Collin

WO
ATLAS

MINI EDITION

Collins

An imprint of HarperCollins Publishers

Westerhill Road, Bishopbriggs,

Glasgow, G64 2QT

www.harpercollins.co.uk

First Published as Collins Mini Atlas of the World 1999

Second edition 2004

Third Edition 2007

Fourth Edition 2009

Fifth Edition 2013

Sixth Edition 2016

A catalogue record for this book is available from the British Library

ISBN 978-0-00-813665-9

10 9 8 7 6 5 4 3 2

Printed in Hong Kong

All mapping in this atlas is generated from Collins Bartholomew™ digital databases. Collins Bartholomew™, the UK's leading independent geographical information supplier, can provide a digital, custom, and premium mapping service to a variety of markets. For further information:

Tel: +44 (0) 208 307 4515

e-mail: collinsbartholomew@harpercollins.co.uk

or visit our website at: www.collinsbartholomew.com

 facebook.com/collinsmaps

 @collinsmaps

Pages	Title	Scale

GEOGRAPHICAL INFORMATION

6–19	**COUNTRIES OF THE WORLD**	
20–21	**OCEANIA** Physical Features	
22–23	**ASIA** Physical Features	
24–25	**EUROPE** Physical Features	
26–27	**AFRICA** Physical Features	
28–29	**NORTH AMERICA** Physical Features	
30–31	**SOUTH AMERICA** Physical Features	
32–33	**OCEANS AND POLES** Physical Features	
34–35	**CLIMATE**	
36–37	**LAND COVER**	
38–39	**POPULATION**	
40–41	**URBANIZATION**	
42–43	**SYMBOLS AND ABBREVIATIONS**	

WORLD

44–45	**WORLD** Physical Features	1:170 000 000
46–47	**WORLD** Countries	1:170 000 000

OCEANIA

48–49	OCEANIA	1:70 000 000
50–51	**AUSTRALIA**	1:25 000 000
52–53	**AUSTRALIA** Southeast	1:10 000 000
54	**NEW ZEALAND**	1:10 000 000

ANTARCTICA

55	ANTARCTICA	1:60 000 000

ASIA

56–57	ASIA	1:70 000 000
58–59	**SOUTHEAST ASIA**	1:30 000 000
60–61	**MALAYSIA AND INDONESIA** West	1:15 000 000
62–63	**CONTINENTAL SOUTHEAST ASIA**	1:15 000 000
64	**PHILIPPINES**	1:15 000 000

CONTENTS

Pages	Title	Scale
65	**NORTH KOREA AND SOUTH KOREA**	1:9 000 000
66–67	**JAPAN**	1:10 000 000
68–69	**CHINA AND MONGOLIA**	1:30 000 000
70–71	**CHINA** Central	1:15 000 000
72–73	**SOUTH ASIA**	1:20 000 000
74–75	**PAKISTAN, INDIA AND BANGLADESH**	1:15 000 000
76–77	**CENTRAL ASIA**	1:20 000 000
78–79	**ARABIAN PENINSULA**	1:15 000 000
80–81	**EAST MEDITERRANEAN**	1:15 000 000
82–83	**RUSSIA**	1:42 000 000
84–85	EUROPE	1:40 000 000
86–87	**EUROPEAN RUSSIA**	1:20 000 000
88–89	**NORTHEAST EUROPE**	1:8 000 000
90–91	**UKRAINE AND MOLDOVA**	1:8 000 000
92–93	**SCANDINAVIA AND ICELAND**	1:10 000 000
94–95	**BRITISH ISLES**	1:8 000 000
96	**SCOTLAND**	1:4 000 000
97	**IRELAND**	1:4 000 000
98–99	**ENGLAND AND WALES**	1:4 000 000
100–101	**NORTHWEST EUROPE**	1:4 000 000
102–103	**CENTRAL EUROPE**	1:8 000 000
104–105	**FRANCE AND SWITZERLAND**	1:8 000 000
106–107	**SPAIN AND PORTUGAL**	1:8 000 000
108–109	**ITALY AND THE BALKANS**	1:8 000 000
110–111	**GREECE, ROMANIA AND BULGARIA**	1:8 000 000
112–113	AFRICA	1:60 000 000
114–115	**NORTHWEST AFRICA**	1:26 000 000

CONTENTS

Pages	Title	Scale
116–117	**NORTHEAST AFRICA**	1:26 000 000
118–119	**CENTRAL AFRICA**	1:20 000 000
120–121	**SOUTHERN AFRICA**	1:20 000 000
122–123	**SOUTH AFRICA**	1:10 000 000
124–125	NORTH AMERICA	1:70 000 000
126–127	**CANADA**	1:30 000 000
128–129	**CANADA** West	1:15 000 000
130–131	**CANADA** East	1:15 000 000
132–133	**UNITED STATES OF AMERICA**	1:25 000 000
134–135	**USA** West	1:11 000 000
136–137	**USA** North Central	1:11 000 000
138–139	**USA** South Central	1:11 000 000
140–141	**USA** Northeast	1:11 000 000
142–143	**USA** Southeast	1:11 000 000
144–145	**MEXICO**	1:15 000 000
146–147	**CARIBBEAN**	1:20 000 000
148–149	SOUTH AMERICA	1:70 000 000
150–151	**SOUTH AMERICA** North	1:25 000 000
152–153	**SOUTH AMERICA** South	1:25 000 000
154–155	**BRAZIL** Southeast	1:10 000 000
	OCEANS	
156–157	**PACIFIC OCEAN**	1:120 000 000
158	**ATLANTIC OCEAN**	1:120 000 000
159	**INDIAN OCEAN**	1:120 000 000
160	**ARCTIC OCEAN**	1:60 000 000
161–256	INDEX	

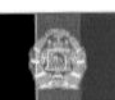

AFGHANISTAN
Islamic Republic of Afghanistan
Capital Kābul

Area sq km	652 225	**Currency** Afghani
Area sq miles	251 825	**Languages** Dari, Pashto
Population	30 552 000	(Pashtu), Uzbek, Turkmen

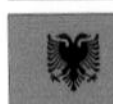

ALBANIA
Republic of Albania
Capital Tirana (Tiranë)

Area sq km	28 748	**Currency** Lek
Area sq miles	11 100	**Languages** Albanian, Greek
Population	3 173 000	

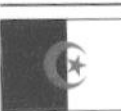

ALGERIA
People's Democratic Republic of Algeria
Capital Algiers (Alger)

Area sq km	2 381 741	**Currency** Algerian dinar
Area sq miles	919 595	**Languages** Arabic, French,
Population	39 208 000	Berber

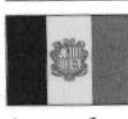

ANDORRA
Principality of Andorra
Capital Andorra la Vella

Area sq km	465	**Currency** Euro
Area sq miles	180	**Languages** Catalan, Spanish,
Population	79 000	French

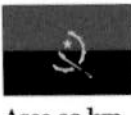

ANGOLA
Republic of Angola
Capital Luanda

Area sq km	1 246 700	**Currency** Kwanza
Area sq miles	481 354	**Languages** Portuguese,
Population	21 472 000	Bantu, other local lang.

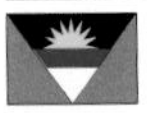

ANTIGUA AND BARBUDA
Capital St John's

Area sq km	442	**Currency** East Caribbean
Area sq miles	171	dollar
Population	90 000	**Languages** English, creole

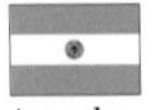

ARGENTINA
Argentine Republic
Capital Buenos Aires

Area sq km	2 766 889	**Currency** Argentinian peso
Area sq miles	1 068 302	**Languages** Spanish, Italian,
Population	41 446 000	Amerindian lang.

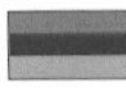

ARMENIA
Republic of Armenia
Capital Yerevan (Erevan)

Area sq km	29 800	**Currency** Dram
Area sq miles	11 506	**Languages** Armenian,
Population	2 977 000	Kurdish

AUSTRALIA
Commonwealth of Australia
Capital Canberra

Area sq km	7 692 024	**Currency** Australian dollar
Area sq miles	2 969 907	**Languages** English, Italian,
Population	23 343 000	Greek

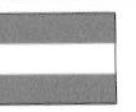

AUSTRIA
Republic of Austria
Capital Vienna (Wien)

Area sq km	83 855	**Currency** Euro
Area sq miles	32 377	**Languages** German,
Population	8 495 000	Croatian, Turkish

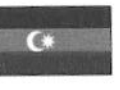

AZERBAIJAN
Republic of Azerbaijan
Capital Baku (Bakı)

Area sq km	86 600	**Currency** Azerbaijani manat
Area sq miles	33 436	**Languages** Azeri, Armenian,
Population	9 413 000	Russian, Lezgian

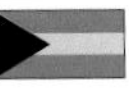

THE BAHAMAS
Commonwealth of The Bahamas
Capital Nassau

Area sq km	13 939	**Currency** Bahamian dollar
Area sq miles	5 382	**Languages** English, creole
Population	377 000	

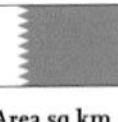

BAHRAIN
Kingdom of Bahrain
Capital Manama (Al Manāmah)

Area sq km	691	**Currency** Bahraini dinar
Area sq miles	267	**Languages** Arabic, English
Population	1 332 000	

BANGLADESH
People's Republic of Bangladesh
Capital Dhaka (Dacca)

Area sq km	143 998	**Currency** Taka
Area sq miles	55 598	**Languages** Bengali, English
Population	156 595 000	

BARBADOS
Capital Bridgetown

Area sq km	430	**Currency** Barbadian dollar
Area sq miles	166	**Languages** English, creole
Population	285 000	

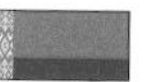

BELARUS
Republic of Belarus
Capital Minsk

Area sq km	207 600	**Currency**	Belarusian rouble
Area sq miles	80 155	**Languages**	Belarusian,
Population	9 357 000		Russian

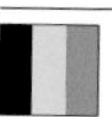

BELGIUM
Kingdom of Belgium
Capital Brussels (Brussel/Bruxelles)

Area sq km	30 520	**Currency**	Euro
Area sq miles	11 784	**Languages**	Dutch (Flemish),
Population	11 104 000		French (Walloon),
			German

BELIZE
Capital Belmopan

Area sq km	22 965	**Currency**	Belize dollar
Area sq miles	8 867	**Languages**	English, Spanish,
Population	332 000		Mayan, creole

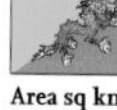

BENIN
Republic of Benin
Capital Porto-Novo

Area sq km	112 620	**Currency**	CFA franc*
Area sq miles	43 483	**Languages**	French, Fon,
Population	10 323 000		Yoruba, Adja,
			other local lang.

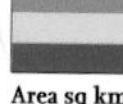

BHUTAN
Kingdom of Bhutan
Capital Thimphu

Area sq km	46 620	**Currency**	Ngultrum,
Area sq miles	18 000		Indian rupee
Population	754 000	**Languages**	Dzongkha,
			Nepali, Assamese

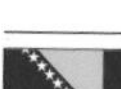

BOLIVIA
Plurinational State of Bolivia
Capital La Paz/Sucre

Area sq km	1 098 581	**Currency**	Boliviano
Area sq miles	424 164	**Languages**	Spanish, Quechua,
Population	10 671 000		Aymara

BOSNIA AND HERZEGOVINA
Capital Sarajevo

Area sq km	51 130	**Currency**	Convertible mark
Area sq miles	19 741	**Languages**	Bosnian, Serbian,
Population	3 829 000		Croatian

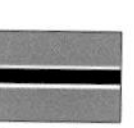

BOTSWANA
Republic of Botswana
Capital Gaborone

Area sq km	581 370	**Currency**	Pula
Area sq miles	224 468	**Languages**	English, Setswana,
Population	2 021 000		Shona, other local
			lang.

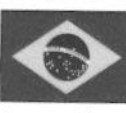

BRAZIL
Federative Republic of Brazil
Capital Brasília

Area sq km	8 514 879	**Currency**	Real
Area sq miles	3 287 613	**Languages**	Portuguese
Population	200 362 000		

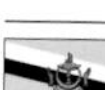

BRUNEI
Brunei Darussalam
Capital Bandar Seri Begawan

Area sq km	5 765	**Currency**	Bruneian dollar
Area sq miles	2 226	**Languages**	Malay, English,
Population	418 000		Chinese

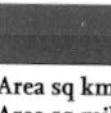

BULGARIA
Republic of Bulgaria
Capital Sofia

Area sq km	110 994	**Currency**	Lev
Area sq miles	42 855	**Languages**	Bulgarian,
Population	7 223 000		Turkish, Romany,
			Macedonian

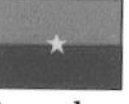

BURKINA FASO
Capital Ouagadougou

Area sq km	274 200	**Currency**	CFA franc*
Area sq miles	105 869	**Languages**	French, Moore
Population	16 935 000		(Mossi), Fulani,
			other local lang.

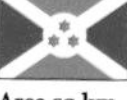

BURUNDI
Republic of Burundi
Capital Bujumbura

Area sq km	27 835	**Currency**	Burundian franc
Area sq miles	10 747	**Languages**	Kirundi (Hutu,
Population	10 163 000		Tutsi), French

CAMBODIA
Kingdom of Cambodia
Capital Phnom Penh

Area sq km	181 035	**Currency**	Riel
Area sq miles	69 884	**Languages**	Khmer,
Population	15 135 000		Vietnamese

* CFA Communauté Financière Africaine

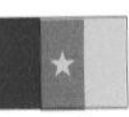

CAMEROON
Republic of Cameroon
Capital Yaoundé

Area sq km	475 442	**Currency**	CFA franc*
Area sq miles	183 569	**Languages**	French, English,
Population	22 254 000		Fang, Bamileke,
			other local lang.

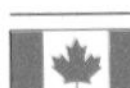

CANADA
Capital Ottawa

Area sq km	9 984 670	**Currency**	Canadian dollar
Area sq miles	3 855 103	**Languages**	English, French,
Population	35 182 000		other local lang.

CAPE VERDE (CABO VERDE)
Republic of Cabo Verde
Capital Praia

Area sq km	4 033	**Currency**	Cape Verdean
Area sq miles	1 557		escudo
Population	499 000	**Languages**	Portuguese, creole

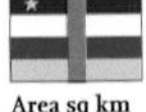

CENTRAL AFRICAN REPUBLIC
Capital Bangui

Area sq km	622 436	**Currency**	CFA franc*
Area sq miles	240 324	**Languages**	French, Sango,
Population	4 616 000		Banda, Baya,
			other local lang.

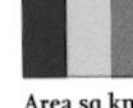

CHAD
Republic of Chad
Capital Ndjamena

Area sq km	1 284 000	**Currency**	CFA franc*
Area sq miles	495 755	**Languages**	Arabic, French,
Population	12 825 000		Sara, other local
			lang.

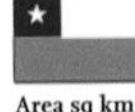

CHILE
Republic of Chile
Capital Santiago

Area sq km	756 945	**Currency**	Chilean peso
Area sq miles	292 258	**Languages**	Spanish,
Population	17 620 000		Amerindian lang.

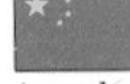

CHINA
People's Republic of China
Capital Beijing (Peking)

Area sq km	9 606 802	**Currency**	Yuan, HK dollar,
Area sq miles	3 709 186		Macao pataca
Population	1 369 993 000	**Languages**	Mandarin
			(Putonghua), Wu,
			Cantonese, Hsiang,
			regional lang.

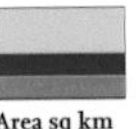

COLOMBIA
Republic of Colombia
Capital Bogotá

Area sq km	1 141 748	**Currency**	Colombian peso
Area sq miles	440 831	**Languages**	Spanish,
Population	48 321 000		Amerindian lang.

COMOROS
Union of the Comoros
Capital Moroni

Area sq km	1 862	**Currency**	Comorian franc
Area sq miles	719	**Languages**	Shikomor
Population	735 000		(Comorian),
			French, Arabic

CONGO
Republic of the Congo
Capital Brazzaville

Area sq km	342 000	**Currency**	CFA franc*
Area sq miles	132 047	**Languages**	French, Kongo,
Population	4 448 000		Monokutuba,
			other local lang.

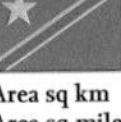

CONGO, DEMOCRATIC REPUBLIC OF THE
Capital Kinshasa

Area sq km	2 345 410	**Currency**	Congolese franc
Area sq miles	905 568	**Languages**	French, Lingala,
Population	67 514 000		Swahili, Kongo,
			other local lang.

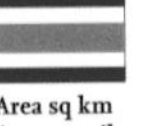

COSTA RICA
Republic of Costa Rica
Capital San José

Area sq km	51 100	**Currency**	Costa Rican colón
Area sq miles	19 730	**Languages**	Spanish
Population	4 872 000		

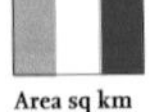

CÔTE D'IVOIRE (IVORY COAST)
Republic of Côte d'Ivoire
Capital Yamoussoukro

Area sq km	322 463	**Currency**	CFA franc*
Area sq miles	124 504	**Languages**	French, creole,
Population	20 316 000		Akan, other local
			lang.

CROATIA
Republic of Croatia
Capital Zagreb

Area sq km	56 538	**Currency**	Kuna
Area sq miles	21 829	**Languages**	Croatian, Serbian
Population	4 290 000		

CUBA
Republic of Cuba
Capital Havana (La Habana)

Area sq km	110 860	**Currency**	Cuban peso
Area sq miles	42 803	**Languages**	Spanish
Population	11 266 000		

CYPRUS
Republic of Cyprus
Capital Nicosia (Lefkosia)

Area sq km	9 251	**Currency**	Euro
Area sq miles	3 572	**Languages**	Greek, Turkish,
Population	1 141 000		English

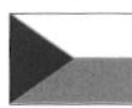

CZECH REPUBLIC
Capital Prague (Praha)

Area sq km	78 864	**Currency**	Czech koruna
Area sq miles	30 450	**Languages**	Czech, Moravian,
Population	10 702 000		Slovakian

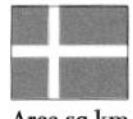

DENMARK
Kingdom of Denmark
Capital Copenhagen (København)

Area sq km	43 075	**Currency**	Danish krone
Area sq miles	16 631	**Languages**	Danish
Population	5 619 000		

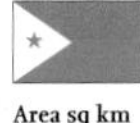

DJIBOUTI
Republic of Djibouti
Capital Djibouti

Area sq km	23 200	**Currency**	Djiboutian franc
Area sq miles	8 958	**Languages**	Somali, Afar,
Population	873 000		French, Arabic

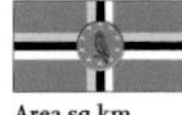

DOMINICA
Commonwealth of Dominica
Capital Roseau

Area sq km	750	**Currency**	East Caribbean
Area sq miles	290		dollar
Population	72 000	**Languages**	English, creole

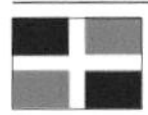

DOMINICAN REPUBLIC
Capital Santo Domingo

Area sq km	48 442	**Currency**	Dominican peso
Area sq miles	18 704	**Languages**	Spanish, creole
Population	10 404 000		

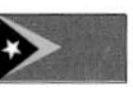

EAST TIMOR (TIMOR-LESTE)
Democratic Republic of Timor-Leste
Capital Dili

Area sq km	14 874	**Currency**	US dollar
Area sq miles	5 743	**Languages**	Portuguese, Tetun,
Population	1 133 000		English

ECUADOR
Republic of Ecuador
Capital Quito

Area sq km	272 045	**Currency**	US dollar
Area sq miles	105 037	**Languages**	Spanish, Quechua,
Population	15 738 000		Amerindian lang.

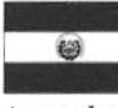

EGYPT
Arab Republic of Egypt
Capital Cairo (Al Qāhirah)

Area sq km	1 000 250	**Currency**	Egyptian pound
Area sq miles	386 199	**Languages**	Arabic
Population	82 056 000		

EL SALVADOR
Republic of El Salvador
Capital San Salvador

Area sq km	21 041	**Currency**	US dollar
Area sq miles	8 124	**Languages**	Spanish
Population	6 340 000		

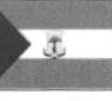

EQUATORIAL GUINEA
Republic of Equatorial Guinea
Capital Malabo

Area sq km	28 051	**Currency**	CFA franc*
Area sq miles	10 831	**Languages**	Spanish, French,
Population	757 000		Fang

ERITREA
State of Eritrea
Capital Asmara

Area sq km	117 400	**Currency**	Nakfa
Area sq miles	45 328	**Languages**	Tigrinya, Tigre
Population	6 333 000		

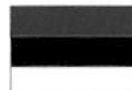

ESTONIA
Republic of Estonia
Capital Tallinn

Area sq km	45 200	**Currency**	Euro
Area sq miles	17 452	**Languages**	Estonian, Russian
Population	1 287 000		

ETHIOPIA
Federal Democratic Republic of Ethiopia
Capital Addis Ababa (Ādīs Ābeba)

Area sq km	1 133 880	Currency	Birr
Area sq miles	437 794	Languages	Oromo, Amharic,
Population	94 101 000		Tigrinya, other
			local lang.

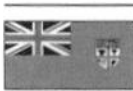

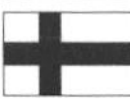

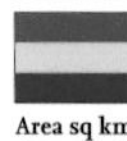

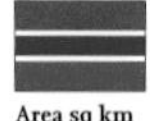

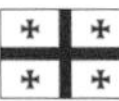

GEORGIA
Capital Tbilisi

Area sq km	69 700	Currency	Lari
Area sq miles	26 911	Languages	Georgian, Russian,
Population	4 341 000		Armenian, Azeri,
			Ossetian, Abkhaz

FIJI
Republic of Fiji
Capital Suva

Area sq km	18 330	Currency	Fijian dollar
Area sq miles	7 077	Languages	English, Fijian,
Population	881 000		Hindi

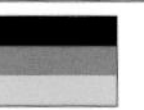

GERMANY
Federal Republic of Germany
Capital Berlin

Area sq km	357 022	Currency	Euro
Area sq miles	137 849	Languages	German, Turkish
Population	82 727 000		

FINLAND
Republic of Finland
Capital Helsinki (Helsingfors)

Area sq km	338 145	Currency	Euro
Area sq miles	130 559	Languages	Finnish, Swedish,
Population	5 426 000		Sami

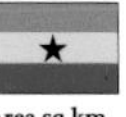

GHANA
Republic of Ghana
Capital Accra

Area sq km	238 537	Currency	Cedi
Area sq miles	92 100	Languages	English, Hausa,
Population	25 905 000		Akan, other local
			lang.

FRANCE
French Republic
Capital Paris

Area sq km	543 965	Currency	Euro
Area sq miles	210 026	Languages	French, German
Population	64 291 000		dialects, Italian,
			Arabic, Breton

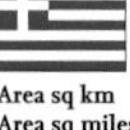

GREECE
Hellenic Republic
Capital Athens (Athina)

Area sq km	131 957	Currency	Euro
Area sq miles	50 949	Languages	Greek
Population	11 128 000		

GABON
Gabonese Republic
Capital Libreville

Area sq km	267 667	Currency	CFA franc*
Area sq miles	103 347	Languages	French, Fang,
Population	1 672 000		other local lang.

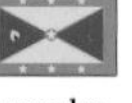

GRENADA
Capital St George's

Area sq km	378	Currency	East Caribbean
Area sq miles	146		dollar
Population	106 000	Languages	English, creole

THE GAMBIA
Republic of The Gambia
Capital Banjul

Area sq km	11 295	Currency	Dalasi
Area sq miles	4 361	Languages	English, Malinke,
Population	1 849 000		Fulani, Wolof

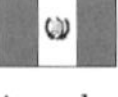

GUATEMALA
Republic of Guatemala
Capital Guatemala City

Area sq km	108 890	Currency	Quetzal
Area sq miles	42 043	Languages	Spanish,
Population	15 468 000		Mayan lang.

Gaza
Disputed territory
Capital Gaza

Area sq km	363	Currency	Israeli shekel
Area sq miles	140	Languages	Arabic
Population	1 701 437		

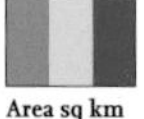

GUINEA
Republic of Guinea
Capital Conakry

Area sq km	245 857	Currency	Guinean franc
Area sq miles	94 926	Languages	French, Fulani,
Population	11 745 000		Malinke, other
			local lang.

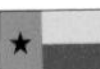

GUINEA-BISSAU
Republic of Guinea-Bissau
Capital Bissau

Area sq km	36 125	**Currency**	CFA franc*
Area sq miles	13 948	**Languages**	Portuguese, crioulo, other local lang.
Population	1 704 000		

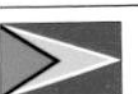

GUYANA
Co-operative Republic of Guyana
Capital Georgetown

Area sq km	214 969	**Currency**	Guyana dollar
Area sq miles	83 000	**Languages**	English, creole, Amerindian lang.
Population	800 000		

HAITI
Republic of Haiti
Capital Port-au-Prince

Area sq km	27 750	**Currency**	Gourde
Area sq miles	10 714	**Languages**	French, creole
Population	10 317 000		

HONDURAS
Republic of Honduras
Capital Tegucigalpa

Area sq km	112 088	**Currency**	Lempira
Area sq miles	43 277	**Languages**	Spanish, Amerindian lang.
Population	8 098 000		

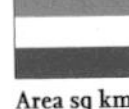

HUNGARY
Capital Budapest

Area sq km	93 030	**Currency**	Forint
Area sq miles	35 919	**Languages**	Hungarian
Population	9 955 000		

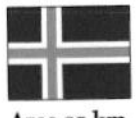

ICELAND
Republic of Iceland
Capital Reykjavík

Area sq km	102 820	**Currency**	Icelandic króna
Area sq miles	39 699	**Languages**	Icelandic
Population	330 000		

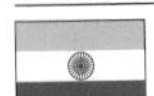

INDIA
Republic of India
Capital New Delhi

Area sq km	3 166 620	**Currency**	Indian rupee
Area sq miles	1 222 632	**Languages**	Hindi, English, many regional lang.
Population	1 252 140 000		

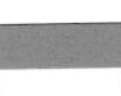

INDONESIA
Republic of Indonesia
Capital Jakarta

Area sq km	1 919 445	**Currency**	Rupiah
Area sq miles	741 102	**Languages**	Indonesian, other local lang.
Population	249 866 000		

IRAN
Islamic Republic of Iran
Capital Tehrān

Area sq km	1 648 000	**Currency**	Iranian rial
Area sq miles	636 296	**Languages**	Farsi, Azeri, Kurdish, regional lang.
Population	77 447 000		

IRAQ
Republic of Iraq
Capital Baghdād

Area sq km	438 317	**Currency**	Iraqi dinar
Area sq miles	169 235	**Languages**	Arabic, Kurdish, Turkmen
Population	33 765 000		

IRELAND
Capital Dublin (Baile Átha Cliath)

Area sq km	70 282	**Currency**	Euro
Area sq miles	27 136	**Languages**	English, Irish
Population	4 627 000		

ISRAEL
State of Israel
Capital Jerusalem* (Yerushalayim) (El Quds)

Area sq km	22 072	**Currency**	Shekel
Area sq miles	8 522	**Languages**	Hebrew, Arabic
Population	7 733 000		

* De facto capital. Disputed.

ITALY
Italian Republic
Capital Rome (Roma)

Area sq km	301 245	**Currency**	Euro
Area sq miles	116 311	**Languages**	Italian
Population	60 990 000		

JAMAICA
Capital Kingston

Area sq km	10 991	**Currency**	Jamaican dollar
Area sq miles	4 244	**Languages**	English, creole
Population	2 784 000		

JAPAN
Capital Tōkyō

Area sq km	377 727	Currency	Yen
Area sq miles	145 841	Languages	Japanese
Population	127 144 000		

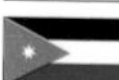

JORDAN
Hashemite Kingdom of Jordan
Capital 'Ammān

Area sq km	89 206	Currency	Jordanian dinar
Area sq miles	34 443	Languages	Arabic
Population	7 274 000		

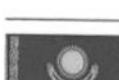

KAZAKHSTAN
Republic of Kazakhstan
Capital Astana (Akmola)

Area sq km	2 717 300	Currency	Tenge
Area sq miles	1 049 155	Languages	Kazakh, Russian,
Population	16 441 000		Ukrainian, German,
			Uzbek, Tatar

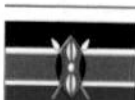

KENYA
Republic of Kenya
Capital Nairobi

Area sq km	582 646	Currency	Kenyan shilling
Area sq miles	224 961	Languages	Swahili, English,
Population	44 354 000		other local lang.

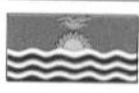

KIRIBATI
Republic of Kiribati
Capital Bairiki

Area sq km	717	Currency	Australian dollar
Area sq miles	277	Languages	Gilbertese,
Population	102 000		English

KOSOVO
Republic of Kosovo
Capital Prishtinë (Priština)

Area sq km	10 908	Currency	Euro
Area sq miles	4 212	Languages	Albanian, Serbian
Population	1 815 606		

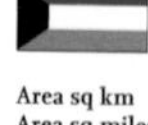

KUWAIT
State of Kuwait
Capital Kuwait (Al Kuwayt)

Area sq km	17 818	Currency	Kuwaiti dinar
Area sq miles	6 880	Languages	Arabic
Population	3 369 000		

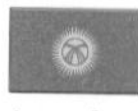

KYRGYZSTAN
Kyrgyz Republic
Capital Bishkek (Frunze)

Area sq km	198 500	Currency	Kyrgyz som
Area sq miles	76 641	Languages	Kyrgyz, Russian,
Population	5 548 000		Uzbek

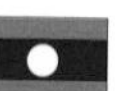

LAOS
Lao People's Democratic Republic
Capital Vientiane (Viangchan)

Area sq km	236 800	Currency	Kip
Area sq miles	91 429	Languages	Lao, other local
Population	6 770 000		lang.

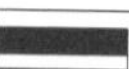

LATVIA
Republic of Latvia
Capital Rīga

Area sq km	64 589	Currency	Euro
Area sq miles	24 938	Languages	Latvian, Russian
Population	2 050 000		

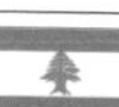

LEBANON
Lebanese Republic
Capital Beirut (Beyrouth)

Area sq km	10 452	Currency	Lebanese pound
Area sq miles	4 036	Languages	Arabic, Armenian,
Population	4 822 000		French

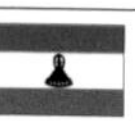

LESOTHO
Kingdom of Lesotho
Capital Maseru

Area sq km	30 355	Currency	Loti,
Area sq miles	11 720		S. African rand
Population	2 074 000	Languages	Sesotho, English,
			Zulu

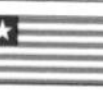

LIBERIA
Republic of Liberia
Capital Monrovia

Area sq km	111 369	Currency	Liberian dollar
Area sq miles	43 000	Languages	English, creole,
Population	4 294 000		other local lang.

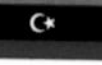

LIBYA
State of Libya
Capital Tripoli (Ṭarābulus)

Area sq km	1 759 540	Currency	Libyan dinar
Area sq miles	679 362	Languages	Arabic, Berber
Population	6 202 000		

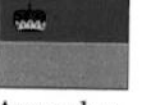

LIECHTENSTEIN
Principality of Liechtenstein
Capital Vaduz

Area sq km	160	Currency	Swiss franc
Area sq miles	62	Languages	German
Population	37 000		

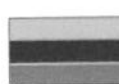

LITHUANIA
Republic of Lithuania
Capital Vilnius

Area sq km	65 200	**Currency** Euro
Area sq miles	25 174	**Languages** Lithuanian,
Population	3 017 000	Russian, Polish

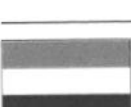

LUXEMBOURG
Grand Duchy of Luxembourg
Capital Luxembourg

Area sq km	2 586	**Currency** Euro
Area sq miles	998	**Languages** Letzeburgish,
Population	530 000	German, French

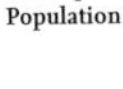

MACEDONIA (F.Y.R.O.M.)
Republic of Macedonia
Capital Skopje

Area sq km	25 713	**Currency** Macedonian denar
Area sq miles	9 928	**Languages** Macedonian,
Population	2 107 000	Albanian, Turkish

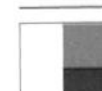

MADAGASCAR
Republic of Madagascar
Capital Antananarivo

Area sq km	587 041	**Currency** Ariary
Area sq miles	226 658	**Languages** Malagasy, French
Population	22 925 000	

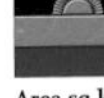

MALAWI
Republic of Malawi
Capital Lilongwe

Area sq km	118 484	**Currency** Malawian kwacha
Area sq miles	45 747	**Languages** Chichewa,
Population	16 363 000	English, other
		local lang.

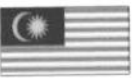

MALAYSIA
Capital Kuala Lumpur/Putrajaya

Area sq km	332 965	**Currency** Ringgit
Area sq miles	128 559	**Languages** Malay, English,
Population	29 717 000	Chinese, Tamil,
		other local lang.

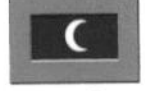

MALDIVES
Republic of the Maldives
Capital Male

Area sq km	298	**Currency** Rufiyaa
Area sq miles	115	**Languages** Divehi
Population	345 000	(Maldivian)

MALI
Republic of Mali
Capital Bamako

Area sq km	1 240 140	**Currency** CFA franc*
Area sq miles	478 821	**Languages** French, Bambara,
Population	15 302 000	other local lang.

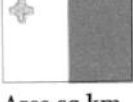

MALTA
Republic of Malta
Capital Valletta

Area sq km	316	**Currency** Euro
Area sq miles	122	**Languages** Maltese, English
Population	429 000	

MARSHALL ISLANDS
Republic of the Marshall Islands
Capital Delap-Uliga-Djarrit

Area sq km	181	**Currency** US dollar
Area sq miles	70	**Languages** English,
Population	53 000	Marshallese

MAURITANIA
Islamic Republic of Mauritania
Capital Nouakchott

Area sq km	1 030 700	**Currency** Ouguiya
Area sq miles	397 955	**Languages** Arabic, French,
Population	3 890 000	other local lang.

MAURITIUS
Republic of Mauritius
Capital Port Louis

Area sq km	2 040	**Currency** Mauritius rupee
Area sq miles	788	**Languages** English, creole,
Population	1 244 000	Hindi, Bhojpūrī,
		French

MEXICO
United Mexican States
Capital Mexico City

Area sq km	1 972 545	**Currency** Mexican peso
Area sq miles	761 604	**Languages** Spanish,
Population	122 332 000	Amerindian lang.

MICRONESIA, FEDERATED STATES OF
Capital Palikir

Area sq km	701	**Currency** US dollar
Area sq miles	271	**Languages** English, Chuukese,
Population	104 000	Pohnpeian, other
		local lang.

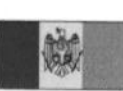

MOLDOVA
Republic of Moldova
Capital Chişinău (Kishinev)

Area sq km	33 700	**Currency** Moldovan leu
Area sq miles	13 012	**Languages** Romanian,
Population	3 487 000	Ukrainian,
		Gagauz, Russian

NAMIBIA
Republic of Namibia
Capital Windhoek

Area sq km	824 292	**Currency** Namibian dollar
Area sq miles	318 261	**Languages** English, Afrikaan
Population	2 303 000	German, Ovamb
		other local lang.

MONACO
Principality of Monaco
Capital Monaco-Ville

Area sq km	2	**Currency** Euro
Area sq miles	1	**Languages** French,
Population	38 000	Monégasque,
		Italian

NAURU
Republic of Nauru
Capital Yaren (de facto)

Area sq km	21	**Currency** Australian dollar
Area sq miles	8	**Languages** Nauruan, Englisl
Population	10 000	

MONGOLIA
Capital Ulan Bator (Ulaanbaatar)

Area sq km	1 565 000	**Currency** Tugrik (tögrög)
Area sq miles	604 250	**Languages** Khalka
Population	2 839 000	(Mongolian),
		Kazakh, other
		local lang.

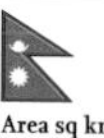

NEPAL
Federal Democratic Republic of Nepal
Capital Kathmandu

Area sq km	147 181	**Currency** Nepalese rupee
Area sq miles	56 827	**Languages** Nepali, Maithili,
Population	27 797 000	Bhojpuri, Englisl
		other local lang.

MONTENEGRO
Republic of Montenegro
Capital Podgorica

Area sq km	13 812	**Currency** Euro
Area sq miles	5 333	**Languages** Serbian
Population	621 000	(Montenegrin),
		Albanian

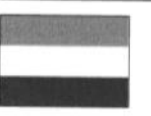

NETHERLANDS
Kingdom of the Netherlands
Capital Amsterdam/The Hague ('s-Gravenhaç

Area sq km	41 526	**Currency** Euro
Area sq miles	16 033	**Languages** Dutch, Frisian
Population	16 759 000	

MOROCCO
Kingdom of Morocco
Capital Rabat

Area sq km	446 550	**Currency** Moroccan dirham
Area sq miles	172 414	**Languages** Arabic, Berber,
Population	33 008 000	French

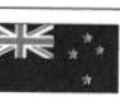

NEW ZEALAND
Capital Wellington

Area sq km	270 534	**Currency** New Zealand
Area sq miles	104 454	dollar
Population	4 506 000	**Languages** English, Maori

MOZAMBIQUE
Republic of Mozambique
Capital Maputo

Area sq km	799 380	**Currency** Metical
Area sq miles	308 642	**Languages** Portuguese,
Population	25 834 000	Makua, Tsonga,
		other local lang.

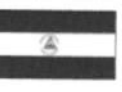

NICARAGUA
Republic of Nicaragua
Capital Managua

Area sq km	130 000	**Currency** Córdoba
Area sq miles	50 193	**Languages** Spanish,
Population	6 080 000	Amerindian lang.

MYANMAR (Burma)
Republic of the Union of Myanmar
Capital Nay Pyi Taw

Area sq km	676 577	**Currency** Kyat
Area sq miles	261 228	**Languages** Burmese, Shan,
Population	53 259 000	Karen, other
		local lang.

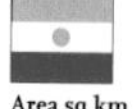

NIGER
Republic of Niger
Capital Niamey

Area sq km	1 267 000	**Currency** CFA franc*
Area sq miles	489 191	**Languages** French, Hausa,
Population	17 831 000	Fulani, other loca
		lang.

NIGERIA
Federal Republic of Nigeria
Capital Abuja

Area sq km	923 768	**Currency**	Naira
Area sq miles	356 669	**Languages**	English, Hausa, Yoruba, Ibo, Fulani, other local lang.
Population	173 615 000		

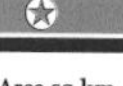

NORTH KOREA
Democratic People's Republic of Korea
Capital P'yŏngyang

Area sq km	120 538	**Currency**	North Korean won
Area sq miles	46 540	**Languages**	Korean
Population	24 895 000		

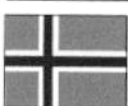

NORWAY
Kingdom of Norway
Capital Oslo

Area sq km	323 878	**Currency**	Norwegian krone
Area sq miles	125 050	**Languages**	Norwegian, Sami
Population	5 043 000		

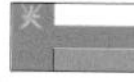

OMAN
Sultanate of Oman
Capital Muscat (Masqaţ)

Area sq km	309 500	**Currency**	Omani rial
Area sq miles	119 499	**Languages**	Arabic, Baluchi, Indian lang.
Population	3 632 000		

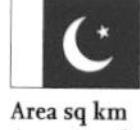

PAKISTAN
Islamic Republic of Pakistan
Capital Islamabad

Area sq km	881 888	**Currency**	Pakistani rupee
Area sq miles	340 497	**Languages**	Urdu, Punjabi, Sindhi, Pashto (Pashtu), English, Balochi
Population	182 143 000		

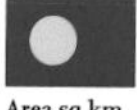

PALAU
Republic of Palau
Capital Melekeok (Ngerulmud)

Area sq km	497	**Currency**	US dollar
Area sq miles	192	**Languages**	Palauan, English
Population	21 000		

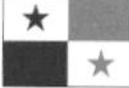

PANAMA
Republic of Panama
Capital Panama City

Area sq km	77 082	**Currency**	Balboa
Area sq miles	29 762	**Languages**	Spanish, English, Amerindian lang.
Population	3 864 000		

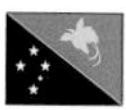

PAPUA NEW GUINEA
Independent State of Papua New Guinea
Capital Port Moresby

Area sq km	462 840	**Currency**	Kina
Area sq miles	178 704	**Languages**	English, Tok Pisin (creole), other local lang.
Population	7 321 000		

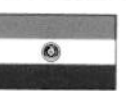

PARAGUAY
Republic of Paraguay
Capital Asunción

Area sq km	406 752	**Currency**	Guaraní
Area sq miles	157 048	**Languages**	Spanish, Guaraní
Population	6 802 000		

PERU
Republic of Peru
Capital Lima

Area sq km	1 285 216	**Currency**	Nuevo sol
Area sq miles	496 225	**Languages**	Spanish, Quechua, Aymara
Population	30 376 000		

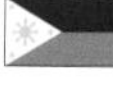

PHILIPPINES
Republic of the Philippines
Capital Manila

Area sq km	300 000	**Currency**	Philippine peso
Area sq miles	115 831	**Languages**	English, Filipino, Tagalog, Cebuano, other local lang.
Population	98 394 000		

POLAND
Republic of Poland
Capital Warsaw (Warszawa)

Area sq km	312 683	**Currency**	Złoty
Area sq miles	120 728	**Languages**	Polish, German
Population	38 217 0000		

PORTUGAL
Portuguese Republic
Capital Lisbon (Lisboa)

Area sq km	88 940	**Currency**	Euro
Area sq miles	34 340	**Languages**	Portuguese
Population	10 608 000		

QATAR
State of Qatar
Capital Doha (Ad Dawḩah)

Area sq km	11 437	**Currency**	Qatari riyal
Area sq miles	4 416	**Languages**	Arabic
Population	2 169 000		

COUNTRIES OF THE WORLD

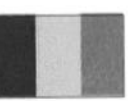

ROMANIA
Capital Bucharest (Bucureşti)

Area sq km	237 500	**Currency**	Romanian leu
Area sq miles	91 699	**Languages**	Romanian,
Population	21 699 000		Hungarian

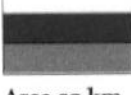

RUSSIA
Capital Moscow (Moskva)

Area sq km	17 075 400	**Currency**	Russian rouble
Area sq miles	6 592 849	**Languages**	Russian, Tatar,
Population	142 834 000		Ukrainian, other
			local lang.

RWANDA
Republic of Rwanda
Capital Kigali

Area sq km	26 338	**Currency**	Rwandan franc
Area sq miles	10 169	**Languages**	Kinyarwanda,
Population	11 777 000		French, English

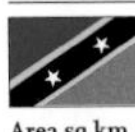

ST KITTS AND NEVIS
Federation of St Kitts and Nevis
Capital Basseterre

Area sq km	261	**Currency**	East Caribbean
Area sq miles	101		dollar
Population	54 000	**Languages**	English, creole

ST LUCIA
Capital Castries

Area sq km	616	**Currency**	East Caribbean
Area sq miles	238		dollar
Population	182 000	**Languages**	English, creole

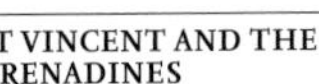

ST VINCENT AND THE GRENADINES
Capital Kingstown

Area sq km	389	**Currency**	East Caribbean
Area sq miles	150		dollar
Population	109 000	**Languages**	English, creole

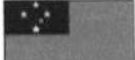

SAMOA
Independent State of Samoa
Capital Apia

Area sq km	2 831	**Currency**	Tala
Area sq miles	1 093	**Languages**	Samoan, English
Population	190 000		

SAN MARINO
Republic of San Marino
Capital San Marino

Area sq km	61	**Currency**	Euro
Area sq miles	24	**Languages**	Italian
Population	31 000		

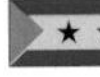

SÃO TOMÉ AND PRÍNCIPE
Democratic Rep. of São Tomé and Príncipe
Capital São Tomé

Area sq km	964	**Currency**	Dobra
Area sq miles	372	**Languages**	Portuguese, creole
Population	193 000		

SAUDI ARABIA
Kingdom of Saudi Arabia
Capital Riyadh (Ar Riyāḍ)

Area sq km	2 200 000	**Currency**	Saudi Arabian
Area sq miles	849 425		riyal
Population	28 829 000	**Languages**	Arabic

SENEGAL
Republic of Senegal
Capital Dakar

Area sq km	196 720	**Currency**	CFA franc*
Area sq miles	75 954	**Languages**	French, Wolof,
Population	14 133 000		Fulani, other local
			lang.

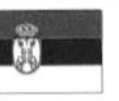

SERBIA
Republic of Serbia
Capital Belgrade (Beograd)

Area sq km	77 453	**Currency**	Serbian dinar,
Area sq miles	29 904	**Languages**	Serbian,
Population	7 181 505		Hungarian

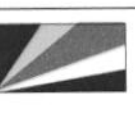

SEYCHELLES
Republic of Seychelles
Capital Victoria

Area sq km	455	**Currency**	Seychelles rupee
Area sq miles	176	**Languages**	English, French,
Population	93 000		creole

SIERRA LEONE
Republic of Sierra Leone
Capital Freetown

Area sq km	71 740	**Currency**	Leone
Area sq miles	27 699	**Languages**	English, creole,
Population	6 092 000		Mende, Temne,
			other local lang.

SINGAPORE
Republic of Singapore
Capital Singapore

Area sq km	639	**Currency**	Singapore dollar
Area sq miles	247	**Languages**	Chinese, English,
Population	5 412 000		Malay, Tamil

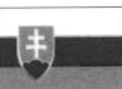

SLOVAKIA
Slovak Republic
Capital Bratislava

Area sq km	49 035	**Currency** Euro
Area sq miles	18 933	**Languages** Slovak, Hungarian, Czech
Population	5 450 000	

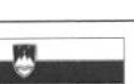

SLOVENIA
Republic of Slovenia
Capital Ljubljana

Area sq km	20 251	**Currency** Euro
Area sq miles	7 819	**Languages** Slovene, Croatian, Serbian
Population	2 072 000	

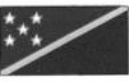

SOLOMON ISLANDS
Capital Honiara

Area sq km	28 370	**Currency** Solomon Islands dollar
Area sq miles	10 954	
Population	561 000	**Languages** English, creole, other local lang.

SOMALIA
Federal Republic of Somalia
Capital Mogadishu (Muqdisho)

Area sq km	637 657	**Currency** Somali shilling
Area sq miles	246 201	**Languages** Somali, Arabic
Population	10 496 000	

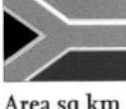

SOUTH AFRICA
Capital Pretoria (Tshwane)/ Cape Town/Bloemfontein

Area sq km	1 219 090	**Currency** Rand
Area sq miles	470 693	**Languages** Afrikaans, English, nine official local lang.
Population	52 776 000	

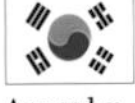

SOUTH KOREA
Republic of Korea
Capital Seoul (Sŏul)

Area sq km	99 274	**Currency** South Korean won
Area sq miles	38 330	
Population	49 263 000	**Languages** Korean

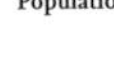

SOUTH SUDAN
Republic of South Sudan
Capital Juba

Area sq km	644 329	**Currency** South Sudan pound
Area sq miles	248 775	
Population	11 296 000	**Languages** English, Arabic, Dinka, Nuer, other local lang.

SPAIN
Kingdom of Spain
Capital Madrid

Area sq km	504 782	**Currency** Euro
Area sq miles	194 897	**Languages** Spanish (Castilian), Catalan, Galician, Basque
Population	46 927 000	

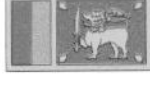

SRI LANKA
Democratic Socialist Republic of Sri Lanka
Capital Sri Jayewardenepura Kotte

Area sq km	65 610	**Currency** Sri Lankan rupee
Area sq miles	25 332	**Languages** Sinhalese, Tamil, English
Population	21 273 000	

SUDAN
Republic of the Sudan
Capital Khartoum

Area sq km	1 861 484	**Currency** Sudanese pound (Sudani)
Area sq miles	718 725	
Population	37 964 000	**Languages** Arabic, English, Nubian, Beja, Fur, other local lang.

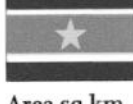

SURINAME
Republic of Suriname
Capital Paramaribo

Area sq km	163 820	**Currency** Surinamese dollar
Area sq miles	63 251	**Languages** Dutch, Surinamese, English, Hindi
Population	539 000	

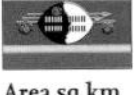

SWAZILAND
Kingdom of Swaziland
Capital Mbabane

Area sq km	17 364	**Currency** Lilangeni, South African rand
Area sq miles	6 704	
Population	1 250 000	**Languages** Swazi, English

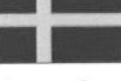

SWEDEN
Kingdom of Sweden
Capital Stockholm

Area sq km	449 964	**Currency** Swedish krona
Area sq miles	173 732	**Languages** Swedish, Sami
Population	9 571 000	

SWITZERLAND
Swiss Confederation
Capital Bern (Berne)

Area sq km	41 293	**Currency** Swiss franc
Area sq miles	15 943	**Languages** German, French, Italian, Romansch
Population	8 078 000	

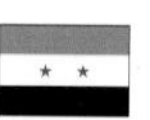

SYRIA
Syrian Arab Republic
Capital Damascus (Dimashq)

Area sq km	184 026	**Currency**	Syrian pound
Area sq miles	71 052	**Languages**	Arabic, Kurdish,
Population	21 898 000		Armenian

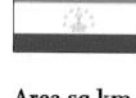

TAIWAN
Republic of China
Capital Taibei

Area sq km	36 179	**Currency**	New Taiwan dollar
Area sq miles	13 969	**Languages**	Mandarin
Population	23 344 000		(Putonghua), Min,
			Hakka, other local
			lang.

The People's Republic of China claims Taiwan as its 23rd province.

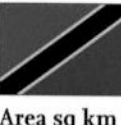

TAJIKISTAN
Republic of Tajikistan
Capital Dushanbe

Area sq km	143 100	**Currency**	Somoni
Area sq miles	55 251	**Languages**	Tajik, Uzbek,
Population	8 208 000		Russian

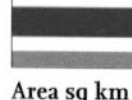

TANZANIA
United Republic of Tanzania
Capital Dodoma

Area sq km	945 087	**Currency**	Tanzanian shilling
Area sq miles	364 900	**Languages**	Swahili, English,
Population	49 253 000		Nyamwezi, other
			local lang.

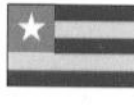

THAILAND
Kingdom of Thailand
Capital Bangkok (Krung Thep)

Area sq km	513 115	**Currency**	Baht
Area sq miles	198 115	**Languages**	Thai, Lao,
Population	67 011 000		Chinese, Malay,
			Mon-Khmer lang.

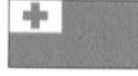

TOGO
Togolese Republic
Capital Lomé

Area sq km	56 785	**Currency**	CFA franc*
Area sq miles	21 925	**Languages**	French, Ewe,
Population	6 817 000		Kabre, other local
			lang.

TONGA
Kingdom of Tonga
Capital Nuku'alofa

Area sq km	748	**Currency**	Pa'anga
Area sq miles	289	**Languages**	Tongan, English
Population	105 000		

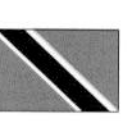

TRINIDAD AND TOBAGO
Republic of Trinidad and Tobago
Capital Port of Spain

Area sq km	5 130	**Currency**	Trinidad and
Area sq miles	1 981		Tobago dollar
Population	1 341 000	**Languages**	English, creole,
			Hindi

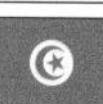

TUNISIA
Republic of Tunisia
Capital Tunis

Area sq km	164 150	**Currency**	Tunisian dinar
Area sq miles	63 379	**Languages**	Arabic, French
Population	10 997 000		

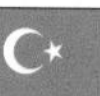

TURKEY
Republic of Turkey
Capital Ankara

Area sq km	779 452	**Currency**	Lira
Area sq miles	300 948	**Languages**	Turkish, Kurdish
Population	74 933 000		

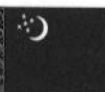

TURKMENISTAN
Capital Aşgabat (Ashkhabad)

Area sq km	488 100	**Currency**	Turkmen manat
Area sq miles	188 456	**Languages**	Turkmen, Uzbek,
Population	5 240 000		Russian

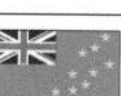

TUVALU
Capital Vaiaku

Area sq km	25	**Currency**	Australian dollar
Area sq miles	10	**Languages**	Tuvaluan, English
Population	10 000		

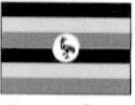

UGANDA
Republic of Uganda
Capital Kampala

Area sq km	241 038	**Currency**	Ugandan shilling
Area sq miles	93 065	**Languages**	English, Swahili,
Population	37 579 000		Luganda, other
			local lang.

UKRAINE
Capital Kiev (Kyiv)

Area sq km	603 700	**Currency**	Hryvnia
Area sq miles	233 090	**Languages**	Ukrainian,
Population	45 239 000		Russian

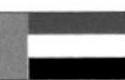

UNITED ARAB EMIRATES
Federation of Emirates
Capital Abu Dhabi (Abū Ẓaby)

Area sq km	77 700	**Currency** UAE dirham
Area sq miles	30 000	**Languages** Arabic, English
Population	9 346 000	

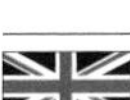

UNITED KINGDOM
United Kingdom of Great Britain and
Northern Ireland
Capital London

Area sq km	243 609	**Currency** Pound sterling
Area sq miles	94 058	**Languages** English, Welsh,
Population	63 136 000	Gaelic

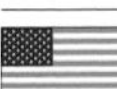

UNITED STATES OF AMERICA
Capital Washington D.C.

Area sq km	9 826 635	**Currency** US dollar
Area sq miles	3 794 085	**Languages** English, Spanish
Population	320 051 000	

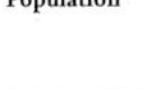

URUGUAY
Oriental Republic of Uruguay
Capital Montevideo

Area sq km	176 215	**Currency** Uruguayan peso
Area sq miles	68 037	**Languages** Spanish
Population	3 407 000	

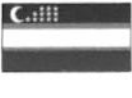

UZBEKISTAN
Republic of Uzbekistan
Capital Tashkent

Area sq km	447 400	**Currency** Uzbek som
Area sq miles	172 742	**Languages** Uzbek, Russian,
Population	28 934 000	Tajik, Kazakh

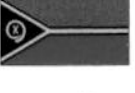

VANUATU
Republic of Vanuatu
Capital Port Vila

Area sq km	12 190	**Currency** Vatu
Area sq miles	4 707	**Languages** English,
Population	253 000	Bislama (creole), French

VATICAN CITY
Vatican City State or Holy See
Capital Vatican City

Area sq km	0.5	**Currency** Euro
Area sq miles	0.2	**Languages** Italian
Population	800	

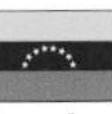

VENEZUELA
Bolivarian Republic of Venezuela
Capital Caracas

Area sq km	912 050	**Currency** Bolívar
Area sq miles	352 144	**Languages** Spanish,
Population	30 405 000	Amerindian lang.

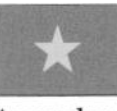

VIETNAM
Socialist Republic of Vietnam
Capital Ha Nôi (Hanoi)

Area sq km	329 565	**Currency** Dong
Area sq miles	127 246	**Languages** Vietnamese, Thai,
Population	91 680 000	Khmer, Chinese, other local lang.

West Bank
Disputed territory

Area sq km	5 860	**Currency** Jordanian dinar,
Area sq miles	2 263	Israeli shekel
Population	2 719 112	**Languages** Arabic, Hebrew

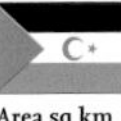

Western Sahara
Disputed territory (Morocco)
Capital Laâyoune

Area sq km	266 000	**Currency** Moroccan dirham
Area sq miles	102 703	**Languages** Arabic
Population	567 000	

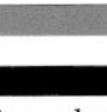

YEMEN
Republic of Yemen
Capital Şan'ā'

Area sq km	527 968	**Currency** Yemeni riyal
Area sq miles	203 850	**Languages** Arabic
Population	24 407 000	

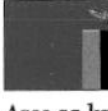

ZAMBIA
Republic of Zambia
Capital Lusaka

Area sq km	752 614	**Currency** Zambian kwacha
Area sq miles	290 586	**Languages** English, Bemba,
Population	14 539 000	Nyanja, Tonga, other local lang.

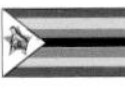

ZIMBABWE
Republic of Zimbabwe
Capital Harare

Area sq km	390 759	**Currency** US dollar and
Area sq miles	150 873	other currencies
Population	14 150 000	**Languages** 16 local languages including English, Shona, Ndebele

Total Land Area 8 844 516 sq km / 3 414 868 sq miles
(includes New Guinea and Pacific Island nations)

HIGHEST MOUNTAIN
Puncak Jaya
4 884 m / 16 023 feet

Oceania cross section

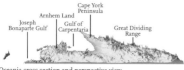

Joseph
Bonaparte Gulf

Arnhem Land

Gulf of
Carpentaria

Cape York
Peninsula

Great Dividing
Range

Oceania cross section and perspective view

Cook Strait

North Island

North Cape

Tasman Sea

HIGHEST MOUNTAINS	metres	feet	Map page
Puncak Jaya, Indonesia	4 884	16 023	59 D3
Puncak Trikora, Indonesia	4 730	15 518	59 D3
Puncak Mandala, Indonesia	4 700	15 420	59 D3
Puncak Yamin, Indonesia	4 595	15 075	—
Mt Wilhelm, Papua New Guinea	4 509	14 793	59 D3
Mt Kubor, Papua New Guinea	4 359	14 301	—

LARGEST ISLAND
New Guinea
808 510 sq km /
312 166 sq miles

LARGEST ISLANDS	sq km	sq miles	Map page
New Guinea	808 510	312 166	59 D3
South Island (Te Waipounamu)	151 215	58 384	54 B2
North Island (Te Ika-a-Māui)	115 777	44 701	54 B1
Tasmania	67 800	26 178	51 D4

LONGEST RIVERS	km	miles	Map page
Murray-Darling	3 672	2 282	52 B2
Darling	2 844	1 767	52 B2
Murray	2 375	1 476	52 B3
Murrumbidgee	1 485	923	52 B2
Lachlan	1 339	832	53 C2
Cooper Creek	1 113	692	52 B1

LARGEST LAKES	sq km	sq miles	Map page
Kati Thanda-Lake Eyre	0–8 900	0–3 436	52 A1
Lake Torrens	0–5 780	0–2 232	52 A1

LARGEST LAKE AND LOWEST POINT
Kati Thanda-Lake Eyre
0-8 900 sq km / 0-3 436 sq miles
16 m / 52 feet below sea level

LONGEST RIVER AND
LARGEST DRAINAGE BASIN
Murray-Darling
3 672 km / 2 282 miles
1 058 000 sq km / 409 000 sq miles

Total Land Area 45 036 492 sq km / 17 388 590 sq miles

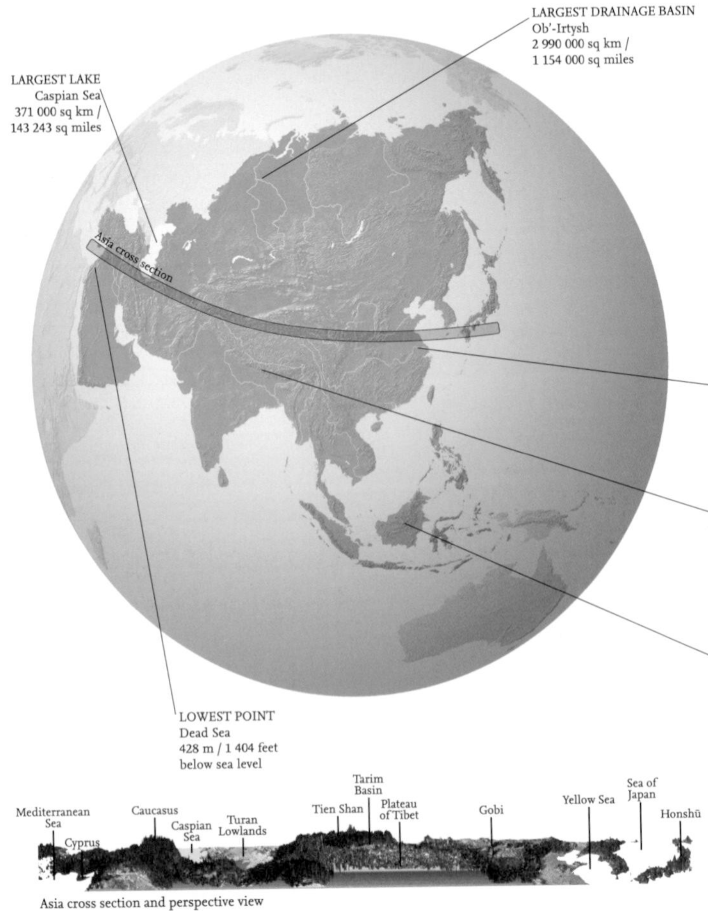

LARGEST DRAINAGE BASIN
Ob'-Irtysh
2 990 000 sq km /
1 154 000 sq miles

LARGEST LAKE
Caspian Sea
371 000 sq km /
143 243 sq miles

Asia cross section

LOWEST POINT
Dead Sea
428 m / 1 404 feet
below sea level

Mediterranean Sea
Cyprus
Caucasus
Caspian Sea
Turan Lowlands
Tien Shan
Tarim Basin
Plateau of Tibet
Gobi
Yellow Sea
Sea of Japan
Honshū

Asia cross section and perspective view

HIGHEST MOUNTAINS	metres	feet	Map page
Mt Everest (Sagarmatha/ Qomolangma Feng), China/Nepal	8 848	29 028	75 C2
K2 (Qogir Feng), China/Pakistan	8 611	28 251	74 B1
Kangchenjunga, India/Nepal	8 586	28 169	75 C2
Lhotse, China/Nepal	8 516	27 939	—
Makalu, China/Nepal	8 463	27 765	—
Cho Oyu, China/Nepal	8 201	26 906	—

LARGEST ISLANDS	sq km	sq miles	Map page
Borneo	745 561	287 861	61 C1
Sumatra (Sumatera)	473 606	182 859	60 A1
Honshū	227 414	87 805	67 B3
Celebes (Sulawesi)	189 216	73 056	58 C3
Java (Jawa)	132 188	51 038	61 B2
Luzon	104 690	40 421	64 B1

LONGEST RIVER
Yangtze (Chang Jiang)
6 380 km /
3 965 miles

LONGEST RIVERS	km	miles	Map page
Yangtze (Chang Jiang)	6 380	3 965	70 C2
Ob'-Irtysh	5 568	3 460	86 F2
Yenisey-Angara-Selenga	5 550	3 449	83 H3
Yellow (Huang He)	5 464	3 395	70 B2
Irtysh	4 440	2 759	86 F2
Mekong	4 425	2 750	63 B2

HIGHEST MOUNTAIN
Mt Everest
8 848 m / 29 028 feet

LARGEST LAKES	sq km	sq miles	Map page
Caspian Sea	371 000	143 243	81 C1
Lake Baikal (Ozero Baykal)	30 500	11 776	69 D1
Lake Balkhash (Ozero Balkash)	17 400	6 718	77 D2
Aral Sea (Aral'skoye More)	17 158	6 625	76 B2
Ysyk-Köl	6 200	2 394	77 D2

LARGEST ISLAND
Borneo
745 561 sq km /
287 861 sq miles

Total Land Area 9 908 599 sq km / 3 825 710 sq miles

LARGEST ISLAND
Great Britain
218 476 sq km /
84 354 sq miles

Europe cross section

HIGHEST MOUNTAIN
El'brus
5 642 m / 18 510 feet

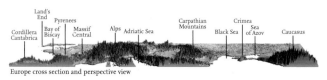

Cordillera
Cantabrica

Land's
End

Bay of
Biscay

Pyrenees

Massif
Central

Alps

Adriatic Sea

Carpathian
Mountains

Black Sea

Crimea

Sea
of Azov

Caucasus

Europe cross section and perspective view

HIGHEST MOUNTAINS	metres	feet	Map pages
El'brus, Russia	5 642	18 510	87 D4
Gora Dykh-Tau, Russia	5 204	17 073	—
Shkhara, Georgia/Russia	5 201	17 063	—
Kazbek, Georgia/Russia	5 047	16 558	76 A2
Mont Blanc, France/Italy	4 810	15 781	105 D2
Dufourspitze, Italy/Switzerland	4 634	15 203	—

LARGEST ISLANDS	sq km	sq miles	Map pages
Great Britain	218 476	84 354	95 C3
Iceland	102 820	39 699	92 A3
Ireland	83 045	32 064	97 C2
Ostrov Severnyy (part of Novaya Zemlya)	47 079	18 177	86 E1
Spitsbergen	37 814	14 600	82 C1

LONGEST RIVER AND
LARGEST DRAINAGE BASIN
Volga
3 688 km / 2 292 miles
1 380 000 sq km / 533 000 sq miles

LONGEST RIVERS	km	miles	Map pages
Volga	3 688	2 292	89 F2
Danube	2 850	1 771	110 A1
Dnieper	2 285	1 420	91 C2
Kama	2 028	1 260	86 E3
Don	1 931	1 200	89 E3
Pechora	1 802	1 120	86 E2

LARGEST LAKE AND LOWEST POINT
Caspian Sea
371 000 sq km / 143 243 sq miles
28m / 92 feet below sea level

LARGEST LAKES	sq km	sq miles	Map pages
Caspian Sea	371 000	143 243	81 C1
Lake Ladoga (Ladozhskoye Ozero)	18 390	7 100	86 C2
Lake Onega (Onezhskoye Ozero)	9 600	3 707	86 C2
Vänern	5 585	2 156	93 F4
Rybinskoye Vodokhranilishche	5 180	2 000	89 E2

Total Land Area 30 343 578 sq km / 11 715 655 sq miles

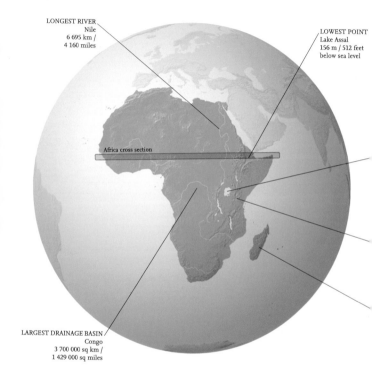

LONGEST RIVER
Nile
6 695 km /
4 160 miles

LOWEST POINT
Lake Assal
156 m / 512 feet
below sea level

Africa cross section

LARGEST DRAINAGE BASIN
Congo
3 700 000 sq km /
1 429 000 sq miles

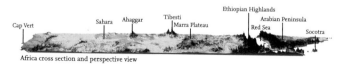

Cap Vert   Sahara   Ahaggar   Tibesti   Marra Plateau   Ethiopian Highlands   Red Sea   Arabian Peninsula   Socotra

Africa cross section and perspective view

HIGHEST MOUNTAINS	metres	feet	Map page
Kilimanjaro, Tanzania	5 892	19 330	119 D3
Mt Kenya (Kirinyaga), Kenya	5 199	17 057	119 D3
Margherita Peak, Democratic Republic of the Congo/Uganda	5 110	16 765	119 C2
Meru, Tanzania	4 565	14 977	119 D3
Ras Dejen, Ethiopia	4 533	14 872	117 B3
Mt Karisimbi, Rwanda	4 510	14 796	—

LARGEST ISLANDS	sq km	sq miles	Map page
Madagascar	587 040	226 656	121 D3

LONGEST RIVERS	km	miles	Map page
Nile	6 695	4 160	116 B1
Congo	4 667	2 900	118 B3
Niger	4 184	2 600	115 C4
Zambezi	2 736	1 700	120 C2
Wabē Shebelē Wenz	2 490	1 547	117 C4
Ubangi	2 250	1 398	118 B3

LARGEST LAKES	sq km	sq miles	Map page
Lake Victoria	68 870	26 591	52 B2
Lake Tanganyika	32 600	12 587	119 C3
Lake Nyasa (Lake Malawi)	29 500	11 390	121 C1
Lake Volta	8 482	3 275	114 C4
Lake Turkana	6 500	2 510	119 D2
Lake Albert	5 600	2 162	119 D2

LARGEST LAKE
Lake Victoria
68 870 sq km /
26 591 sq miles

HIGHEST MOUNTAIN
Kilimanjaro
5 892 m / 19 330 feet

LARGEST ISLAND
Madagascar
587 040 sq km /
226 656 sq miles

Total Land Area 24 680 331 sq km / 9 529 076 sq miles
(including Hawaiian Islands)

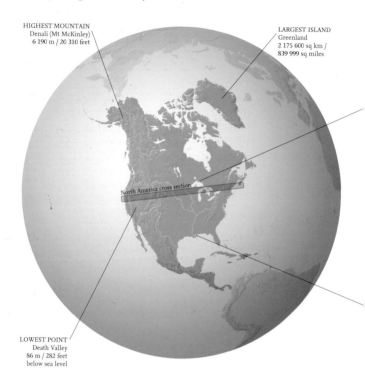

HIGHEST MOUNTAIN
Denali (Mt McKinley)
6 190 m / 20 310 feet

LARGEST ISLAND
Greenland
2 175 600 sq km /
839 999 sq miles

North America cross section

LOWEST POINT
Death Valley
86 m / 282 feet
below sea level

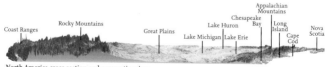

Coast Ranges — Rocky Mountains — Great Plains — Lake Michigan — Lake Huron — Lake Erie — Chesapeake Bay — Appalachian Mountains — Long Island — Cape Cod — Nova Scotia

North America cross section and perspective view

HIGHEST MOUNTAINS	metres	feet	Map page
Denali, USA	6 190	20 310	124 F2
Mt Logan, Canada	5 959	19 550	126 B2
Pico de Orizaba, Mexico	5 610	18 405	145 C3
Mt St Elias, USA	5 489	18 008	126 B2
Volcán Popocatépetl, Mexico	5 452	17 887	145 C3
Mt Foraker, USA	5 303	17 398	—

LARGEST LAKE
Lake Superior
82 100 sq km /
31 699 sq miles

LARGEST ISLANDS	sq km	sq miles	Map page
Greenland	2 175 600	839 999	127 I2
Baffin Island	507 451	195 927	127 G2
Victoria Island	217 291	83 896	126 D2
Ellesmere Island	196 236	75 767	127 F1
Cuba	110 860	42 803	146 B2
Newfoundland	108 860	42 031	131 E2
Hispaniola	76 192	29 418	147 C2

LONGEST RIVERS	km	miles	Map page
Mississippi-Missouri	5 969	3 709	133 D3
Mackenzie-Peace-Finlay	4 241	2 635	126 C2
Missouri	4 086	2 539	137 E3
Mississippi	3 765	2 340	142 C3
Yukon	3 185	1 979	126 A2
St Lawrence	3 058	1 900	131 D2

LONGEST RIVER AND
LARGEST DRAINAGE BASIN
Mississippi-Missouri
5 969 km / 3 709 miles
3 250 000 sq km / 1 255 000
sq miles

LARGEST LAKES	sq km	sq miles	Map page
Lake Superior	82 100	31 699	140 B1
Lake Huron	59 600	23 012	140 C2
Lake Michigan	57 800	22 317	140 B2
Great Bear Lake	31 328	12 096	126 C2
Great Slave Lake	28 568	11 030	128 C1
Lake Erie	25 700	9 923	140 C2
Lake Winnipeg	24 387	9 416	129 E2
Lake Ontario	18 960	7 320	141 D2

Total Land Area 17 815 420 sq km / 6 878 534 sq miles

LARGEST LAKE
Lake Titicaca
8 340 sq km /
3 220 sq miles

South America cross section

LARGEST ISLAND
Isla Grande de Tierra del Fuego
47 000 sq km / 18 147 sq miles

Andes

Selvas

Bahia de
São Marcos

Cabo de
São Roque

South America cross section and perspective view

HIGHEST MOUNTAINS	metres	feet	Map page
Cerro Aconcagua, Argentina	6 959	22 831	153 B4
Nevado Ojos del Salado, Argentina/Chile	6 908	22 664	152 B3
Cerro Bonete, Argentina	6 872	22 546	—
Cerro Pissis, Argentina	6 858	22 500	—
Cerro Tupungato, Argentina/Chile	6 800	22 309	—
Cerro Mercedario, Argentina	6 770	22 211	—

LARGEST ISLANDS	sq km	sq miles	Map page
Isla Grande de Tierra del Fuego	47 000	18 147	153 B6
Isla de Chiloé	8 394	3 241	153 A5
East Falkland	6 760	2 610	153 C6
West Falkland	5 413	2 090	153 B6

LONGEST RIVER AND
LARGEST DRAINAGE BASIN
Amazon
8 516 km / 4 049 miles
7 050 000 sq km / 2 722 000 sq miles

LONGEST RIVERS	km	miles	Map page
Amazon (Amazonas)	6 516	4 049	150 C1
Río de la Plata-Paraná	4 500	2 796	153 C4
Purus	3 218	2 000	150 B2
Madeira	3 200	1 988	150 C2
São Francisco	2 900	1 802	151 E3
Tocantins	2 750	1 709	151 D2

HIGHEST MOUNTAIN
Cerro Aconcagua
6 959 m / 22 831 feet

LARGEST LAKES	sq km	sq miles	Map page
Lake Titicaca	8 340	3 220	152 B2

LOWEST POINT
Laguna del Carbón
105 m / 344 feet below sea level

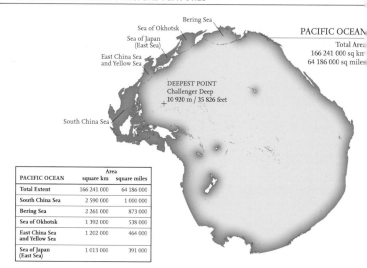

Bering Sea

Sea of Okhotsk

Sea of Japan
(East Sea)

East China Sea
and Yellow Sea

South China Sea

DEEPEST POINT
Challenger Deep
10 920 m / 35 826 feet

PACIFIC OCEAN

Total Area
166 241 000 sq km
64 186 000 sq miles

PACIFIC OCEAN	Area	
	square km	square miles
Total Extent	166 241 000	64 186 000
South China Sea	2 590 000	1 000 000
Bering Sea	2 261 000	873 000
Sea of Okhotsk	1 392 000	538 000
East China Sea and Yellow Sea	1 202 000	464 000
Sea of Japan (East Sea)	1 013 000	391 000

ANTARCTICA

Total Land Area 12 093 000 sq km /
4 669 107 sq miles (excluding ice shelves)

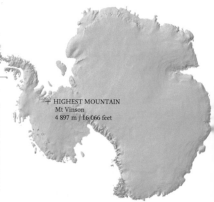

HIGHEST MOUNTAIN
Mt Vinson
4 897 m / 16 066 feet

HIGHEST MOUNTAINS	Height	
	metres	feet
Mt Vinson	4 897	16 066
Mt Tyree	4 852	15 918
Mt Kirkpatrick	4 528	14 855
Mt Markham	4 351	14 275
Mt Sidley	4 285	14 058
Mt Minto	4 165	13 665

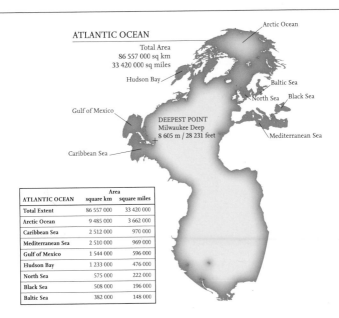

ATLANTIC OCEAN

Total Area
86 557 000 sq km
33 420 000 sq miles

Arctic Ocean

Hudson Bay

Baltic Sea

North Sea   Black Sea

Gulf of Mexico

DEEPEST POINT
Milwaukee Deep
8 605 m / 28 231 feet

Mediterranean Sea

Caribbean Sea

| ATLANTIC OCEAN | Area | |
	square km	square miles
Total Extent	86 557 000	33 420 000
Arctic Ocean	9 485 000	3 662 000
Caribbean Sea	2 512 000	970 000
Mediterranean Sea	2 510 000	969 000
Gulf of Mexico	1 544 000	596 000
Hudson Bay	1 233 000	476 000
North Sea	575 000	222 000
Black Sea	508 000	196 000
Baltic Sea	382 000	148 000

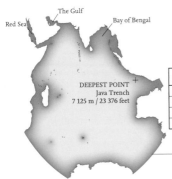

The Gulf

Bay of Bengal

Red Sea

DEEPEST POINT
Java Trench
7 125 m / 23 376 feet

| INDIAN OCEAN | Area | |
	square km	square miles
Total Extent	73 427 000	28 350 000
Bay of Bengal	2 172 000	839 000
Red Sea	453 000	175 000
The Gulf	238 000	92 000

INDIAN OCEAN

Total Area
73 427 000 sq km
28 350 000 sq miles

MAJOR CLIMATIC REGIONS AND SUB-TYPES
Köppen classification system
Winkel Tripel Projection
scale 1:200 000 000

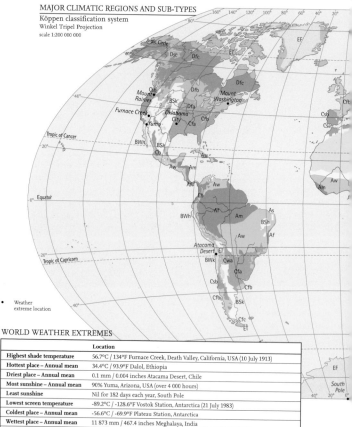

• Weather extreme location

WORLD WEATHER EXTREMES

	Location
Highest shade temperature	56.7°C / 134°F Furnace Creek, Death Valley, California, USA (10 July 1913)
Hottest place – Annual mean	34.4°C / 93.9°F Dalol, Ethiopia
Driest place – Annual mean	0.1 mm / 0.004 inches Atacama Desert, Chile
Most sunshine – Annual mean	90% Yuma, Arizona, USA (over 4 000 hours)
Least sunshine	Nil for 182 days each year, South Pole
Lowest screen temperature	-89.2°C / -128.6°F Vostok Station, Antarctica (21 July 1983)
Coldest place – Annual mean	-56.6°C / -69.9°F Plateau Station, Antarctica
Wettest place – Annual mean	11 873 mm / 467.4 inches Meghalaya, India
Highest surface wind speed	
- High altitude	372 km per hour/231 miles per hour Mount Washington, New Hampshire, USA, (12 April 1934)
- Low altitude	408 km per hour/254 miles per hour Barrow Island, Australia (10 April 1996)
- Tornado	512 km per hour / 318 miles per hour in a tornado, Oklahoma City, Oklahoma, USA (3 May 1999)
Greatest snowfall	31 102 mm / 1 224.5 inches Mount Rainier, Washington, USA (19 February 1971 – 18 February 1972)

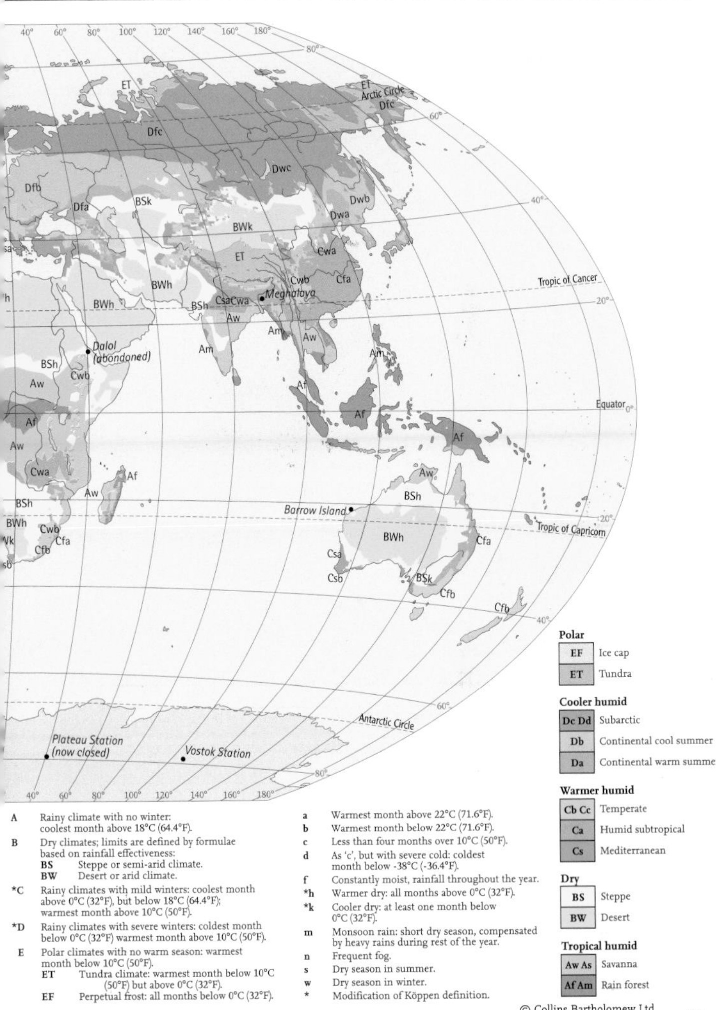

A   Rainy climate with no winter:
     coolest month above 18°C (64.4°F).

B   Dry climates; limits are defined by formulae
     based on rainfall effectiveness:
     BS   Steppe or semi-arid climate.
     BW   Desert or arid climate.

*C   Rainy climates with mild winters: coolest month
     above 0°C (32°F), but below 18°C (64.4°F);
     warmest month above 10°C (50°F).

*D   Rainy climates with severe winters: coldest month
     below 0°C (32°F) warmest month above 10°C (50°F).

E   Polar climates with no warm season: warmest
     month below 10°C (50°F).
     ET   Tundra climate: warmest month below 10°C
          (50°F) but above 0°C (32°F).
     EF   Perpetual frost: all months below 0°C (32°F).

a   Warmest month above 22°C (71.6°F).
b   Warmest month below 22°C (71.6°F).
c   Less than four months over 10°C (50°F).
d   As 'c', but with severe cold: coldest
     month below -38°C (-36.4°F).
f   Constantly moist, rainfall throughout the year.
*h   Warmer dry: all months above 0°C (32°F).
*k   Cooler dry: at least one month below
     0°C (32°F).
m   Monsoon rain: short dry season, compensated
     by heavy rains during rest of the year.
n   Frequent fog.
s   Dry season in summer.
w   Dry season in winter.
*   Modification of Köppen definition.

Polar

EF	Ice cap
ET	Tundra

Cooler humid

Dc Dd	Subarctic
Db	Continental cool summer
Da	Continental warm summer

Warmer humid

Cb Cc	Temperate
Ca	Humid subtropical
Cs	Mediterranean

Dry

BS	Steppe
BW	Desert

Tropical humid

Aw As	Savanna
Af Am	Rain forest

© Collins Bartholomew Ltd

35

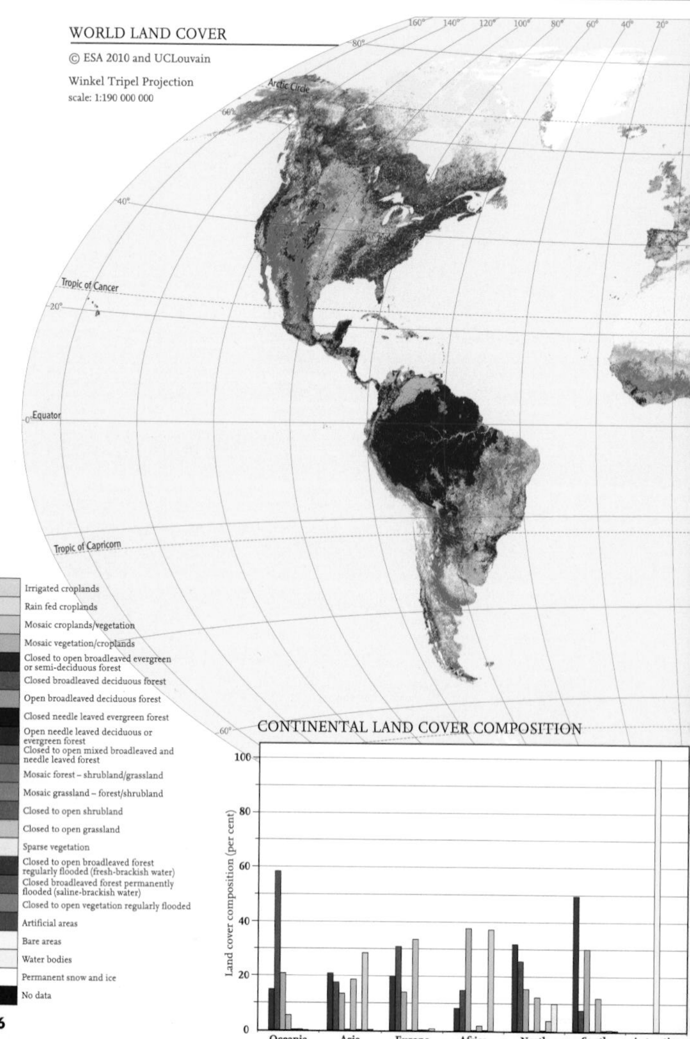

WORLD LAND COVER

© ESA 2010 and UCLouvain

Winkel Tripel Projection
scale: 1:190 000 000

Arctic Circle

Tropic of Cancer

Equator

Tropic of Capricorn

Irrigated croplands
Rain fed croplands
Mosaic croplands/vegetation
Mosaic vegetation/croplands
Closed to open broadleaved evergreen
or semi-deciduous forest
Closed broadleaved deciduous forest
Open broadleaved deciduous forest
Closed needle leaved evergreen forest
Open needle leaved deciduous or
evergreen forest
Closed to open mixed broadleaved and
needle leaved forest
Mosaic forest – shrubland/grassland
Mosaic grassland – forest/shrubland
Closed to open shrubland
Closed to open grassland
Sparse vegetation
Closed to open broadleaved forest
regularly flooded (fresh-brackish water)
Closed broadleaved forest permanently
flooded (saline-brackish water)
Closed to open vegetation regularly flooded
Artificial areas
Bare areas
Water bodies
Permanent snow and ice
No data

CONTINENTAL LAND COVER COMPOSITION

Land cover composition [per cent]

100

80

60

40

20

0

Oceania   Asia   Europe   Africa   North   South   Antarctica
                                              America America

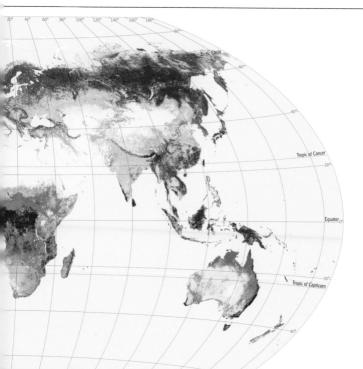

LAND COVER GRAPHS - CLASSIFICATION

Class description	Map classes
Forest/Woodland	Evergreen needleleaf forest
	Evergreen broadleaf forest
	Deciduous needleleaf forest
	Deciduous broadleaf forest
	Mixed forest
Shrubland	Closed shrublands
	Open shrublands
Grass/Savanna	Woody savannas
	Savannas
	Grasslands
Wetland	Permanent wetlands
Crops/Mosaic	Croplands
	Cropland/Natural vegetation mosaic
Urban	Urban and built-up
Snow/Ice	Snow and Ice
Barren	Barren or sparsely vegetated

GLOBAL LAND COVER COMPOSITION

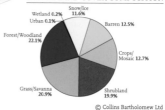

Wetland 0.2%
Urban 0.1%
Snow/Ice 11.6%
Barren 12.5%
Forest/Woodland 22.1%
Crops/Mosaic 12.7%
Grass/Savanna 20.9%
Shrubland 19.9%

© Collins Bartholomew Ltd

WORLD POPULATION DISTRIBUTION

Population Density
Winkel Tripel Projection
scale 1:190 000 000

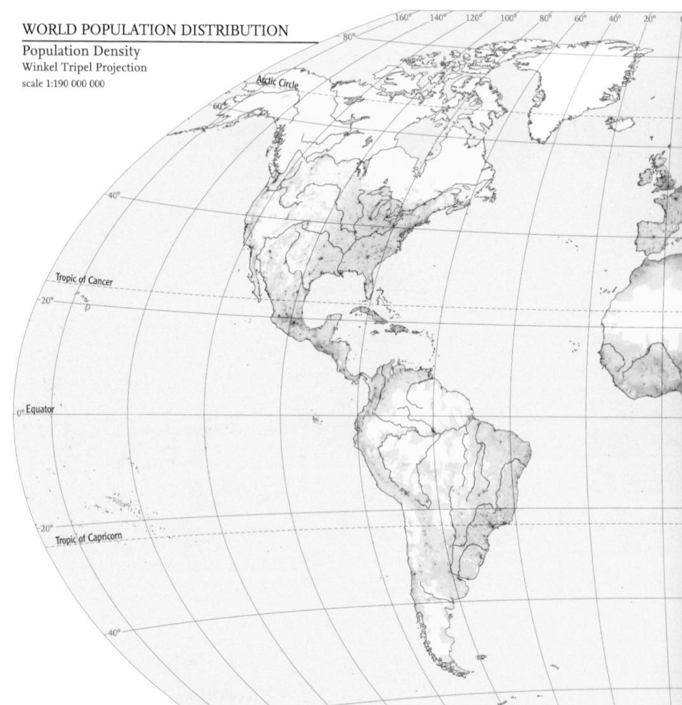

KEY POPULATION STATISTICS FOR MAJOR REGIONS

	Population 2013 (millions)	Growth (per cent)	Infant mortality rate	Total fertility rate	Life expectancy (years)
World	7 162	1.1	37	2.5	70
More developed regions[1]	1 253	0.3	6	1.7	78
Less developed regions[2]	5 909	1.3	40	2.6	67
Africa	1 111	2.5	64	4.7	58
Asia	4 299	1.0	31	2.2	71
Europe[3]	742	0.1	6	1.6	76
Latin America and the Caribbean[4]	617	1.1	18	2.2	75
North America	355	0.9	6	1.9	79
Oceania	38	1.4	20	2.4	78

1. Europe, North America, Australia, New Zealand and Japan.
2. Africa, Asia (excluding Japan), Latin America and the Caribbean, and Oceania (excluding Australia and New Zealand).
3. Includes Russia.
4. South America, Central America (including Mexico) and all Caribbean Islands.

Except for population (2013) the data are annual averages projected for the period 2010–2015.

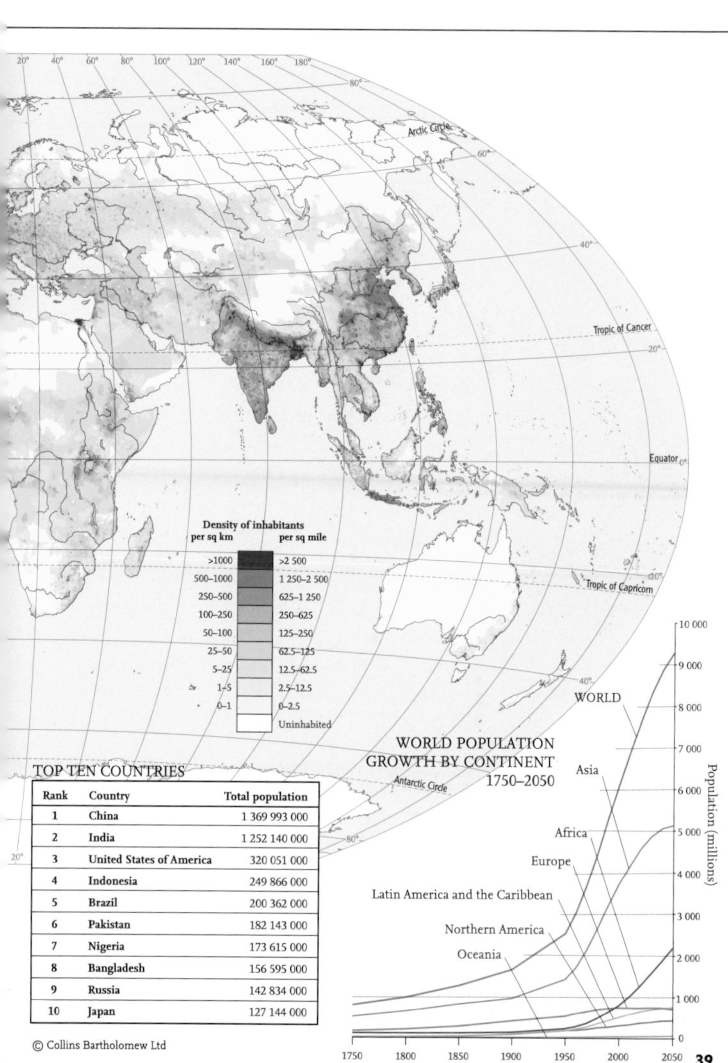

Density of inhabitants

per sq km	per sq mile
>1000	>2 500
500–1000	1 250–2 500
250–500	625–1 250
100–250	250–625
50–100	125–250
25–50	62.5–125
5–25	12.5–62.5
1–5	2.5–12.5
0–1	0–2.5
	Uninhabited

TOP TEN COUNTRIES

Rank	Country	Total population
1	China	1 369 993 000
2	India	1 252 140 000
3	United States of America	320 051 000
4	Indonesia	249 866 000
5	Brazil	200 362 000
6	Pakistan	182 143 000
7	Nigeria	173 615 000
8	Bangladesh	156 595 000
9	Russia	142 834 000
10	Japan	127 144 000

© Collins Bartholomew Ltd

WORLD POPULATION GROWTH BY CONTINENT 1750–2050

WORLD

Asia

Africa

Europe

Latin America and the Caribbean

Northern America

Oceania

Population (millions)

Year

THE WORLD'S MAJOR CITIES

Urban agglomerations with over
1 million inhabitants.
Winkel Tripel Projection
scale 1:190 000 000

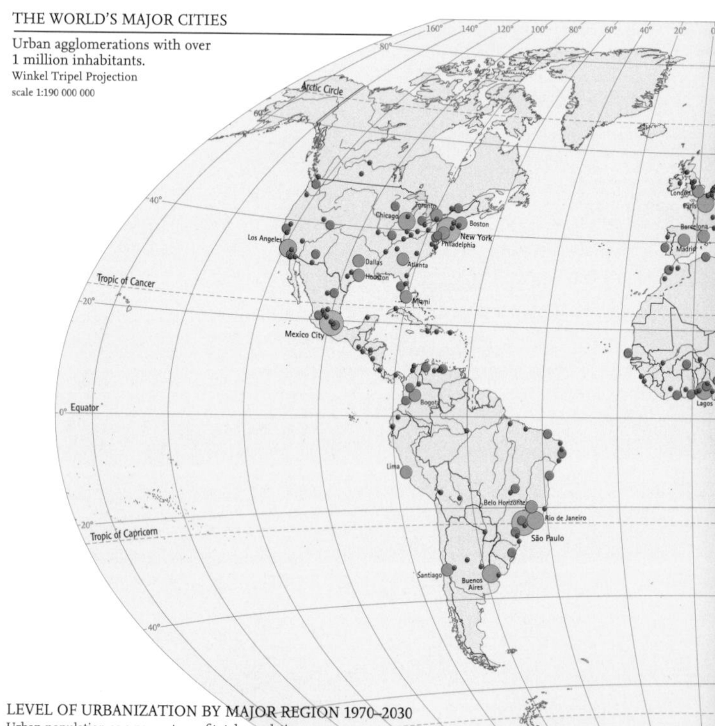

LEVEL OF URBANIZATION BY MAJOR REGION 1970–2030

Urban population as a percentage of total population

	1970	2010	2030
World	36.6	51.6	59.9
More developed regions[1]	66.6	77.5	82.1
Less developed regions[2]	25.3	46.0	55.8
Africa	23.5	39.2	47.7
Asia	23.7	44.4	55.5
Europe[3]	62.8	72.7	77.4
Latin America and the Caribbean[4]	57.1	78.8	83.4
Northern America	73.8	82.0	85.8
Oceania	71.2	70.7	71.4

1. Europe, North America, Australia,
New Zealand and Japan.

2. Africa, Asia (excluding Japan), Latin
America and the Caribbean, and
Oceania (excluding Australia and
New Zealand).

3. Includes Russia.

4. South America, Central America
(including Mexico) and all Caribbean
Islands.

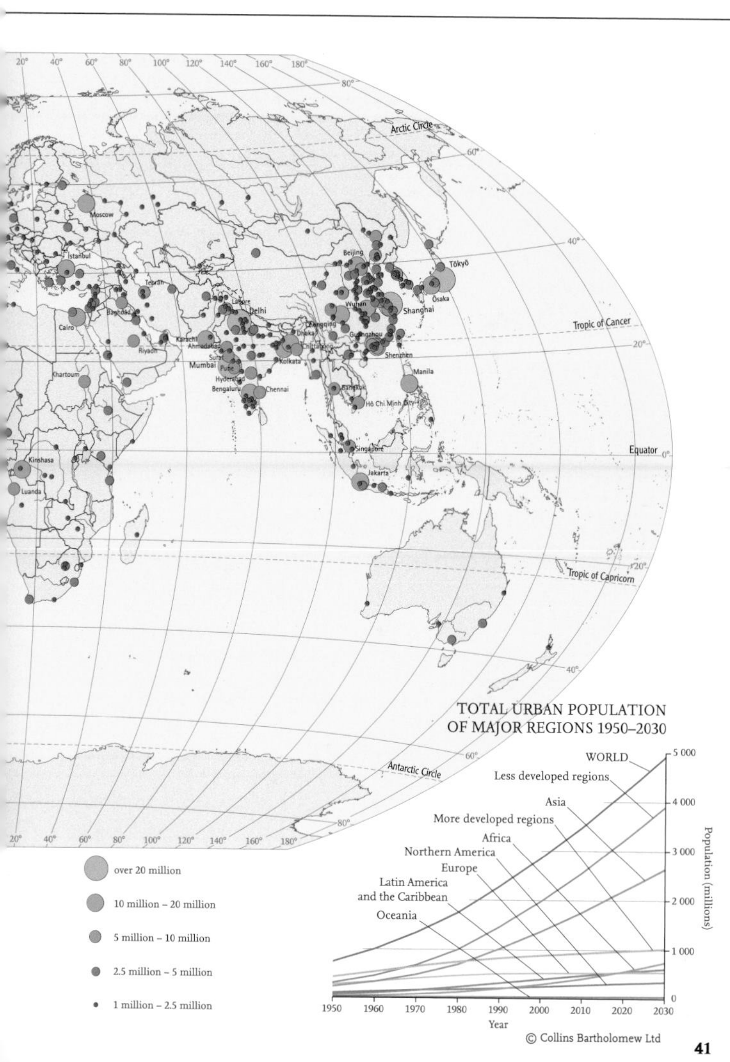

TOTAL URBAN POPULATION
OF MAJOR REGIONS 1950–2030

over 20 million

10 million – 20 million

5 million – 10 million

2.5 million – 5 million

1 million – 2.5 million

© Collins Bartholomew Ltd

SYMBOLS AND ABBREVIATIONS

SYMBOLS

Map symbols used on the map pages are explained here. The status of nations and their boundaries are shown in this atlas as they are in reality at time of going to press, as far as can be ascertained. Where international boundaries are subject of disputes the aim is to take a strictly neutral viewpoint, based on advice from expert consultants. Settlements are classified in terms of both population and administrative significance. The abbreviations listed are those used in place names on the map pages and within the index.

BOUNDARIES

International boundary

Disputed international boundary or alignment unconfirmed

Undefined international boundary in the sea. All land within this boundary is part of state or territory named.

Disputed territory boundary

Administrative boundary Shown for selected countries only.

Ceasefire line or other boundary described on the map

TRANSPORT

Motorway

Main road

Track

Main railway

Canal

Main airport

LAND AND WATER FEATURES

Lake

Impermanent lake

Salt lake or lagoon

Impermanent salt lake

Dry salt lake or salt pan

River

Impermanent river

Ice cap / Glacier

123 Pass
height in metres

123 Summit
△ height in metres

1234 Volcano
▲ Height in metres

∴ Site of special interest

ᴖᴖᴖᴖ Wall

CITIES AND TOWNS

Built-up area
SCALE 1:4 000 000 only

Population	National Capital	Administrative Capital Shown for selected countries only	Other City or Town
over 10 million	**BEIJING** ▣	**São Paulo** ◉	**New York** ◉
5 to 10 million	**MADRID** ▣	**Toronto** ◉	**Philadelphia** ◉
1 to 5 million	**KUWAIT** ☐	**Sydney** ○	**Seattle** ○
500 000 to 1 million	**BANGUI** ☐	**Winnipeg** ○	**Warangal** ○
100 000 to 500 000	WELLINGTON ☐	Edinburgh ○	Apucarana ○
50 000 to 100 000	PORT OF SPAIN ☐	Bismarck ○	Invercargill ○
under 50 000	MALABO ▫	Charlottetown ○	Ceres ○

STYLES OF LETTERING

Cities and towns are explained separately

		Physical features	
Country	**FRANCE**	Island	*Gran Canaria*
Overseas Territory/Dependency	**Guadeloupe**	Lake	*Lake Erie*
Disputed Territory	WESTERN SAHARA	Mountain	*Mt Blanc*
Administrative name Shown for selected countries only.	**SCOTLAND**	River	*Thames*
Area name	PATAGONIA	Region	*LAPPLAND*

CONTINENTAL MAPS

BOUNDARIES

——— International boundary

------ Disputed international boundary

········ Ceasefire line

CITIES AND TOWNS

National capital	Other city or town
Kuwait □	Seattle ○

ABBREVIATIONS

Arch.	Archipelago			
B.	Bay			
	Bahia, Baía	Portuguese	bay	
	Bahía	Spanish	bay	
	Baie	French	bay	
C.	Cape			
	Cabo	Portuguese, Spanish	cape, headland	
	Cap	French	cape, headland	
Co	Cerro	Spanish	hill, peak, summit	
E.	East, Eastern			
Est.	Estrecho	Spanish	strait	
Gt	Great			
I.	Island, Isle			
	Ilha	Portuguese	island	
	Isla	Spanish	island	
Is	Islands, Isles			
	Islas	Spanish	islands	
Khr.	Khrebet	Russian	mountain range	
L.	Lake			
	Loch	(Scotland)	lake	
	Lough	(Ireland)	lake	
	Lac	French	lake	
	Lago	Portuguese, Spanish	lake	
M.	Mys	Russian	cape, point	
Mt	Mount			
	Mont	French	hill, mountain	
Mt.	Mountain			

Mts	Mountains			
	Monts	French	hills, mountains	
N.	North, Northern			
O.	Ostrov	Russian	island	
Pt	Point			
Pta	Punta	Italian, Spanish	cape, point	
R.	River			
	Rio	Portuguese	river	
	Río	Spanish	river	
	Rivière	French	river	
Ra.	Range			
S.	South, Southern			
	Salar, Salina, Salinas	Spanish	saltpan, saltpans	
Sa	Serra	Portuguese	mountain range	
	Sierra	Spanish	mountain range	
Sd	Sound			
S.E.	Southeast, Southeastern			
St	Saint			
	Sankt	German		
	Sint	Dutch	saint	
Sta	Santa	Italian, Portuguese, Spanish	saint	
Ste	Sainte	French	saint	
Str.	Strait			
W.	West, Western			
	Wadi, Wādī	Arabic	watercourse	

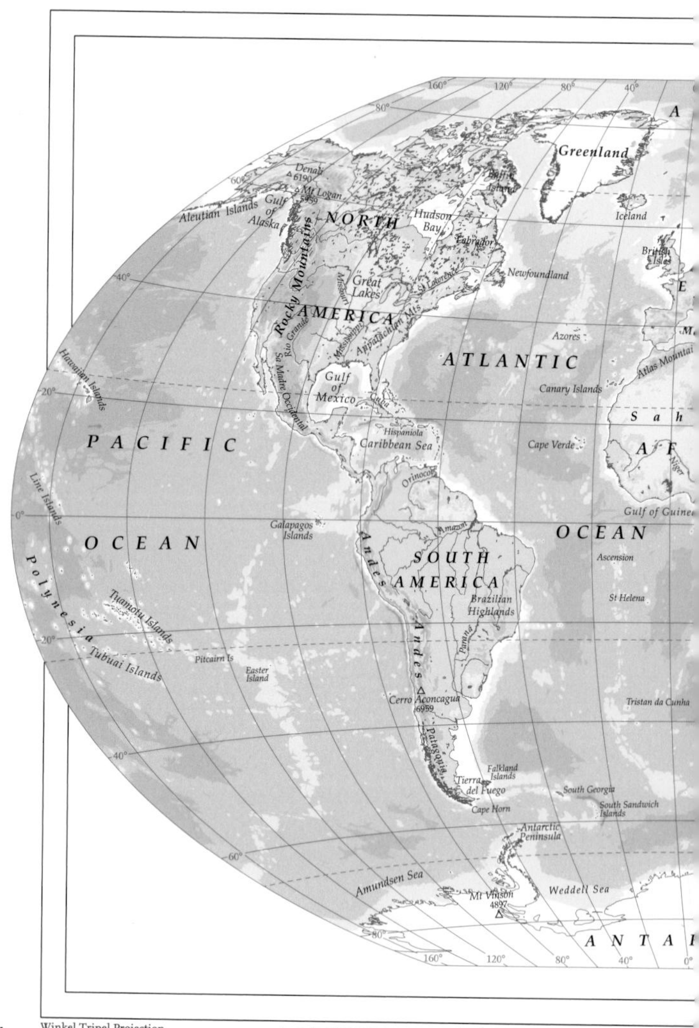

NORTH AMERICA

Denali 6190▲
Mt Logan
Aleutian Islands
Gulf of Alaska
Rocky Mountains
Great Lakes
Rio Grande
Sierra Madre Occidental
Appalachian Mts
Mississippi
Gulf of Mexico
Hispaniola
Caribbean Sea
Orinoco

Hawaiian Islands

Greenland
Baffin Island
Iceland
British Isles
Newfoundland
Labrador
Hudson Bay

Azores
Canary Islands
Cape Verde

ATLANTIC

Atlas Mountains
Sahara
AFRICA
Gulf of Guinea

PACIFIC

Line Islands

Polynesia

OCEAN

Galapagos Islands

Amazon

Tuamotu Islands
Tubuai Islands
Pitcairn Is
Easter Island

SOUTH AMERICA

Andes
Brazilian Highlands
Pampas
Cerro Aconcagua 6959▲
Tierra del Fuego
Cape Horn
Falkland Islands

OCEAN

Ascension
St Helena
Tristan da Cunha

South Georgia
South Sandwich Islands

Antarctic Peninsula
Amundsen Sea
Mt Vinson 4897△
Weddell Sea

ANTARCTICA

Winkel Tripel Projection

1 : 170 000 000

MILES 0   1000   2000   3000

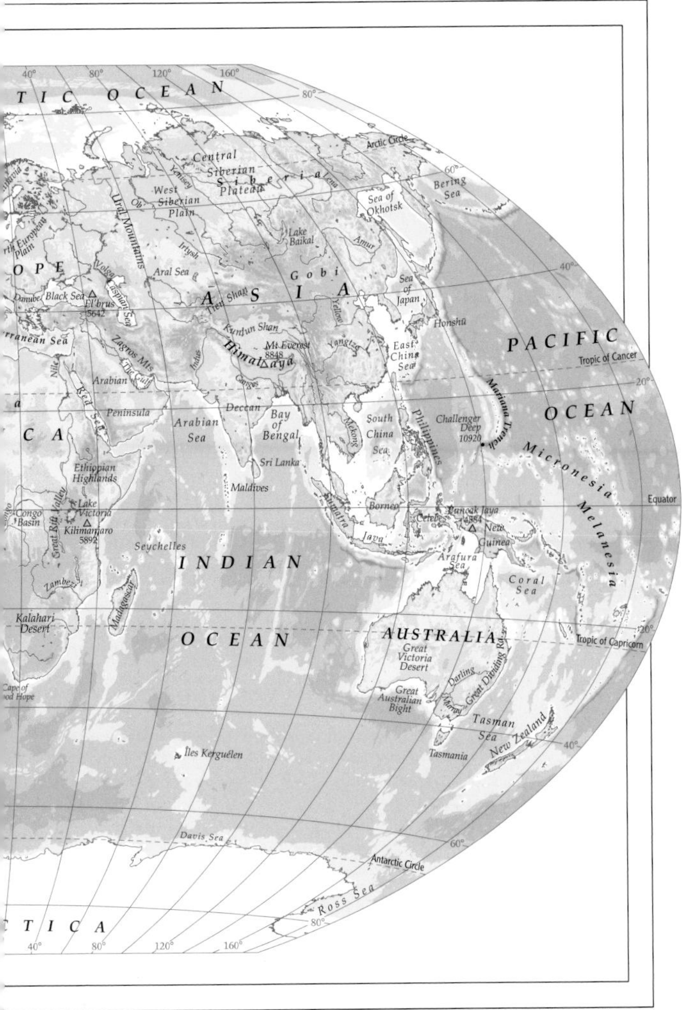

ARCTIC OCEAN

Arctic Circle

Central
Siberian
Plateau

West
Siberian
Plain

Siberian Lands

Bering
Sea

Sea of
Okhotsk

Ural Mountains

EUROPE

Great European
Plain

Volga

Ob

Irtysh

Aral Sea

Lake
Baikal

Amur

Danube

Black Sea

Elbrus
5642

Gobi

ASIA

Sea
of
Japan

Honshū

PACIFIC

Mediterranean Sea

Zagros Mts

Red Sea

Tien Shan

Kunlun Shan

Mt Everest
8848

Himalaya

Yangtze

East
China
Sea

Tropic of Cancer

Nile

Arabian
Peninsula

Deccan

Bay
of
Bengal

Mekong

South
China
Sea

Challenger
Deep
10920

Mariana Trench

OCEAN

AFRICA

Arabian
Sea

Sri Lanka

Philippines

Micronesia

Maldives

Sumatra

Borneo

Celebes

Equator

Congo
Basin

Lake
Victoria

Kilimanjaro
5892

Great Rift Valley

Ethiopian
Highlands

Seychelles

INDIAN

Java

Bangkok

New
Guinea

Melanesia

Zambezi

Arafura
Sea

Coral
Sea

Madagascar

Kalahari
Desert

OCEAN

AUSTRALIA

Great
Victoria
Desert

Darling

Great Dividing Range

Tropic of Capricorn

Cape of
Good Hope

Great
Australian
Bight

Murray

Tasman
Sea

New Zealand

Tasmania

40°

Îles Kerguélen

Davis Sea

Antarctic Circle

ANTARCTICA

Ross Sea

80°

40°   80°   120°   160°

| 0 | 1000 | 2000 | 3000 | 4000 | 5000 KILOMETRES |

© Collins Bartholomew Ltd

45

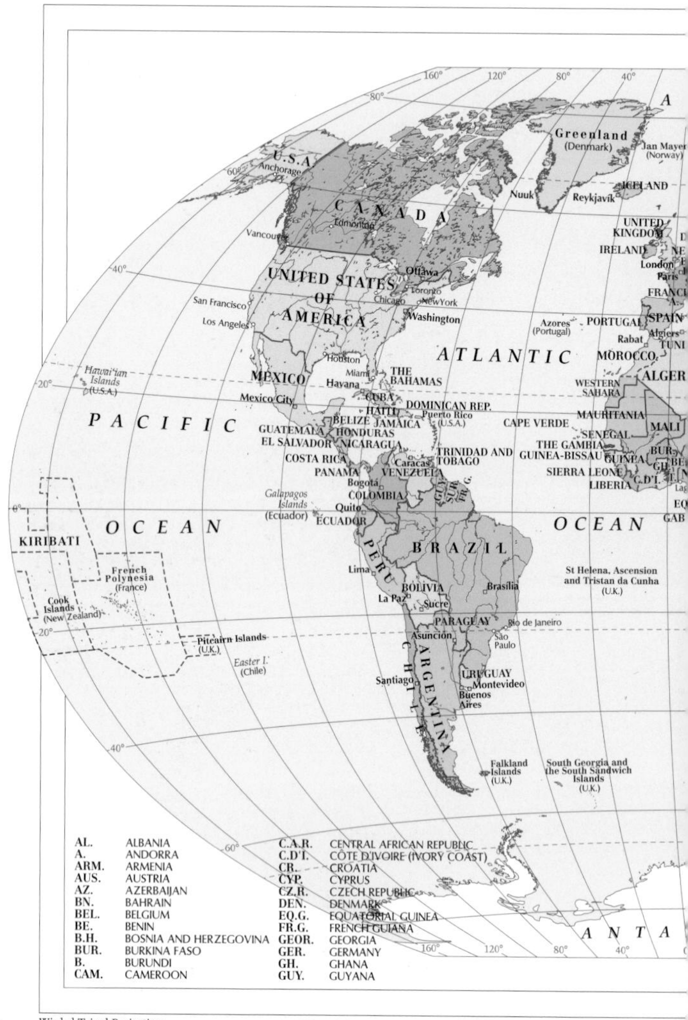

AL.	ALBANIA	C.A.R.	CENTRAL AFRICAN REPUBLIC
A.	ANDORRA	C.D'I.	CÔTE D'IVOIRE (IVORY COAST)
ARM.	ARMENIA	CR.	CROATIA
AUS.	AUSTRIA	CYP.	CYPRUS
AZ.	AZERBAIJAN	CZ.R.	CZECH REPUBLIC
BN.	BAHRAIN	DEN.	DENMARK
BEL.	BELGIUM	EQ.G.	EQUATORIAL GUINEA
BE.	BENIN	FR.G.	FRENCH GUIANA
B.H.	BOSNIA AND HERZEGOVINA	GEOR.	GEORGIA
BUR.	BURKINA FASO	GER.	GERMANY
B.	BURUNDI	GH.	GHANA
CAM.	CAMEROON	GUY.	GUYANA

Winkel Tripel Projection   1 : 170 000 000   MILES 0   1000   2000   3000

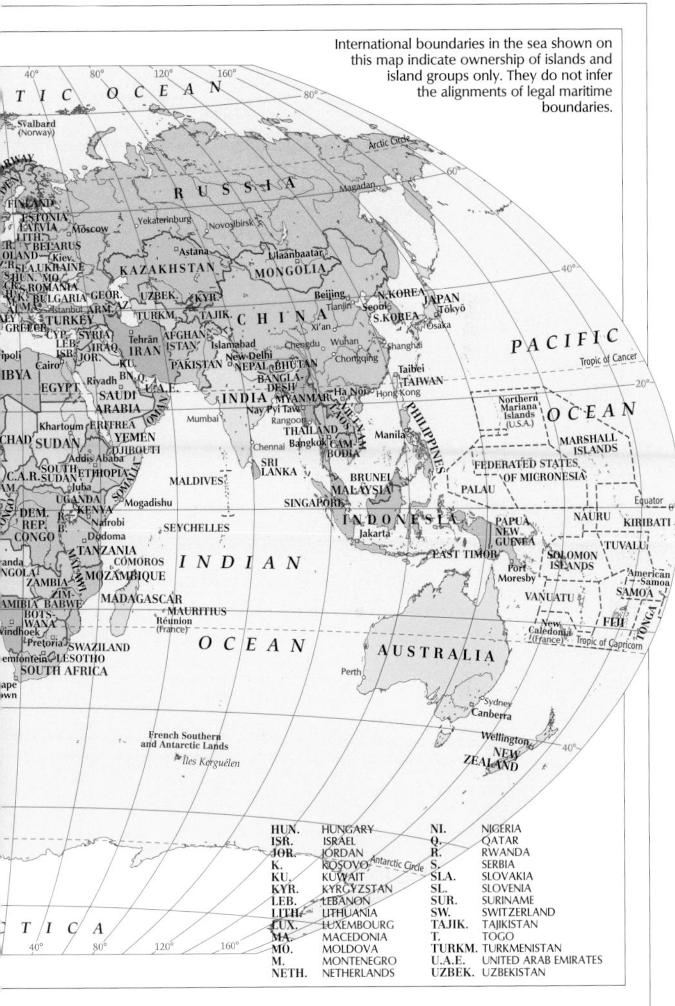

International boundaries in the sea shown on this map indicate ownership of islands and island groups only. They do not infer the alignments of legal maritime boundaries.

HUN.	HUNGARY	NI.	NIGERIA
ISR.	ISRAEL	Q.	QATAR
JOR.	JORDAN	R.	RWANDA
K.	KOSOVO	S.	SERBIA
KU.	KUWAIT	SLA.	SLOVAKIA
KYR.	KYRGYZSTAN	SL.	SLOVENIA
LEB.	LEBANON	SUR.	SURINAME
LITH.	LITHUANIA	SW.	SWITZERLAND
LUX.	LUXEMBOURG	TAJIK.	TAJIKISTAN
MA.	MACEDONIA	T.	TOGO
MO.	MOLDOVA	TURKM.	TURKMENISTAN
M.	MONTENEGRO	U.A.E.	UNITED ARAB EMIRATES
NETH.	NETHERLANDS	UZBEK.	UZBEKISTAN

0   1000   2000   3000   4000   5000 KILOMETRES

© Collins Bartholomew Ltd

47

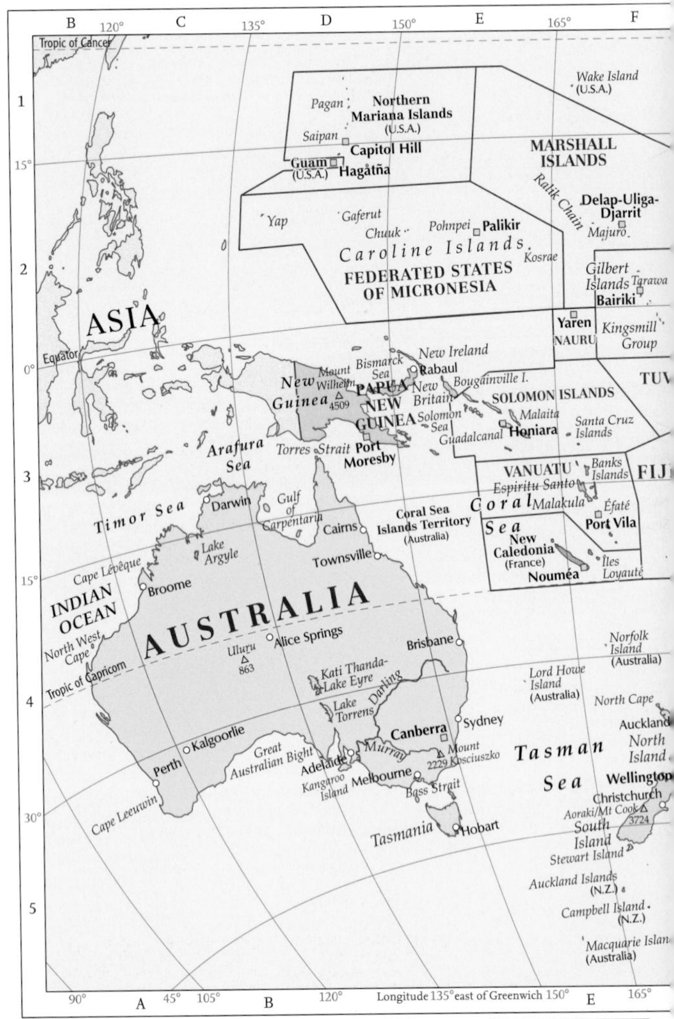

B
C
D
E
F
120°
135°
150°
165°
Tropic of Cancer
1
2
3
4
5
15°
0°
30°
Wake Island
(U.S.A.)
Pagan
Northern
Mariana Islands
(U.S.A.)
Saipan
Capitol Hill
Guam
(U.S.A.)
Hagåtña
MARSHALL
ISLANDS
Ralik Chain
Delap-Uliga-
Djarrit
Majuro
Yap
Gaferut
Chuuk
Pohnpei
Palikir
Caroline Islands
Kosrae
FEDERATED STATES
OF MICRONESIA
Gilbert
Islands
Tarawa
Bairiki
Yaren
NAURU
Kingsmill
Group
ASIA
Equator
New Ireland
Bismarck
Sea
Rabaul
Mount
Wilhelm
4509
New
Guinea
PAPUA
NEW
GUINEA
New
Britain
Bougainville I.
SOLOMON ISLANDS
TUV
Solomon
Sea
Malaita
Honiara
Guadalcanal
Santa Cruz
Islands
Arafura
Sea
Torres Strait
Port
Moresby
VANUATU
Banks
Islands
FIJI
Espiritu Santo
Malakula
Éfaté
Port Vila
Timor Sea
Darwin
Gulf
of
Carpentaria
Cairns
Coral Sea
Islands Territory
(Australia)
Coral
Sea
New
Caledonia
(France)
Nouméa
Îles
Loyauté
Lake
Argyle
Townsville
Cape Lévêque
Broome
INDIAN
OCEAN
AUSTRALIA
North West
Cape
Alice Springs
Uluru
863
Brisbane
Norfolk
Island
(Australia)
Tropic of Capricorn
Kati Thanda-
Lake Eyre
Darling
Lord Howe
Island
(Australia)
Lake
Torrens
North Cape
Sydney
Auckland
Canberra
North
Island
Kalgoorlie
Great
Australian Bight
Murray
Mount
2229
Kosciuszko
Tasman
Sea
Perth
Adelaide
Kangaroo
Island
Melbourne
Bass Strait
Wellington
Christchurch
Aoraki/Mt Cook
3724
South
Island
Cape Leeuwin
Tasmania
Hobart
Stewart Island
Auckland Islands
(N.Z.)
Campbell Island
(N.Z.)
Macquarie Island
(Australia)
90°
A
45°
105°
B
120°
Longitude 135° east of Greenwich 150°
E
165°

1 : 72 000 000

MILES 0 500 1000

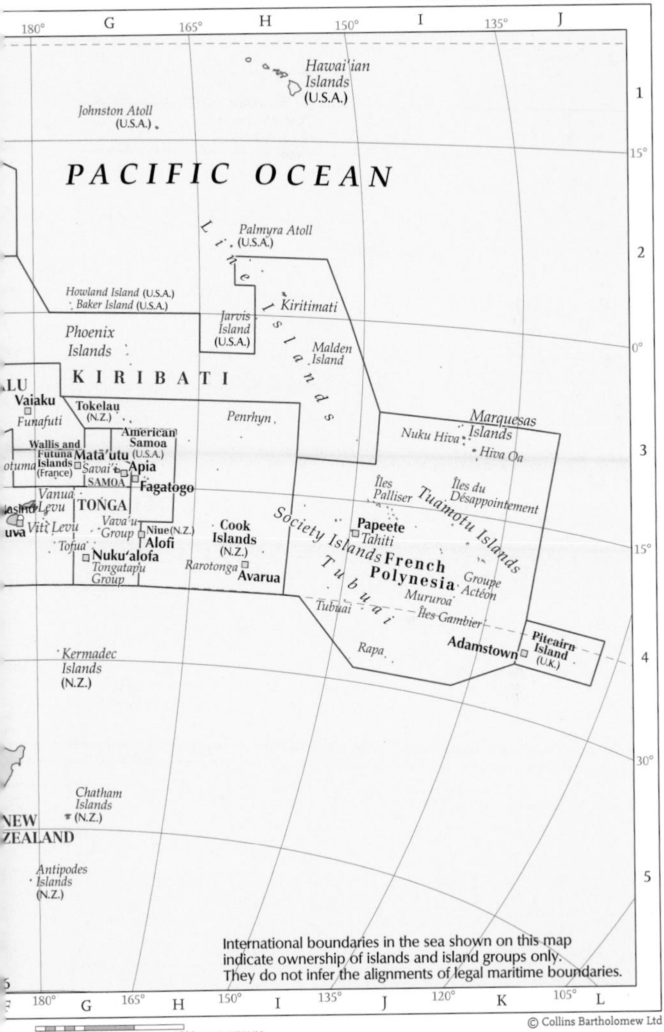

PACIFIC OCEAN

Hawai'ian
Islands
(U.S.A.)

Johnston Atoll
(U.S.A.)

Palmyra Atoll
(U.S.A.)

Howland Island (U.S.A.)
Baker Island (U.S.A.)

Jarvis
Island
(U.S.A.)

Kiritimati

Phoenix
Islands

Malden
Island

KIRIBATI

LU
Vaiaku
Funafuti

Tokelau
(N.Z.)
American
Samoa

Wallis and
Futuna
Islands
(France)

Mata'utu
Savai'i
SAMOA
(U.S.A.)
Apia

Marquesas
Islands
Nuku Hiva

Hiva Oa

Penrhyn

Vanua
Levu

otuma

oshin
uva

Viti Levu

TONGA
Vava'u
Group

Fagatogo

Niue(N.Z.)
Alofi

Tofua

Nuku'alofa
Tongatapu
Group

Cook
Islands
(N.Z.)

Rarotonga
Avarua

Society Islands

Papeete
Tahiti

French
Polynesia

T u b u a i

Îles
Palliser

Îles du
Désappointement

Tuamotu Islands

Groupe
Actéon

Mururoa

Tubuai

Îles Gambier

Rapa

Adamstown

Pitcairn
Island
(U.K.)

Kermadec
Islands
(N.Z.)

Chatham
Islands
(N.Z.)

NEW
ZEALAND

Antipodes
Islands
(N.Z.)

International boundaries in the sea shown on this map
indicate ownership of islands and island groups only.
They do not infer the alignments of legal maritime boundaries.

© Collins Bartholomew Ltd

49

0   500   1000   1500 KILOMETRES

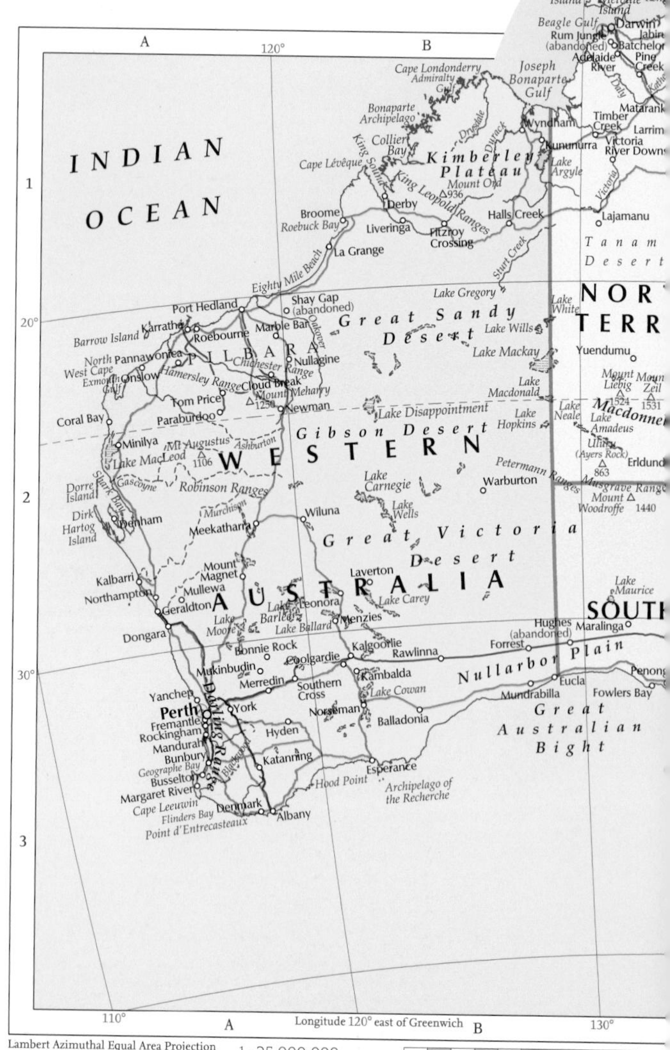

INDIAN

OCEAN

Cape Londonderry
Admiralty

Bonaparte
Archipelago
Collier
Bay
Cape Lévêque
King George River
Kimberley
Plateau
Mount Ord
936

Joseph
Bonaparte
Gulf

Bathurst
Island   Melville Isl'd
Beagle Gulf   Darwin
Rum Jungle   Jabiru
(abandoned)   Batchelor
Adelaide   Pine
River   Creek
Matarank
Timber   Larrima
Creek   Victoria
River Downs
Wyndham   Kununurra   Lake
Argyle

A   120°   B

1

Broome
Roebuck Bay
Liveringa
La Grange

Derby
Fitzroy
Crossing
Halls Creek

Lajamanu

Tanami
Desert

NOR
TERR

Eighty Mile Beach

20°

Port Hedland
Barrow Island
North Pannawonica
West Cape
Exmouth Gulf
Onslow
Coral Bay

Shay Gap
(abandoned)
Marble Bar
Roebourne
Karratha
PILBARA
Hamersley Range
Tom Price
Paraburdoo
Nullagine
Cloud break
Mount Meharry
1251
Newman

Great Sandy
Desert

Lake White
Lake Wills
Lake Mackay
Lake
Macdonald

Yuendumu

Mount   Moun
Liebig   Zeil
1524   1531
Macdonne
Lake
Neale   Amadeus
Uluru   Erldund
(Ayers Rock)
863

2

Minilya
Lake MacLeod
Mt Augustus
1106
Robinson Ranges

Gibson Desert

Lake Disappointment

Lake
Hopkins

Petermann Ranges

Grave Range
Mount △
Woodroffe 1440

WESTERN

Dorre
Island
Dirk
Hartog
Island
Denham
Gascoyne
Murchison

Lake
Carnegie
Lake
Wells

Warburton

Meekatharra

AUSTRALIA

Great Victoria
Desert

Kalbarri
Northampton
Dongara

Mount
Magnet
Mullewa
Geraldton
Lake
Moore
Lake
Barlee
Lake Ballard

Wiluna

Laverton
Leonora
Menzies

Lake Carey

30°

Bonnie Rock
Mukinbudin
Merredin
York
Southern
Cross
Kalgoorlie
Coolgardie
Kambalda

Rawlinna
Forrest
(abandoned)

Hughes
(abandoned)

Maralinga

SOUTH

Lake
Maurice

Yanchep
Perth
Fremantle
Rockingham
Mandurah
Bunbury
Geographe Bay
Busselton
Margaret River
Cape Leeuwin
Flinders Bay
Point d'Entrecasteaux

Hyden
Katanning
Denmark
Albany

Noseman
Lake Cowan
Balladonia
Esperance
Hood Point
Archipelago of
the Recherche

Nullarbor   Plain
Mundrabilla
Eucla

Great
Australian
Bight

Penon
Fowlers Bay

3

A   120°   B   130°

110°   Longitude 120° east of Greenwich

Lambert Azimuthal Equal Area Projection   1 : 25 000 000   MILES 0   250   500

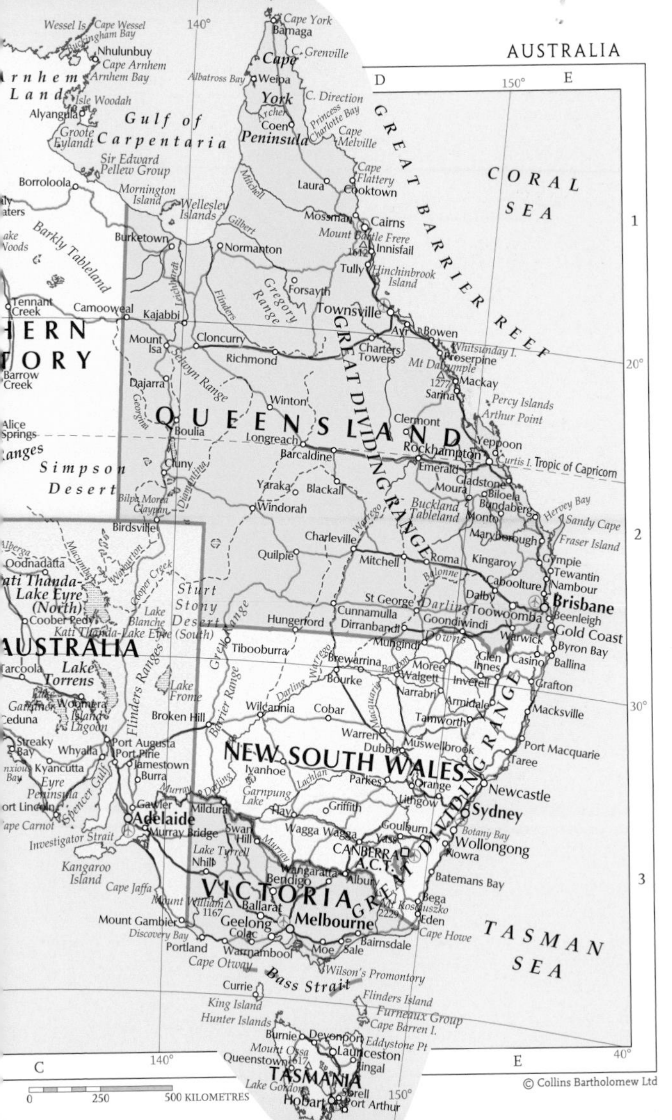

CORAL
SEA

Gulf of
Carpentaria

Cape York
Peninsula

GREAT BARRIER REEF

QUEENSLAND

GREAT DIVIDING RANGE

Tropic of Capricorn

Simpson
Desert

Sturt
Stony
Desert

SOUTH
AUSTRALIA

NEW SOUTH WALES

GREAT DIVIDING RANGE

VICTORIA

Bass Strait

TASMAN
SEA

TASMANIA

© Collins Bartholomew Ltd

0   250   500 KILOMETRES

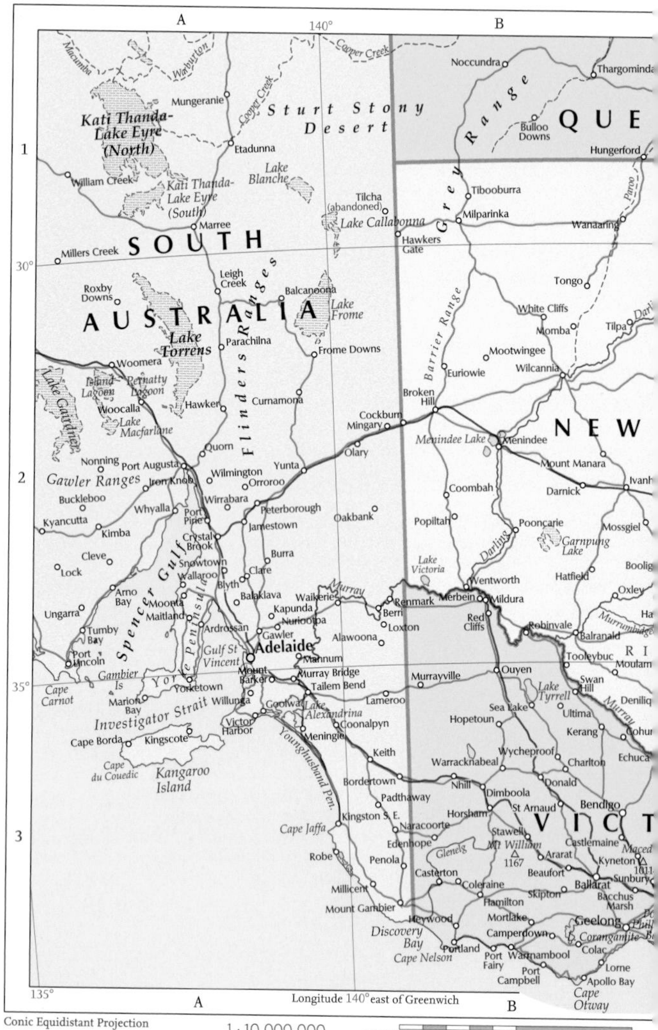

Conic Equidistant Projection

1 : 10 000 000

Longitude 140° east of Greenwich

MILES 0   100   200

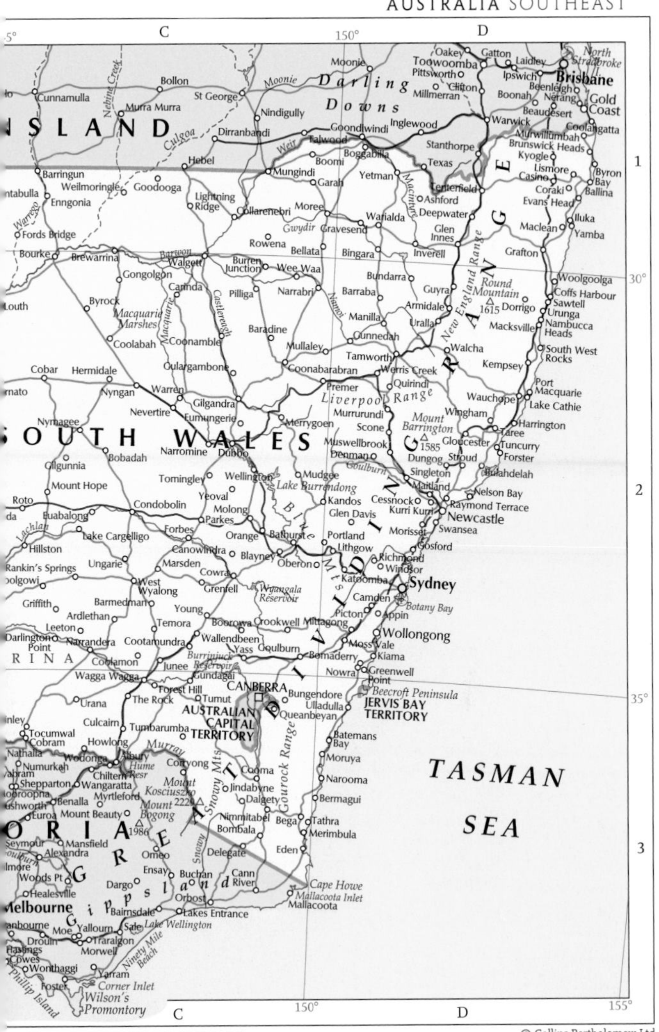

TASMAN

SEA

© Collins Bartholomew Ltd

0   100   200   300 KILOMETRES

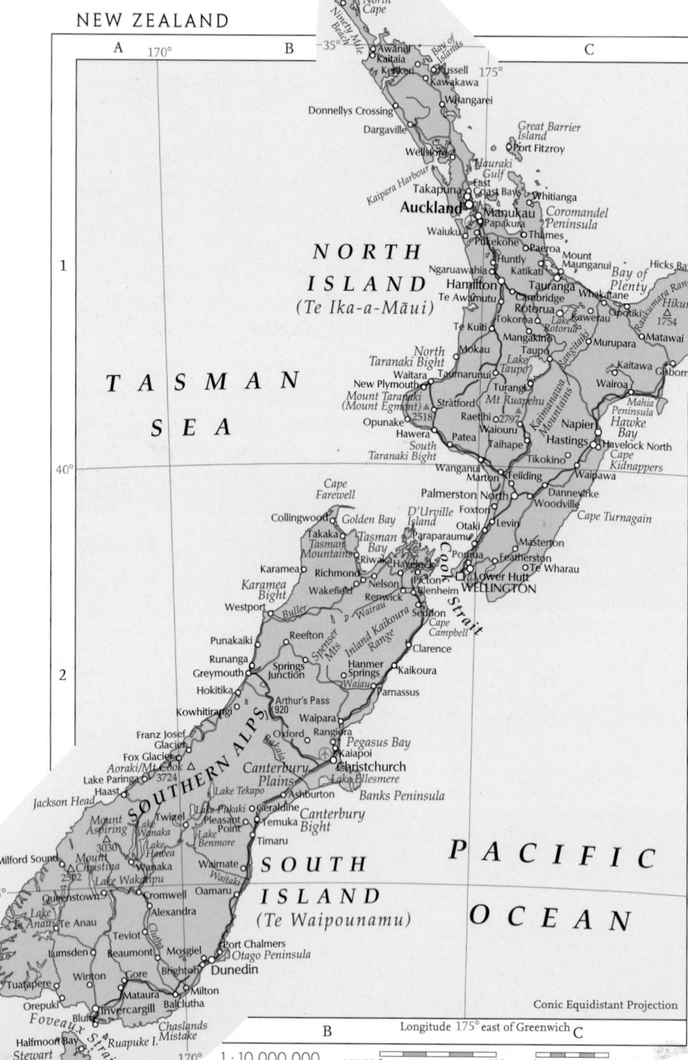

NEW ZEALAND
A
B
C
170°
175°
-35°
40°
-45°
1
2
Te Paki
North Cape
Ninety Mile Beach
Awanui
Kaitaia
Bay of Islands
Kerikeri
Russell
Kawakawa
Whangarei
Donnellys Crossing
Dargaville
Great Barrier Island
Port Fitzroy
Wellsford
Kaipara Harbour
Hauraki Gulf
Takapuna
East Coast Bays
Whitianga
Auckland
Manukau
Coromandel Peninsula
Papakura
Waiuku
Thames
Pukekohe
Paeroa
NORTH ISLAND
(Te Ika-a-Māui)
Huntly
Mount Maunganui
Hicks Bay
Ngaruawahia
Katikati
Bay of Plenty
Hamilton
Tauranga
Whakatane
Raukumara Range
Hikurangi
1754
Te Awamutu
Cambridge
Rotorua
Opotiki
Kawerau
Lake Rotorua
Tokoroa
Te Kuiti
Mangakino
Murupara
Matawai
North Taranaki Bight
Mokau
Taupo
Rangitaiki
Lake Taupo
Kaitawa
Gisborne
Waitara
Taumarunui
Turangi
Wairoa
TASMAN SEA
New Plymouth
Mount Taranaki (Mount Egmont)
Stratford
Mt Ruapehu
Kaimanawa Mountains
Mahia Peninsula
2518
Raetihi
2797
Opunake
Waiouru
Napier
Hawke Bay
Hawera
Patea
Taihape
Hastings
Havelock North
South Taranaki Bight
Cape Kidnappers
Tikokino
Wanganui
Marton
Feilding
Waipawa
Cape Farewell
Palmerston North
Dannevirke
Woodville
D'Urville Island
Foxton
Collingwood
Golden Bay
Cape Turnagain
Otaki
Levin
Takaka
Tasman Mountains
Tasman Bay
Paraparaumu
Masterton
Cook Strait
Porirua
Featherston
Te Wharau
Karamea
Riwaka
Havelock
Richmond
Picton
Lower Hutt
WELLINGTON
Karamea Bight
Wakefield
Nelson
Blenheim
Renwick
Westport
Buller
Wairau
Seddon
Inland Kaikoura Range
Cape Campbell
Punakaiki
Reefton
Spenser Mts
Clarence
Runanga
Hanmer Springs
Kaikoura
Greymouth
Springs Junction
Waiau
Hokitika
Parnassus
Arthur's Pass
920
Kowhitirangi
Waipara
Franz Josef Glacier
Oxford
Rangiora
Pegasus Bay
Kaiapoi
Rakaia
Fox Glacier
Aoraki/Mt Cook
3724
Christchurch
Canterbury Plains
Lake Ellesmere
Lake Paringa
Haast
SOUTHERN ALPS
Lake Tekapo
Ashburton
Banks Peninsula
Jackson Head
Lake Pukaki
Geraldine
Canterbury Bight
Mount Aspiring
3030
Twizel
Lake Wanaka
Pleasant Point
Temuka
Lake Benmore
Lake Hawea
Timaru
PACIFIC OCEAN
Milford Sound
Mount Christina
2502
Wanaka
Waimate
SOUTH ISLAND
(Te Waipounamu)
Lake Wakatipu
Waitaki
Queenstown
Cromwell
Oamaru
Lake Te Anau
Te Anau
Alexandra
Teviot
Clutha
Port Chalmers
Lumsden
Beaumont
Mosgiel
Otago Peninsula
Brighton
Dunedin
Tuatapere
Winton
Gore
Milton
Mataura
Orepuki
Balclutha
Invercargill
Conic Equidistant Projection
Bluff
Foveaux Strait
Chaslands Mistake
Longitude 175° east of Greenwich
Halfmoon Bay
Ruapuke I.
Stewart Island
1 : 10 000 000
MILES 0
100
0
100 KILOMETRES

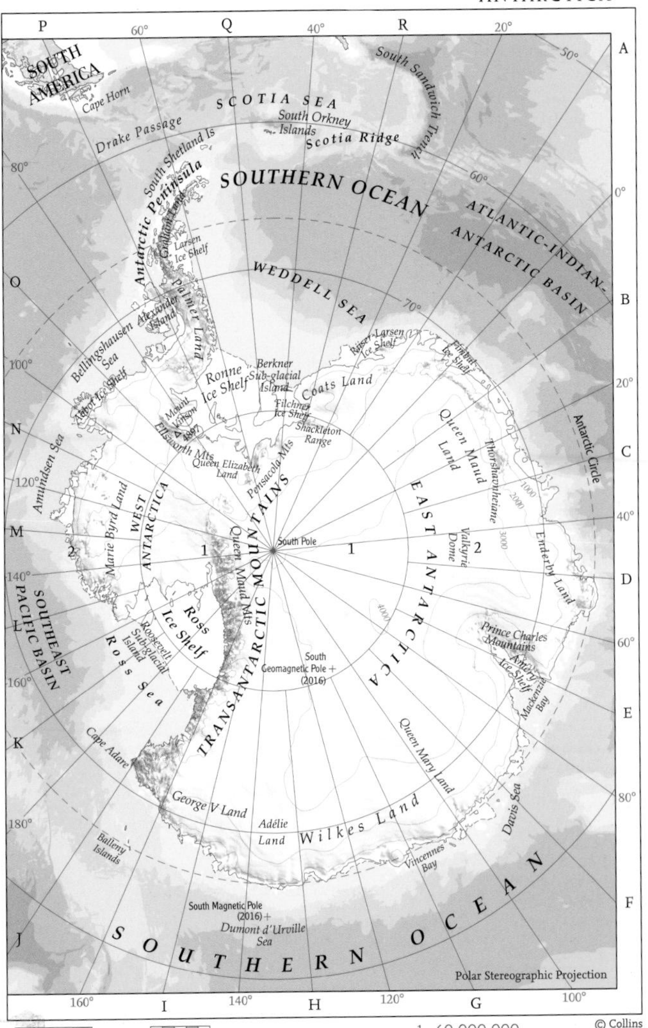

P   60°   Q   40°   R   20°   50°   A

SOUTH AMERICA
Cape Horn
Drake Passage
SCOTIA SEA
South Orkney Islands
Scotia Ridge
South Shetland Is
South Sandwich Trench
SOUTHERN OCEAN
ATLANTIC-INDIAN-ANTARCTIC BASIN
Antarctic Peninsula
Larsen Ice Shelf
Graham Land
WEDDELL SEA
Palmer Land
Bellingshausen Sea
Alexander Island
Larsen Ice Shelf
Filchner Ice Shelf
Abbot Ice Shelf
Ronne Ice Shelf
Berkner Sub-glacial Island
Coats Land
Queen Maud Land
Mount Vinson
Ellsworth Mts
Queen Elizabeth Land
Shackleton Range
Pensacola Mts
Thorshavnheane
EAST ANTARCTICA
Amundsen Sea
Marie Byrd Land
WEST ANTARCTICA
Rockefeller Plateau
South Pole
Valkyrie Dome
Enderby Land
SOUTHEAST PACIFIC BASIN
Ross Ice Shelf
Roosevelt Island
ROSS SEA
TRANSANTARCTIC MOUNTAINS
South Geomagnetic Pole (2016)
Prince Charles Mountains
Amery Ice Shelf
Mackenzie Bay
Cape Adare
Queen Mary Land
Davis Sea
Balleny Islands
George V Land
Adélie Land
Wilkes Land
Vincennes Bay
South Magnetic Pole (2016)
Dumont d'Urville Sea
SOUTHERN OCEAN
Polar Stereographic Projection

Antarctic Circle

O   80°   100°   120°   140°   160°   180°
N   M   L   K   J
I   H   G
140°   120°   100°
0°   60°   70°   20°   40°   60°   80°
B   C   D   E   F

0   400 MILES   0   500 KILOMETRES   1 : 60 000 000   © Collins Bartholomew Ltd   55

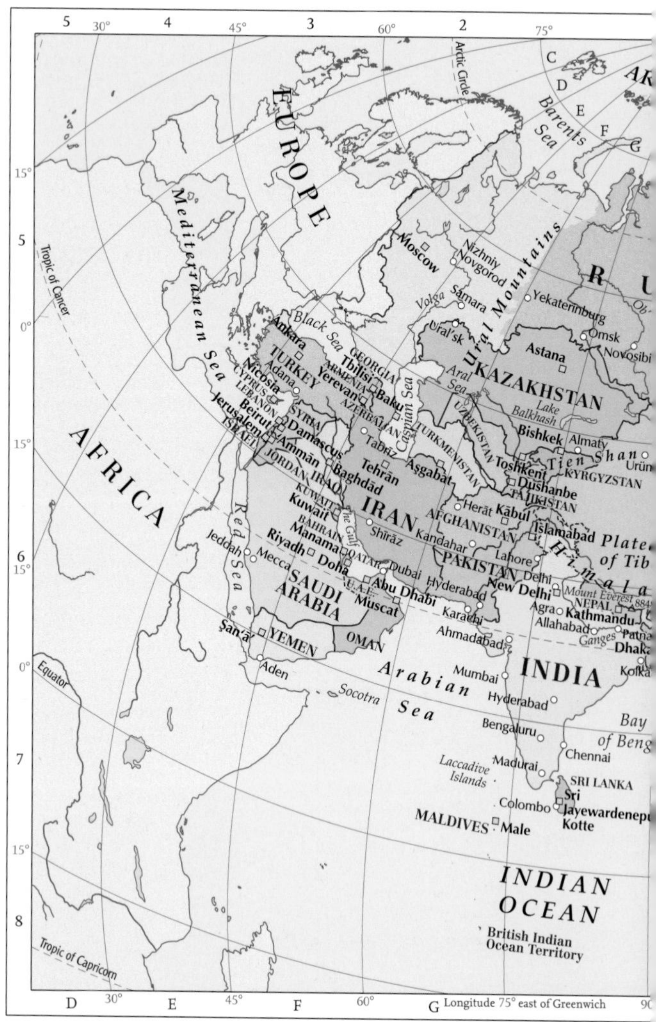
EUROPE
AFRICA
Mediterranean Sea
Black Sea
Caspian Sea
Aral Sea
Red Sea
The Gulf
Arabian Sea
Barents Sea
Bay of Beng
INDIAN OCEAN
Ural Mountains
Arctic Circle
Tropic of Cancer
Equator
Tropic of Capricorn
Moscow
Nizhniy Novgorod
Samara
Volga
Ural'sk
Yekaterinburg
Omsk
Novosibi
Astana
KAZAKHSTAN
Lake Balkhash
Bishkek
Almaty
Tien Shan
Urün
KYRGYZSTAN
UZBEKISTAN
Toshkent
Dushanbe
TAJIKISTAN
TURKMENISTAN
Aşgabat
GEORGIA
Tbilisi
ARMENIA
Yerevan
AZERBAIJAN
Baku
Ankara
TURKEY
Adana
Nicosia
CYPRUS
LEBANON
Beirut
SYRIA
Damascus
Jerusalem
ISRAEL
Ammān
JORDAN
IRAQ
Baghdād
Tabrīz
Tehrān
IRAN
Shīrāz
KUWAIT
Kuwait
BAHRAIN
Manama
QATAR
Doha
Riyadh
SAUDI ARABIA
Jeddah
Mecca
U.A.E.
Abu Dhabi
Dubai
Muscat
OMAN
Şan'ā'
YEMEN
Aden
Socotra
Herāt
Kābul
AFGHANISTAN
Kandahar
Islamabad
PAKISTAN
Lahore
Hyderabad
Karachi
Delhi
New Delhi
Plate of Tib
Himala
Mount Everest 884
NEPAL
Kathmandu
Agra
Allahabad
Ganges
Patna
Dhaka
Kolka
Ahmadabad
INDIA
Mumbai
Hyderabad
Bengaluru
Chennai
Madurai
Laccadive Islands
SRI LANKA
Colombo
Sri Jayewardenepu Kotte
MALDIVES
Male
British Indian Ocean Territory
R
AR
C
D
E
F
G
5
4
3
2
6
7
8
D
E
F
G
30°
45°
60°
75°
90
15°
0°
Longitude 75° east of Greenwich

1 : 86 000 000

MILES 0 500 1000 1500

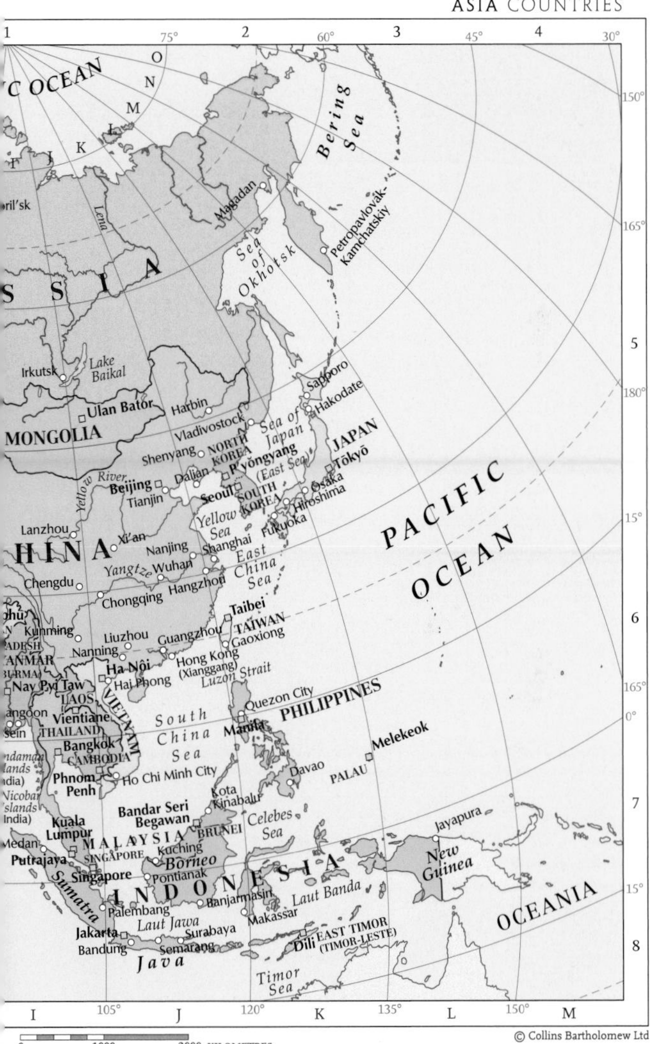

1   75°   2   60°   3   45°   4   30°

150°

C OCEAN

I K I L M N O

ril'sk

Lena

165°

Sea of Okhotsk

Magadan

Bering Sea

Petropavlovsk-Kamchatskiy

R U S S I A

Irkutsk   Lake Baikal

5   180°

Ulan Bator   Harbin   Sapporo   Hakodate

MONGOLIA   Vladivostock   Sea of Japan

Shenyang   NORTH   (East Sea)   JAPAN

Yellow River   Beijing   Dalian   P'yongyang   Tōkyō

Lanzhou   Tianjin   Seoul   Ōsaka

C H I N A   Xi'an   Yellow   SOUTH   Hiroshima

Nanjing   Sea   KOREA   Fukuoka

Chengdu   Yangtze   Wuhan   Shanghai   East

Chongqing   Hangzhou   China

Kunming   Liuzhou   Taibei   Sea   15°

Nanning   Guangzhou   TAIWAN

BANGLADESH   Gaoxiong

MYANMAR   Hong Kong

(BURMA)   Ha Nôi   (Xianggang)

Nay Pyi Taw   Hai Phong   Luzon Strait

ngoon   VIETNAM

sein   Vientiane   South   Quezon City   165°

THAILAND   China   PHILIPPINES   0°

Bangkok   LAOS   Sea   Manila

CAMBODIA   Melekeok

Andaman   Phnom   Ho Chi Minh City   Davao

Islands   Penh   Kota   PALAU

(India)   Bandar Seri   Kinabalu

Nicobar   Begawan   BRUNEI   Celebes

Islands   Kuala   Kuching   Sea   Jayapura

(India)   Lumpur   MALAYSIA   Borneo

Medan   Putrajaya   SINGAPORE   Pontianak   New

Singapore   I N D O N E S I A   Guinea

Sumatra   Palembang   Banjarmasin   Laut Banda

Jakarta   Laut Jawa   Surabaya   Makassar   Dili   EAST TIMOR   OCEANIA   15°

Bandung   Semarang   (TIMOR-LESTE)

Java   Timor Sea

MONGOLIA

PACIFIC

OCEAN

Lake Baikal

Sea of Okhotsk

I   105°   J   120°   K   135°   L   150°   M

0   1000   2000 KILOMETRES

© Collins Bartholomew Ltd   **57**

A   105°   B   120°

Pyinmana Louangphabang Nam Đinh Suwen Luzon
Taung-ngu Chiang Rai Gulf of Haikou Batan Islands
Phayao Tongking Wencheng Babuyan Islands
Chiang Mai Nan Phônsavan Thanh Hoa Chengmai Qionghai Wanning
Phrae Uttaradit don Vinh Ha Tinh Dongfang Wanning Hainan Dao Laoag City Tugueg
Thaton Tak Phitsanulok Dong Hoi (China) Vigan Bontoc

Moulmein VIENTIANE Khon Kaen Savannakhet Huế Đà Nẵng San Fernando Luz
(Mawlamyine) (Viangchan) alavan Da Nang Dagupan
Ye Ayutthaya THAILAND Surin Ubon Ratchathani Quang Ngai Quezon City

Tavoy BANGKOK Nakhon Ratchasima Quezon City MANILA
(Dawei) (Krung Thep) Lucena
Myeik Palaw Pattaya CAMBODIA Buôn Mê Thuột Quy Nhon Batangas
Tenasserim Chanthaburi Battambang Nha Trang Mindoro Calamian Romble
Padaung Kompong Som PHNOMPENH Da Lat Group Cuyo

Khlong Yai Gulf of Don Keo Phan Thiết SOUTH Taytay Ilo
Chumphon Thailand Kep Long Xuyên Biên Hoa Ho Chi Minh City (Saigon) Puerto Pana
Ranong Rach Gia Princesa Palawan
Takua Pa Can Tho Bac Liêu CHINA Brooke's Point

Nakhon Si Bac Liêu Mui Ca Mau Point Sulu
Krabi Thammarat Mouths of the Mekong SEA Negr
Phuket Balabac Strait
Banda Hat Yai Songkhla Kudat Banggi Zamboang
Aceh Yala Kota Bharu Kota Kinabalu Sandakan Basilan
Sigli George Town Alor Star Pasir Putih Gunung Kinabalu Lahad Datu Archipela
Bireun Sungai BANDAR SERI SABAH Semporna Ce
Langsa Taiping Petani Kuala Terengganu BEGAWAN Tawau Tarakan
Pangkalansusu Ipoh Kuala Lipis BRUNEI Miri Tanjungredeb
Medan KUALA Natuna Besar Igan Bintulu Tanjungselor
Simeulue LUMPUR Kepulauan Mukah Sangkulirang
Gunungsitoli PUTRAJAYA Anambas Kuching Sibu Sri Aman Sangkulirang Semenanju
Labuhanbilik Melaka Kepulauan Likup Belaga Lubok Antu Tanjungredeb Montong
Sibolga Muar Tambelan Serian Teluk
Pulau-pulau Dumai Kabupaten Sambas Delok Samarinda Tomini
Batu Minas JOHOR BAHRU Riau Singkawang Mempawah Donggala
Payakumbuh SINGAPORE Pontianak BORNEO Palu
Bukittinggi Kuala Belinyu Ketapang Sukadana CELEBES
Padang Jambi Pangkalpinang Sungaliat Balikpapan (SULAWES
Sijunjung Bangka Ketapang Kendawangan Amuntai Parepare
Kepulauan Belitung Pangkalanbun Sampit Kotabaru Watampone
Mentawai Bengkulu Tebingtinggi Toboali Banjarmasin Makale
Bintuhan Lahat Martapura Makassar Bontosunggu Bente
Enggano Krui Tg Selatan Laut Bulukum
Bandar INDON Kepulauan
Lampung JAKARTA Kangean
Sukabumi Cirebon Semarang Laut Jawa Madura Laut Bali Laut Flores
Bandung Surabaya (Java Sea) (Bali Sea)
Cilacap Surakarta Malang Jember Mataram Sumbawa
JAVA Denpasar Sumba
(JAWA) Selat Lombok Waikabubak Wangar
Christmas I. Lesser Sunda Islands Sumba Timo
(Australia)

INDIAN
OCEAN

A Longitude 105° east of Greenwich B 120°

Albers Equal Area Conic Projection 1 : 30 000 000 MILES 0   200   400   600

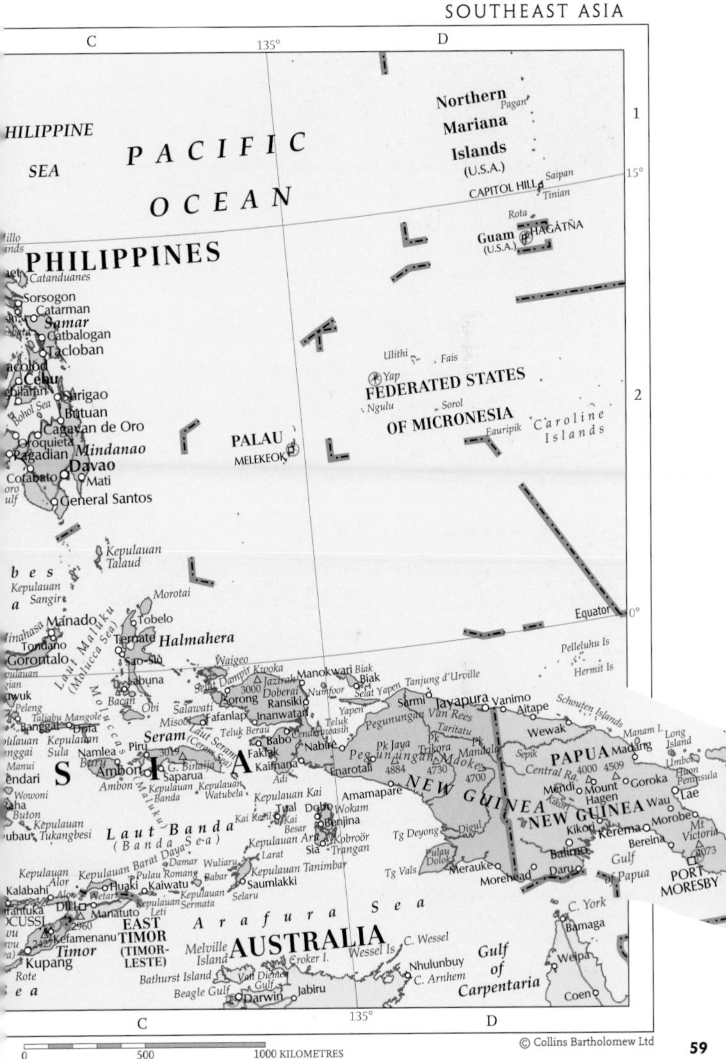

PACIFIC OCEAN

PHILIPPINE SEA

PHILIPPINES

Catanduanes
Sorsogon
Catarman
Catbalogan
Samar
Tacloban
Cebu
Surigao
Butuan
Cagayan de Oro
Mindanao
Davao
General Santos
Mati
Cotabato

Northern Mariana Islands (U.S.A.)

Pagan

Saipan
CAPITOL HILL
Tinian
Rota

Guam HAGÅTÑA
(U.S.A.)

Ulithi · Fais

Yap

FEDERATED STATES OF MICRONESIA

Ngulu · Sorol
Eauripik

Caroline Islands

PALAU
MELEKEOK

Kepulauan Talaud

Kepulauan Sangir

Morotai

Equator

Pelleluhu Is
Hermit Is

Manado
Tondano
Gorontalo
Ternate
Halmahera
Sao–Sio
Tobelo

Laut Maluku
(Molucca Sea)

Waigeo

Manokwari
Biak
Jayapura
Vanimo

Salawati
Sorong
Afanlap
Inanwatan
Faktak
Nabire
Enarotali
Wamena
Sarmi
Wewak

Schouten Islands
Tanjung d'Urville
Manam I.
Long Island

Seram

Piru
Ambon
Saparua

Teluk Berau
Teluk Cenderawasih

Pegunungan Van Rees
Taritatu

PAPUA
Madang

Pk Jaya
Trikora

Central Ra.
Mt Hagen
Goroka
Wau
Lae

Namlea
Buru

Kepulauan Watubela
Kepulauan Kai

Kaimana

NEW GUINEA

Kikori
Kerema

Laut Banda
(Banda Sea)

Kepulauan Banda

Kai Kecil
Kai Besar
Dobo
Wokam

Kepulauan Aru
Kobroor
Trangan

Merauke
Morehead
Daru

Gulf of Papua

PORT MORESBY

Kepulauan Tanimbar
Larat
Saumlakki

Tg Deyong
Tg Vals

Pulau Dolok

Kepulauan Sermata
Wuliaru
Babar

Kalabahi
Alor
Leti
Huaki
Kaiwatu
Pulau Romang
Wetar

Damar

Kefamenanu
DILI
Manatuto
Timor

EAST TIMOR (TIMOR-LESTE)

Arafura Sea

C. Wessel
Wessel Is

Melville Island
Bathurst Island
Van Diemen
Croker I.
C. Arnhem

Nhulunbuy

Gulf of Carpentaria

OECUSSI
Kupang
Rote

Kelamenanu

Beagle Gulf
Darwin
Jabiru

AUSTRALIA

Weipa
Coen

C. York
Bamaga

© Collins Bartholomew Ltd

59

0   500   1000 KILOMETRES

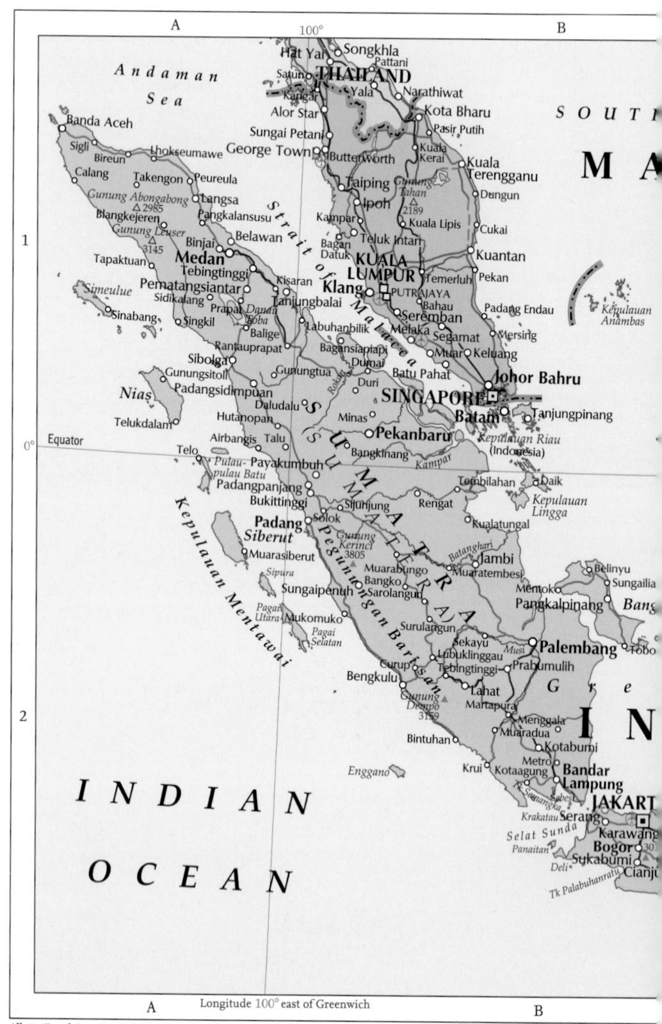

Andaman Sea

Banda Aceh
Sigli
Bireun
Lhokseumawe
Calang
Takengon
Peureula
Gunung Abongabong
△ 2985
Langsa
Blangkejeren
Pangkalansusu
Gunung Leuser
△ 3145
Belawan
Tapaktuan
Medan
Binjai
Tebingtinggi
Simeulue
Pematangsiantar
Sidikalang
Kisaran
Prapat *Danau*
Tanjungbalai
Sinabang
Singkil
Toba
Balige
Labuhanbilik
Rantauprapat
Bagansiapiapi
Sibolga
Gunungsitoli
Gunungtua
Dumai
Padangsidimpuan
Daludalu
Duri
Nias
Hutanopan
SINGAPORE
Telukdalam
Airbangis
Talu
Minas
Pekanbaru
0° Equator
Telo
Pulau Payakumbuh
Bangkinang
Pulau Batu
Padangpanjang
Kampar
Bukittinggi
Solok
Padang
Sijunjung
Rengat
Siberut
Muarasiberut
Gunung
Kerinci
Sipura
3805
Muarabungo
Batanghari
Sungaipenuh
Bangko
Sarolangun
Pagai
Mukomuko
Utara
Surulangun
Pagai
Sekayu
Selatan
Lubuklinggau
Curup
Tebingtinggi
Prabumulih
Bengkulu
Lahat
Martapura
Gunung
Dempo
3159
Menggala
Muaradua
Bintuhan
Kotabumi
Enggano
Metro
Krui
Kotaagung
Bandar
Lampung
Krakatau
Serang
Panaitan
Selat Sunda
JAKART
Deli
Karawang
Tk Pulahuhmratu
Bogor
Sukabumi
Cianju

Hat Yai
Songkhla
Pattani
Sadao
THAILAND
Yala
Narathiwat
Kangar
Alor Star
Pasir Putih
Kota Bharu
Sungai Petani
Butterworth
Kuala
Kerai
Kuala
George Town
Terengganu
Taiping
Gunung
Tahan
Ipoh
△ 2189
Dungun
Kampar
Kuala Lipis
Teluk Intan
Cukai
Bagan
Datuk
KUALA
Kuantan
LUMPUR
Temerluh
Klang
Pekan
PUTRAJAYA
Bahau
Padang Endau
Seremban
Melaka
Segamat
Mersing
Muar
Keluang
Batu Pahat
Johor Bahru
Tanjungpinang
Batam
Kepulauan Riau
(Indonesia)
Tembilahan
Daik
Kepulauan
Lingga
Kualatungal
Jambi
Belinyu
Muaratembesi
Mentok
Sungailia
Pangkalpinang
Bang
Palembang
Tobo
Gre
IN
Kepulauan Anambas

SUMATERA
Pegunungan Barisan

S O U T H
M A

Kepulauan Mentawai

I N D I A N

O C E A N

Albers Equal Area Conic Projection
1 : 15 000 000
MILES 0   100   200   300

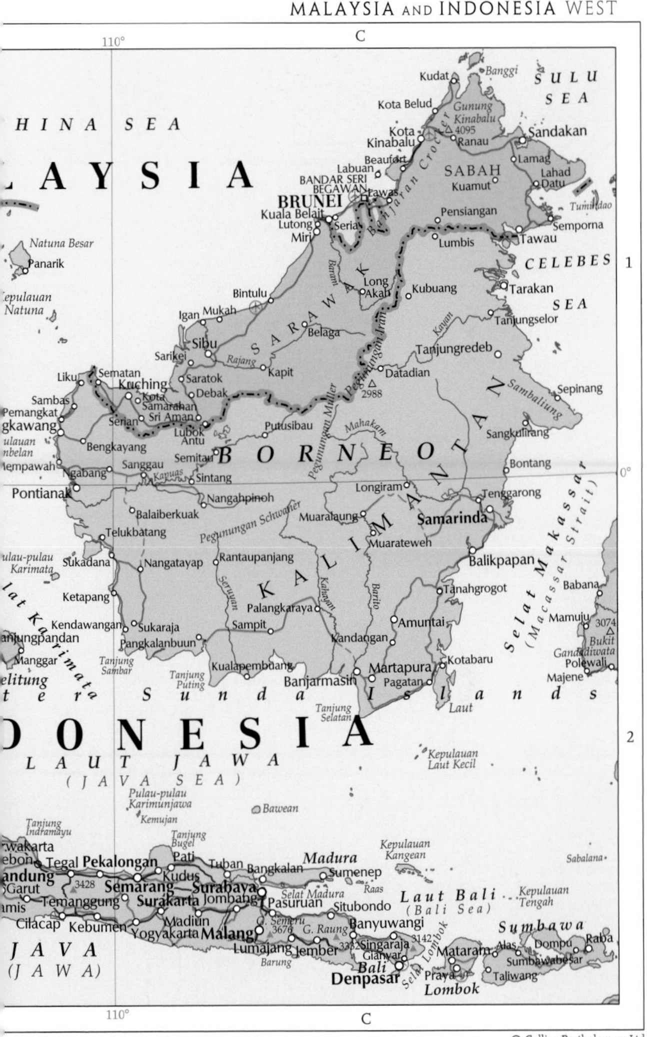

110°

SULU SEA

Banggi

Kudat

Kota Belud

SOUTH CHINA SEA

Gunung
Kinabalu
4095
Kota
Kinabalu
Beaufort   Ranau   Sandakan

Labuan   Lamag   Lahad
BANDAR SERI   SABAH   Kuamut   Datu
BEGAWAN   SARAWAK
BRUNEI   Pensiangan   Semporna
Kuala Belait   Tumindao
Lutong   Serian   CELEBES   1
Miri   Lumbis   Tawau

LAYSIA

Natuna Besar

Panarik

Long   Kubuang   Tarakan
Pakah   SEA

Bintulu   Tanjungselor

*Kepulauan
Natuna*

Igan   Mukah
Sarikei   Belaga   Tanjungredeb
Sibu   Kapit   Datadian   Sepinang
Liku   Sematan   *Rajang*   Sangkulirang
Sambas   Kuching   Saratok
Pemangkat   Kota   Debak   Putusibau   Bontang
gkawang   Samarahan   Sri Aman   *Mahakam*
*ualuan   Serian   Lubok   BORNEO   0°
nbelan   Bengkayang   Antu
empawah   Sanggau   Semitau   Tenggarong
Pontianak   Sintang   Longiram   **Samarinda**
Ngabang   Nangahpinoh   Muaralaung   *Selat Makassar*
Balaiberkuak   Muarateweh   Balikpapan   (Macassar Strait)
Telukbatang   Rantaupanjang   Babana
*ulau-pulau
Karimata*   Sukadana   Nangatayap   Tanahgrogot
Ketapang   Palangkaraya   Amuntai   Mamuju
lat Karimata   Kendawangan   Sukaraja   Sampit   Bukit
Pangkalanbuun   Kualapembuang   Kandangan   Gandadiwata
njungpandan   Kotabaru   Polewali
Manggar   Tanjung   Martapura   Majene
elitung   Sambar   Banjarmasin   Pagatan
Puting   Sunda   Islands
*era   Tanjung   Laut
Selatan
DONESIA   2

LAUT JAWA
(JAVA SEA)   *Kepulauan
Laut Kecil*

Pulau-pulau
Karimunjawa   Laut Bali
*Tanjung
Kemujan*   *Bawean*   *(Bali Sea)*   *Sabalana*

Kepulauan   *Kepulauan
Kangean   Tengah*

*Tanjung
Indramayu*
wakarta   Tegal Pekalongan   Pati   Madura   *Sumbawa*
ebong   Kudus   Tuban   Bangkalan   Sumenep
andung   3428   Semarang   Surabaya   Raas
Garut   Temanggung   Surakarta   Jombang   Pasuruan   Situbondo   Sabalana
amis   Cilacap   Madiun   Yogyakarta   **Malang**   G. Raung   Banyuwangi   **Sumbawa**
Kebumen   *Semeru*   Lumajang   Jember   3142   Alas   Dompu   Raba
3676   Singaraja   Mataram   Sumbawabesar
JAVA   *Barung*   **Bali**   Praya   Taliwang
(JAWA)   Glagah   **Denpasar**   *Selat Lombok*   **Lombok**

110°
C

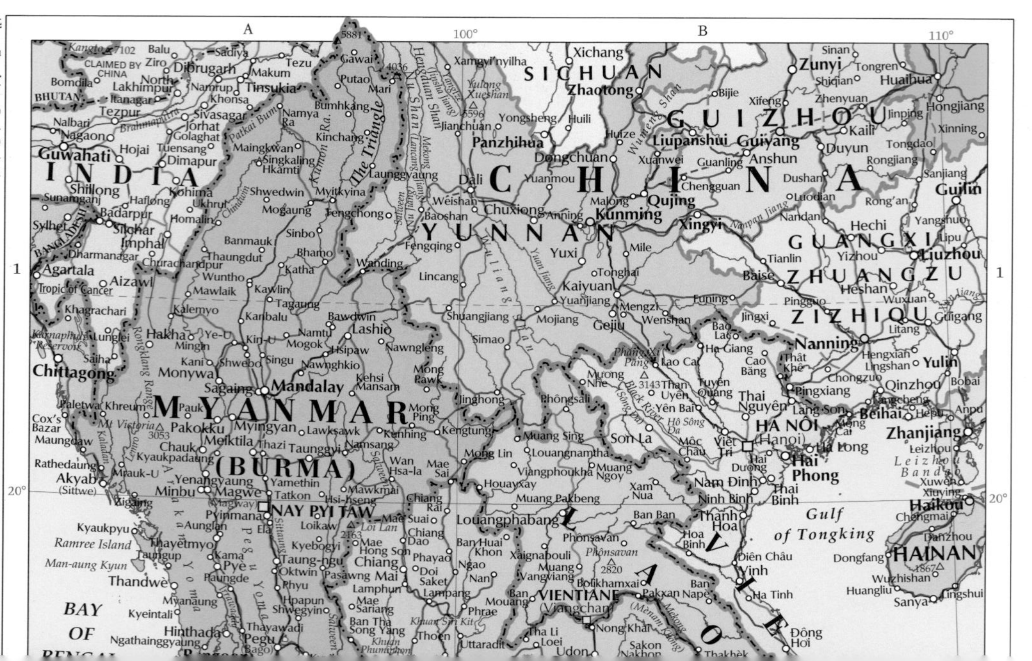

Albers Equal Area Conic Projection

1 : 15 000 000

MILES 0   100   200   300

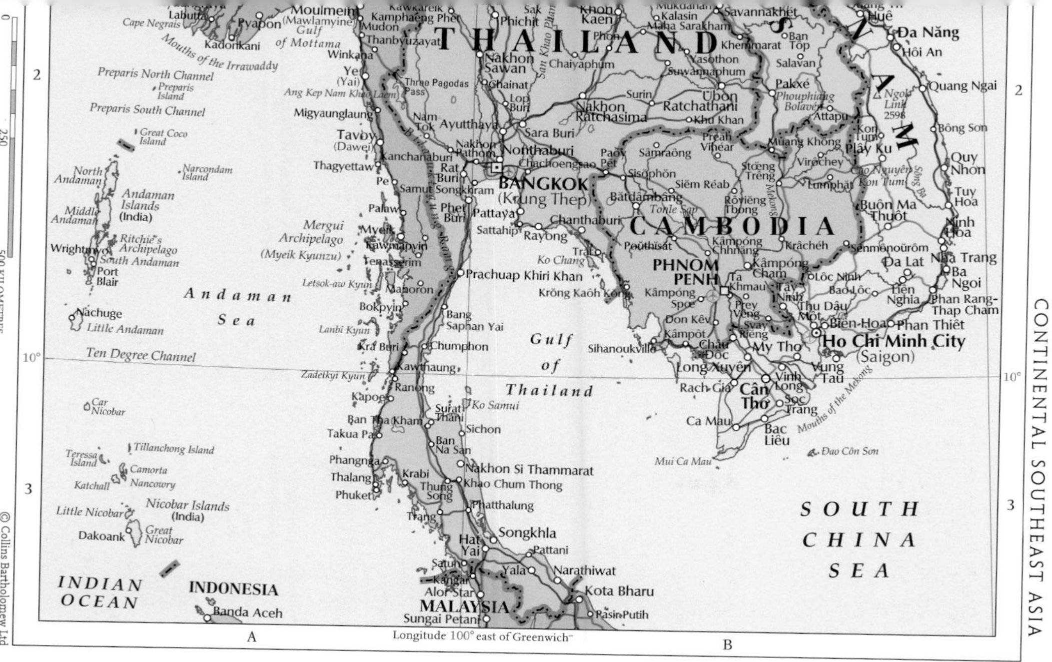

© Collins Bartholomew Ltd

0   250   500 KILOMETRES

THAILAND
CAMBODIA
MALAYSIA
INDONESIA
BANGKOK
(Krung Thep)
PHNOM PENH
Ho Chi Minh City
(Saigon)
Gulf of Thailand
Andaman Sea
SOUTH CHINA SEA
INDIAN OCEAN
Gulf of Mottama
Mouths of the Irrawaddy
Mouths of the Mekong
Mergui Archipelago
(Myeik Kyunzu)
Andaman Islands
(India)
Nicobar Islands
(India)
Ten Degree Channel
Preparis North Channel
Preparis South Channel
Moulmein
(Mawlamyine)
Tavoy
(Dawei)
Nakhon Sawan
Nakhon Ratchasima
Ubon Ratchathani
Khon Kaen
Songkhla
Hat Yai
Phuket
Surat Thani
Nakhon Si Thammarat
Kota Bharu
Banda Aceh
Port Blair
Đa Năng
Huê
Hôi An
Quang Ngai
Quy Nhon
Nha Trang
Đa Lat
Cân Tho
Ca Mau
Savannakhét
Pakxé
Sihanoukville
Kâmpóng Cham
Siëm Réab
Bătdâmbâng
Mekong
Longitude 100° east of Greenwich

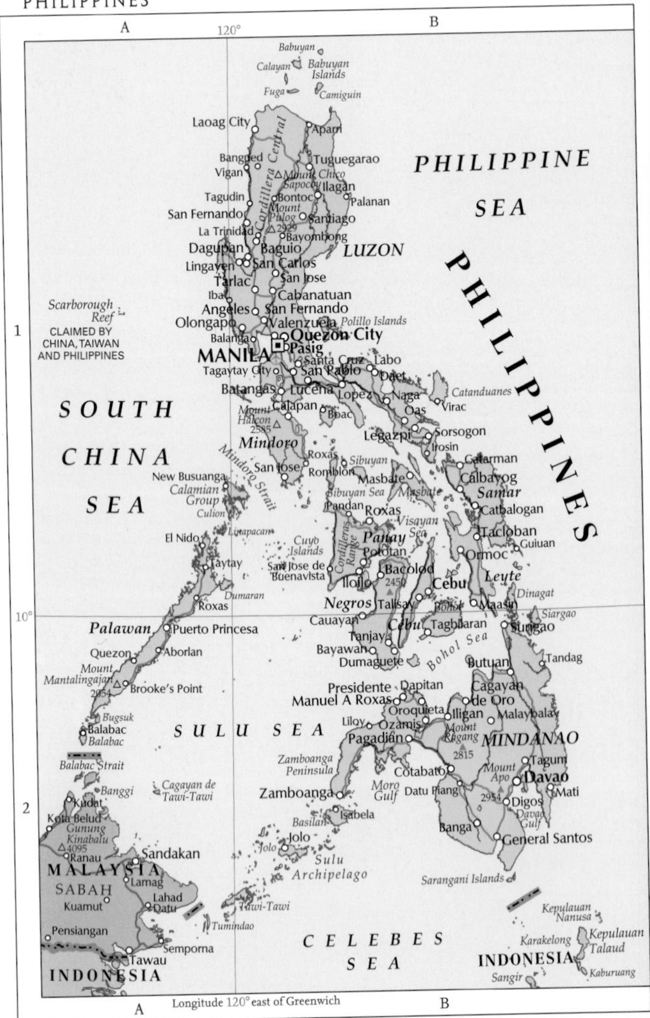

A   120°   B

Babuyan
Calayan *Babuyan*
Islands
Fuga *Camiguin*

Laoag City
Aparri
Bangged   Tuguegarao
Vigan   *Bapoch* Ilagan
Tagudin   *Mount* Palanan
San Fernando   Sagrada
La Trinidad   Bayombong
Dagupan   Baguio
Lingayen   San Carlos   **LUZON**
Tarlac   San Jose
Cabanatuan
Iba   San Fernando
Angeles   Valenzuela   Polillo Islands
Olongapo   Balanga
MANILA   Quezon City
Pasig
Tagaytay City   Santa Cruz   Labo
Batangas   San Pablo   Daet
  Lucena   Lopez   Naga   Catanduanes
Mount Calapan   Bhac   Virac
Halcon   Legazpi   Sorsogon
Mindoro   Roxas   Irosin
New Busuanga   San Jose   *Sibuyan*   Calamian
Calamian   Romblon   Masbate   **Samar**
Group   *Sibuyan Sea*   Calbayog
Culion   Pandan   Masbate   Catbalogan
El Nido   *Cuyo*   Roxas   *Visayan*   Tacloban
Islands   *Sea*   Guiuan
Taytay   San Jose de   Pototan   Ormoc
Lutapacan   Buenavista   Bacolod   **Leyte**
Iloilo   **Cebu**
Dumaran   **Negros**   Talisay   **Cebu**   Dinagat
Roxas   Cauayan   Tagbilaran   Siargao
Palawan   Puerto Princesa   Tanjay   Surigao
Bayawan   *Bohol Sea*
Quezon   Aborlan   Dumaguete   Butuan   Tandag
Mount   Presidente   Dapitan   Cagayan
Mantalingajan   Manuel A Roxas   Oroquieta   de Oro
Brooke's Point   Liloy   Iligan   Malaybalay
Bugsuk   Ozamis   *Mount*
Balabac   **SULU SEA**   Pagadian   *Kitanglad*   **MINDANAO**
Balabac   *Zamboanga*   2815
Balabac Strait   *Peninsula*   Cotabato   *Mount*   Tagum   **Davao**
Cagayan de   *Apo*   2954   Mati
Tawi-Tawi   Zamboanga   Datu Piang   Digos
Moro   *Davao*
Banggi   Isabela   *Gulf*   Banga   *Gulf*
Kudat   Jolo   General Santos
Kota Belud   *Basilan*
Gunung   *Sulu*
Kinabalu   *Archipelago*
4095   *Sarangani Islands*
Ranau
Sandakan   *Kepulauan*
Nanusa
MALAYSIA   Lamag
SABAH   Lahad   *Tawi-Tawi*
Kuamut   Datu   *Karakelong*   *Kepulauan*
Talaud
Pensiangan   *Tumindao*   **C E L E B E S**   *Sangir*   *Kaburuang*
Semporna   **S E A**
Tawau   **INDONESIA**
INDONESIA

PHILIPPINE

S E A

P H I L I P P I N E S

Scarborough
Reef
CLAIMED BY
CHINA, TAIWAN
AND PHILIPPINES

S O U T H

C H I N A

S E A

Mindoro Strait

Cordillera Central

1

10°

2

Longitude 120° east of Greenwich

A   B

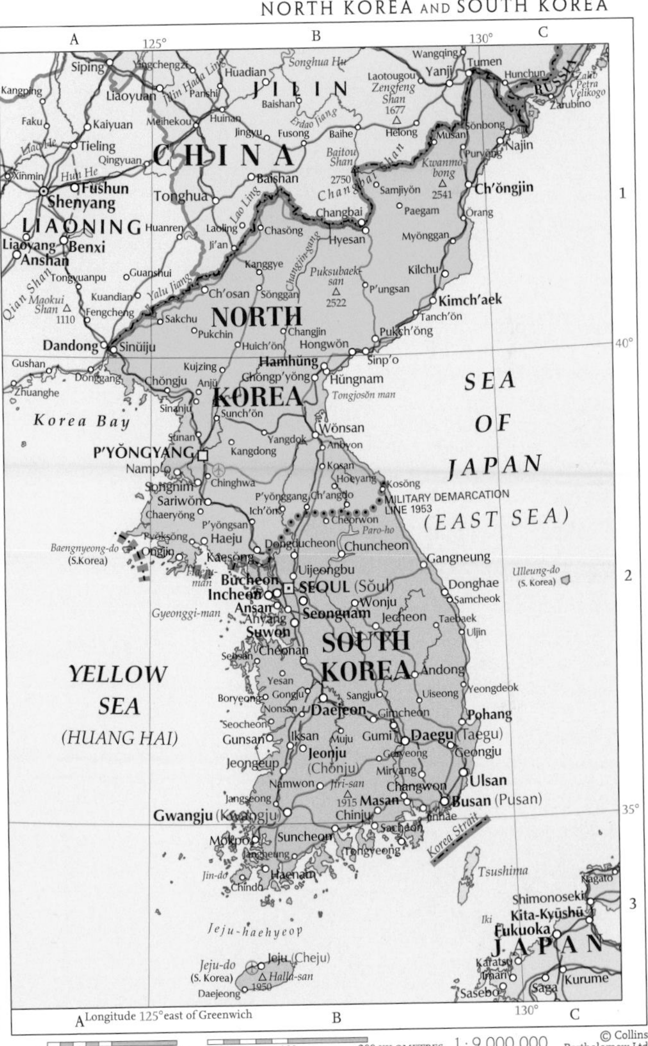

Siping · 125° · Mingchengzi · Huadian · Songhua Hu · Wangqing · 130° · C
Kangping · Liaoyuan · JILIN · Baishan · Laotougou · Zengfeng Shan 1677 · Yanji · Tumen · Hunchun · RUSSIA
Faku · Kaiyuan · Meihekou · Huinan · Jingyu · Fusong · Baihe · Helong · Musan · Sonbong · Najin
Fushun · Tonghua · Baishan · Changbai · Hyesan · Samjiyön · Kwanmo bong 2541 · Ch'ŏngjin
Tieling · Qingyuan · CHINA · Puksubaek san 2522 · Paegam · Orang
Shenyang · LIAONING · Huanren · Ji'an · Chasŏng · Kanggye · P'ungsan · Kilchu
Liaoyang · Benxi · Tongyuanpu · Guanshui · Ch'osan · Sŏnggan · Myŏnggan
Anshan · Kuandian · Fengcheng · Pukchin · Changjin · Hongwŏn · Tanch'ŏn · Kimch'aek
Qian Shan 1110 · Maokui Shan · Sakchu · Huich'ŏn · Pukch'ŏng
Gushan · Dandong · Sinŭiju · Kujzing · Hamhŭng · Sinp'o · 40°
Zhuanghe · Donggang · Chŏngju · Anju · Chŏngp'yŏng · Hŭngnam · SEA
Korea Bay · Sinanju · Kangdong · Wŏnsan · OF
P'YŎNGYANG · Nampo · Chinhwa · Yangdŏk · Kosan · JAPAN
Songnim · Sariwŏn · P'yŏngsan · Ich'ŏn · Ch'angdo · Kosŏng · MILITARY DEMARCATION LINE 1953 (EAST SEA)
Chaeryŏng · Anyon · Paro-ho · Ch'ŏrwon
Baengnyeong-do (S.Korea) · Ongjin · Haeju · Kaesŏng · Dongducheon · Chuncheon · Gangneung · Ulleung-do (S. Korea)
Gyeonggi-man · Bucheon · Uijeongbu · Donghae
Incheon · SEOUL (Sŏul) · Wonju · Samcheok
Ansan · Seongnam · Jecheon · Uljin
Anyang · Suwon · Cheonan · SOUTH KOREA · Andong · Yeongdeok
YELLOW SEA (HUANG HAI) · Yesan · Gongju · Sangju · Uiseong · Pohang
Boryeong · Nonsan · Gimcheon · Gumi · Gyeongju · Daegu (Taegu)
Seocheon · Daejeon · Gyeongsan
Gunsan · Iksan · Muju · Gimhae · Miryang · Changwon · Ulsan
Jeongeup · Jeonju (Chŏnju) · Jiri-san 1915 · Masan · Busan (Pusan)
Namwon · Jangseong · Chinju · Jinhae
Gwangju (Kwangju) · Naju · Suncheon · Sacheon · Tongyeong · Korea Strait · Tsushima
Mokpo · Jangheung · Iki · Shimonoseki · Kita-Kyūshū · Kagoto
Jin-do · Haenam · Fukuoka · JAPAN
Jeju-haehyeop · Jeju (Cheju) · Karatsu · Imari · Saga · Kurume
Jeju-do (S. Korea) · Halla-san 1950 · Daejeong · Sasebo

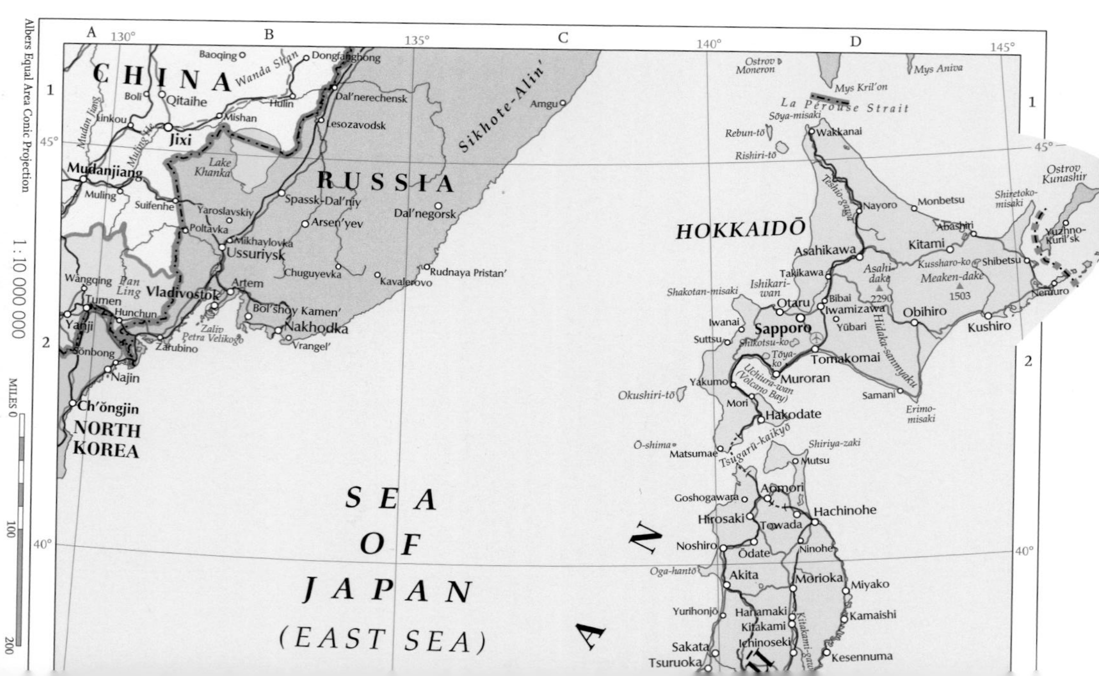

Albers Equal Area Conic Projection

1 : 10 000 000

MILES 0   100   200

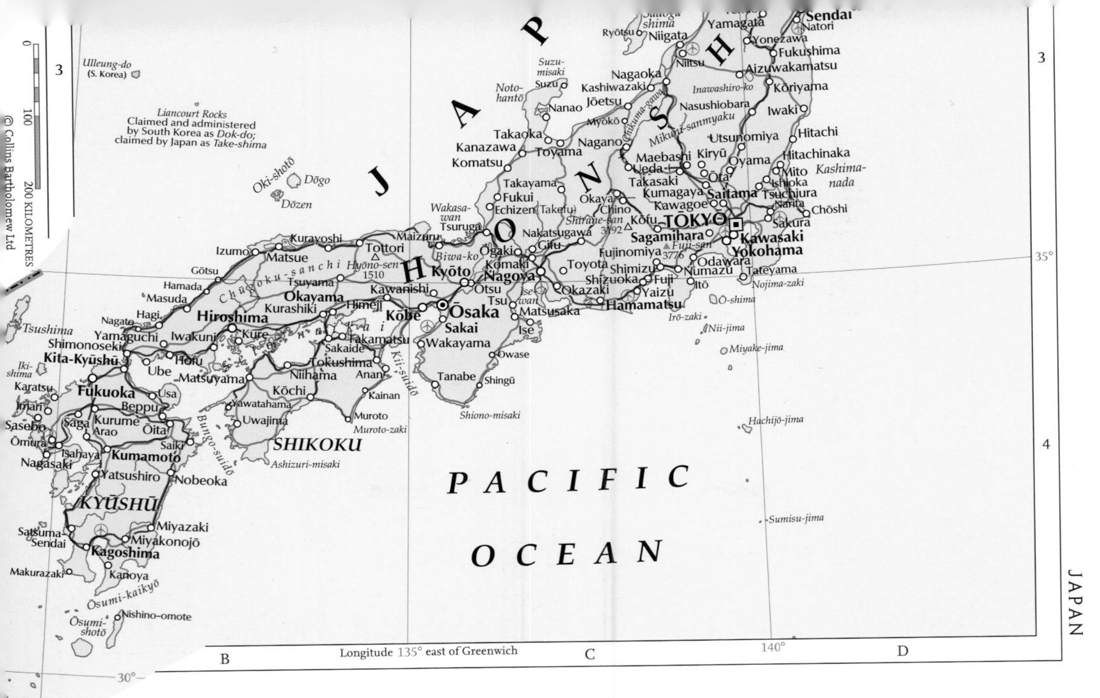

3

35°

140°

Longitude 135° east of Greenwich

30°

130°

PACIFIC

OCEAN

Sendai
Natori
Yamagata
Kaminoyama
Fukushima
Yonezawa
Iwakamatsu
Kōriyama
Hitachi
Kashima-nada
Kitaibaraki
Mito
Hitachinaka
Tsukuba
Chōshi
Tsuchiura
Sakura
Kawagoe
Utsunomiya
Kiryū
Ashikaga
Maebashi
TOKYO
Chiba
Kawasaki
Yokohama
Nasushiobara
Takasaki
Kumagaya
Ōmiya
Urawa
Saitama
Odawara
Nojima-zaki
Nagaoka
Kashiwazaki
Joetsu
Nagano
Ueda
Kōfu
Fuji
Numazu
Atami
Ō-shima
Inubō-zaki
Niijima
Nii-jima
Mikura-jima
Hachijō-jima
Miyake-jima
Nagano
Suzaka
Matsumoto
Kōfu
Shizuoka
Hamamatsu
Toyohashi
Iida
Ina
Fujinomiya
Mt Fuji 3776
Shizuoka
Irō-zaki
Ryōtsu
Sado-shima
Niigata
Shibata
Suzu-misaki
Noto-hantō
Nanao
Takaoka
Toyama
Kurobe
Kanazawa
Komatsu
Takayama
Takefu
Echizen
Fukui
Gifu
Ichinomiya
Nagoya
Okazaki
Ōgaki
Kuwana
Toyota
Owase
Shingū
Tsu
Matsusaka
Ise
Kanazawa
Tsuruga
Wakasa-wan
Maizuru
Toyooka
Amino
Tottori
Kurayoshi
Matsue
Izumo
Yonago
Gōtsu
Hamada
Masuda
Hagi
Yamaguchi
Shimonoseki
Ube
Nagato
KYŪSHŪ
Kita-Kyūshū
Fukuoka
Kurume
Ōmuta
Saga
Ōita
Beppu
Kumamoto
Nobeoka
Miyazaki
Miyakonojō
Kagoshima
Sendai
Satsuma
Nagasaki
Sasebo
Karatsu
Unzen
Isahaya
Yatsushiro
Minamata
Akune
Kanoya
Makurazaki
Ōsumi-shotō
Ōsumi-kaikyō
Tanega-shima
Bungo-suidō
Kōchi
Nankoku
Sukumo
Uwajima
Nakamura
Ashizuri-misaki
Muroto-zaki
Muroto
Anan
Komatsushima
Tokushima
Naruto
Kainan
Wakayama
Shiono-misaki
Tanabe
Kii-suidō
Osaka
Sakai
Kishiwada
Kōbe
Kōya-san 1510
Hashimoto
Kyōto
Ōtsu
Hikone
Nara
Yao
Suzuka
Yokkaichi
Nabari
KYŪSHŪ
SHIKOKU
Ōkayama
Kurashiki
Fukuyama
Onomichi
Hiroshima
Kure
Iwakuni
Tokuyama
Matsuyama
Imabari
Saijō
Niihama
Sakaide
Marugame
Takamatsu
Tsuyama
Biwa-ko
Hyōno-sen
Chūgoku-sanchi
Naka-umi
Daisen 1710
Oki-shotō
Dōgo
Dōzen
Ulleung-do
(S. Korea)
Liancourt Rocks
Claimed and administered
by South Korea as Dok-do;
claimed by Japan as Take-shima
Tsushima

H O N S H Ū

J A P A N

3
D
C
B

0   100   200 KILOMETRES

© Collins Bartholomew Ltd

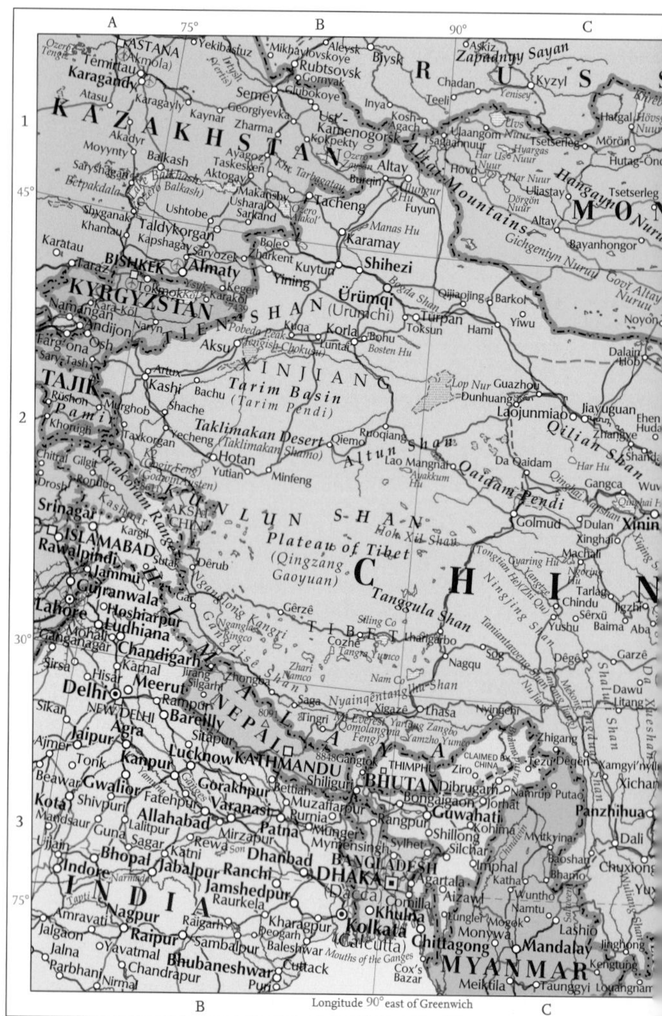

Albers Equal Area Conic Projection

1 : 30 000 000

MILES 0   200   400   600

Longitude 90° east of Greenwich

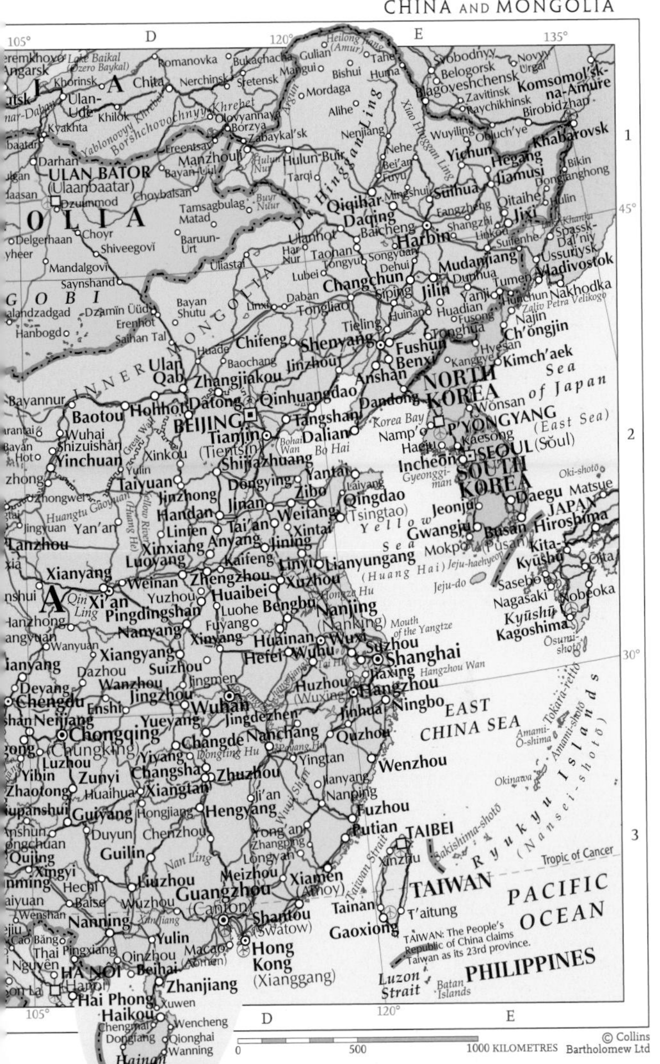

PACIFIC OCEAN
EAST CHINA SEA
Sea of Japan (East Sea)
Yellow Sea (Huang Hai)
NORTH KOREA
SOUTH KOREA
JAPAN
TAIWAN
PHILIPPINES
BEIJING
SHANGHAI
P'YONGYANG
SEOUL (Sŏul)
ULAN BATOR (Ulaanbaatar)
TAIBEI
HA NOI (Hanoi)
Hong Kong (Xianggang)
Ryukyu Islands (Nansei-shotō)
Tropic of Cancer
TAIWAN: The People's Republic of China claims Taiwan as its 23rd province.
0 500 1000 KILOMETRES
© Collins Bartholomew Ltd

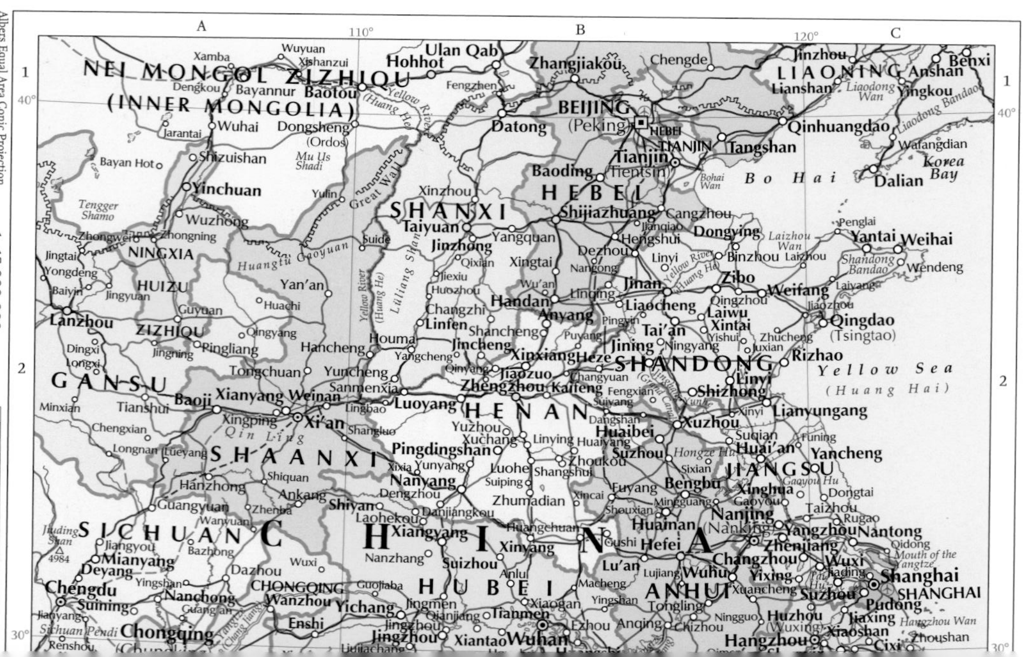

Albers Equal Area Conic Projection

1 : 15 000 000

MILES 0   100   200   300

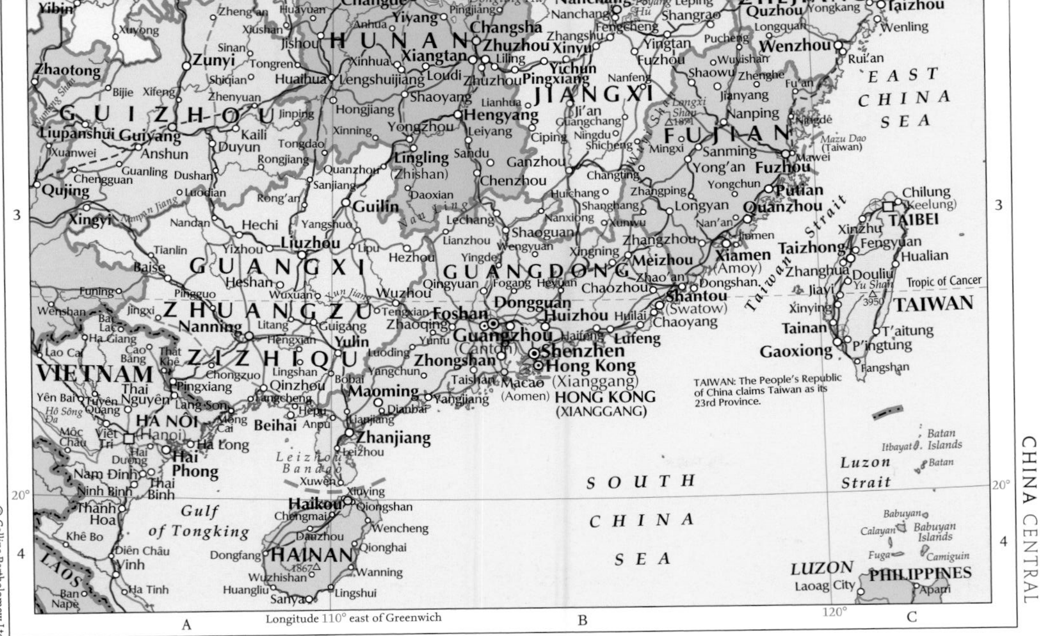

C H I N A   S E A

E A S T

Zhenghe
Wenling
Qingyuan
Yongkang
Longquan
Qingzhou
Quzhou
Leping
Nanchang
Changde
Yiyang
Changsha
Pingxiang
Zhuzhou
Xinyu
Yichun

H U N A N

Xiangtan
Lengshuijiang
Loudi
Zhuzhou
Pingxiang

Shaoyang
Hengyang
Yong'an

J I A N G X I

Ji'an

Taizhou
Chilung (Keelung)
TAIBEI
Hualian

T A I W A N

Taitung

Fuzhou

F U J I A N

Quanzhou
Xiamen (Amoy)
Putian

Longyan
Zhangzhou

Nan'an

Taizhong

Zhanghua

Nantou

Xinying

T A I N A N

Gaoxiong

P'ingtung
Fangshan

Luzon   Strait

Batan
Is. Batan Islands

Ibuhat Is.
Y'ami I.

120°

4

L U Z O N

Laoag City

P H I L I P P I N E S

Pagudpud

Vigan

Bangui

Calayan
Babuyan
Islands

Fuga
Camiguin

TAIWAN: The People's Republic
of China claims Taiwan as its
23rd Province.

Tropic of Cancer

20°

S O U T H

C H I N A

S E A

Guangzhou
Shenzhen
Hong Kong
(Xianggang)
Macao
(Aomen)

HONG KONG
(XIANGGANG)

G U A N G D O N G

Shaoguan
Qingyuan
Foshan
Dongguan
Huizhou
Chaozhou
Shantou (Swatow)
Jieyang
Lufeng

Meizhou
Zhanzhou
Zhao'an Dongshan

Chenzhou

G U A N G X I

Guilin
Liuzhou
Hechi

Nanning

Z H U A N G Q U

Wuzhou
Yulin
Beihai
Qinzhou
Fangcheng

Zhanjiang
Maoming
Yangjiang

Leizhou
Bandao

HAIKOU

H A I N A N

Sanya

Gulf

of

Tongking

V I E T N A M

HANOI

Hai Phong

L A O S

G U I Z H O U

Guiyang
Anshun
Duyun
Kaili
Zunyi

Qujing

Longitude 110° east of Greenwich

© Collins Bartholomew Ltd

0   250   500 KILOMETRES

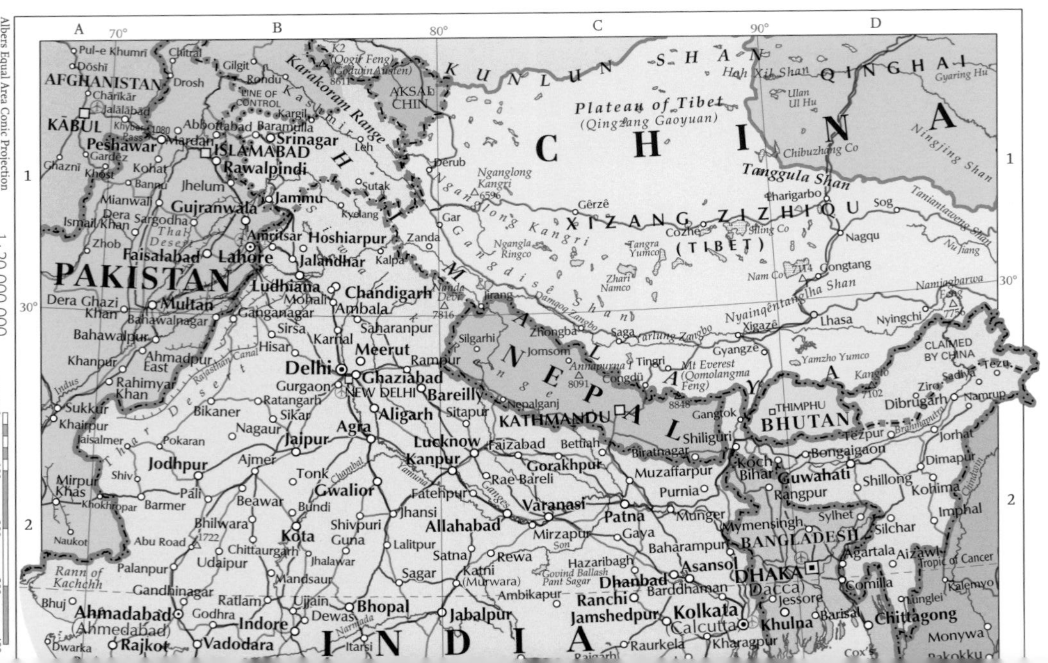

Albers Equal Area Conic Projection   1 : 20 000 000   MILES 0   100   200   300   400

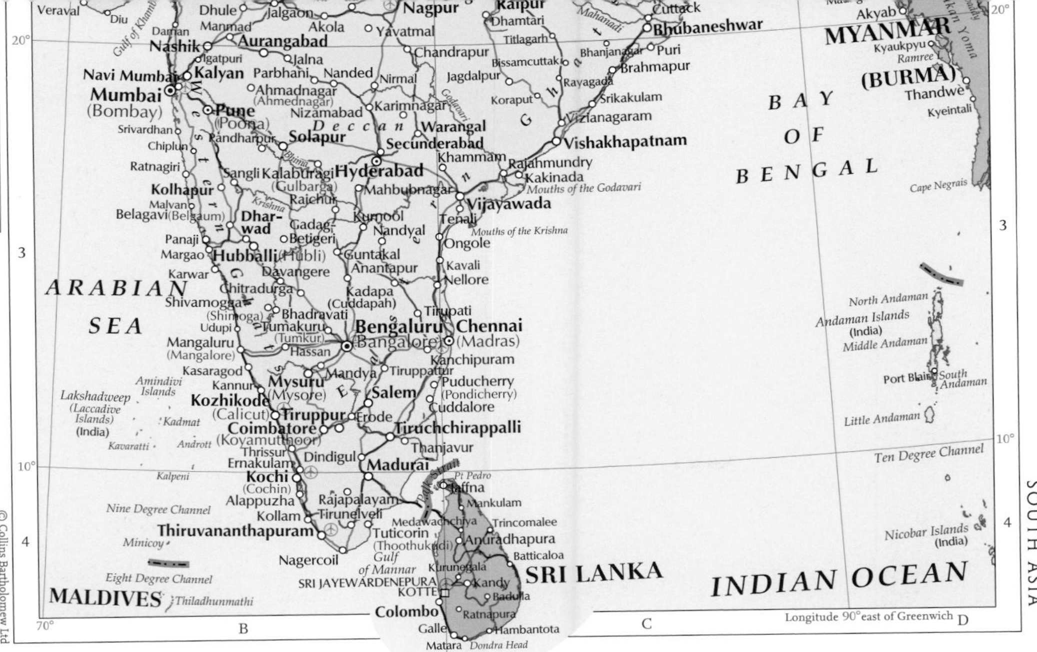

INDIAN OCEAN

BAY

OF

BENGAL

MYANMAR
(BURMA)

Akyab

Kyaukpyu
Ramree
Thandwe
Kyeintali

Cape Negrais

North Andaman
South Andaman
Andaman Islands
(India)
Middle Andaman
Little Andaman
Port Blair

Ten Degree Channel

Nicobar Islands
(India)

ARABIAN

SEA

SRI LANKA

Jaffna
Vavuniya
Trincomalee
Anuradhapura
Batticaloa
Negombo
Kandy
Badulla
COLOMBO
SRI JAYEWARDENEPURA KOTTE
Ratnapura
Galle
Matara   Dondra Head
Hambantota

Gulf of
Mannar
Palk Strait
Adam's
Bridge
Cape Comorin

Cuttack
Bhubaneshwar
Puri
Brahmapur
Srikakulam
Vizianagaram
Vishakhapatnam
Kakinada
Rajahmundry
Mouths of the Godavari
Vijayawada
Mouths of the Krishna
Machilipatnam
Ongole
Kavali
Nellore
Tirupati
Kanchipuram
Chennai (Madras)
Puducherry (Pondicherry)
Cuddalore
Tiruchchirappalli
Thanjavur
Karaikal
Madurai
Tuticorin (Thoothukudi)
Tirunelveli
Nagercoil

Nagpur
Kanpur
Dhamtari
Chandrapur
Yavatmal
Akola
Amravati
Aurangabad
Jalgaon
Dhule
Malegaon
Manmad
Nashik
Kalyan
Navi Mumbai
Mumbai (Bombay)
Srivardhan
Chiplun
Ratnagiri
Malvan
Vengurla
Panaji (Panjim)
Margao
Karwar
Udupi
Mangaluru (Mangalore)
Kasaragod
Kannur
Kozhikode (Calicut)
Thrissur
Ernakulam
Kochi (Cochin)
Alappuzha
Kollam (Quilon)
Thiruvananthapuram (Trivandrum)

Pune (Poona)
Solapur
Kolhapur
Sangli
Ichalkaranji
Nanded
Parbhani
Bidar
Kalaburagi (Gulbarga)
Raichur
Bagalkot
Belagavi (Belgaum)
Hubballi-Dharwad
Gadag
Davangere
Chitradurga
Shivamogga (Shimoga)
Tumakuru (Tumkur)
Hassan
Bengaluru (Bangalore)
Mysuru (Mysore)
Mandya
Salem
Erode
Tiruppur
Coimbatore
Dindigul
Rajapalayam

Hyderabad
Secunderabad
Warangal
Khammam
Nalgonda
Mahbubnagar
Kurnool
Nandyal
Kadapa (Cuddapah)
Anantapur
Guntakal

Nanded
Karimnagar
Nizamabad
Nirmal

Veraval
Diu
Daman
Gulf of Khambhat

Ahmednagar (Ahmadnagar)

MALDIVES

Lakshadweep (Laccadive Islands) (India)
Amindivi Islands
Minicoy
Kavaratti
Androth
Kadmat
Kalpeni

Eight Degree Channel
Nine Degree Channel

D e c c a n

Godavari
Krishna

20°

10°

10°

20°

70°

80°

90°

1

2

3

4

B

C

3

4

0   200   400   600 KILOMETRES

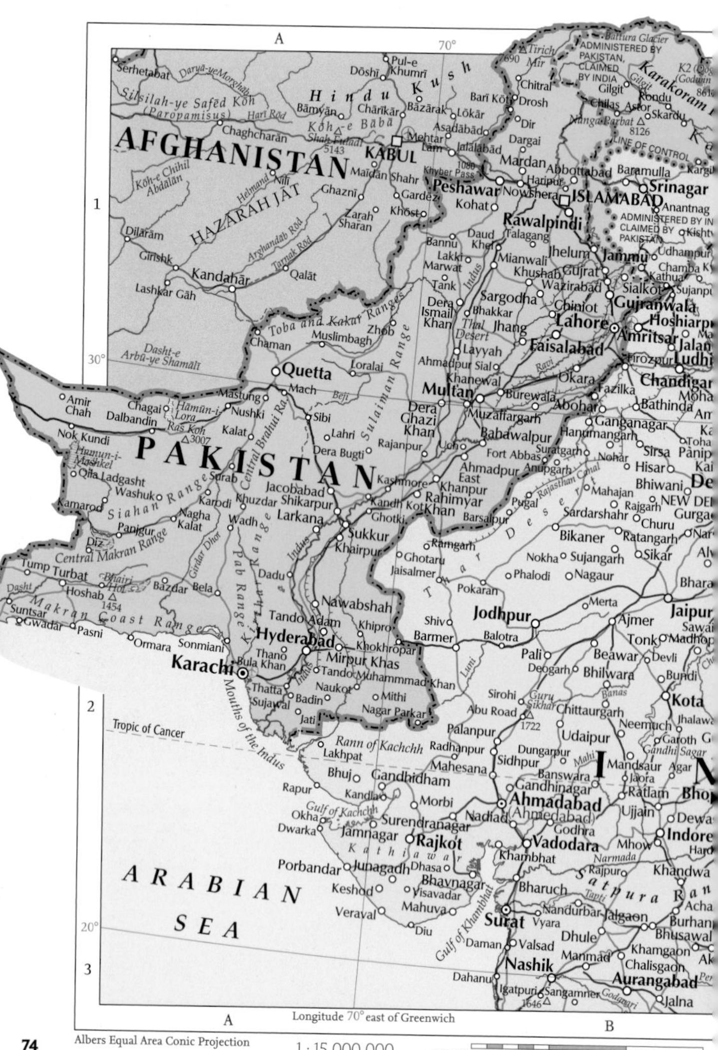

Albers Equal Area Conic Projection

1 : 15 000 000

MILES 0   100   200   300

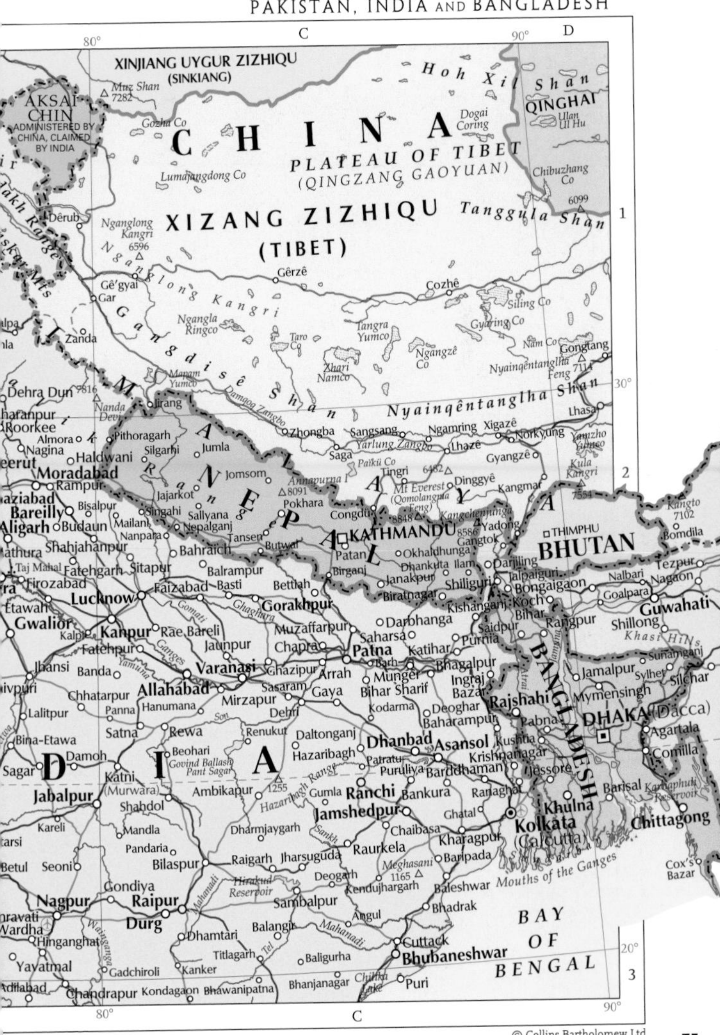

C
D
80°
90°
XINJIANG UYGUR ZIZHIQU
(SINKIANG)
Hoh Xil Shan
QINGHAI
AKSAI CHIN
ADMINISTERED BY CHINA, CLAIMED BY INDIA
CHINA
PLATEAU OF TIBET
(QINGZANG GAOYUAN)
XIZANG ZIZHIQU
(TIBET)
Tanggula Shan
Nganglong Kangri
Gangdisê Shan
Nyainqêntanglha Shan
NEPAL
KATHMANDU
BHUTAN
THIMPHU
BANGLADESH
DHAKA (Dacca)
INDIA
Mt Everest (Qomolangma Feng)
8848
Kangchenjunga
8586
Annapurna I
8091
Lucknow
Kanpur
Varanasi
Allahabad
Patna
Gorakhpur
Jabalpur
Nagpur
Raipur
Durg
Ranchi
Jamshedpur
Dhanbad
Asansol
Kolkata (Calcutta)
Bhubaneshwar
Cuttack
Guwahati
Chittagong
Khulna
Rajshahi
Mouths of the Ganges
BAY OF BENGAL
1
2
3
30°
20°
© Collins Bartholomew Ltd
0
250
500 KILOMETRES

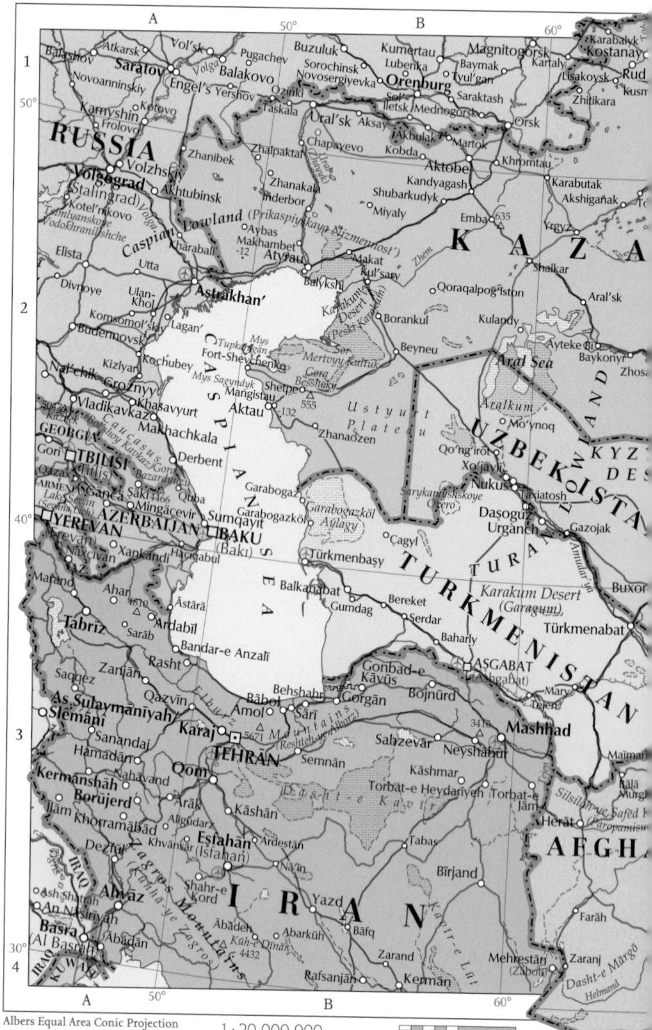

Albers Equal Area Conic Projection

1 : 20 000 000

MILES 0   100   200

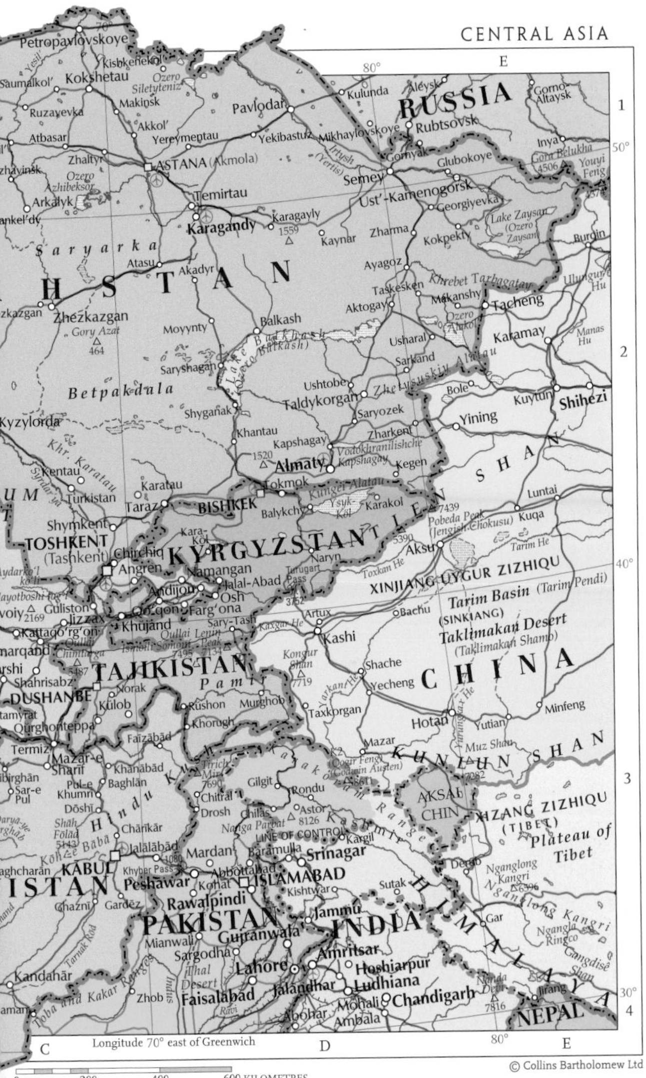

Petropavlovskoye
Kokshetau
Kiskenekol'
Saumalkol'
Ozero
Siletyteniz
Makiysk
Kulunda
Aleysk
Gorno-
Altaysk
RUSSIA
Ruzayevka
Atbasar
Akkol'
Yereymentau
Yekibastuz
Mikhaylovskoye
Rubtsovsk
Inya
Gora Belukha
4506
Yunyi
Feng
Zhaltyr
ASTANA (Akmola)
Tiemirtau
Semey
Ust'-Kamenogorsk
Georgiyevka
zhinsk
Akkol'
Arkalyk
Tengiz
Zhibekszol
ankel'dy

Karagandy
Karagayly
1559
Kaynar
Zharma
Kokpekty
Burqin

K A Z A K H S T A N

Saryarka

zkazgan
Zhezkazgan
Gory Azat
464
Atasu
Akadyr
Moyynty
Balkash
Ayagoz
Taskesken
Khrebet Tarbagatay
Makanshy
Ozero
Ala-Köl'
Tacheng
Ulungur
Ulungur He

Kyzylorda
Betpakdala
Saryshagan
Ushtobe
Usharal
Sarkand
Karamay
Manas
He

UM
Kentau
Khr. Karatau
Shyganak
Taldykorgan
Saryozek
Bole
Kuytun
Shihezi

TOSHKENT
(Tashkent)
Chirchiq
Khantau
Kapshagay
Zharkent
Yining
T I E N S H A N

Almaty
Tokmok
Kegen

BISHKEK
Balykchy
Ysyk-Köl'
Karakol
Pobeda Peak
(Jengish Chokusu)
7439
Luntai
Kuqa
Tarim He

Shymkent
Taraz
Kara-Köl'
KYRGYZSTAN
Naryn
Aksu

Angren
Namangan
Jalal-Abad
XINJIANG UYGUR ZIZHIQU
(SINKIANG)

voiy
Gulistan
Andijon
Osh
Artux
Bachu
Tarim Basin (Tarim Pendi)

Jizzax
Farg'ona
Kashi
Taklimakan Desert
(Taklimakan Shamo)

Khujand
TAJIKISTAN
Shache
Yecheng
C H I N A

DUSHANBE
Pamir
Murghob
Taxkorgan
Hotan
Yutian
Minfeng

Termiz
Kulob
Khorugh
Mazar
K U N L U N   S H A N

Mazar-e
Sharif
Faizabad
Gilgit
AKSAI
CHIN
XIZANG ZIZHIQU
(TIBET)

I S T A N
KABUL
Peshawar
Kohat
ISLAMABAD
Srinagar
Plateau of
Tibet

PAKISTAN
Rawalpindi
Jammu
I N D I A
H I M A L A Y A

Kandahar
Lahore
Amritsar
Hoshiarpur
Chandigarh
NEPAL

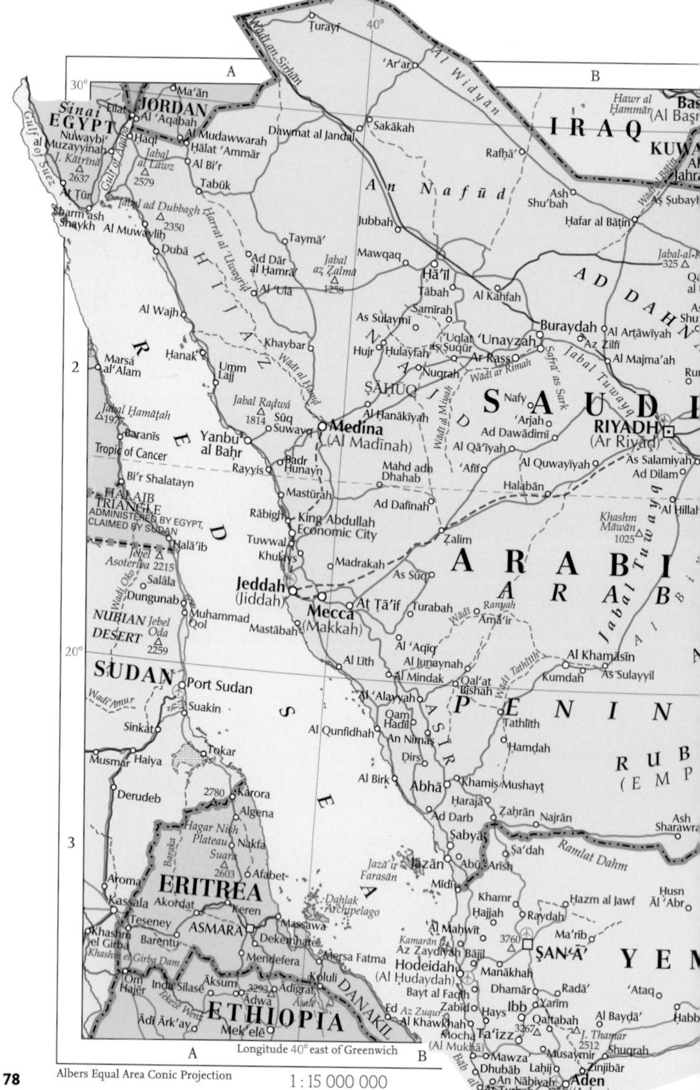

JORDAN
EGYPT
Sinai
IRAQ
KUWAIT
SAUDI ARABIA
ARABIAN PENINSULA
RIYADH
(Ar Riyāḍ)
Medina
(Al Madīnah)
Mecca
(Makkah)
Jeddah
(Jiddah)
An Nafūd
AD DAHNĀ'
HIJAZ
NAJD
ASIR
ṢAḤŪQ
RED SEA
Gulf of Suez
Gulf of Aqaba
Ma'ān
Eilat
Al 'Aqabah
Al Mudawwarah
Ḥālat 'Ammār
Al Bi'r
Tabūk
Ṭurayf
'Ar'ar
Al Widyān
Wādī an Sirḥān
Sakākah
Dawmat al Jandal
Rafḥā'
Hawr al Ḥammār
Al Jahra
Ash Shu'bah
Ḥafar al Bāṭin
Aṣ Ṣubayḥ
Wādī al Bāṭin
Jubbah
Mawqaq
Ḥā'il
Ṭābah
Al Kahfah
Taymā'
Ad Dār al Ḥamrā'
Jabal az Zalmā
1258
Al 'Ulā
Ḥarrat al 'Uwayriḍ
Jabal ad Dubbagh
2350
Nuwaybi'
al Muzayyinah
J. Kātrīnā
2637
Ḥaql
Jabal al Lawz
2579
Aṭ Ṭūr
Sharm ash Shaykh
Al Muwayliḥ
Ḍubā
Al Wajh
Khaybar
Wādī al Ḥamḍ
Ḥanak
Umm Lajj
Marsá al'Alam
Jabal Raḍwá
1814
Sūq Suwayq
Yanbu' al Baḥr
Jabal Ḥamāṭah
1977
Baranīs
Tropic of Cancer
Rayyis
Badr Ḥunayn
Bi'r Shalatayn
Mastūrah
HALAIB TRIANGLE
ADMINISTERED BY EGYPT, CLAIMED BY SUDAN
Rābigh
King Abdullah Economic City
Tuwwal
Khulays
Ḥalā'ib
Jebel Asoteriba
2215
Salâla
Dungunab
Wadi Oko
Muhammad Qol
NUBIAN DESERT
Jebel Oda
2259
SUDAN
Port Sudan
Suakin
Wadi Amur
Sinkat
Tokar
Musmar
Haiya
Derudeb
2780
Karora
Algena
Hagar Nish Plateau
Nakfa
Suara
2603
Afabet
Baraka
ERITREA
Aroma
Kassala
Akordat
Keren
Teseney
ASMARA
Massawa
Khashm el Girba
Barentu
Dekemhare
Khashm el Girba Dam
Mendefera
Mersa Fatma
Om Hajer
Inda Silase
Āksum
Ādwa
3293
Ādigrat
Koluli
Tekezē Wenz
DANAKIL
ETHIOPIA
Ādī Ārk'ay
Mek'elē
Dahlak Archipelago
Jazā'ir Farasān
Jāzān
Medina
Mahd adh Dhahab
Ad Dafīnah
Madrakah
As Sūq
Aṭ Ṭā'if
Turabah
Mastābah
Al 'Aqīq
Al Līth
Al Junaynah
Al Mindak
Al 'Alayyah
Qam Hadīl
Al Qunfidhah
An Nimāṣ
Dirs
Al Birk
Abhā
Khamīs Mushayṭ
Ḥarajā
Ad Darb
Ẓahrān
Najrān
Ṣabyā
Ṣa'dah
Abū 'Arīsh
Mīdī
Ramlat Dahm
Buraydah
Al Arṭāwīyah
Az Zilfī
Al Majma'ah
'Unayzah
'Uqlat aṣ Ṣuqūr
Ar Rass
Wādī ar Rimah
Nuqrah
As Sulaymī
Hujr
Ḥulayfah
Samīrah
Al Ḥanākīyah
Wādī al Miyah
Ṣafrā' as Sark
Jabal Ṭuwayq
Nafy
'Arjah
Ad Dawādimī
Al Qā'īyah
'Afīf
Al Quwayīyah
As Salamiyah
Ad Dilam
Ḥalabān
Al Hillah
Khashm Māwān
1025
Ẓalim
Ranyah
Wādī 'Amā'ir
Wādī Tathlīth
Al Khamāsīn
Kumdah
As Sulayyil
Qal'at Bīshah
Tathlīth
Ḥamḍah
RUB' AL KHALI
(EMPTY QUARTER)
Ash Sharawrah
Jabal-al-Ḥ
325
Khamr
Ḥajjah
Raydah
Ḥazm al Jawf
Ḥusn Āl 'Abr
Al Maḥwīt
Ma'rib
Kamarān
Az Zaydīyah
Bājil
3760
ṢAN'Ā'
YEMEN
Hodeidah
(Al Ḥudaydah)
Manākhah
Bayt al Faqīh
Dhamār
Radā'
'Ataq
Zabīd
Ḥays
Ibb
Yarīm
Ed
Az Zuqur
Al Khawkhah
Qa'ṭabah
Al Bayḍā'
Ḥabbān
Mocha
(Al Mukhā)
Ta'izz
3267
J. Thamar
2512
Mawza'
Musaymīr
Shuqrah
Dhubāb
Laḥij
Zinjibār
Bāb al Mandab
An Nābiyah
Aṭ Ṭurbah
Aden
('Adan)
30°
20°
40°
A
B
2
3
Longitude 40° east of Greenwich
Albers Equal Area Conic Projection
1 : 15 000 000

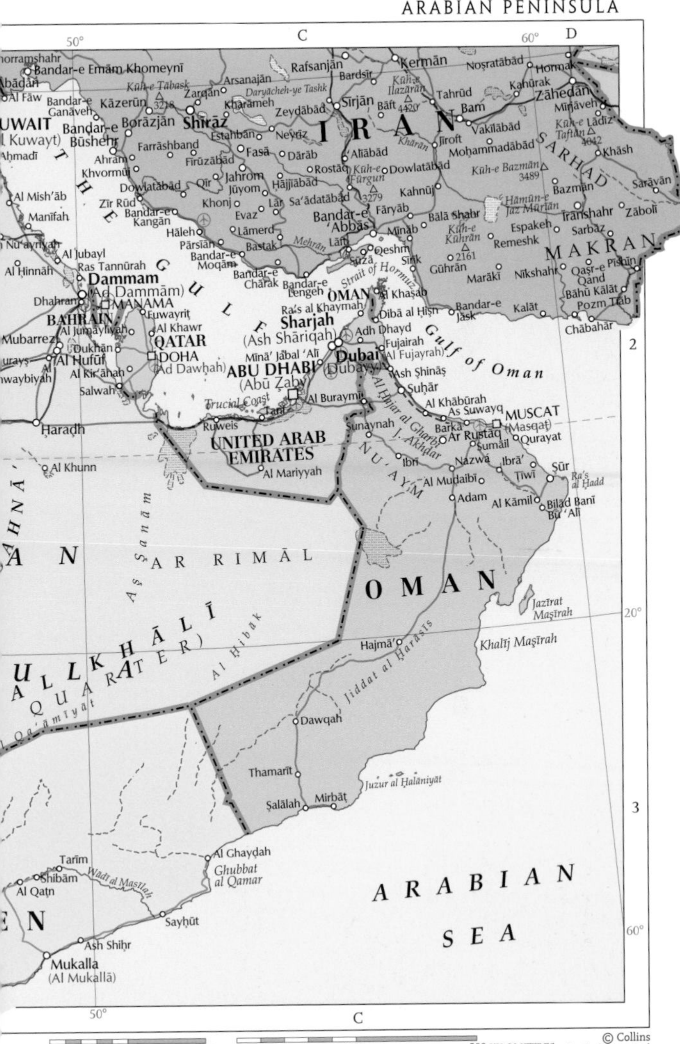
IRAN
Kermān
Rafsanjān
Shīrāz
Bandar-e Būshehr
Bandar-e 'Abbās
Zāhedān
MAKRAN
SARHAD
Dammam (Ad Dammām)
MANAMA
BAHRAIN
QATAR
DOHA (Ad Dawḩah)
ABU DHABI (Abū Ẓaby)
Dubai (Dubayy)
Sharjah (Ash Shāriqah)
UNITED ARAB EMIRATES
Fujairah (Al Fujayrah)
Ṣuḩār
MUSCAT (Masqaţ)
Nazwá
Ṣūr
OMAN
Strait of Hormuz
Gulf of Oman
THE GULF
AR RIMĀL
AL RUB' AL KHĀLĪ (EMPTY QUARTER)
Dawqah
Thamarīt
Şalālah
Mirbāţ
Jazīrat Maşīrah
Khalīj Maşīrah
Juzur al Ḩalānīyāt
Al Ghaydah
Mukalla (Al Mukallā)
ARABIAN SEA
Al Hufūf
Haraḑh
Al Buraymī
Ruweis
Ibrī
Hajmā'
Jiddat al Ḩarāsīs
Al Jubayl
Ras Tannūrah
Chābahār
MILES 0 100 200
0 250 500 KILOMETRES

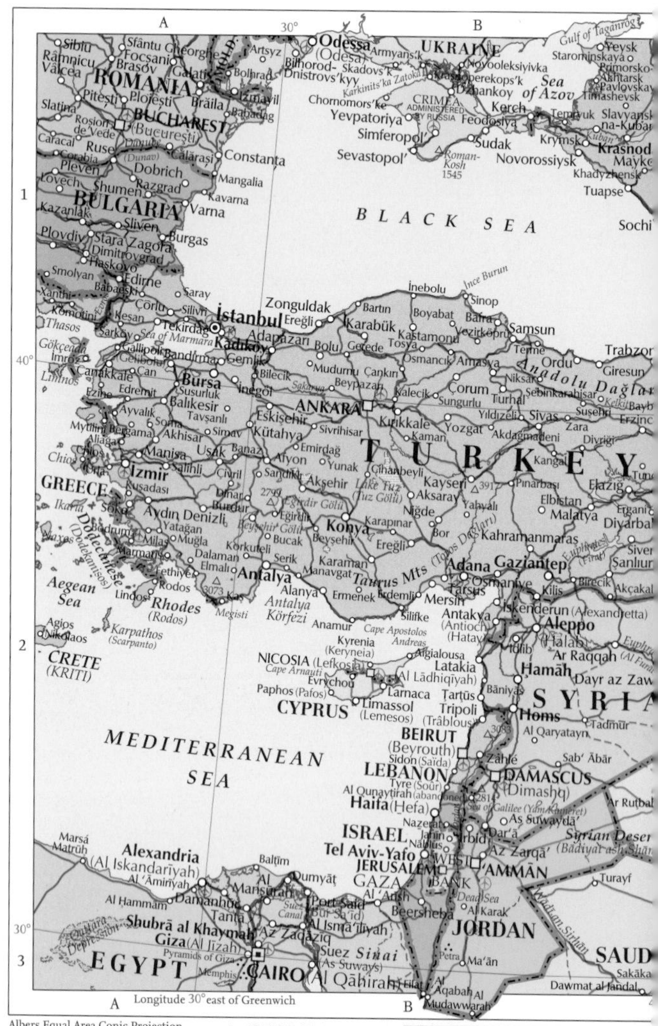

Albers Equal Area Conic Projection   1 : 15 000 000   MILES 0   100   200   300

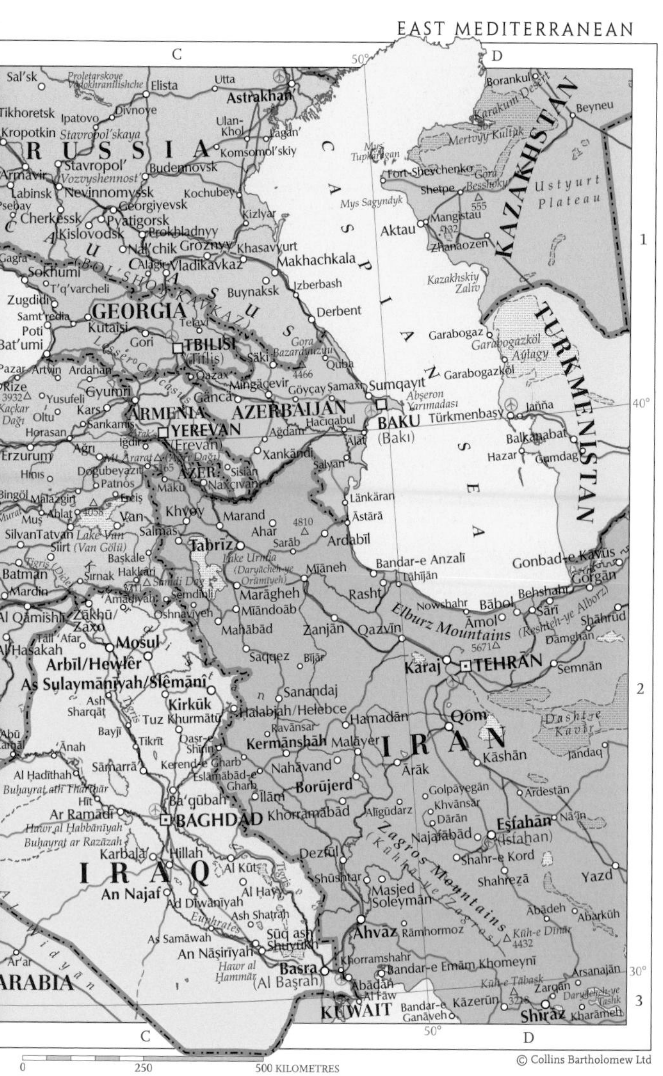

0   250   500 KILOMETRES

© Collins Bartholomew Ltd

40°
30°
1
3
RUSSIA
GEORGIA
ARMENIA
AZERBAIJAN
AZER.
KAZAKHSTAN
TURKMENISTAN
IRAN
IRAQ
KUWAIT
ARABIA
CASPIAN SEA
CAUCASUS
BOL'SHOY KAVKAZ
Lesser Caucasus
Elburz Mountains
Reshteh-ye Alborz
Zagros Mountains
Kūhhā-ye Zāgros
Ustyurt Plateau
Karakum Desert
Dasht-e Kavīr
Al Widyān
Sal'sk
Proletarskoye Vodokhranilishche
Elista
Utta
Astrakhan'
Tikhoretsk
Ipatovo
Divnoye
Ulan-Khol
Lagan'
Kropotkin
Stavropol'skaya Vozvyshennost'
Komsomol'skiy
Stavropol'
Armavir
Budennovsk
Labinsk
Nevinnomyssk
Kochubey
Psebay
Georgiyevsk
Cherkessk
Pyatigorsk
Kizlyar
Kislovodsk
Prokhladnyy
Nal'chik
Groznyy
Khasavyurt
Gagra
Alagir
Vladikavkaz
Makhachkala
Sokhumi
T'q'varcheli
Buynaksk
Izberbash
Zugdidi
Samt'redia
Derbent
Poti
Kutaisi
Telavi
Bat'umi
Gori
TBILISI
(Tiflis)
Gora Bazardyuzyu
4466
Şäki
Quba
Pazar
Artvin
Ardahan
Qazax
Rize
Mingäçevir
Göyçay
Şamaxı
Sumqayıt
3932
Yusufeli
Gyumri
Gäncä
Abşeron Yarımadası
Kaçkar Dağı
Oltu
Kars
BAKU
(Bakı)
Hacıqabul
Sarıkamış
Ağdam
Horasan
YEREVAN
(Erevan)
Ālāt
Iğdır
Xankändi
Erzurum
Ağrı
Mt Ararat
(Ağrı Dağı)
5165
Salyan
Doğubeyazıt
Sisian
Hınıs
Patnos
Naxçıvan
Mākū
Bingöl
Malazgirt
Erciş
Länkäran
Murat
Muş
Ahlat
4058
Van
Khvoy
Marand
Āstārā
SilvanTatvan
Lake Van
(Van Gölü)
Salmās
Ahar
4810
Siirt
Sarāb
Ardabīl
Başkale
Tabrīz
Tigris (Dicle)
Lake Urmia
(Daryācheh-ye Orūmīyeh)
Batman
Şırnak
Hakkâri
3811
Samdi Dag
Mīāneh
Mardin
Āmādīyah
Semdinli
Marāgheh
Rasht
Al Qāmishlī
Zākhū/Zaxo
Oshnavīyeh
Mīāndoāb
Tall 'Afar
Mosul
Mahābād
Zanjān
Qazvīn
Al Ḩasakah
Arbīl/Hewlêr
Saqqez
Bījār
As Sulaymānīyah/Slêmānî
Sanandaj
Ash Sharqāţ
Kirkūk
Ḩalabjah/Helebce
Tuz Khurmātū
Hamadān
Bayjī
Ravānsar
Tikrīt
Qasr-e Shīrīn
Kermānshāh
Malāyer
Abū Kamāl
'Ānah
Kerend-e Gharb
Sāmarrā'
Nahāvand
Al Ḩadīthah
Eslāmābād-e Gharb
Buḩayrat ath Tharthār
Borūjerd
Hīt
Ba'qūbah
Īlām
Ar Ramādī
BAGHDĀD
Khorramābād
Hawr al Ḩabbānīyah
Buḩayrat ar Razāzah
Karbalā'
Hillah
Al Kūt
Dezfūl
An Najaf
Al Ḩayy
Shūshtar
Masjed Soleymān
Ad Dīwānīyah
Ash Shaţrah
Euphrates
Ahvāz
Rāmhormoz
As Samāwah
Sūq ash Shuyūkh
An Nāşirīyah
Hawr al Ḩammār
Basra
(Al Başrah)
Khorramshahr
Ābādān
Al Fāw
Ar'ar
Bandar-e Emām Khomeynī
Bandar-e Ganāveh
Kāzerūn
Mys Tupkaragan
Fort-Shevchenko
Gora Besshoky
Shetpe
555
Mys Sagyndyk
Mangistau
Aktau
332
Zhanaozen
Kazakhskiy Zaliv
Borankul
Beyneu
Sor Mertvyy Kultuk
Garabogaz
Garabogazköl Aýlagy
Garabogazköl
Türkmenbaşy
Jaňňa
Balkanabat
Hazar
Gumdag
Bandar-e Anzalī
Lāhījān
Gonbad-e Kāvūs
Gorgān
Behshahr
Nowshahr
Bābol
Sārī
Āmol
Shāhrūd
Dāmghān
5671
Karaj
TEHRĀN
Semnān
Qom
Kāshān
Jandaq
Arāk
Golpāyegān
Ardestān
Alīgūdarz
Khvānsār
Dārān
Eşfahān
(Isfahan)
Nā'īn
Najafābād
Shahr-e Kord
Shahreẕā
Yazd
Ābādeh
Abarkūh
Kūh-e Dīnār
4432
Kūh-e Tābask
3218
Arsanajān
Zarqān
Daryācheh-ye Tashk
Shīrāz
Kharāmeh

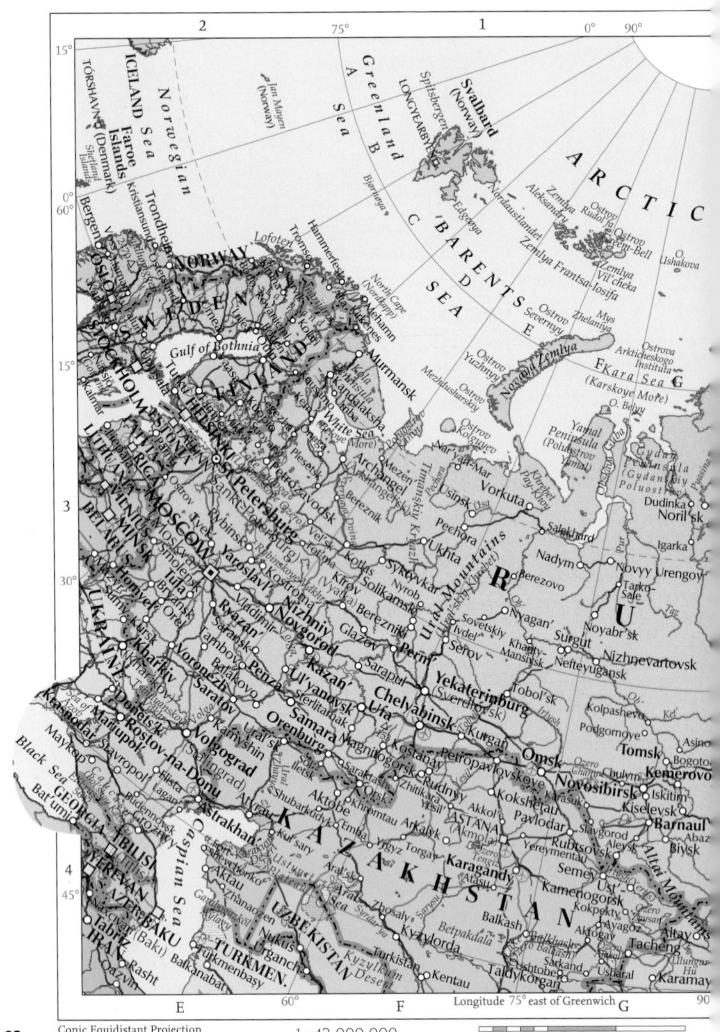

Conic Equidistant Projection

1 : 42 000 000

MILES 0   250   500   750

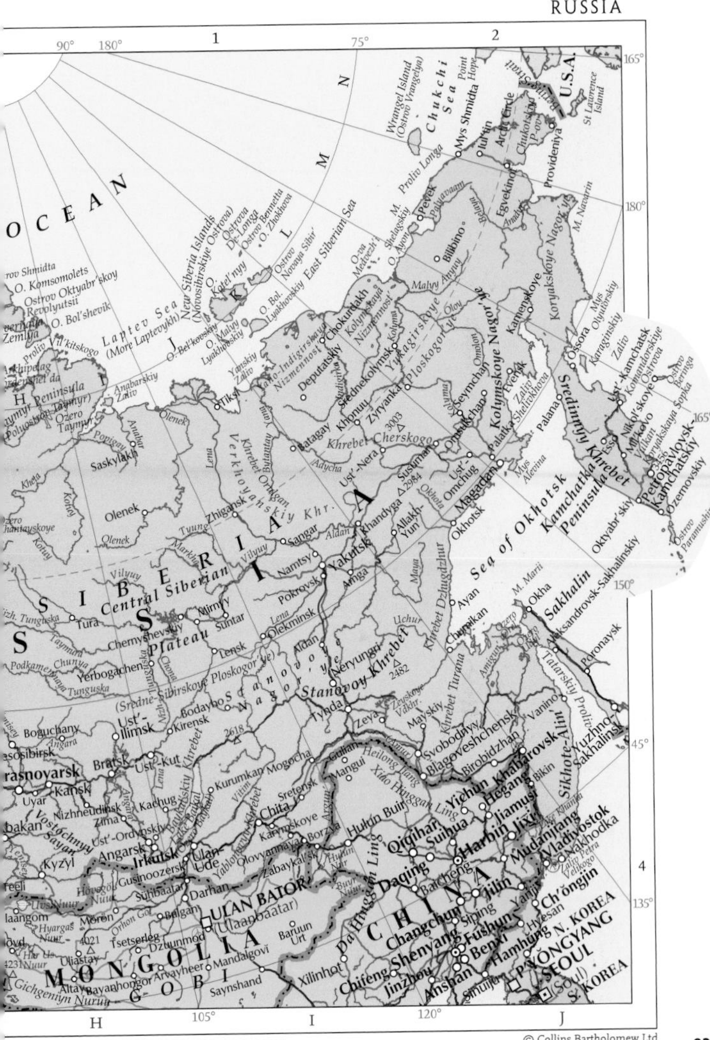

© Collins Bartholomew Ltd

0   500   1000   1500 KILOMETRES

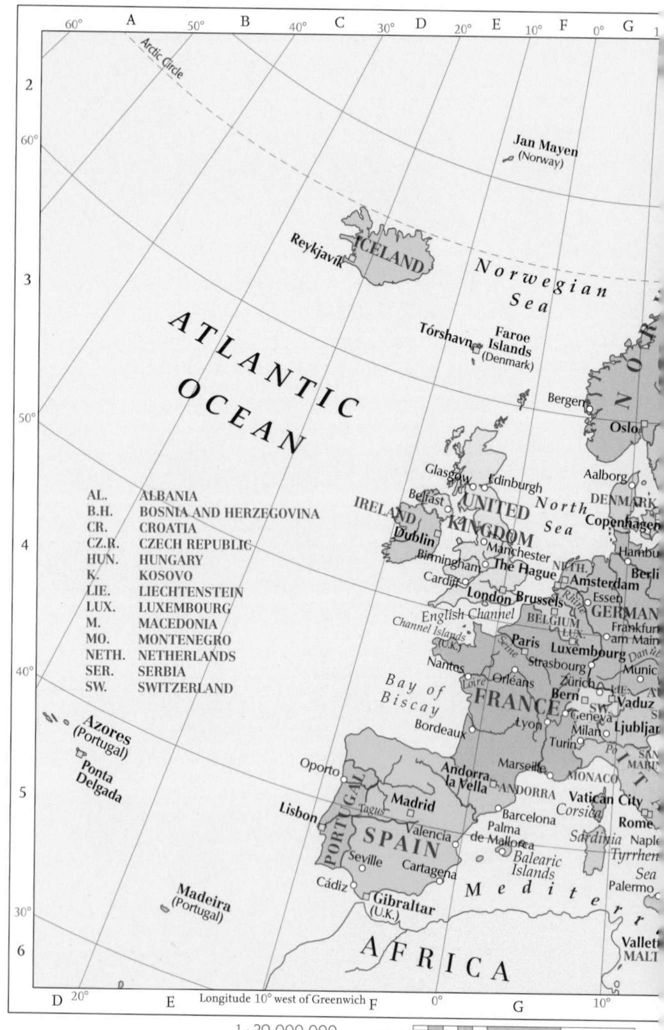

84

2

60°

Arctic Circle

Jan Mayen
(Norway)

Reykjavík   ICELAND

*Norwegian
Sea*

3

Tórshavn   Faroe
Islands
(Denmark)

A T L A N T I C

Bergen

N O R

O C E A N

Oslo

50°

Glasgow   Edinburgh

Aalborg

Belfast   UNITED   *North*   DENMARK

IRELAND   KINGDOM   *Sea*   Copenhagen

AL.	ALBANIA
B.H.	BOSNIA AND HERZEGOVINA
CR.	CROATIA
CZ.R.	CZECH REPUBLIC
HUN.	HUNGARY
K.	KOSOVO
LIE.	LIECHTENSTEIN
LUX.	LUXEMBOURG
M.	MACEDONIA
MO.	MONTENEGRO
NETH.	NETHERLANDS
SER.	SERBIA
SW.	SWITZERLAND

4

Dublin   Manchester   The Hague

Birmingham   NETH.

Cardiff   London   Brussels   Essen   GERMAN

Hamburg

Berlin

English Channel   BELGIUM   LUX.

Channel Islands   Paris   Luxembourg   Frankfurt
(U.K.)   am Main

40°

Strasbourg   Dan

Nantes   *Loire*   Orleans   Zürich   Munic

Bay of   FRANCE   Bern   Vaduz
Biscay   Geneva   Ljubljar

Bordeaux   Lyon   Milan   Turin   *Po*   SAN
MARIN

Marseille   MONACO

Azores
(Portugal)

Oporto   Andorra   Corsica   Vatican City

la Vella   ANDORRA   Rome

Ponta
Delgada

5

Lisbon   Madrid   Barcelona   *Sardinia*   Naple

Tagus   Valencia   Palma   *Tyrrhen*
de Mallorca   *Sea*

SPAIN   Palermo

Seville   *Balearic
Islands*

Madeira
(Portugal)

Cádiz   Cartagena   M e d i t e r r

Gibraltar
(U.K.)

Valleta
MALT

30°

6

A F R I C A

1 : 39 000 000   MILES 0   250   500   750

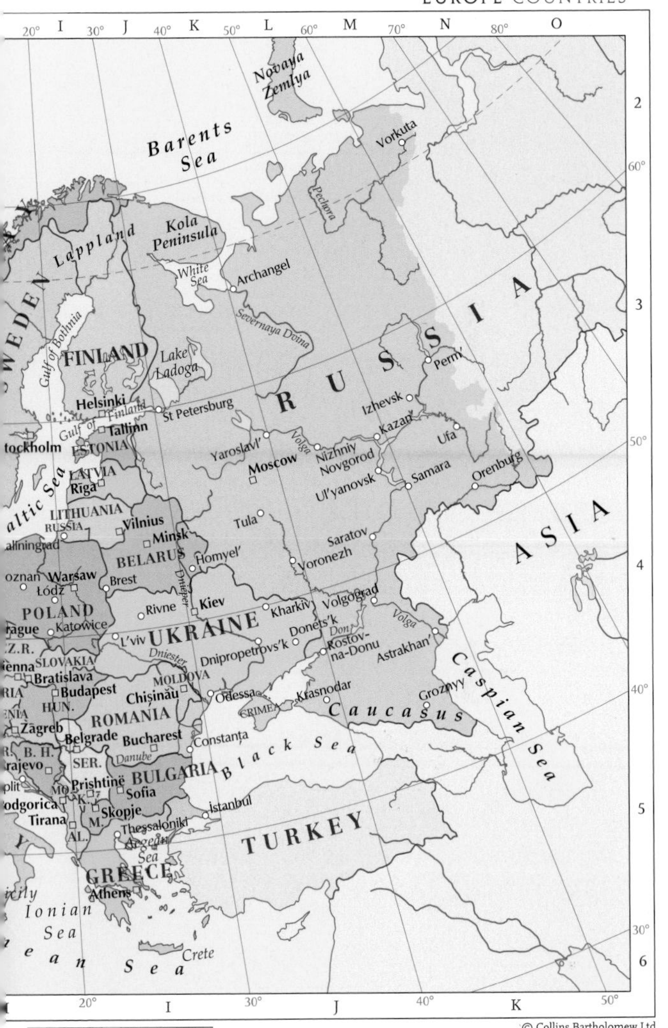

© Collins Bartholomew Ltd

Novaya Zemlya
Barents Sea
Vorkuta
Pechora
Kola Peninsula
Lappland
White Sea
Archangel
SWEDEN
Gulf of Bothnia
Severnaya Dvina
FINLAND
Lake Ladoga
R U S S I A
Perm'
Helsinki
Gulf of Finland
St Petersburg
Izhevsk
Tallinn
Kazan'
Stockholm
ESTONIA
Yaroslavl'
Volga
Nizhniy Novgorod
Ufa
Moscow
Baltic Sea
LATVIA
Riga
Ul'yanovsk
Samara
Orenburg
LITHUANIA
RUSSIA
Kaliningrad
Vilnius
Tula
Minsk
Saratov
ASIA
BELARUS
Homyel'
Voronezh
Poznan
Warsaw
Brest
Łódź
Dnieper
POLAND
Rivne
Kiev
Kharkiv
Volgograd
Prague
Katowice
UKRAINE
Donets'k
Volga
CZ.R.
L'viv
Don
Dniester
Dnipropetrovs'k
Rostov-na-Donu
Astrakhan'
Vienna
SLOVAKIA
Bratislava
MOLDOVA
Caspian Sea
AUSTRIA
Budapest
Chișinău
Odessa
Krasnodar
Groznyy
HUN.
CRIMEA
Caucasus
ROMANIA
Zagreb
Belgrade
Bucharest
Constanța
Black Sea
B. H.
Sarajevo
SER.
Danube
BULGARIA
Split
MO.
Prishtinë
K.
Sofia
Podgorica
Skopje
Istanbul
Tirana
AL.
Thessaloniki
Aegean Sea
TURKEY
GREECE
Sicily
Athens
Ionian Sea
Crete
Mediterranean Sea
I J K L M N O
20° 30° 40° 50° 60° 70° 80°
2 3 4 5 6
60° 50° 40° 30°
0 500 1000 KILOMETRES

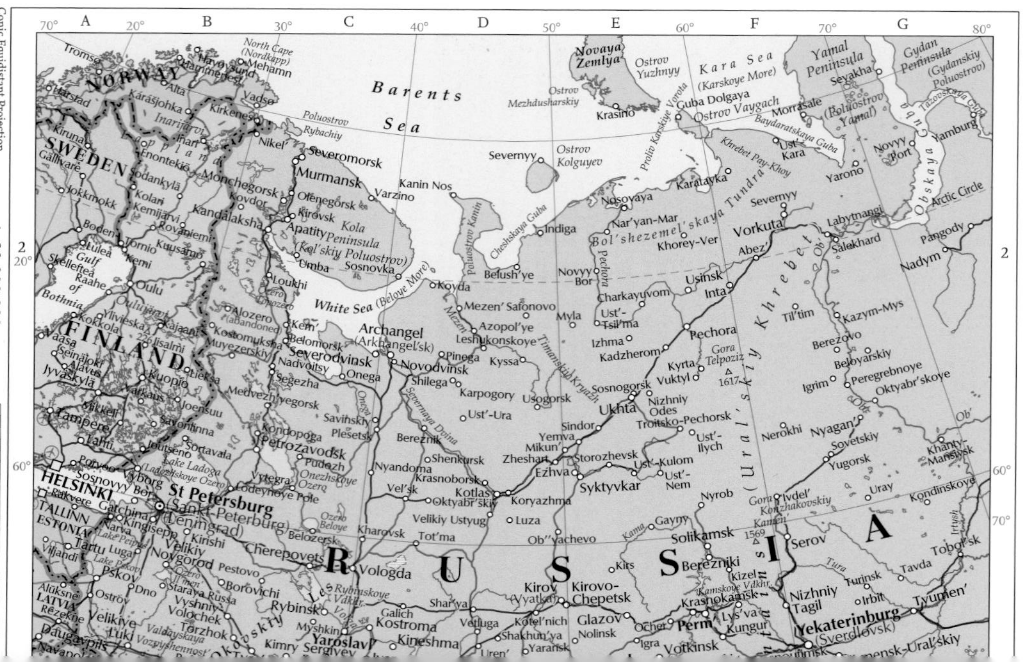

Conic Equidistant Projection

1 : 20 000 000

MILES 0   100   200   300   400

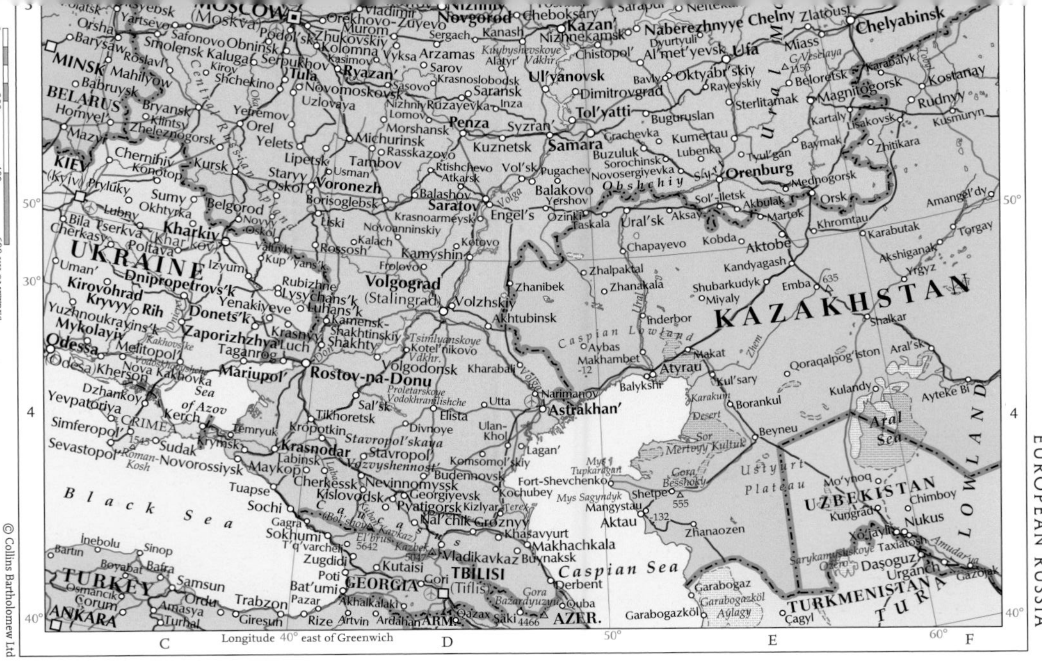
KAZAKHSTAN
UKRAINE
BELARUS
GEORGIA
TURKEY
UZBEKISTAN
TURKMENISTAN
AZER.
ARM.
Caspian Sea
Black Sea
Sea of Azov
Aral Sea
Caspian Lowland
Ustyurt Plateau
Central Russian Upland
Obshchiy Syrt
MOSCOW (Moskva)
KIEV (Kyiv)
MINSK
TBILISI (Tiflis)
ANKARA
Volgograd (Stalingrad)
Rostov-na-Donu
Saratov
Samara
Kazan'
Ufa
Orenburg
Chelyabinsk
Astrakhan'
Atyrau
Aktobe
Kharkiv (Khar'kov)
Dnipropetrovs'k
Donets'k
Zaporizhzhya
Odesa (Odessa)
Voronezh
Krasnodar
Stavropol'
Makhachkala
Longitude 40° east of Greenwich

0 200 400 600 KILOMETRES

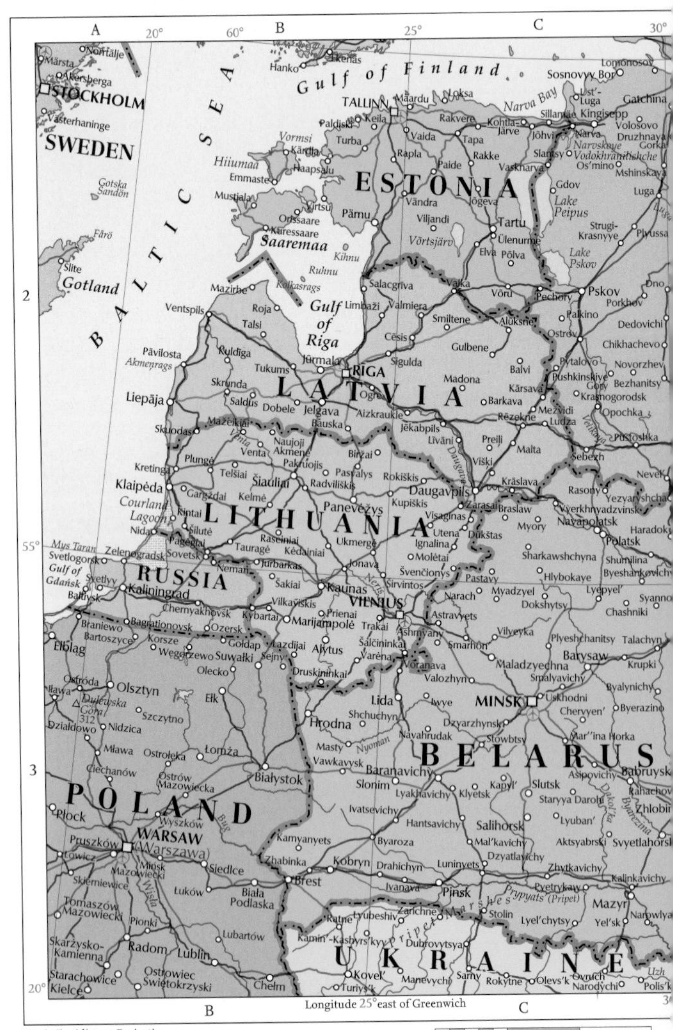

Conic Equidistant Projection   1 : 8 000 000   MILES 0   50   100   150

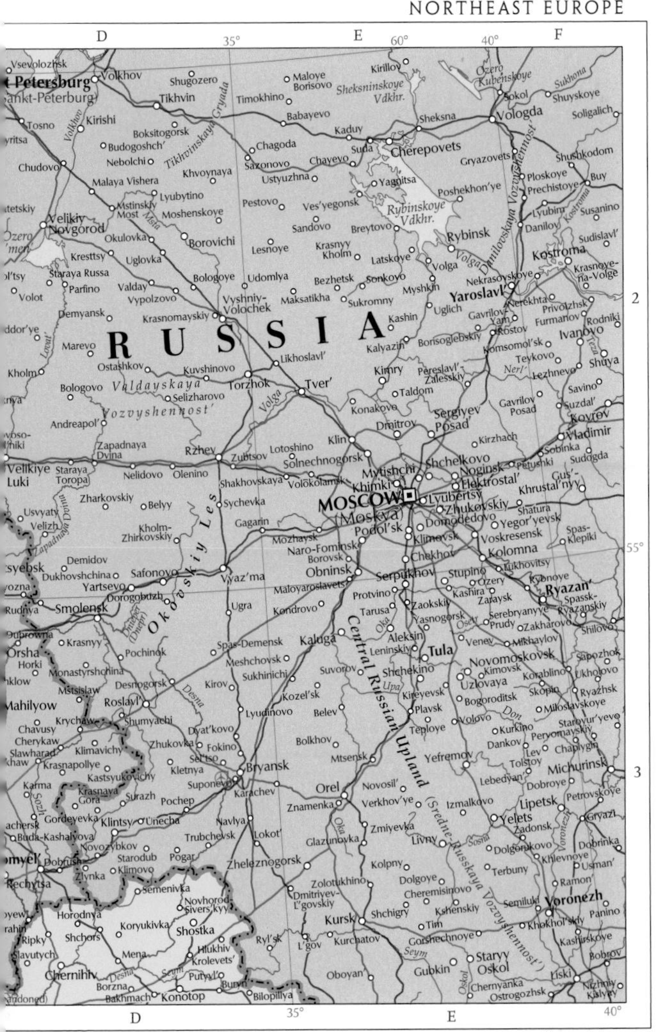

D   35°   E   60°   40°   F

Vsevolozhsk
Volkhov   Shugozero   Maloye   Kirillov   Ozero   Sukhona   Soligalich
Petersburg   Borisovo   Sheksninskoye   Kubenskoye
ankt-Peterburg   Tikhvin   Timokhino   Vdkhr.   Sokol   Vologda   Shuyskoye
Tosno   Kirishi   Babayevo   Sheksna
yritsa   Boksitogorsk   Chagoda   Suda   Cherepovets   Gryazovets   Ploskove   Shunkodom
Chudovo   Budogoshch'   Khvoynaya   Sazonovo   Chayevo   Buy
Nebolchi   Ustyuzhna   Yagnitsa   Poshekhon'ye   Prechistoye
Malaya Vishera   Lyubytino   Pestovo   Ves'yegonsk   Rybinskoye
atetskiy   Moshenskoye   Sandovo   Breytovo   Vdkhr.   Lyubim   Susanino
Velikiy   Mstinskiy   Borovichi   Krasny   Rybinsk   Danilov
Novgorod   Most   Okulovka   Lesnoye   Kholm   Latskoye   Volga   Sudislavl'
'tsy   Kresttsy   Uglovka   Bologoye   Udomlya   Bezhetsk   Skromny   Myshkin   Yaroslavl'   Kostroma
Staraya Russa   Vyshniy-   Maksatikha   Nekrasovskoye   Krasnoye-
Parfino   Valday   Volochek   Kashin   Uglich   Gavrilov   na-Volge
'mon   Volot   Vypolzovo   Demyansk   Krasnomayskiy   Likhoslavl'   Kalyazin   Borisoglebsk   Yam   Furmanov   Rodniki
ddor'   R U S S I A   Komsomol'sk   Ivanovo
Kholm   Ostashkov   Kuvshinovo   Kimry   Pereslavl'-   Teykovo   Shuya
Bologovo   Valdayskaya   Torzhok   Tver'   Taldom   Zalesskiy   Savino
unya   Selizharovo   Vozvyshennost'   Gavrilov   Suzdal'
Andreapol'   Zapadnaya   Rzhev   Zubtsov   Konakovo   Dmitrov   Sergiyev   Posad   Kirzhach   Kovrov
Dvina   Lotoshino   Klin   Vladimir
yoso-   Nelidovo   Olenino   Solnechnogorsk   Shchelkovo   Petushki   Sobinka
thiki   Zapadnaya   Shakhovskaya   Mytishchi   Pokrov   Sudogda
Velikiye   Staraya   Volokolamsk   Khimki   Elektrostal'   Gus'-   Gus'-
Luki   Toropa   Belyy   Sychevka   MOSCOW   Lyubertsy   Khrustal'ny   ny
Usvyaty   Zharkovskiy   Gagarin   (Moskva)   Zhukovskiy   Yegor'yevsk
Velizh   Kholm-   Safonovo   Vyaz'ma   Naro-Fominsk   Podol'sk   Domodedovo   Voskresensk   Spas-
yozna   Zhirkovskiy   Ugra   Borovsk   Klimovsk   Shatura   Klepiki
syelsk   Dukhovshchina   Obninsk   Serpukhov   Stupino   Ozery   Kolomna
Demidov   Yartsevo   Maloyaroslavets   Protvino   Kashira   Kuznove
Smolensk   Dorogobuzh   Kondrovo   Tarusa   Zaoksky   Zaraysk   Ryazan'
Roslavl'   Spas-Demensk   Kaluga   Serebryanyye   Spassk-
Orsha   Krasnyy   Pochinok   Meshchovsk   Aleksin   Prudy   Zakharovo   Shilovo
Horki   Monastyrshchina   Sukhinichi   Leninskiy   Tula   Novomoskovsk   Sapozhok
klov   Mstislaw   Desnogorsk   Kirov   Suvorov   Shchekino   Uzlovaya   Kimovsk   Skopin   Ukholovo
Mahilyow   Roslavl'   Lyudinovo   Belev   Plavsk   Bogoroditsk   Ryazhsk
Chavusy   Krychaw   Shumyachi   Dyat'kovo   Bolkhov   Mtsensk   Yefremov   Dankov   Peryomyshl'   Chaplygin
Cherykaw   Klimavichy   Zhukovka   Fokino   Kletnya   Bryansk   Orel   Verkhov'ye   Izmalkovo   Lebedyan   Michurinsk
Slawharad   Krasnapollye   Kastsyukovichy   Suponevo   Karachev   Znamenka   Livny   Dolgorukovo   Yelets   Zadonsk   Ramon   Voronezh
Karma   Gordeyevka   Surazh   Pochep   Navlya   Glazunovka   Zmiyevka   Terbuny   Semiluki
chersk   Klintsy   Unecha   Trubchevsk   Lokot'   Kolpny   Cheremisinovo   Khokhol'skiy
mych   Dobrush   Starodub   Pogar   Zheleznogorsk   Dolgoye   Shchigry   Gubkin   Staryy   Oskol   Chernyanka   Ostrogozhsk
Rechitsa   Zlynka   Klimovo   Semenivka   Zolotukhino   Kshenskiy   Tim   Ostrogozhsk
yew   Horodnya   Novhorod-   Koryukivka   Kursk   Kurchatov   Gorshechnoye
rahin   Ripky   Shchors   Sivers'kyy   Shostka   Ryl'sk   L'gov   Oboyan   Gubkin

0   100   200 KILOMETRES   © Collins Bartholomew Ltd   89

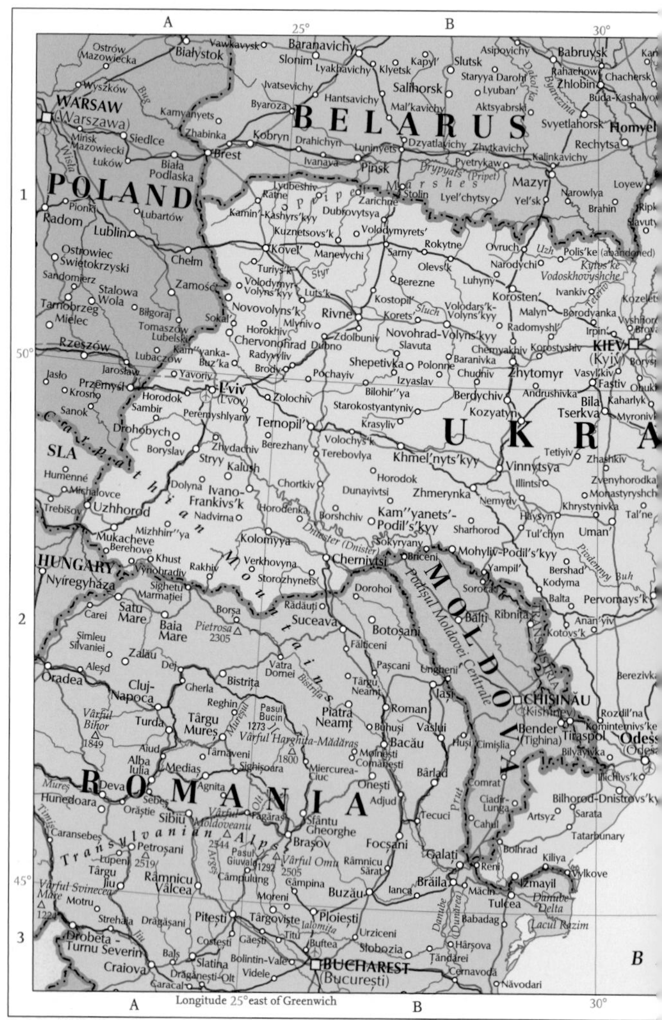

1 : 8 000 000

MILES 0   50   100   150

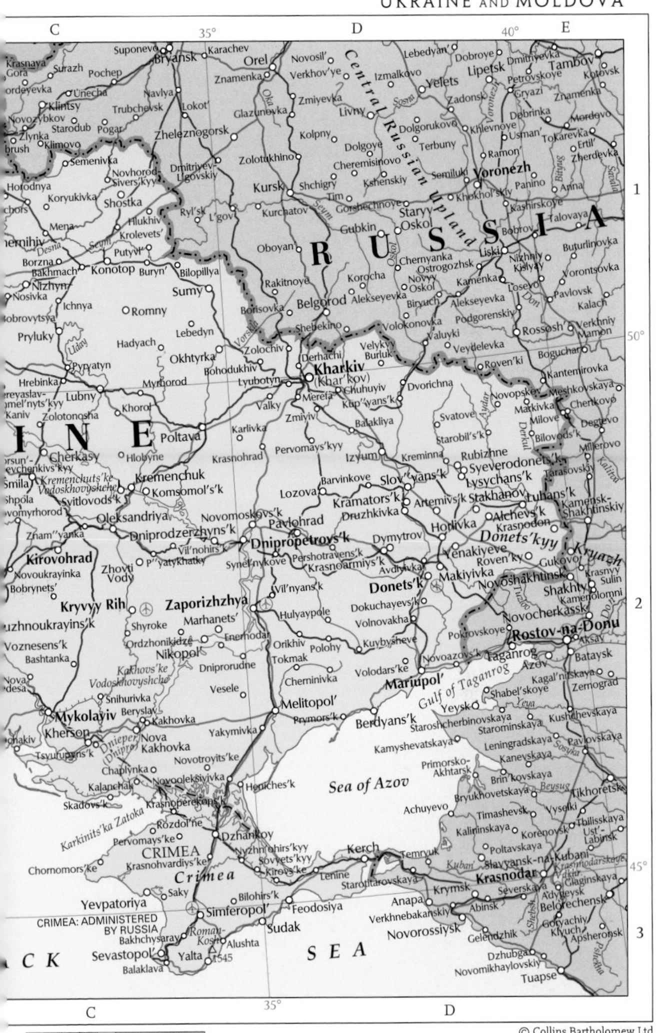
C
35°
D
40°
E
1
2
3
50°
45°
RUSSIA
Central Russian Upland
UKRAINE
Bryansk
Karachev
Orel
Suponevo
Surazh
Pochep
Unecha
Navlya
Klintsy
Trubchevsk
Lokot'
Novozybkov
Starodub
Pogar
Zheleznogorsk
Zlynka
Klimovo
Semenivka
Novhorod-Sivers'kyy
Dmitriyev-L'govskiy
Horodnya
Koryukivka
Shostka
Ryl'sk
L'gov
Hlukhiv
Krolevets'
Mena
Putyvl'
Borzna
Bakhmach
Konotop
Buryn'
Bilopillya
Nizhyn
Nosivka
Ichnya
Romny
Sumy
Pryluky
Lebedyn
Hadyach
Okhtyrka
Pyryatyn
Bohodukhiv
Hrebinka
Lubny
Myrhorod
Khorol
Zolotonosha
Kaniv
Poltava
Cherkasy
Hlobyne
Kremenchuk
Kremenchuts'ke Vodoskhovyshche
Komsomol's'k
Smila
Svitlovods'k
Shpola
Oleksandriya
Znam"yanka
Kirovohrad
Novoukrayinka
Bobrynets'
Zhovti Vody
Kryvyy Rih
Voznesens'k
Bashtanka
Mykolayiv
Kherson
Nova Kakhovka
Kakhovka
Beryslav
Snihurivka
Tsyurupyns'k
Chaplynka
Kalanchak
Skadovs'k
Karkinits'ka Zatoka
Krasnoperekops'k
Rozdol'ne
Pervomays'ke
CRIMEA
Crimea
Chornomors'ke
Krasnohvardiys'ke
Yevpatoriya
Saky
Simferopol'
Bakhchysaray
Sevastopol'
Balaklava
Yalta
Alushta
Roman-Kosh
1545
CRIMEA: ADMINISTERED BY RUSSIA
Dzhankoy
Nyzhn'ohirs'kyy
Sovyets'kyy
Kirovs'ke
Bilohirs'k
Feodosiya
Sudak
Lenine
Kerch
Henichesk
Novooleksiyivka
Novotroyits'ke
Yakymivka
Melitopol'
Prymors'k
Berdyans'k
Sea of Azov
Gulf of Taganrog
Mariupol'
Volodars'ke
Tokmak
Cherninivka
Vesele
Dniprorudne
Kakhovs'ke Vodoskhovyshche
Nikopol'
Marhanets'
Enerhodar
Ordzhonikidze
Shyroke
Zaporizhzhya
Vil'nyans'k
Hulyaypole
Orikhiv
Polohy
Synel'nykove
Dnipropetrovs'k
Dniprodzerzhyns'k
Novomoskovs'k
Pavlohrad
Vil'nohirs'k
P"yatykhatky
Pershotravens'k
Krasnoarmiys'k
Donets'k
Makiyivka
Horlivka
Yenakiyeve
Dymytrov
Avdiyivka
Dokuchayevs'k
Volnovakha
Kuybysheve
Novoazovs'k
Kramators'k
Druzhkivka
Slov"yans'k
Artemivs'k
Lozova
Barvinkove
Izyum
Krasnohrad
Pervomays'kyy
Karlivka
Zmiyiv
Balakliya
Valky
Merefa
Kharkiv
(Khar'kov)
Chuhuyiv
Kup"yans'k
Lyubotyn
Zolochiv
Derhachi
Velykyy Burluk
Dvorichna
Svatove
Starobil's'k
Kreminna
Rubizhne
Syeverodonets'k
Lysychans'k
Stakhanov
Luhans'k
Alchevs'k
Krasnodon
Roven'ky
Donets'kyy Kryazh
Novopskov
Markivka
Milove
Bilovods'k
Aydar
Derkul
Belgorod
Shebekino
Borisovka
Rakitnoye
Oboyan'
Korocha
Alekseyevka
Novyy Oskol
Biryuch
Volokonovka
Valuyki
Veydelevka
Roven'ki
Kursk
Kurchatov
Shchigry
Tim
Gorshechnoye
Staryy Oskol
Gubkin
Chernyanka
Ostrogozhsk
Liski
Voronezh
Khokhol'skiy
Semiluki
Kashirskoye
Bobrov
Talovaya
Anna
Panino
Ramon'
Usman'
Khlevnoye
Dolgorukovo
Terbuny
Yelets
Lipetsk
Zadonsk
Livny
Zmiyevka
Glazunovka
Kolpny
Dolgoye
Cheremisinovo
Kshenskiy
Zolotukhino
Novosil'
Verkhov'ye
Znamenka
Izmalkovo
Lebedyan'
Dobroye
Dmitriyevka
Tambov
Petrovskoye
Gryazi
Kotovsk
Znamenka
Dobrinka
Mordovo
Tokarevka
Ertil'
Zherdevka
Nizhniy Kislyay
Buturlinovka
Vorontsovka
Pavlovsk
Kalach
Losevo
Kamenka
Podgorenskiy
Rossosh'
Verkhniy Mamon
Boguchar
Kantemirovka
Meshkovskaya
Chertkovo
Degtevo
Millerovo
Tarasovskiy
Kamensk-Shakhtinskiy
Gukovo
Krasnyy Sulin
Shakhty
Kamenolomni
Novoshakhtinsk
Novocherkassk
Rostov-na-Donu
Aksay
Bataysk
Azov
Taganrog
Pokrovskoye
Kagal'nitskaya
Zernograd
Shabel'skoye
Yeysk
Yeya
Staroshcherbinovskaya
Starominskaya
Kushchevskaya
Leningradskaya
Pavlovskaya
Kamyshevatskaya
Kanevskaya
Primorsko-Akhtarsk
Brin'kovskaya
Bryukhovetskaya
Beysug
Tikhoretsk
Achuyevo
Timashevsk
Vyselki
Kalininskaya
Korenovsk
Tbilisskaya
Ust'-Labinsk
Poltavskaya
Slavyansk-na-Kubani
Krasnodar
Kuban'
Temryuk
Starotitarovskaya
Krymsk
Severskaya
Adygeysk
Glaginskaya
Belorechensk
Anapa
Abinsk
Verkhnebakanskiy
Novorossiysk
Gelendzhik
Goryachiy Klyuch
Apsheronsk
Dzhubga
Novomikhaylovskiy
Tuapse
Pshekha
BLACK SEA
Desna
Seym
Oka
Sosna
Voronezh
Bityug
Savala
Oskol
Vorskla
Uday
Don
Kalitva
Tuzlov
Dnieper (Dnipro)
Sosyka
Shebsh
0
100
200 KILOMETRES

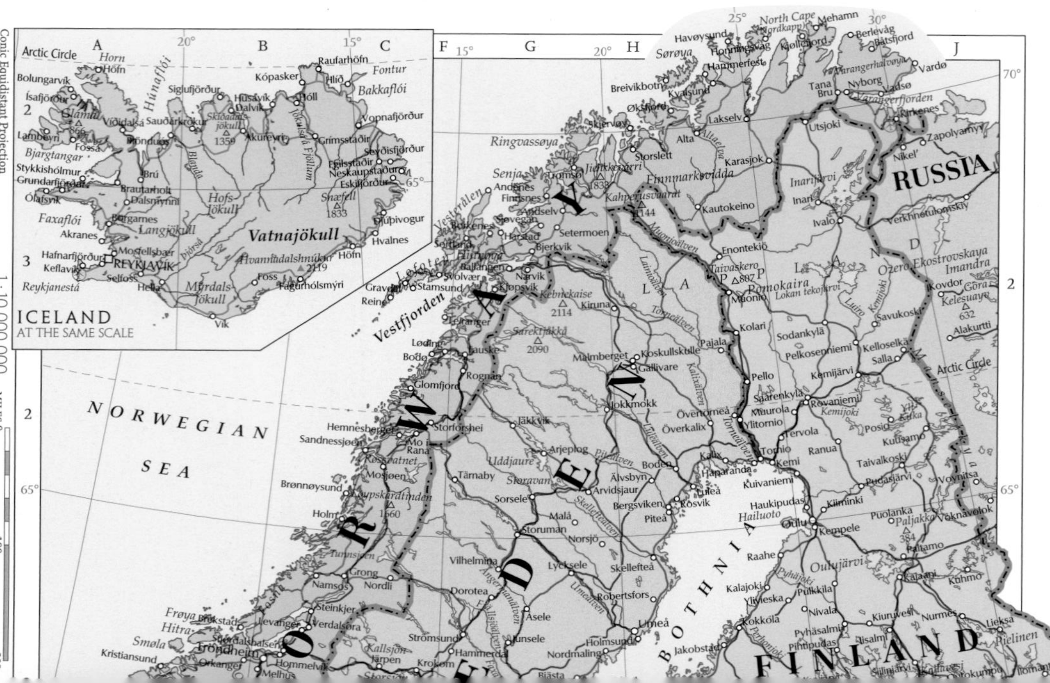

Conic Equidistant Projection

1:10 000 000   MILES 0   100   200

KILOMETRES 0   100   200   300

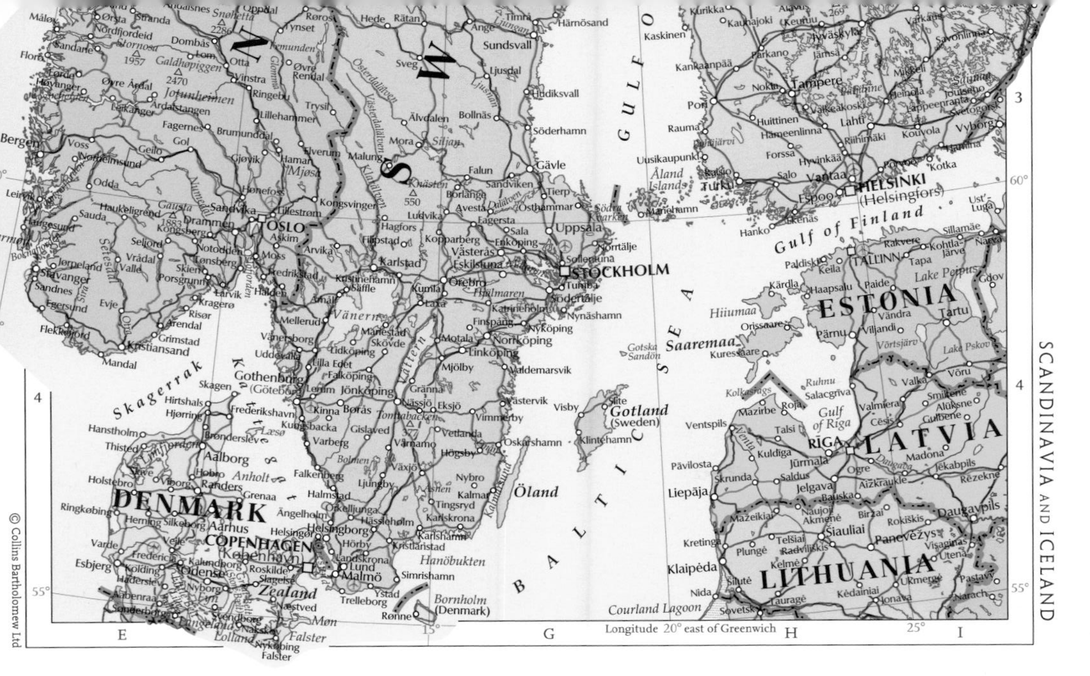
NORWAY
SWEDEN
DENMARK
ESTONIA
LATVIA
LITHUANIA
OSLO
STOCKHOLM
HELSINKI
(Helsingfors)
TALLINN
RĪGA
COPENHAGEN
(København)
Gulf of Finland
Gulf of Riga
GULF OF
BALTIC SEA
Skagerrak
Kattegat
Gotland
(Sweden)
Öland
Saaremaa
Hiiumaa
Åland
Islands
Bornholm
(Denmark)
Zealand
Lake Peipus
Vänern
Vättern
Hanöbukten
Courland Lagoon
Longitude 20° east of Greenwich
Bergen
Stavanger
Kristiansand
Gothenburg
(Göteborg)
Malmö
Uppsala
Tampere
Turku
Tartu
Aalborg
Århus
Odense
Klaipėda
Liepāja
Daugavpils

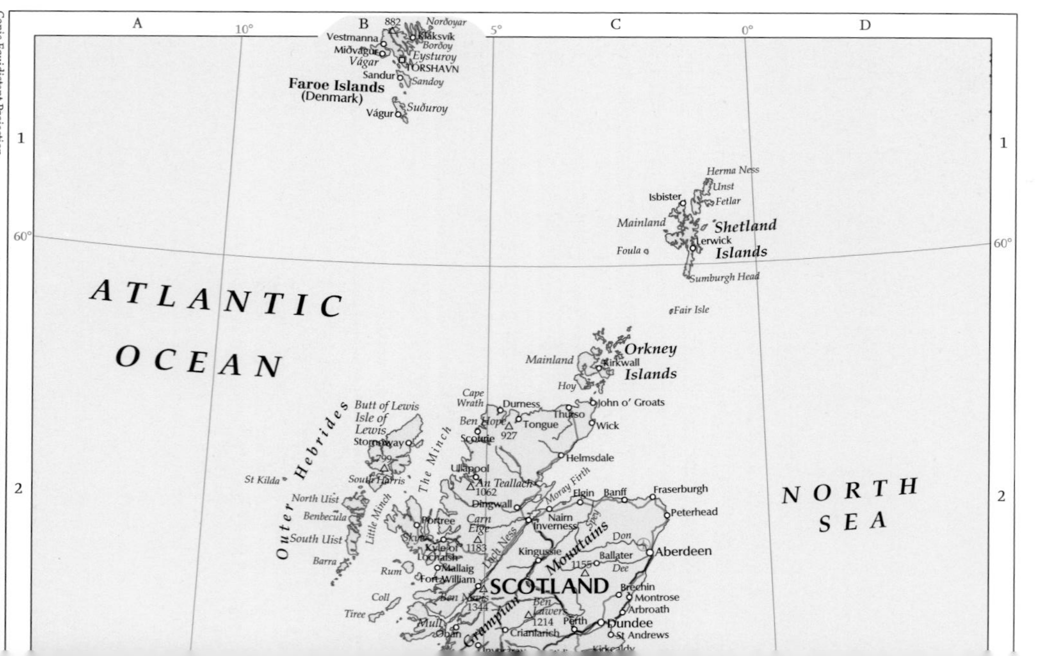

Conic Equidistant Projection

1 : 8 000 000

MILES 0   50   100   150

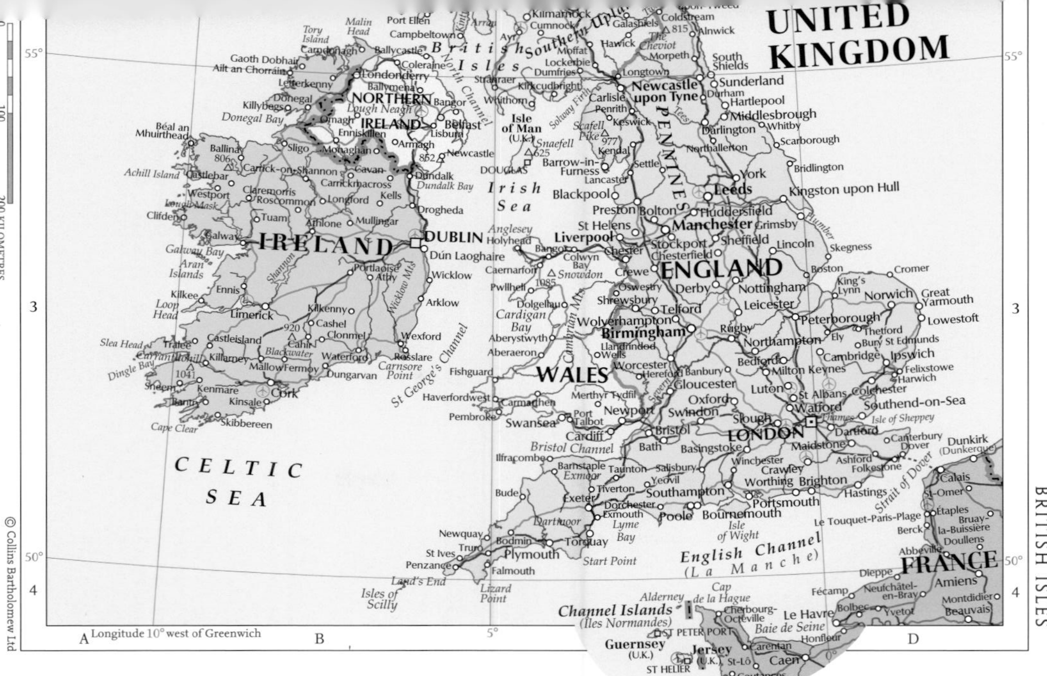
UNITED KINGDOM
IRELAND
NORTHERN IRELAND
ENGLAND
WALES
FRANCE
PENNINES
CELTIC SEA
Irish Sea
English Channel
(La Manche)
North Channel
St George's Channel
Bristol Channel
Strait of Dover
British Isles
Channel Islands
(Îles Normandes)
Isle of Man
(U.K.)
Isles of Scilly
DUBLIN
LONDON
Belfast
Birmingham
Manchester
Liverpool
Leeds
Sheffield
Newcastle upon Tyne
Bristol
Cardiff
Swansea
Plymouth
Southampton
Portsmouth
Brighton
Oxford
Cambridge
Norwich
Kingston upon Hull
York
Nottingham
Leicester
Cork
Limerick
Galway
Waterford
Wexford
Kilkenny
Sligo
Londonderry
Le Havre
Caen
Calais
Amiens
Jersey
Guernsey
Longitude 10° west of Greenwich
55°
50°
5°
0°
A
B
D
3
4

0 100 200 KILOMETRES

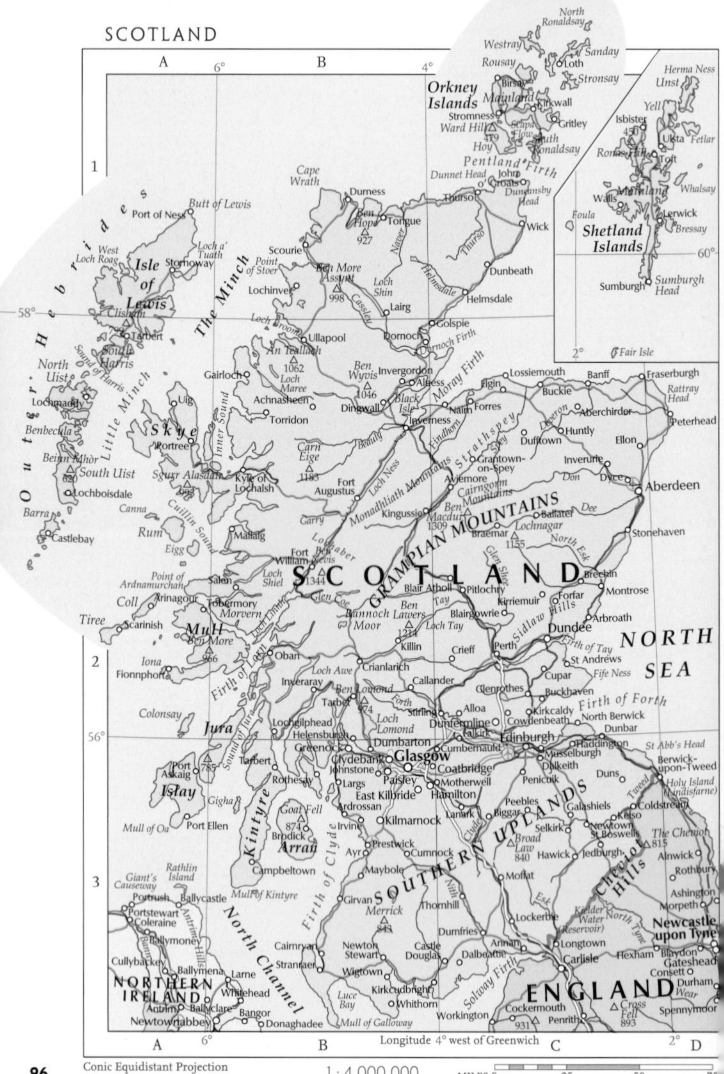

A   6°   B   4°

1

North Ronaldsay
Westray
Sanday
Rousay
Stronsay
Eday
Orkney Islands
Birsay
Mainland
Kirkwall
Stromness
Scapa
Ward Hill
Hoy
South Ronaldsay
Pentland Firth

Herma Ness
Unst
Yell
Isbister
450
Fetlar
Ronas Hill
Mainland
Whalsay
Bressay
Wall
Lerwick
Foula
Shetland Islands
Sumburgh
Sumburgh Head
60°

Cape Wrath
Durness
Dunnet Head
John o'Groats
Dunnimans Head
Thurso
Ben Hope
927
Tongue
Wick

Butt of Lewis
Port of Ness
Stornoway
Point of Stoer
Scourie
Lochinver
Ben More Assynt
998
Loch Shin
Dunbeath
Helmsdale

West Loch Ross
Isle of Lewis
Clisham
Tarbert
Ullapool
Loch Broom
An t-Sàilean
Golspie
Dornoch Firth

58°

North Uist
South Harris
Loch Maree
Gairloch
An Teallach
1062
Loch Maree
1046
Dingwall
Black Isle
Invergordon
Alness
Moray Firth
Lossiemouth
Elgin
Buckie
Banff
Fraserburgh
Rattray Head
Peterhead

Benbecula
Lochmaddy
Torridon
Skye
Portree
Kyle of Lochalsh
Carn Eige
1183
Inverness
Nairn
Forres
Findhorn
Huntly
Aberchirder
Ellon

South Uist
Lochboisdale
Canna
Spùr Alasdair
Cuillin Hills
Kyle of Lochalsh
Fort Augustus
Strathspey
Grantown-on-Spey
Aviemore
Inverurie
Dyce
Aberdeen
Don

Barra
Eigg
Rum
Point of Ardnamurchan
Fort William
Ben Nevis
Loch Shiel
GRAMPIAN MOUNTAINS
Monadhliath Mountains
Kingussie
Cairngorm
Cairngorm Mountains
1309
Ben Macdui
1309
Braemar
1155
Lochnagar
Ballater
Dee
North Esk
Stonehaven

Castlebay
Coll
Arinagour
Tobermory
Morvern
Mull
Ben More
Loch Aber
Glen Coe
Glen Lyon
SCOTLAND
Ben Lawers
1214
Blair Atholl
Pitlochry
Kirriemuir
Forfar
Sidlaw Hills
Brechin
Montrose

NORTH SEA

Tiree
Iona
Fionnphort
Colonsay
Oban
Loch Awe
Crianlarich
Killin
Loch Tay
Rannoch Moor
Crieff
Perth
Firth of Tay
St Andrews
Dundee
Blairgowrie

56°

Scarinish
Jura
Inveraray
Ben Lomond
Callander
Clackmannan
Dunfermline
Alloa
Stirling
Cupar
Ness Fife
Buckhaven

Port Askaig
Islay
Lochgilphead
Tarbert
Loch Lomond
Dumbarton
Helensburgh
Greenock
Cumbernauld
Edinburgh
Musselburgh
Dalkeith
North Berwick
Dunbar
Firth of Forth
St Abb's Head

Port Ellen
Gigha
Rothesay
Arran
Goat Fell
874
Brodick
Paisley
Glasgow
Clydebank
Johnstone
Coatbridge
Motherwell
East Kilbride
Hamilton
Peebles
Penicuik
Galashiels
Duns
Berwick-upon-Tweed
Coldstream

3

Mull of Oa
Campbeltown
Kintyre
Irvine
Kilmarnock
Lanark
Biggar
Selkirk
Newtown St Boswells
Kelso
Jedburgh
The Cheviot
815
Alnwick

Giant's Causeway
Rathlin Island
Mull of Kintyre
Prestwick
Ayr
Cumnock
SOUTHERN UPLANDS
Broad Law
840
Hawick
Cheviot Hills
Rothbury

Portrush
Ballycastle
Girvan
Merrick
Thornhill
Moffat
Ettrick Water (Reservoir)
Ashington
Morpeth

Portstewart
Coleraine
Ballymoney
Maybole
Newton Stewart
Dumfries
Lockerbie
Longtown
Newcastle upon Tyne
Blaydon
Gateshead

Cullybackey
Ballymena
Larne
Stranraer
Castle Douglas
Annan
Carlisle
Hexham
Consett
Durham

NORTHERN IRELAND
Antrim
Ballyclare
Whitehead
Wigtown
Kirkcudbright
Workington
Solway Firth
Cockermouth
931
Penrith
Cross Fell
893
ENGLAND
Wear
Spennymoor

Newtownabbey
Bangor
Donaghadee
Luce Bay
Mull of Galloway
Longitude 4° west of Greenwich

A   6°   B   C   2°   D

Conic Equidistant Projection

1 : 4 000 000

MILES 0   25   50   75

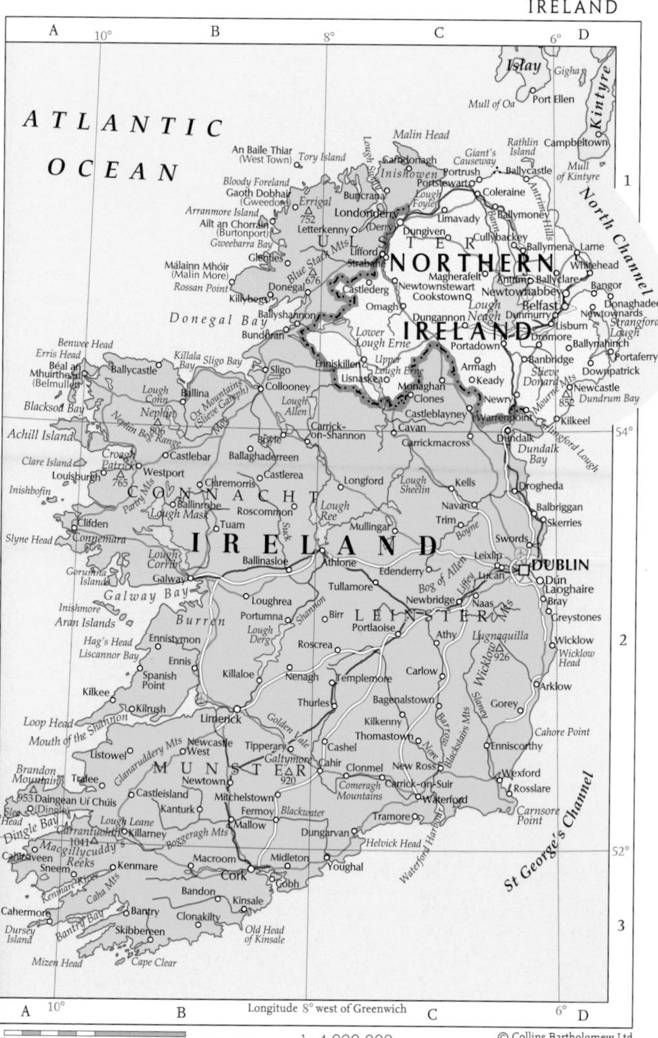

0   50   100 KILOMETRES   1 : 4 000 000   © Collins Bartholomew Ltd

Longitude 8° west of Greenwich

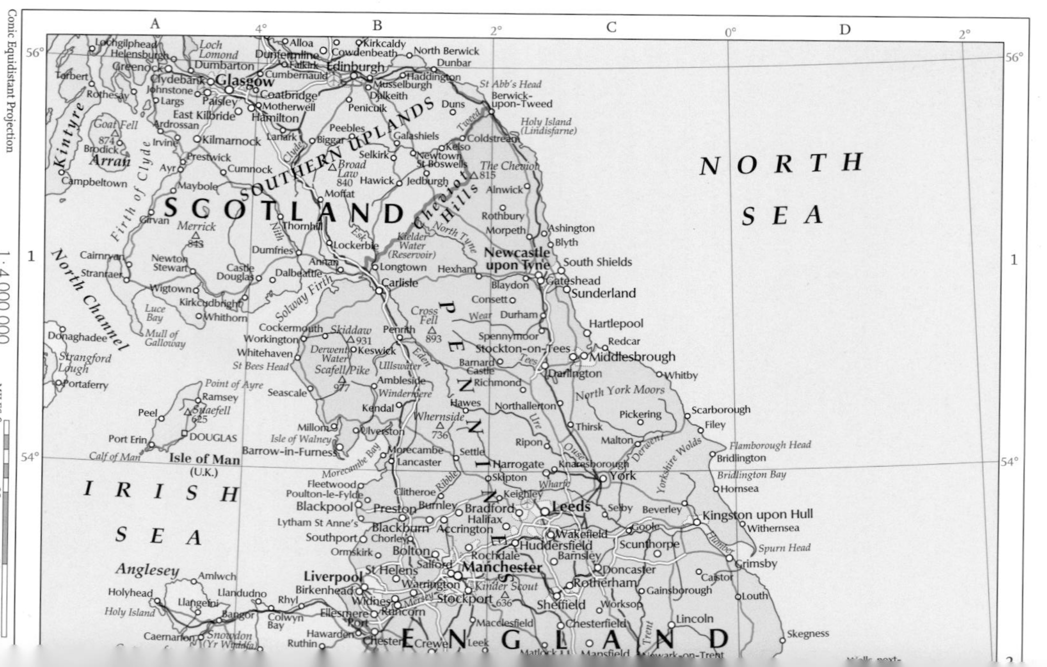

Conic Equidistant Projection

1 : 4 000 000

MILES 0   25   50   75

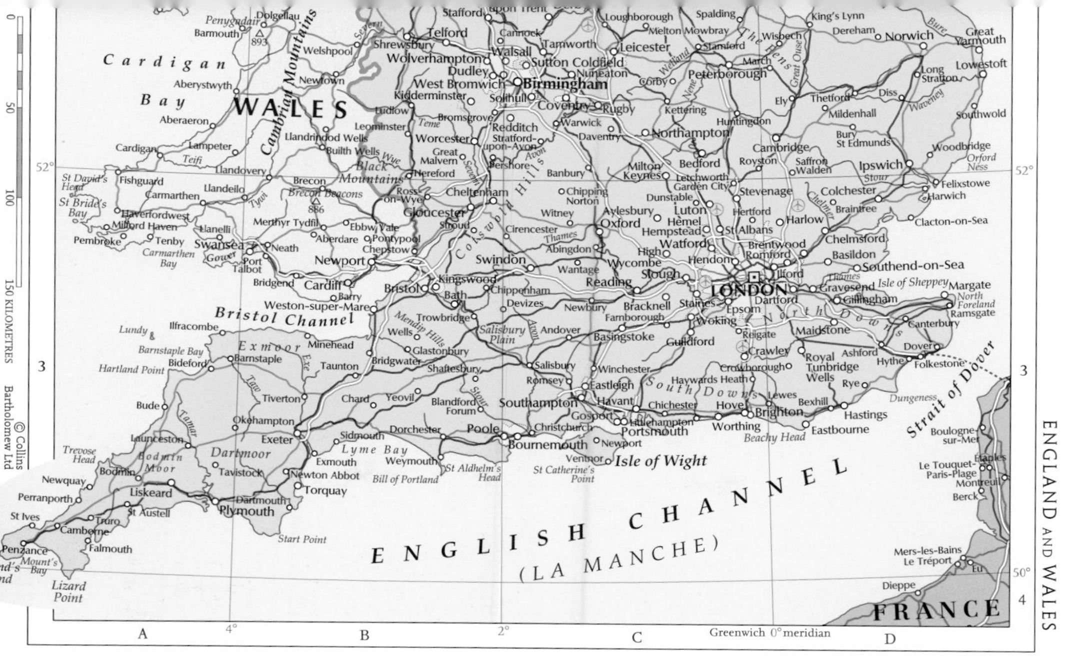
WALES
Cardigan Bay
Cambrian Mountains
Black Mountains
Brecon Beacons
Bristol Channel
Exmoor
Dartmoor
Bodmin Moor
Salisbury Plain
Mendip Hills
Cotswold Hills
North Downs
South Downs
The Fens
ENGLISH CHANNEL
(LA MANCHE)
Strait of Dover
Isle of Wight
FRANCE
LONDON
Birmingham
Norwich
Ipswich
Southend-on-Sea
Margate
Ramsgate
Canterbury
Dover
Folkestone
Hastings
Eastbourne
Brighton
Worthing
Portsmouth
Southampton
Bournemouth
Poole
Weymouth
Exeter
Torquay
Plymouth
Penzance
Truro
Falmouth
Newquay
Bristol
Bath
Cardiff
Newport
Swansea
Gloucester
Cheltenham
Oxford
Reading
Swindon
Coventry
Leicester
Northampton
Peterborough
Cambridge
Wolverhampton
Shrewsbury
Aberystwyth
Land's End
Lizard Point
Start Point
Greenwich 0° meridian
A
B
C
D
0
50
100
150 KILOMETRES
© Collins Bartholomew Ltd

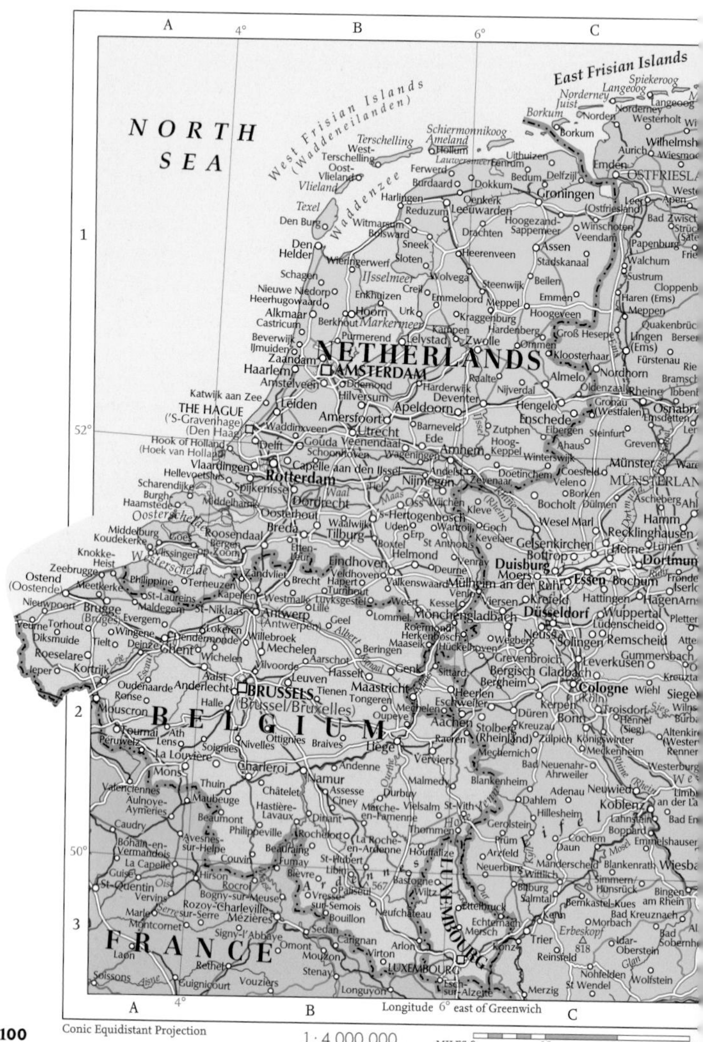

NORTH SEA

East Frisian Islands

NETHERLANDS
■AMSTERDAM

BELGIUM
■BRUSSELS
Brüssel/Bruxelles

FRANCE

LUXEMBOURG

Conic Equidistant Projection

1 : 4 000 000

Longitude 6° east of Greenwich

MILES 0   25   50   75

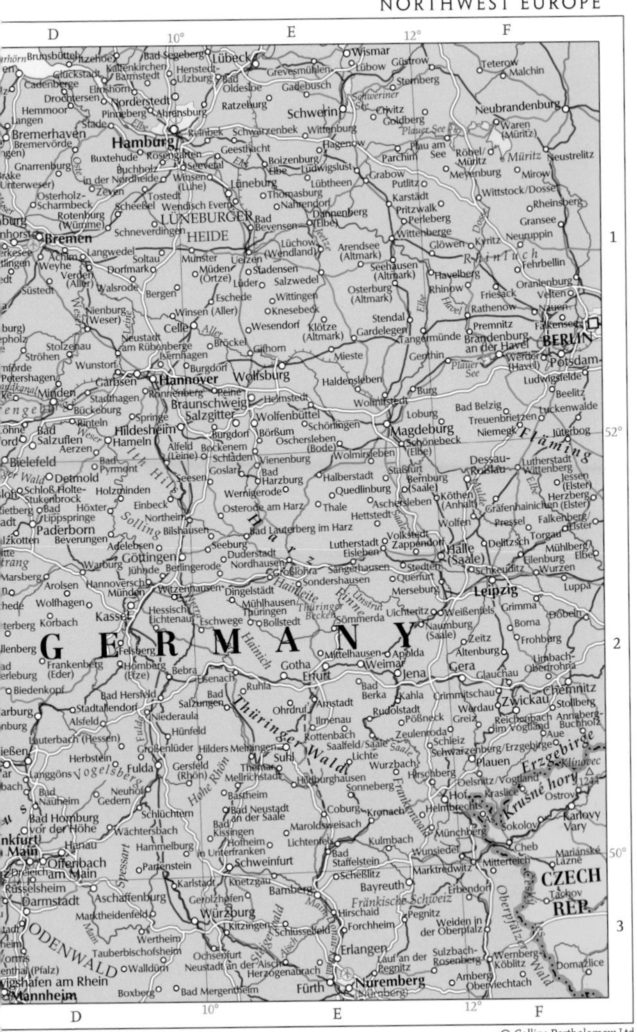
D
E
F
10°
12°
1
2
3
52°
50°
GERMANY
CZECH REP.
BERLIN
Hamburg
Bremen
Bremerhaven
Hannover
Wolfsburg
Braunschweig
Salzgitter
Hildesheim
Magdeburg
Lübeck
Schwerin
Wismar
Güstrow
Neubrandenburg
Lüneburg
LÜNEBURGER HEIDE
Uelzen
Celle
Potsdam
Brandenburg an der Havel
Halle (Saale)
Leipzig
Dessau-Roßlau
Göttingen
Kassel
Paderborn
Bielefeld
Detmold
Hameln
Goslar
Harz
Erfurt
Weimar
Jena
Gera
Gotha
Eisenach
Chemnitz
Zwickau
Plauen
Thüringer Wald
Erzgebirge
Krušné hory
Karlovy Vary
Cheb
Coburg
Bamberg
Bayreuth
Schweinfurt
Würzburg
Erlangen
Nuremberg (Nürnberg)
Fürth
Fulda
Hanau
Offenbach am Main
Darmstadt
Mannheim
Ludwigshafen am Rhein
ODENWALD
Spessart
Hohe Rhön
Vogelsberg
Fränkische Schweiz
Frankenwald
Oberpfälzer Wald
Fläming
Elbe
Weser
Aller
Saale
Main
Havel
Minden
Nienburg (Weser)
Stendal
Wittenberge
Tangermünde
Halberstadt
Quedlinburg
Wernigerode
Nordhausen
Mühlhausen Thüringen
Altenburg
Zeitz
Naumburg (Saale)
Merseburg
Torgau
Lutherstadt Wittenberg
Jüterbog
Luckenwalde
Amberg
Weiden in der Oberpfalz
Tachov
Domažlice
Mariánské Lázně
Sokolov
© Collins Bartholomew Ltd
0
50
100
150 KILOMETRES

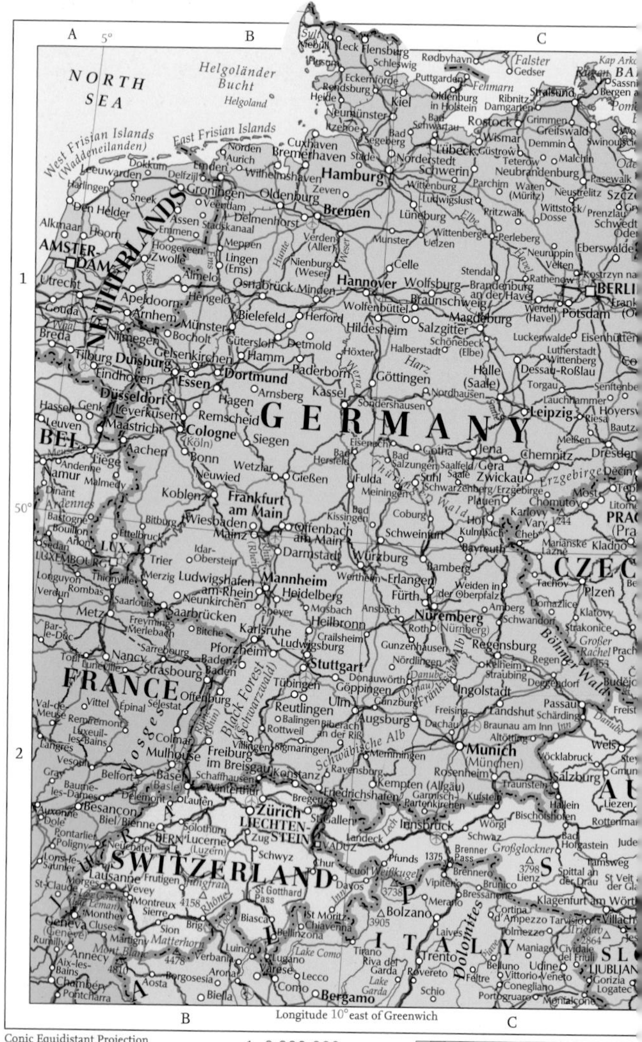

Conic Equidistant Projection

1 : 8 000 000

MILES 0   50   100   150

Longitude 10° east of Greenwich

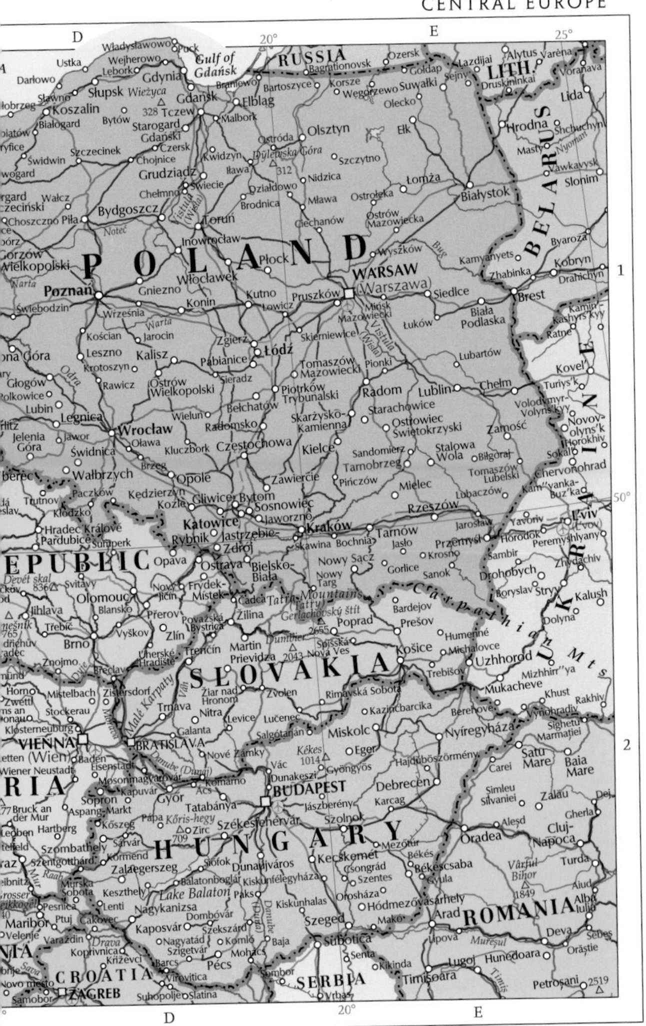
D
E
20°
25°
1
2
50°
RUSSIA
LITH.
BELARUS
POLAND
UKRAINE
SLOVAKIA
HUNGARY
ROMANIA
SERBIA
CROATIA
REPUBLIC
RIA
Gulf of Gdańsk
Władysławowo
Puck
Ustka
Wejherowo
Lebork
Gdynia
Darłowo
Sławno
Słupsk
Wieżyca
328
Gdańsk
Braniewo
Bartoszyce
Korsze
Węgorzewo
Suwałki
Olecko
Ozersk
Bagrationovsk
Goldap
Sejny
Lazdijai
Alytus
Varėna
Druskininkai
Voranava
Lida
Koszalin
Białogard
Bytów
Tczew
Elbląg
Malbork
Starogard Gdański
Czersk
Szczecinek
Świdwin
Chojnice
Kwidzyn
Iława
Ostróda
Olsztyn
Dylewska Góra
312
Szczytno
Ełk
Hrodna
Shchuchyn
Masty
Nyoman
Vawkavysk
Slonim
Grudziądz
Nidzica
Łomża
Białystok
Chełmno
Świecie
Działdowo
Wałcz
Brodnica
Mława
Ostrołęka
Choszczno
Piła
Bydgoszcz
Vistula (Wisła)
Toruń
Ciechanów
Ostrów Mazowiecka
Noteć
Inowrocław
Wyszków
Bug
Byaroza
Gorzów Wielkopolski
Płock
WARSAW (Warszawa)
Kamyanyets
Kobryn
Zhabinka
Drahichyn
Warta
Poznań
Włocławek
Gniezno
Kutno
Pruszków
Siedlce
Brest
Świebodzin
Konin
Łowicz
Mińsk Mazowiecki
Biała Podlaska
Kamin'-Kashyrs'kyy
Września
Łuków
Ratne
Kościan
Jarocin
Zgierz
Skierniewice
Leszno
Kalisz
Łódź
Lubartów
Zielona Góra
Odra
Krotoszyn
Pabianice
Tomaszów Mazowiecki
Pionki
Kovel'
Głogów
Rawicz
Ostrów Wielkopolski
Sieradz
Piotrków Trybunalski
Radom
Lublin
Chełm
Turiys'k
Polkowice
Bełchatów
Starachowice
Volodymyr-Volyns'kyy
Lubin
Legnica
Wieluń
Skarżysko-Kamienna
Ostrowiec Świętokrzyski
Novovolyns'k
Wrocław
Radomsko
Zamość
Jelenia Góra
Jawor
Oława
Częstochowa
Kielce
Horokhiv
Świdnica
Kluczbork
Sandomierz
Stalowa Wola
Biłgoraj
Sokal'
Wałbrzych
Brzeg
Tarnobrzeg
Tomaszów Lubelski
Chervonohrad
Opole
Zawiercie
Pińczów
Mielec
Kam''yanka-Buz'ka
Trutnov
Paczków
Kędzierzyn-Koźle
Gliwice
Bytom
Lubaczów
Kłodzko
Sosnowiec
Rzeszów
Jaworzno
Hradec Králové
Katowice
Kraków
Tarnów
Jarosław
Yavoriv
L'viv (L'vov)
Pardubice
Šumperk
Rybnik
Jastrzębie-Zdrój
Skawina
Bochnia
Jasło
Przemyśl
Horodok
Peremyshlyany
Opava
Ostrava
Bielsko-Biała
Nowy Sącz
Gorlice
Krosno
Sambir
Zhydachiv
Devět skal
836
Svitavy
Nový Jičín
Frýdek-Místek
Nowy Targ
Sanok
Drohobych
Olomouc
Čadca
Tatra Mountains (Tatry)
Boryslav
Stryy
Kalush
Jihlava
Blansko
Přerov
Žilina
Gerlachovský štít
Bardejov
Carpathian Mts
Třebíč
Považská Bystrica
2655
Poprad
Prešov
Dolyna
Vyškov
Zlín
Humenné
Brno
Martin
Ďumbier
Spišská Nová Ves
Košice
Michalovce
Uherské Hradiště
Trenčín
Prievidza
2043
Uzhhorod
Znojmo
Břeclav
Trebišov
Mizhhirr''ya
Dyje
Malé Karpaty
Váh
Žiar nad Hronom
Zvolen
Rimavská Sobota
Mukacheve
Horn
Mistelbach
Zistersdorf
Khust
Rakhiv
Zwettl
Trnava
Kazincbarcika
Berehove
Vynohradiv
Stockerau
Nitra
Levice
Lučenec
Sighetu Marmației
Klosterneuburg
Morava
Galanta
Salgótarján
Miskolc
Nyíregyháza
VIENNA (Wien)
BRATISLAVA
Kékes
1014
Eger
Satu Mare
Baden
Nové Zámky
Hajdúböszörmény
Baia Mare
Wiener Neustadt
Eisenstadt
Danube (Duna)
Vác
Gyöngyös
Carei
Mosonmagyaróvár
Komárno
Dunakeszi
Kapuvár
Győr
Ács
BUDAPEST
Debrecen
Şimleu Silvaniei
Zalău
Dej
Sopron
Tatabánya
Jászberény
Karcag
Bruck an der Mur
Aspang-Markt
Pápa
Kőris-hegy
Székesfehérvár
Szolnok
Aleşd
Gherla
Kőszeg
709
Zirc
Oradea
Cluj-Napoca
Leoben
Hartberg
Sárvár
Mezőtúr
Szombathely
Körmend
Kecskemét
Békés
Turda
Szentgotthárd
Siófok
Dunaújváros
Csongrád
Békéscsaba
Vârful Bihor
Raab
Zalaegerszeg
Balatonboglár
Kiskunfélegyháza
Szentes
Gyula
Mur
Murska Sobota
Keszthely
Lake Balaton
Paks
Orosháza
1849
Aiud
Lenti
Nagykanizsa
Kiskunhalas
Hódmezővásárhely
Alba Iulia
Pesnica
Dombóvár
Szeged
Arad
Maribor
Ptuj
Čakovec
Kaposvár
Szekszárd
Makó
Velenje
Varaždin
Drava
Nagyatád
Komló
Baja
Subotica
Lipova
Mureşul
Deva
Sebeş
Koprivnica
Szigetvár
Mohács
Senta
Orăştie
Križevci
Barcs
Pécs
Kikinda
Lugoj
Hunedoara
Sava
Virovitica
Sombor
Timişoara
Timiş
Novo mesto
ZAGREB
Samobor
Suhopolje
Slatina
Vrbas
Petroşani
2519

0 100 200 KILOMETRES

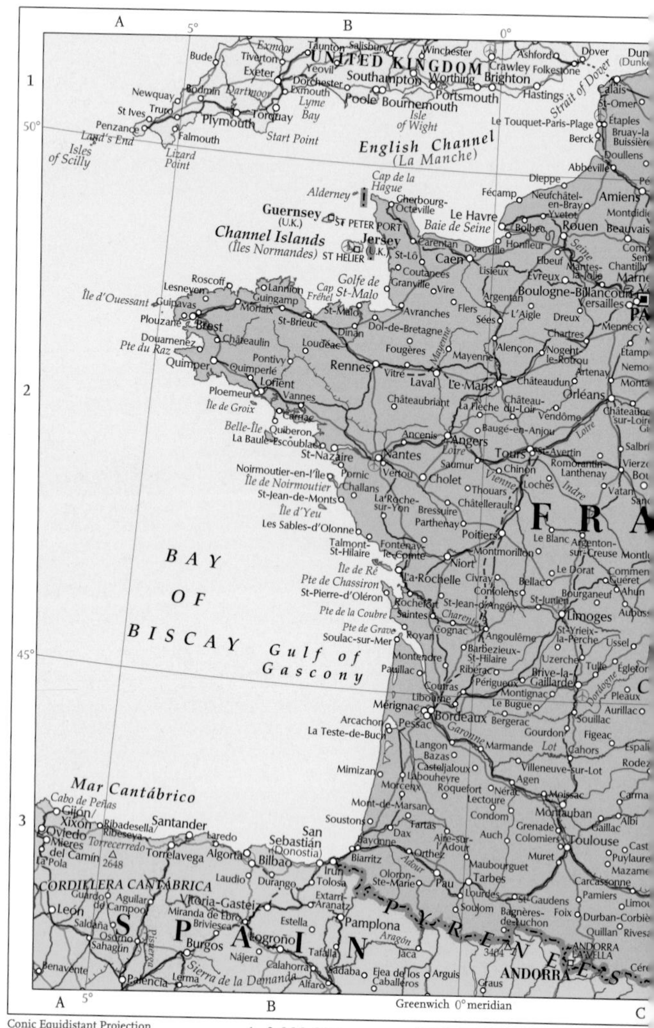

1 : 8 000 000

MILES 0   50   100   150

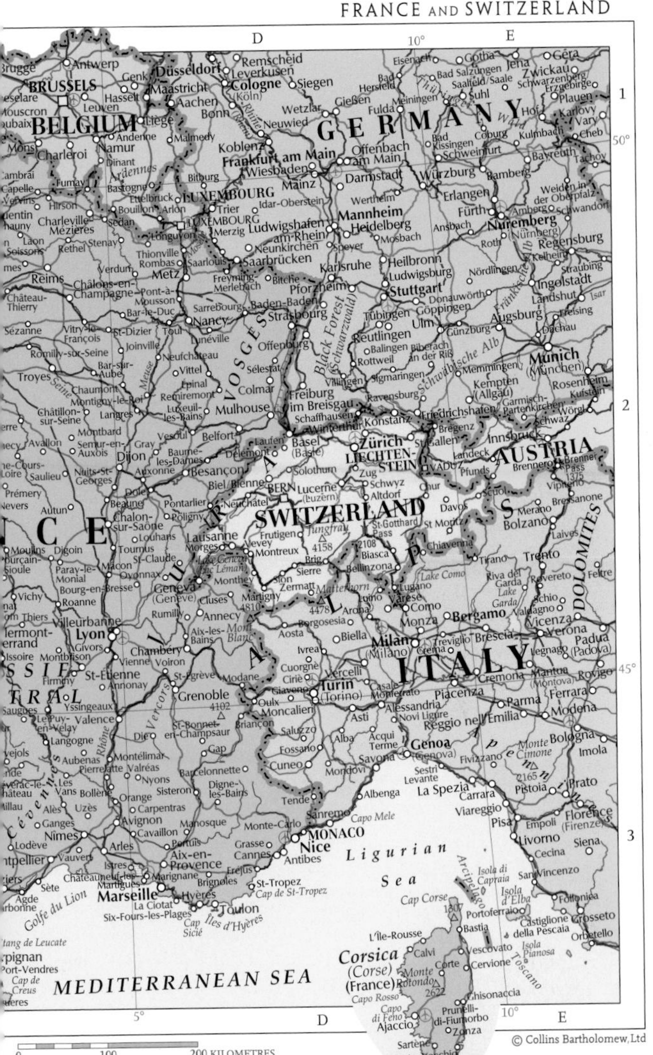

D
E
10°
1
2
3
50°
45°
5°
GERMANY
BELGIUM
LUXEMBOURG
SWITZERLAND
LIECHTENSTEIN
AUSTRIA
ITALY
MONACO
ALPS
VOSGES
JURA
DOLOMITES
Corsica
(Corse)
(France)
MEDITERRANEAN SEA
Ligurian Sea
Golfe du Lion
Black Forest
(Schwarzwald)
Schwäbische Alb
Fränkische Alb
Thüringer Wald
Ardennes
Vercors
Cévennes
Apennines
Lake Geneva
(Lac Léman)
Lake Como
Lake Garda
Arcipelago Toscano
Rhine
Rhône
Meuse
Moselle
Seine
Isar
Jungfrau
4158
Matterhorn
4478
Mont Blanc
4810
Monte Cimone
2165
Monte Rotondo
2622
Brenner Pass
1375
St Gotthard Pass
2108
4102
1307
Cap Corse
Capo Rosso
Capo di Feno
Capo Mele
Cap de St-Tropez
Cap Sicié
Îles d'Hyères
Isola di Capraia
Isola d'Elba
Isola Pianosa
Étang de Leucate
Cap de Creus
BRUSSELS
Antwerp
Brugge
Genk
Hasselt
Leuven
Liège
Namur
Charleroi
Mons
Dinant
Düsseldorf
Remscheid
Leverkusen
Cologne
(Köln)
Siegen
Bonn
Maastricht
Aachen
Koblenz
Neuwied
Wetzlar
Gießen
Frankfurt am Main
Wiesbaden
Mainz
Offenbach am Main
Darmstadt
Mannheim
Heidelberg
Ludwigshafen am Rhein
Karlsruhe
Heilbronn
Stuttgart
Pforzheim
Baden-Baden
Tübingen
Reutlingen
Ulm
Augsburg
Munich
(München)
Nuremberg
(Nürnberg)
Regensburg
Würzburg
Erlangen
Fürth
Bamberg
Bayreuth
Ingolstadt
Freiburg im Breisgau
Konstanz
Friedrichshafen
Kempten (Allgäu)
Saarbrücken
LUXEMBOURG
Trier
Metz
Nancy
Strasbourg
Mulhouse
Colmar
Reims
Troyes
Dijon
Besançon
Belfort
Basel
(Bâle)
Zürich
BERN
Lucerne
(Luzern)
Lausanne
Geneva
(Genève)
Innsbruck
Lyon
Villeurbanne
Grenoble
Chambéry
Annecy
St-Étienne
Avignon
Nîmes
Montpellier
Marseille
Toulon
Aix-en-Provence
Nice
Cannes
Antibes
Turin
(Torino)
Milan
(Milano)
Genoa
(Genova)
Bergamo
Brescia
Verona
Padua
(Padova)
Vicenza
Trento
Bolzano
Parma
Modena
Bologna
Reggio nell'Emilia
Piacenza
Cremona
Mantua
(Mantova)
Ferrara
La Spezia
Pisa
Livorno
Florence
(Firenze)
Prato
Pistoia
Siena
Ajaccio
Bastia
Corte
Bonifacio
Porto-Vecchio
Sartène
Perpignan
Port-Vendres

0 100 200 KILOMETRES

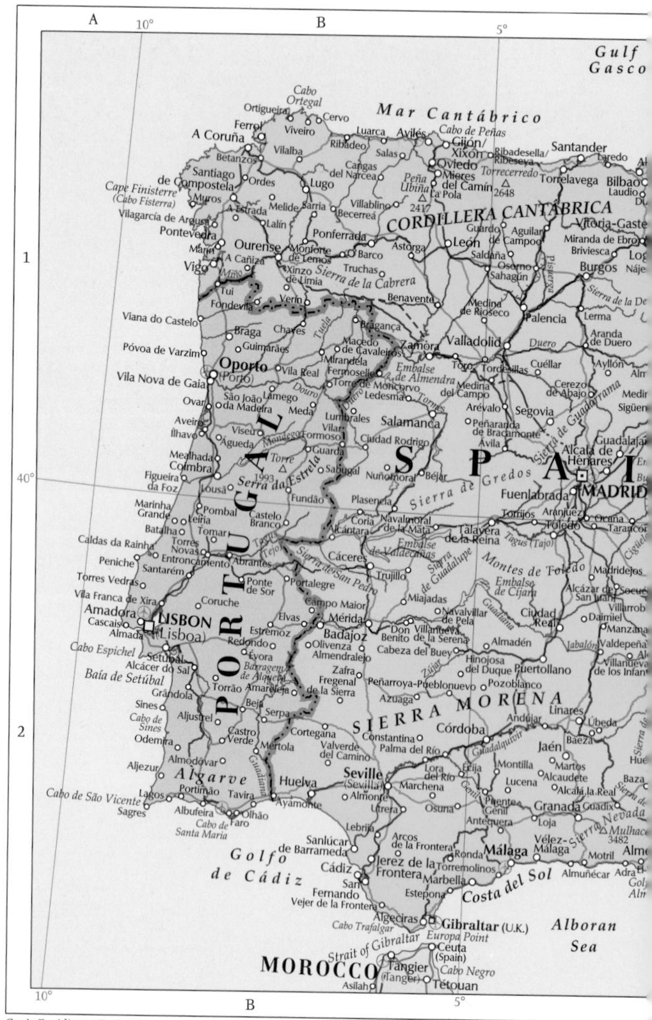

Conic Equidistant Projection   1 : 8 000 000

MILES 0   50   100   150

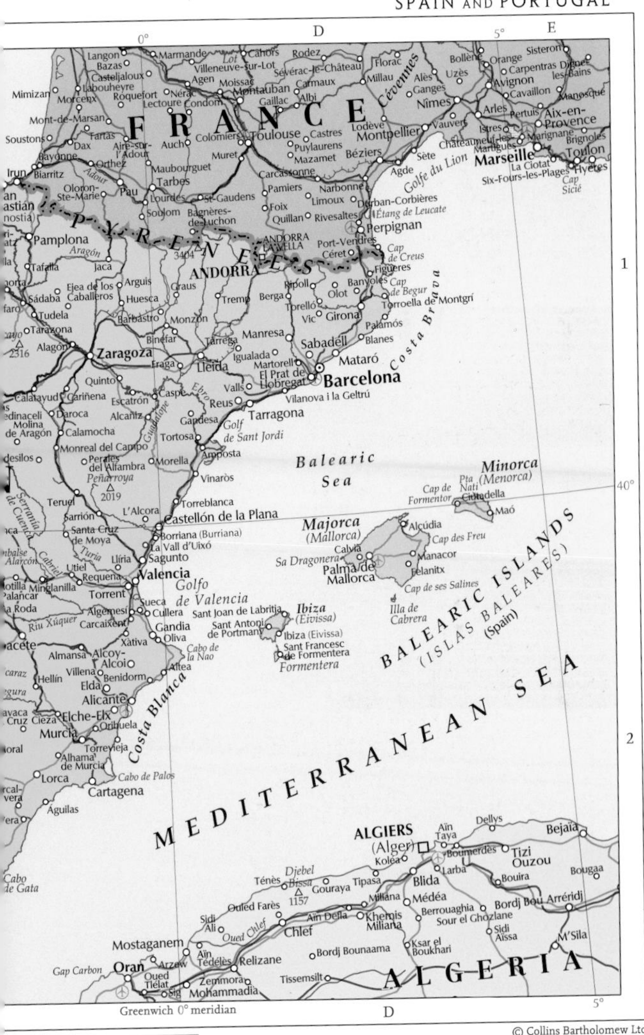
D
E
0°
5°
1
2
40°
FRANCE
PYRENEES
ANDORRA
ANDORRA LA VELLA
Langon
Bazas
Casteljaloux
Labouheyre
Mimizan
Morcenx
Roquefort
Nérac
Lectoure
Condom
Marmande
Villeneuve-sur-Lot
Lot
Cahors
Agen
Moissac
Montauban
Rodez
Sévérac-le-Château
Carmaux
Albi
Gaillac
Florac
Millau
Cévennes
Alès
Uzès
Ganges
Nîmes
Bollène
Orange
Avignon
Carpentras
Cavaillon
Sisteron
Digne-les-Bains
Manosque
Arles
Pertuis
Aix-en-Provence
Mont-de-Marsan
Tartas
Soustons
Dax
Aire-sur-l'Adour
Auch
Colomiers
Toulouse
Castres
Puylaurens
Mazamet
Lodève
Montpellier
Vauvert
Istres
Marignane
Brignoles
Bayonne
Biarritz
Irun
Orthez
Adour
Muret
Maubourguet
Tarbes
Pau
Oloron-Ste-Marie
Lourdes
St-Gaudens
Soulom
Bagnères-de-Luchon
Béziers
Sète
Agde
Golfe du Lion
Châteauneuf-les-Martigues
Marseille
La Ciotat
Toulon
Hyères
Six-Fours-les-Plages
Cap Sicié
Carcassonne
Pamiers
Foix
Narbonne
Limoux
Durban-Corbières
Quillan
Rivesaltes
Étang de Leucate
Perpignan
Port-Vendres
Céret
Cap de Creus
Pamplona
Aragón
Jaca
Tafalla
3404
Figueres
Ejea de los Caballeros
Arguis
Graus
Huesca
Sádaba
Tremp
Berga
Ripoll
Olot
Banyoles
Cap de Begur
Torroella de Montgrí
Costa Brava
Tudela
Barbastro
Monzón
Torelló
Vic
Girona
Palamós
Tarazona
Alagón
2316
Binéfar
Tàrrega
Manresa
Blanes
Sabadell
Zaragoza
Fraga
Lleida
Igualada
Martorell
Mataró
El Prat de Llobregat
Barcelona
Quinto
Calatayud
Cariñena
Escatrón
Caspe
Ebro
Valls
Reus
Vilanova i la Geltrú
Tarragona
Daroca
Alcañiz
Gandesa
Golf de Sant Jordi
Molina de Aragón
Calamocha
Guadalope
Tortosa
Monreal del Campo
Amposta
Peralejos
Perales del Alfambra
Morella
Peñarroya
2019
Vinaròs
Balearic Sea
Minorca (Menorca)
Pta Nati
Cap de Formentor
Ciutadella
Maó
Teruel
Torreblanca
L'Alcora
Serranía de Cuenca
Sarrión
Castellón de la Plana
Santa Cruz de Moya
Borriana (Burriana)
La Vall d'Uixó
Turia
Llíria
Sagunto
Majorca (Mallorca)
Alcúdia
Cap des Freu
Calvià
Manacor
Sa Dragonera
Palma de Mallorca
Felanitx
Cap de ses Salines
Utiel
Cabriel
Requena
Valencia
Golfo de Valencia
Minglanilla
Torrent
Sueca
Cullera
Algemesí
Carcaixent
Riu Xúquer
Sant Joan de Labritja
Ibiza (Eivissa)
Illa de Cabrera
Gandia
Oliva
Sant Antoni de Portmany
Ibiza (Eivissa)
Xàtiva
Cabo de la Nao
Sant Francesc de Formentera
Formentera
BALEARIC ISLANDS
(ISLAS BALEARES)
(Spain)
Almansa
Alcoy-Alcoi
Altea
Hellín
Villena
Benidorm
Elda
Alicante
Costa Blanca
Cieza
Elche-Elx
Orihuela
Murcia
Torrevieja
Alhama de Murcia
Lorca
Cabo de Palos
Cartagena
Águilas
MEDITERRANEAN SEA
Cabo de Gata
ALGIERS
(Alger)
Aïn Taya
Dellys
Bejaïa
Koléa
Boumerdes
Tizi Ouzou
Larba
Djebel Bissa
Ténès
Gouraya
Tipasa
Blida
Bouira
Bougaa
1157
Miliana
Médéa
Bordj Bou Arréridj
Ouled Farès
Berrouaghia
Sour el Ghozlane
Sidi Ali
Aïn Defla
Khemis Miliana
Chlef
Oued Chlef
Sidi Aïssa
M'Sila
Mostaganem
Ksar el Boukhari
Arzew
Aïn Tédélès
Relizane
Bordj Bounaama
Gap Carbon
Oran
Oued Tlélat
Zemmora
Tissemsilt
Sig
Mohammadia
ALGERIA
Greenwich 0° meridian
0
100
200 KILOMETRES

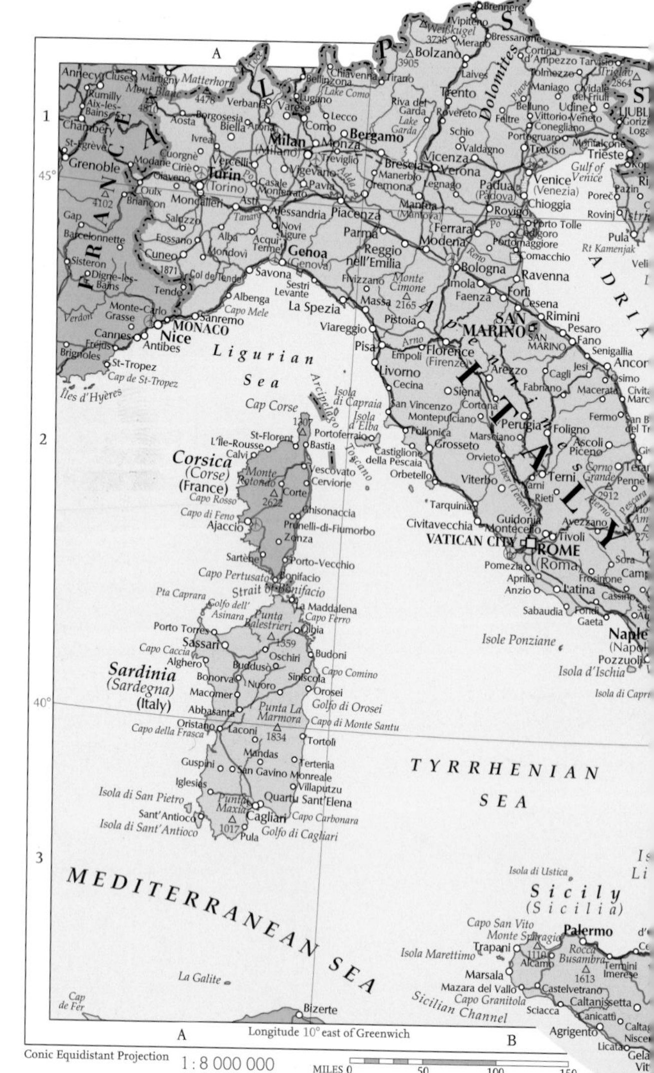

108 Conic Equidistant Projection 1 : 8 000 000 MILES 0   50   100   150

C   20°   D

HUNGARY

Nagyatád Komló
Kaposvár
Pécs Mohács Baja Mako Lipova
Subotica
Senta Timişoara Hunedoara Orăştie Sebeş
Sombor Lugoj Caransebeş Reşiţa Transylvanian Alps
(Carpaţii Meridionali) Petroşani
1

VOJVODINA
Zrenjanin Vršac **ROMANIA** Targu
Novi Sad Jiu
Bačka Palanka Crkva Vârful Srrinecea Motru
Vukovar Pančevo Mehedrevska Palanka Motru Strehaia
Slavonski Brod
Županja **BELGRADE** Obrenovac Požarevac Turnu-Severin
(Beograd) Smederevo
Doboj Brčko Bijeljina Mladenovac Velika Plana Crni Vrh
Tuzla Loznica Lazarevac Bor
Derventa Srebrenica Aranđelovac Gornji 1156 Bailesti
Banja Luka Milanovac Kragujevac Zaječar Montana
BOSNIA AND Valjevo Požega Paraćin Knjaževac
Visoko Čačak **SERBIA** Berkovitsa
HERZEGOVINA Užice Trstenik Kruševac Aleksinac
SARAJEVO Priboj Niš Midžor Iskar
Gornji Pražišta Prokuplje Vlasotince **SOFIA**
Vakuf Novi Leskovac (Sofiya)
Goražde Pazar Sjenica Podujevo
Konjic Bijelo Mitrovica Vushtrri **PRISHTINE** Surdulica Kyustendil
Polje Vranje
Mostar **MONTENEGRO** **KOSOVO** Gjilan Prokhor Blagoevgrad
Gacko Prizren Ferizaj Kumanovo
PODGORICA Rahovec Kriva Sveti
Trebinje Peshkopi **SKOPJE** Nikole Štip Radoviš
Cetinje **ALBANIA** Tetovo Gostivar
Dubrovnik Shkodër Burrel Strumica
Bar Lezhë **MACEDONIA** Kilkis
Lać Debar Kičevo (F.Y.R.O.M.)
Krujë Prilep Bitola **Thessaloniki**
Durrës Struga Edessa
TIRANA Ohrid Veroia Kalamaria
(Tiranë) Elbasan Mesimeri
Lushnjë Pogradec Florina Ptolemaïda
Kuçovë Korçë Kastoria **Katerini**
Berat Kozani Mount
Fier Corovodë Grammos Olympus
Vlorë Tepelenë Ereseke
Girokastër Grevena Elassona Larisa
Delvinë **GREECE**
Sarandë Ioannina Kalambaka Tyrnavos
Corfu Igoumenitsa Trikala Karditsa Almyros
(Kerkyra) Parga Filippiada Domokos
Preveza Arta Lamia
Lefkada Karpenisi
Amfissa Linakia
Lefkada Agrinio Mesolongi Delphi
Cephalonia Nafpaktos Patras
(Kefallonia) Kato
Argostoli Lechaina Achaia Kyllini
Ionian Islands Amaliada Olympia
(Ionioi Nisoi) Pyrgos Tripoli
Zakynthos Zacharo
(Zante) Kyparissia Megalopoli
Kalamata Sparti

ITALY
Vieste
Monte
Calvo Monte Sant'Angelo
1055 Manfredonia
Foggia Barletta
Molfetta
Cerignola Andria Bitonto Monopoli
Melfi Bari
Avellino Altamura Brindisi
Potenza Matera Francavilla
Pisticci Fontana Mesagne Lecce
Marcellana Taranto Copertino
Sapri Policoro Galatina
Scalea Gallipoli
Golfo di
Taranto Casarano
Tricase
Cetraro Punta Alice Capo Sta Maria
Paola Cirò Marina di Leuca
Cosenza San Giovanni in Fiore
Amantea Crotone
Catanzaro Capo Colonna
LA SILA Isola di Capo Rizzuto
Vibo Golfo di Squillace
Valentia Soverato

IONIAN
SEA

Palmi Ionian Islands
Reggio di Calabria (Ionioi Nisoi)
Capo
Spartivento

0   100   200 KILOMETRES

© Collins Bartholomew Ltd   **109**

45°
40°
2
3

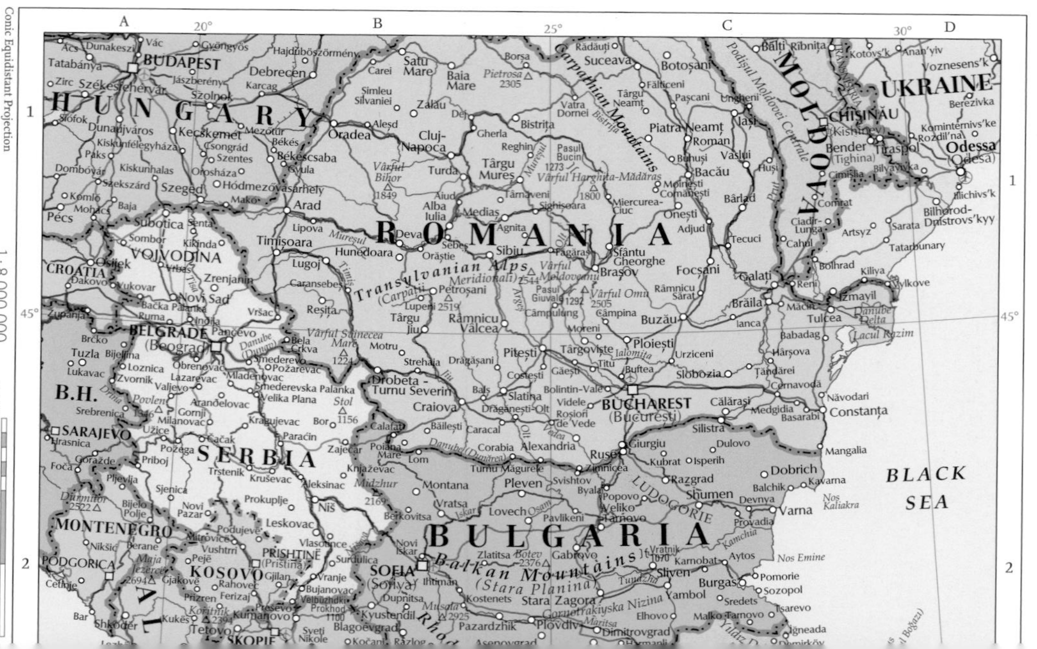

Conic Equidistant Projection

1 : 8 000 000

MILES 0   50   100   150

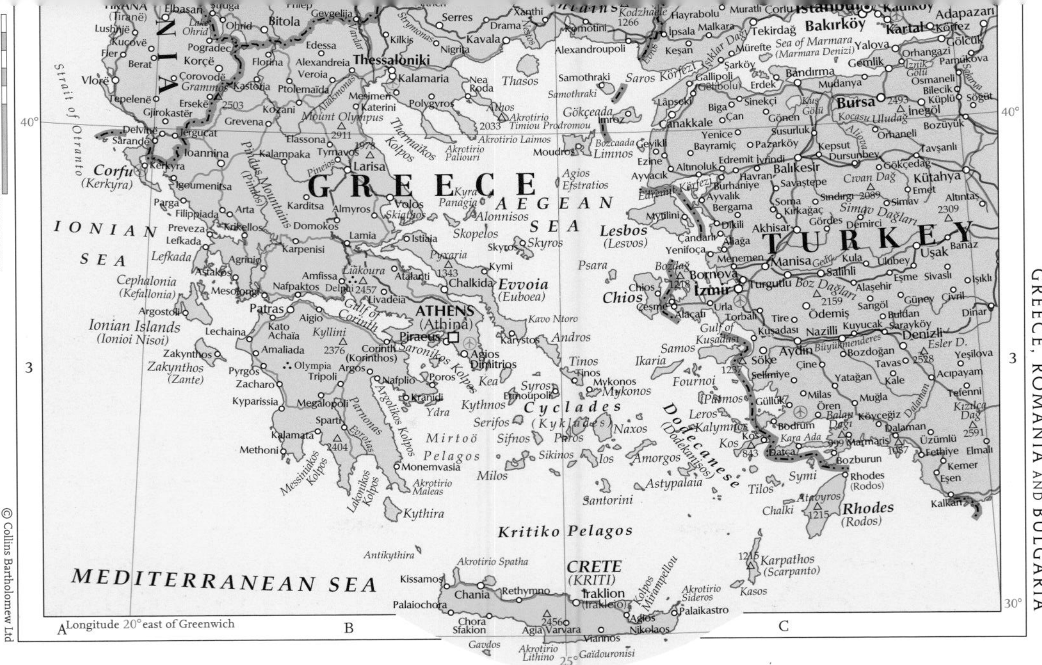

© Collins Bartholomew Ltd

0   100   200 KILOMETRES

GREECE
TURKEY
AEGEAN SEA
IONIAN SEA
MEDITERRANEAN SEA
Kritiko Pelagos
Mirtoö Pelagos
CRETE
(KRITI)
Cyclades
(Kyklades)
Dodecanese
(Dodekanisos)
ATHENS
(Athina)
Thessaloniki
Izmir
Bursa
Rhodes
(Rodos)
Lesbos
(Lesvos)
Chios
Evvoia
(Euboea)
Corfu
(Kerkyra)
Ionian Islands
(Ionioi Nisoi)
Cephalonia
(Kefallonia)
Zakynthos
(Zante)
Strait of Otranto
Sea of Marmara
(Marmara Denizi)
Pindus Mountains
(Pindos)
Mount Olympus
Longitude 20° east of Greenwich

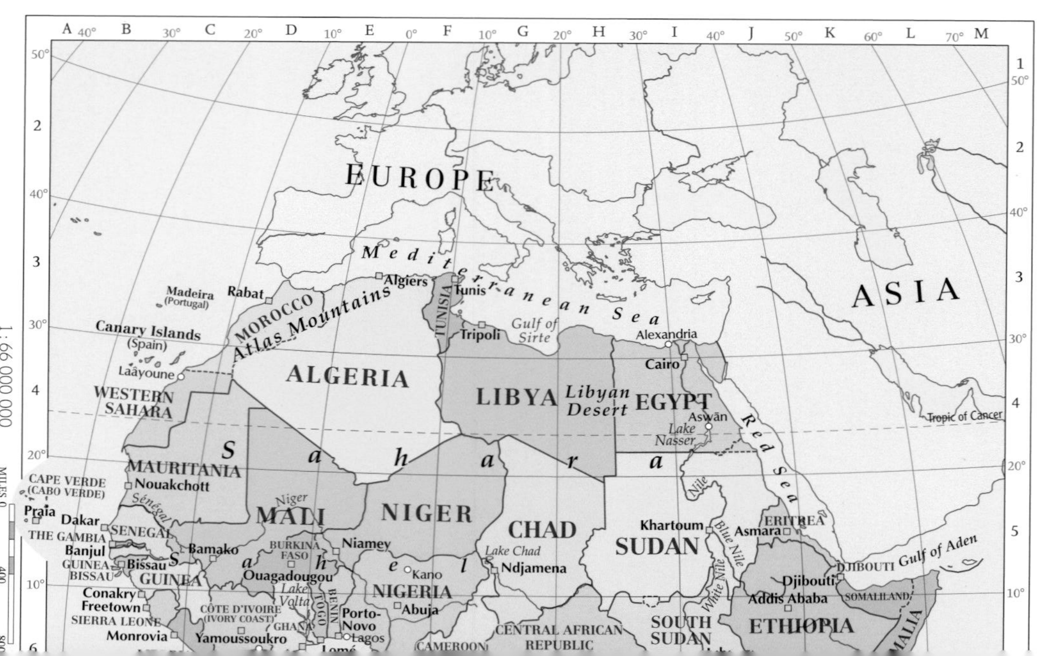

112

1 : 66 000 000

MILES 0   400   800

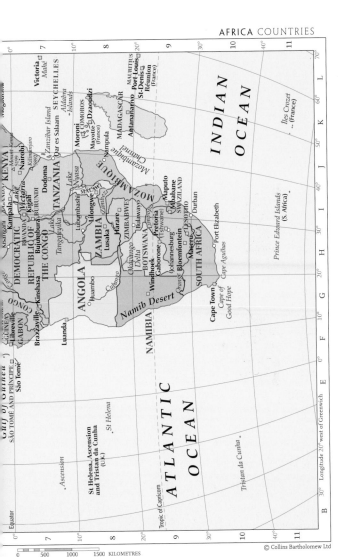

Gulf of Guinea
SÃO TOMÉ AND PRINCIPE
São Tomé

GUINEA
Libreville
GABON
CONGO
Brazzaville
Kinshasa

DEMOCRATIC
REPUBLIC OF
THE CONGO

Luanda

ANGOLA

Cabinda

Cuanza

Okavango
Delta

Namib Desert

NAMIBIA

Windhoek

Orange

Cape Town
Cape of
Good Hope

Cape Agulhas

ATLANTIC
OCEAN

Ascension

St Helena, Ascension
and Tristan da Cunha
(U.K.)

St Helena

Tristan da Cunha

Tropic of Capricorn

Equator

KENYA
Kampala
UGANDA
Nairobi
Mt Kenya
5199

Victoria
Mahé
SEYCHELLES
Aldabra
Islands

Khartoum

RWANDA
Kigali
BURUNDI
Bujumbura

Kilimanjaro
5895

Lake
Victoria

TANZANIA
Dodoma

Zanzibar Island
Dar es Salaam

Moroni
COMOROS
Mayotte
(France)

Dzaoudzi

MADAGASCAR
Antananarivo

MAURITIUS
Port Louis
St-Denis
Réunion
(France)

Lubumbashi

Lake
Nyasa

MALAWI
Lilongwe
Zambezi

ZAMBIA
Lusaka
Harare
ZIMBABWE
Bulawayo

MOZAMBIQUE

Mozambique Channel

Lake
Tanganyika

BOTSWANA
Gaborone
Pretoria
(Tshwane)
Johannesburg
Maputo
Mbabane
SWAZILAND
Maseru
LESOTHO

Bloemfontein

SOUTH AFRICA

Durban

Port Elizabeth

INDIAN
OCEAN

Prince Edward Islands
(S. Africa)

Îles Crozet
(France)

Longitude 20° west of Greenwich

© Collins Bartholomew Ltd

0   500   1000   1500 KILOMETRES

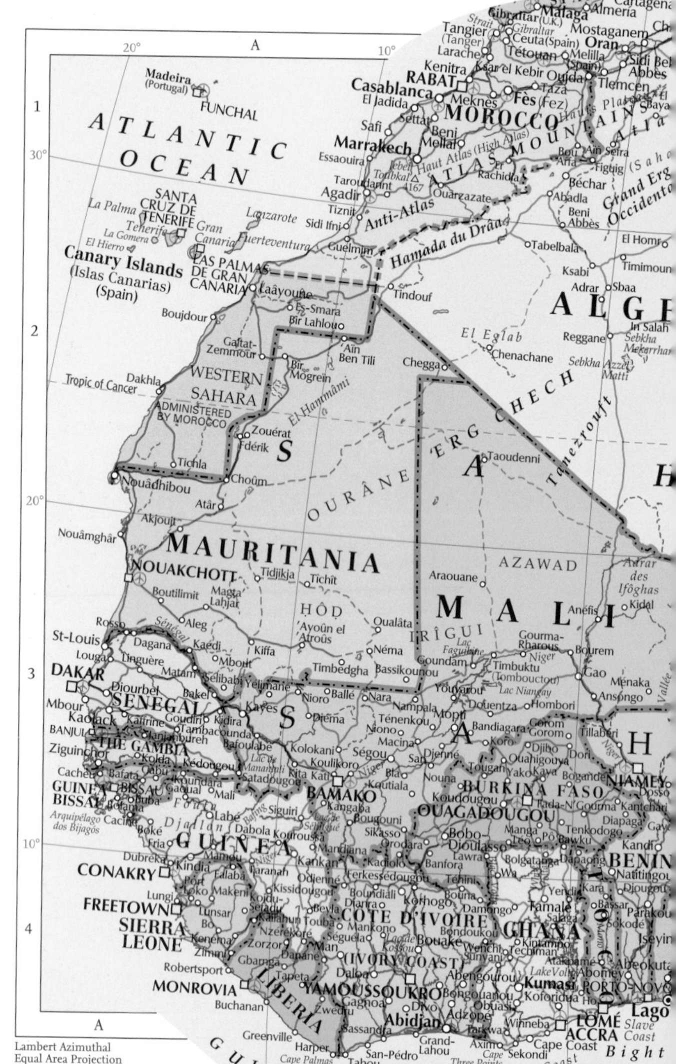
A
20°
10°
1
2
3
4
30°
20°
10°
ATLANTIC OCEAN
Madeira (Portugal)
FUNCHAL
SPAIN
Gibraltar (U.K.)
Málaga
Almería
Cartagena
Strait of Gibraltar
Tangier (Tanger)
Ceuta (Spain)
Mostaganem
Oran
Larache
Tétouan
Melilla (Spain)
Sidi Bel Abbès
Kenitra
Ksar el Kebir
Oujda
RABAT
Tlemcen
Casablanca
Taza
El Jadida
Meknes
Fès (Fez)
Settat
MOROCCO
Hauts Plateaux
Safi
Beni Mellal
ATLAS MOUNTAINS
Marrakech
Essaouira
Haut Atlas (High Atlas)
Toubkal 4167
Er Rachidia
Bou Arfa
Aïn Sefra
Figuig
Taroudannt
Agadir
Ouarzazate
Béchar
Abadla
Beni Abbès
Grand Erg Occidental
Tiznit
Sidi Ifni
Anti-Atlas
SANTA CRUZ DE TENERIFE
La Palma
Tenerife
La Gomera
El Hierro
Gran Canaria
Lanzarote
Fuerteventura
Canary Islands (Islas Canarias) (Spain)
LAS PALMAS DE GRAN CANARIA
Guelmim
Hamada du Drâa
Tabelbala
El Homr
Timimoun
Ksabi
Adrar
Sbaa
Laâyoune
Tindouf
ALG
Es-Smara
Boujdour
Bir Lahlou
In Salah
Sebkha Mekerrhane
Reggane
El Eglab
Gaftat-Zemmour
Aïn Ben Tili
Chenachane
Sebkha Azzel Matti
Bir Mogreïn
Chegga
WESTERN SAHARA
ADMINISTERED BY MOROCCO
Dakhla
Tropic of Cancer
El Hammâmi
ERG CHECH
Tanezrouft
Zouérat
Fdérik
S
Taoudenni
A
Tichla
Nouâdhibou
Choûm
OUARÂNE
Atâr
Akjoujt
Nouâmghâr
MAURITANIA
AZAWAD
NOUAKCHOTT
Tidjikja
Tîchît
Araouane
Adrar des Ifôghas
Boutilimit
Magta Lahjar
HÔD
MALI
Anefis
Kidal
Rosso
Aleg
'Ayoûn el 'Atroûs
Oualâta
IRÎGUI
Sénégal
St-Louis
Dagana
Kaédi
Kiffa
Néma
Gourma-Rharous
Bourem
Louga
Linguère
Mbout
Lac Faguibine
Niger
Matam
Timbédgha
Bassikounou
Goundam
Timbuktu (Tombouctou)
Gao
DAKAR
Sélibabi
Yélimané
Ménaka
Diourbel
Bakel
Nioro
Ballé
Nara
Youvarou
Lac Niangay
Ansongo
Mbour
SENEGAL
Kayes
Nampala
Douentza
Hombori
Kaolack
Kaffrine
Goudiri
Kidira
Diéma
Ténenkou
Mopti
BANJUL
Tambacounda
Niono
Bandiagara
Gorom Gorom
Tillabéri
THE GAMBIA
Bafoulabé
Macina
Koro
Djibo
Ziguinchor
Kolda
Kolokani
Ségou
San
Djenné
Ouahigouya
Dori
Kédougou
Koulikoro
Tougan
Yako
Kaya
Cacheu
Bafatá
Gabú
Kita
Kati
Bla
Nouna
Bogande
NIAMEY
GUINEA-BISSAU
Satadougou
BAMAKO
Koutiala
BURKINA FASO
Dosso
Gaoual
Mali
Kangaba
Koudougou
Fada-N'Gourma
Kantchari
Bolama
Buba
Fouta Djallon
Labé
Siguiri
Bougouni
OUAGADOUGOU
Diapaga
Arquipélago dos Bijagós
Cacine
Boké
Dabola
Kouroussa
Sikasso
Bobo-Dioulasso
Manga
Pô
Tenkodogo
Fria
GUINEA
Mandiana
Orodara
Léo
Bawku
Kandi
Dubréka
Mamou
Kankan
Kadiolo
Banfora
Lawra
Bolgatanga
Dapaong
BENIN
CONAKRY
Kindia
Faranah
Ferkessédougou
Wa
Natitingou
Port Loko
Makeni
Odienné
Tehini
Kara
Djougou
Lungi
Kissidougou
Boundiali
Bouna
Yendi
FREETOWN
Koidu-Sefadu
Dianra
Korhogo
Damongo
Tamale
Bassar
Parakou
SIERRA LEONE
Lunsar
Bo
Kailahun
Beyla
Touba
CÔTE D'IVOIRE
Salaga
Sokodé
Nzérékoré
Mankono
Bondoukou
GHANA
Kenema
Zorzor
Man
Séguéla
Lac de Kossou
Bouaké
Wenchi
Kintampo
Techiman
Iseyin
Zimmi
Gbarnga
Danané
(IVORY COAST)
Sunyani
Atakpamé
Abeokuta
Robertsport
Tapeta
Daloa
Abengourou
Lake Volta
Abomey
MONROVIA
LIBERIA
YAMOUSSOUKRO
Bongouanou
Kumasi
Koforidua
PORTO-NOVO
Buchanan
Gagnoa
Zwedru
Divo
Obuasi
Lago
Adzopé
Abidjan
Tarkwa
Winneba
LOMÉ
ACCRA
Sassandra
Slave Coast
Greenville
Grand-Lahou
Axim
Cape Coast
Harper
San-Pédro
Sekondi
Bight of Benin
Cape Palmas
Tabou
Cape Three Points
Gold Coast
GULF OF GUINEA
0°

Lambert Azimuthal Equal Area Projection

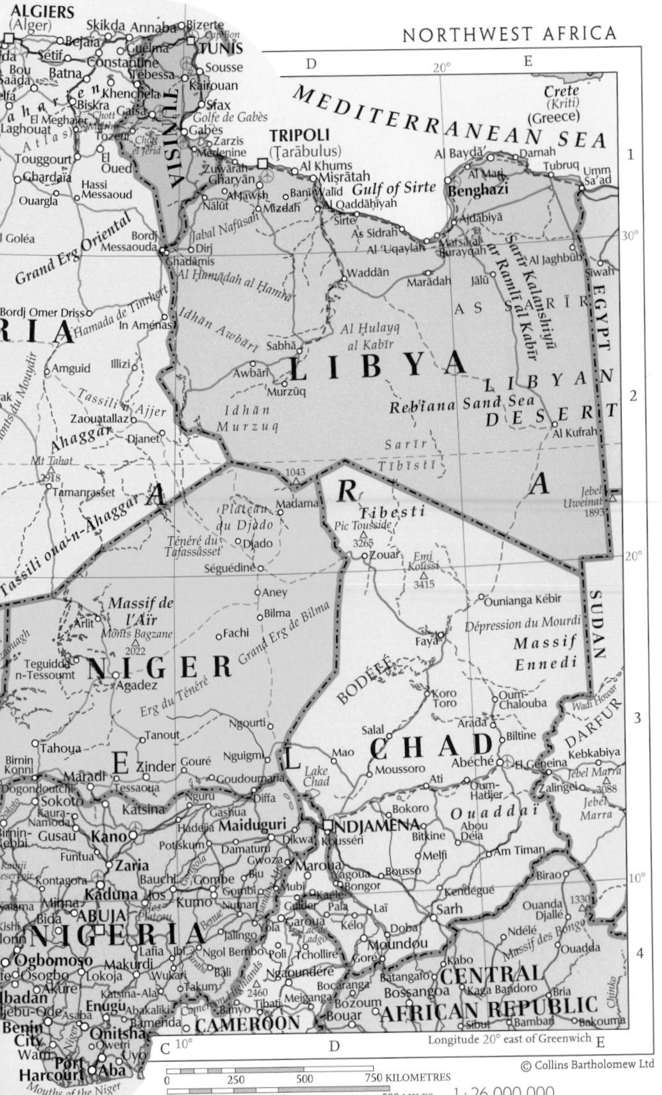

MEDITERRANEAN SEA

ALGIERS
(Alger)

Skikda Annaba
Bizerte

Bejaïa Guelma
TUNIS

Sétif Constantine
Souk Ahras

Batna Tebessa
Sousse

Biskra Kairouan

Khenchela Gafsa
Sfax

El Meghaïer
Tozeur
Chott Golfe de Gabès

Gabès

Oued
Zarzis
Zuwārah **TRIPOLI**
(Ṭarābulus)

Al Khums
Miṣrātah

Al Bayḍā
Darnah
Tubruq

Gharyān
Banī Walīd
Al Qaddāḥīyah

Crete
(Kriti)
(Greece)

Umm
Sa'ad

Bordj
Messaouda
Ghadāmis
Dirj
Jabal Nafūsah
Naūt
Mizdah
Sirte

Benghazi

Ajdābiyā

As Sidrah
Al 'Uqaylah
Marsá al
Burayqah

Al Jaghbūb

Siwah

L I B Y A

Waddān
Marādah

Jālū

A S

Goléa

Grand Erg Oriental

Hamada de Tinrhert

In Amenas

Amguid

Illizi

Al Ḥinnāḥ al al Ḥamrā

Al Ḥulayq
al Kabīr

S A R Ī R

Gharīr Kalanshiyū

Ramlat al Kabīr

EGYPT

Tassili-n-Ajjer

Zaouatallaz

Djanet

Ahaggar

Mt Tahat
2918

Tamanrasset

Tassili oua-n-Ahaggar

Sabhā

Awbārī
Murzuq

Idhān
Awbārī

Idhān
Murzuq

Al Ḥulayq

Rebiana Sand Sea

L I B Y A N

D E S E R T

Al Kufrah

Sarīr
Tībistī

Jebel
Uweinat
1893

S A H A R A

1043

Madama

Plateau
du Djado

Ténéré du
Tafassasset

Djado

T I B E S T I

Pic Toussidé
3265

Zouar

Emi
Koussi
3415

Jebel
Uweinat
1893

Teguidda
n-Tessoumt

Massif de
l'Air

Monts Bagzane
2022

Agadez

Aney

Bilma

Fachi

Erg du Ténéré

Grand Erg de Bilma

Séguédine

Ounianga Kébir

Dépression du Mourdi

SUDAN

Massif
Ennedi

DARFUR

Arlit

Tahoua

Birni
Konni

Maradi

N I G E R

Ngourti

Faya

BODÉLÉ

Koro
Toro

Oum-
Chalouba

Arada

Biltine

Kebkabiya

Dogondoutchi

Sokoto

Gusau

Zinder

Tanout

Tessaoua
Katsina

Goudoumaria

Nguigmi

Mao

Salal

Moussoro

Abéché
El Geneina

Jebel Marra

Zalingei

Jebel
Marra
3088

Kaura
Namoda

Kano

Funtua

Gashua
Hadejia

Nguru

Diffa

Bokoro

Ati

Oum-
Hadjer

Am Zoer
Abou
Deïa

Abou
Deïa

Birnin-
Kebbi

Zaria

Potiskum
Damaturu

Maiduguri

Dikwa
Kousséri

NDJAMENA

Bitkine

Melfi

Am Timan

Biltine

Massif des Bongo

Birao

Minna

Bida

Kaduna
Jos

Bauchi
Gombe

Biu
Mubi

Maroua

Yagoua
Bongor

Bousso

Kélo

Doba

Kendégué

Zaria

ABUJA

Lafia

Makurdi

Wukari

Lake
Chad

Ouaddaï

Pala

Sarh

Ouanda
Djallé

Ouadda

NIGERIA

Osogbo

Lokoja

Jalingo
Bali

Ngol Bembo

Poli

Tcholliré

Moundou

Gore

Ndélé

Batangafo

Kaga Bandoro

Bria

Bambari

Bakouma

Ibadan

Akure

Enugu

Katsina-Ala

Tibati
Meiganga

Ngaoundéré
2460

Bozoum
Bocaranga

Bossangoa

CENTRAL

Benin
City

Onitsha

Oshogbo

CAMEROON

Bouar

Sibut
Bambari

AFRICAN REPUBLIC

Warri

Port
Harcourt
Aba

Owerri
Uyo

Mouths
of the Niger

Longitude 20° east of Greenwich

0   250   500   750 KILOMETRES
0   250   500 MILES

1 : 26 000 000

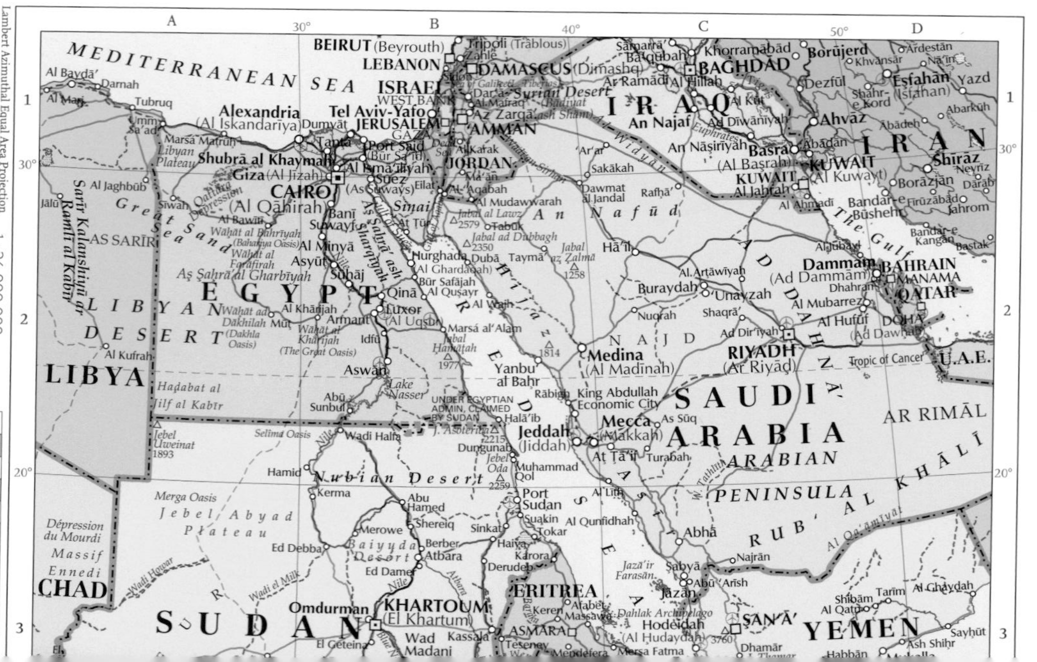

116 Lambert Azimuthal Equal Area Projection   1 : 26 000 000   MILES 0   250   500

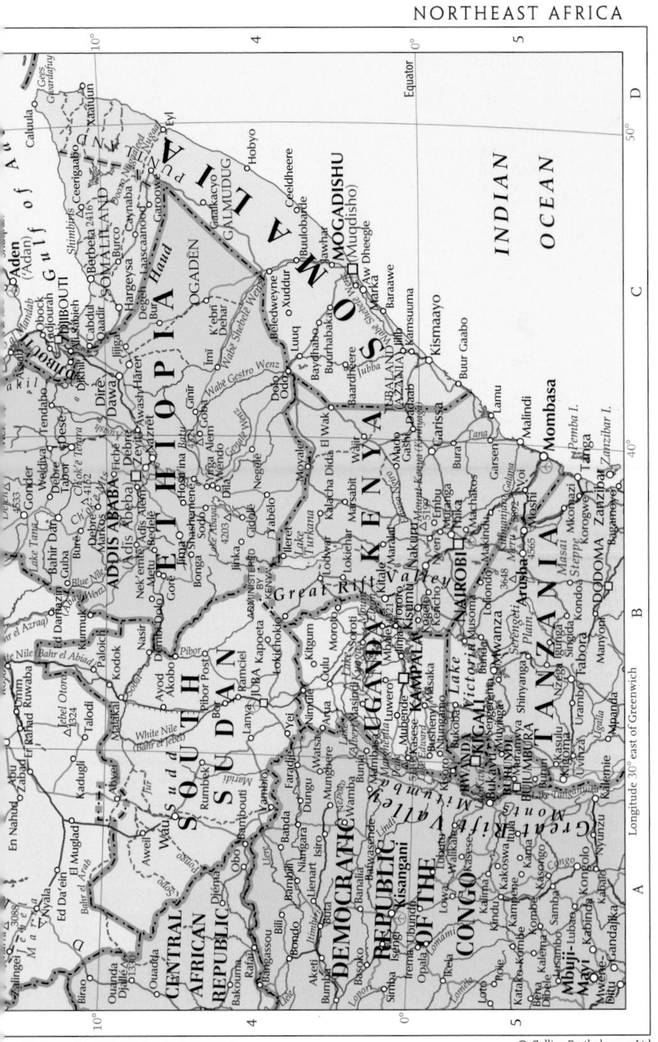

© Collins Bartholomew Ltd

0   250   500   750 KILOMETRES

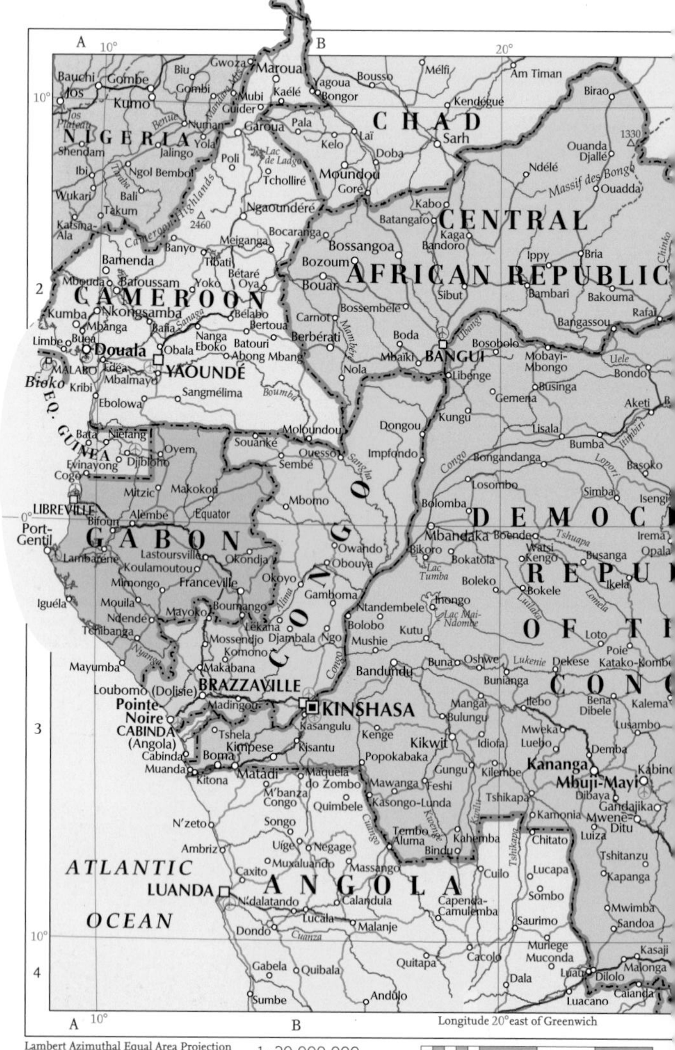

Lambert Azimuthal Equal Area Projection

1 : 20 000 000

MILES 0   100   200   300   400

Longitude 20° east of Greenwich

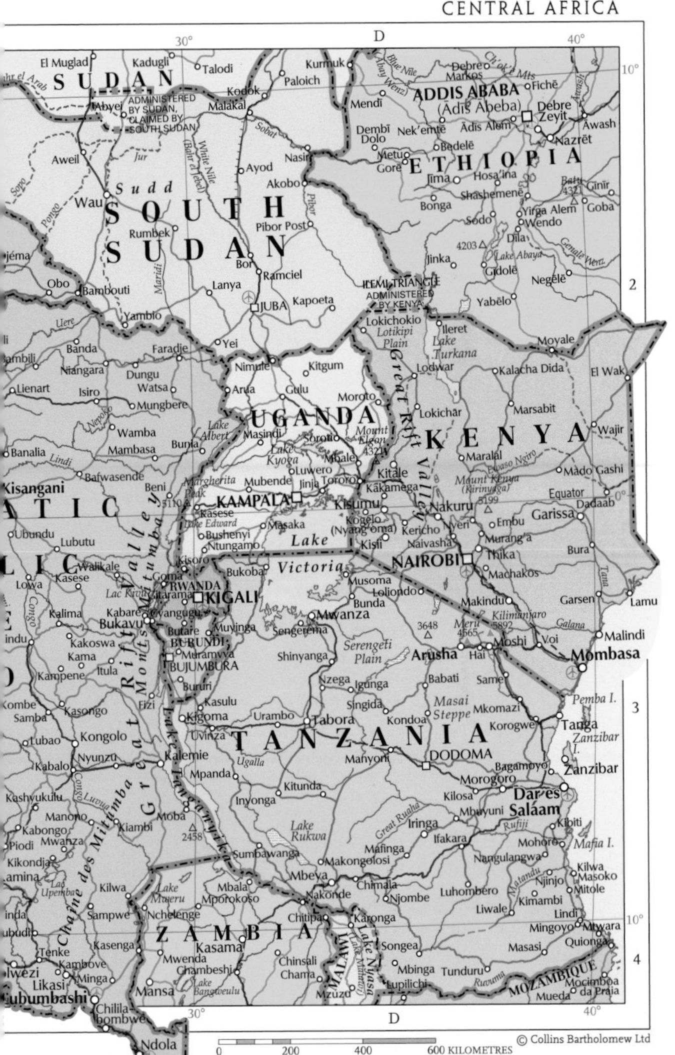
D
30°
40°
10°
2
3
4
0°
SUDAN
SOUTH SUDAN
ETHIOPIA
UGANDA
KENYA
TANZANIA
ZAMBIA
RWANDA
BURUNDI
MALAWI
MOZAMBIQUE
ADDIS ABABA
(Ādīs Ābeba)
JUBA
KAMPALA
NAIROBI
KIGALI
BUJUMBURA
DODOMA
ADMINISTERED BY SUDAN, CLAIMED BY SOUTH SUDAN
ILEMI TRIANGLE ADMINISTERED BY KENYA
El Muglad
Kadugli
Talodi
Kurmuk
Paloich
Kodok
Malakal
Abyei
Aweil
Wau
Sudd
Rumbek
Bor
Ramciel
Lanya
Obo
Bambouti
Yambio
Kapoeta
Yei
Ayod
Akobo
Nasir
Pibor Post
Sobat
White Nile
Jur
Pongo
Sopo
Maridi
Mendī
Dembī Dolo
Nek'emtē
Ādīs Alem
Debre Zeyit
Debre Markos
Fichē
Awash
Nazrēt
Bedelē
Metu
Gorē
Jīma
Hosa'ina
Shashemenē
Bonga
Sodo
Yirga Alem
Wendo
Dila
Batu 4321
Gīnīr
Goba
4203
Lake Abaya
Jinka
Gidolē
Negēlē
Yabēlo
Genale Wenz
Blue Nile
Ch'ok'e Mts
Lokichokio
Lotikipi Plain
Ileret
Lake Turkana
Moyale
Lodwar
Kalacha Dida
El Wak
Lokichar
Marsabit
Wajir
Maralal
Ewaso Ngiro
Mado Gashi
Mount Kenya (Kirinyaga) 5199
Equator
Dadaab
Garissa
Nakuru
Nyeri
Embu
Murang'a
Thika
Machakos
Kericho
Naivasha
Kisii
Kisumu
Kogelo (Nyang'oma)
Kakamega
Kitale
Bura
Tana
Garsen
Lamu
Malindi
Mombasa
Galana
Voi
Kilimanjaro 5892
Meru 4565
Moshi
Hai
Arusha
3648
Makindu
Great Rift Valley
Nimule
Kitgum
Arua
Gulu
Moroto
Lake Albert
Masindi
Soroti
Mount Elgon 4321
Lake Kyoga
Mbale
Luwero
Tororo
Jinja
Mubende
Margherita Peak 5110
Kasese
Lake Edward
Masaka
Bushenyi
Ntungamo
Lake Victoria
Kisoro
Goma
Gitarama
Lac Kivu
Bukavu
Cyangugu
Butare
Muramvya
Muyinga
Bururi
Bukoba
Musoma
Loliondo
Bunda
Mwanza
Sengerema
Shinyanga
Serengeti Plain
Nzega
Igunga
Singida
Babati
Same
Masai Steppe
Mkomazi
Korogwe
Tanga
Pemba I.
Zanzibar I.
Zanzibar
Bagamoyo
Dar es Salâam
Morogoro
Kilosa
Mbuyuni
Kibiti
Rufiji
Mafia I.
Mohoro
Kasulu
Kigoma
Uvinza
Urambo
Tabora
Kondoa
Manyoni
Ugalla
Mpanda
Kitunda
Inyonga
Lake Rukwa
Sumbawanga
Great Ruaha
Iringa
Ifakara
Mafinga
Makongolosi
Mbeya
Chimala
Njombe
Luhombero
Nangulangwa
Matandu
Njinjo
Kilwa Masoko
Mitole
Kimambi
Liwale
Lindi
Mingoyo
Mtwara
Masasi
Quionga
Tunduru
Ruvuma
Mocimboa da Praia
Mueda
Songea
Mbinga
Lupilichi
Karonga
Chitipa
Lake Nyasa (Lake Malawi)
Mzuzu
Chinsali
Chama
Nakonde
Mbala
Mporokoso
Lake Mweru
Nchelenge
Kasama
Mwenda
Chambeshi
Lake Bangweulu
Mansa
Kasenga
Kilwa
Sampwe
Chilila-bombwe
Ndola
Lubumbashi
Likasi
Kolwezi
Kambove
Minga
Tenke
Lake Upemba
Chaîne des Mitumba
Kikondja
Mwanza
Piodi
Kabongo
Manono
Kashyukulu
Kiambi
Moba
2458
Lake Tanganyika
Kalemie
Nyunzu
Kabalo
Luvua
Congo
Kongolo
Lubao
Kasongo
Samba
Kibombe
Kampene
Itula
Kama
Kakoswa
Kalima
Lowa
Kasese
Walikale
Fizi
Ubundu
Lubutu
Kisangani
Bafwasende
Banalia
Lindi
Mambasa
Wamba
Nepoko
Bunia
Beni
Isiro
Lienart
Mungbere
Watsa
Dungu
Faradje
Niangara
Banda
Uele
Monts Mitumba
Great Rift Valley

0 200 400 600 KILOMETRES

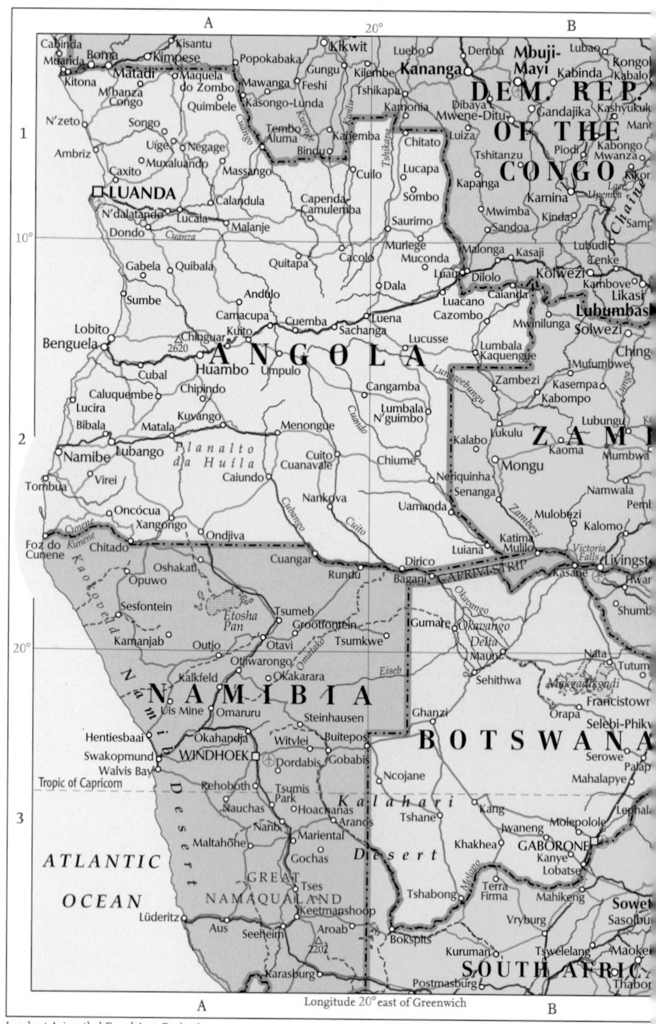

Lambert Azimuthal Equal Area Projection   1 : 20 000 000   MILES 0   100   200   300   400

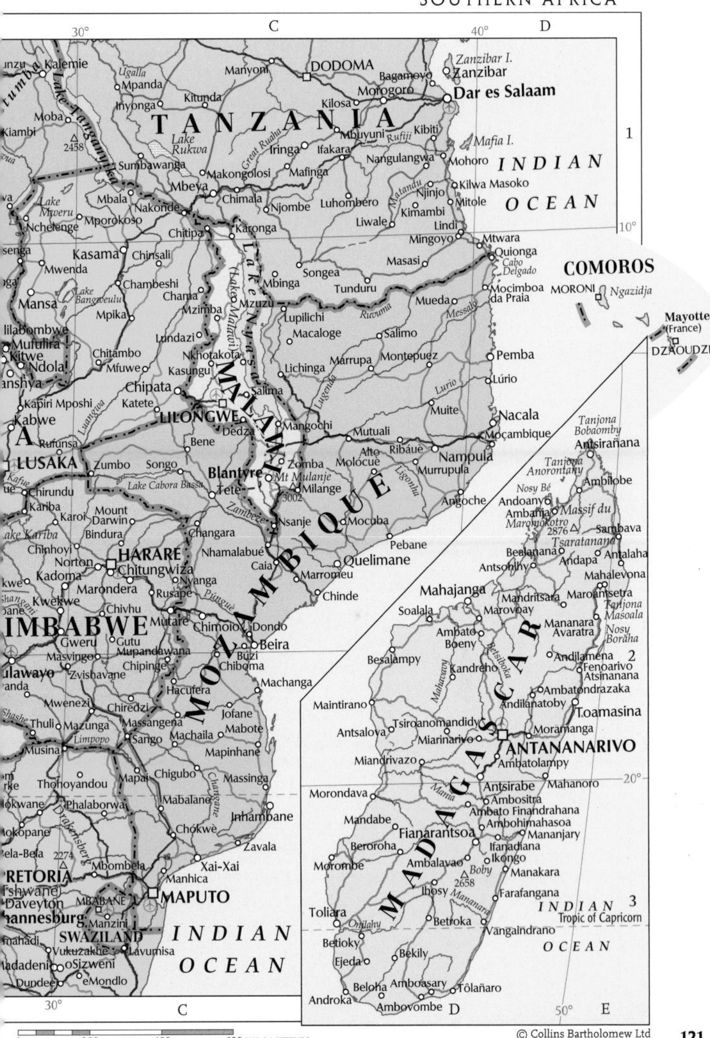

© Collins Bartholomew Ltd

30°
40°
1
2
3
10°
20°
50°
TANZANIA
DODOMA
Zanzibar I.
Zanzibar
Dar es Salaam
Kalemie
Manyoni
Bagamoyo
Morogoro
Ugalla
Mpanda
Kitunda
Inyonga
Kilosa
Lake Tanganyika
Moba
Kiambi
2458
Lake Rukwa
Great Ruaha
Mbuyuni
Rufiji
Kibiti
Mafia I.
Iringa
Ifakara
Sumbawanga
Nangulangwa
Mohoro
INDIAN
OCEAN
Makongolosi
Mafinga
Mbeya
Matandu
Njinjo
Kilwa Masoko
Lake Mweru
Mbala
Nakonde
Chimala
Njombe
Luhombero
Mitole
Kimambi
Nchelenge
Mporokoso
Liwale
Lindi
Chitipa
Karonga
Mingoyo
Mtwara
Quionga
Cabo Delgado
Kasama
Chinsali
Masasi
COMOROS
Mwenda
Lake Nyasa
(Lake Malawi)
Songea
Mbinga
Tunduru
Mocimboa da Praia
MORONI
Ngazidja
Lake Bangweulu
Chambeshi
Chama
Mzuzu
Mueda
Mansa
Mzimba
Lupilichi
Ruvuma
Messalo
Mayotte
(France)
Mpika
Macaloge
Salimo
DZAOUDZI
Mufulira
Kitwe
Ndola
Lundazi
Chitambo
Mfuwe
Nkhotakota
Marrupa
Montepuez
Pemba
Kasungu
Lichinga
Chipata
MALAWI
Lurio
Lúrio
Salima
Lugenda
Kapiri Mposhi
Katete
LILONGWE
Muite
Nacala
Kabwe
Luangwa
Dedza
Mangochi
Tanjona Bobaomby
Bene
Mutuali
Moçambique
Antsirañana
Rufunsa
Alto Molócuè
Ribáuè
Nampula
LUSAKA
Zumbo
Songo
Zomba
Tanjona Anorontany
Kafue
Lake Cabora Bassa
Blantyre
Mt Mulanje
Milange
Murrupula
Ligonha
Ambilobe
Chirundu
Tete
3002
Nosy Bé
Kariba
Karoi
Mount Darwin
Zambeze
Angoche
Andoany
Ambanja
Massif du Maromokotro
2876
Sambava
Tsaratanana
Lake Kariba
Bindura
Nsanje
Mocuba
Chinhoyi
HARARE
Changara
Pebane
Norton
Chitungwiza
Nhamalabué
Befandriana
Antalaha
Andapa
Kadoma
Caia
Quelimane
Antsohihy
Nyanga
Marromeu
Mahalevona
Marondera
Rusape
Mahajanga
Mandritsara
Maroantsetra
Kwekwe
Púnguè
Chinde
Tanjona Masoala
Chivhu
Soalala
Marovoay
ZIMBABWE
Mutare
Chimoio
Dondo
Mananara Avaratra
Nosy Boraha
Ambato Boeny
Gweru
Gutu
Beira
Mupandawana
Masvingo
Búzi
Chiboma
Besalampy
Betsiboka
Andilamena
Chipinge
Fenoarivo Atsinanana
Zvishavane
Kandreho
Machanga
Mahavavy
Hacufera
Ambatondrazaka
Mwenezi
Chiredzi
MOZAMBIQUE
Maintirano
Andilanatoby
Toamasina
Jofane
Thuli
Mazunga
Massangena
Tsiroanomandidy
Mabote
Limpopo
Sango
Machaila
Antsalova
Miarinarivo
Moramanga
Musina
ANTANANARIVO
Mapinhane
Miandrivazo
Ambatolampy
Thohoyandou
Mapai
Chigubo
Changane
Massinga
Mahanoro
Morondava
Antsirabe
Mabalane
Mania
Ambositra
Phalaborwa
Ambato Finandrahana
Drakensberg
Inhambane
Mandabe
Ambohimahasoa
Fianarantsoa
Mananjary
Chókwè
Beroroha
Ifanadiana
Zavala
Ikongo
2274
MADAGASCAR
Mbombela
Morombe
Ambalavao
Boby
Xai-Xai
2658
Manakara
PRETORIA
(Tshwane)
Manhica
Ihosy
MAPUTO
Mananara
Farafangana
MBABANE
INDIAN
Tropic of Capricorn
Johannesburg
Toliara
Manzini
Onilahy
Betroka
SWAZILAND
INDIAN
Vangaindrano
Betioky
OCEAN
Vukuzakhe
Lavumisa
Bekily
OCEAN
Ejeda
Sizweni
Dundee
eMondlo
Beloha
Ambovombe
Tôlañaro
Androka
Ambovombe
0
200
400
600 KILOMETRES

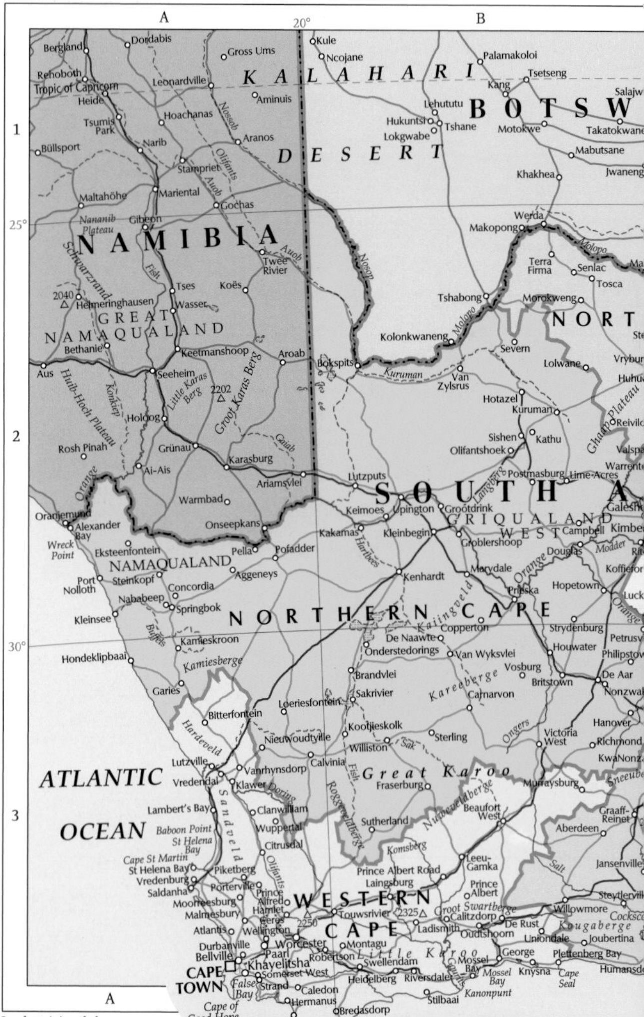

KALAHARI DESERT
BOTSWANA
NAMIBIA
GREAT NAMAQUALAND
NAMAQUALAND
SOUTH AFRICA
NORTHERN CAPE
GRIQUALAND WEST
WESTERN CAPE
ATLANTIC OCEAN
Great Karoo
Little Karoo
Tropic of Capricorn
Windhoek
Rehoboth
Mariental
Keetmanshoop
Karasburg
Upington
Kuruman
Springbok
Calvinia
Beaufort West
Worcester
Paarl
Stellenbosch
CAPE TOWN
Khayelitsha
Oudtshoorn
George
Mossel Bay
Knysna
Cape of Good Hope
Cape Agulhas
Struis Bay

Lambert Azimuthal
Equal Area Projection

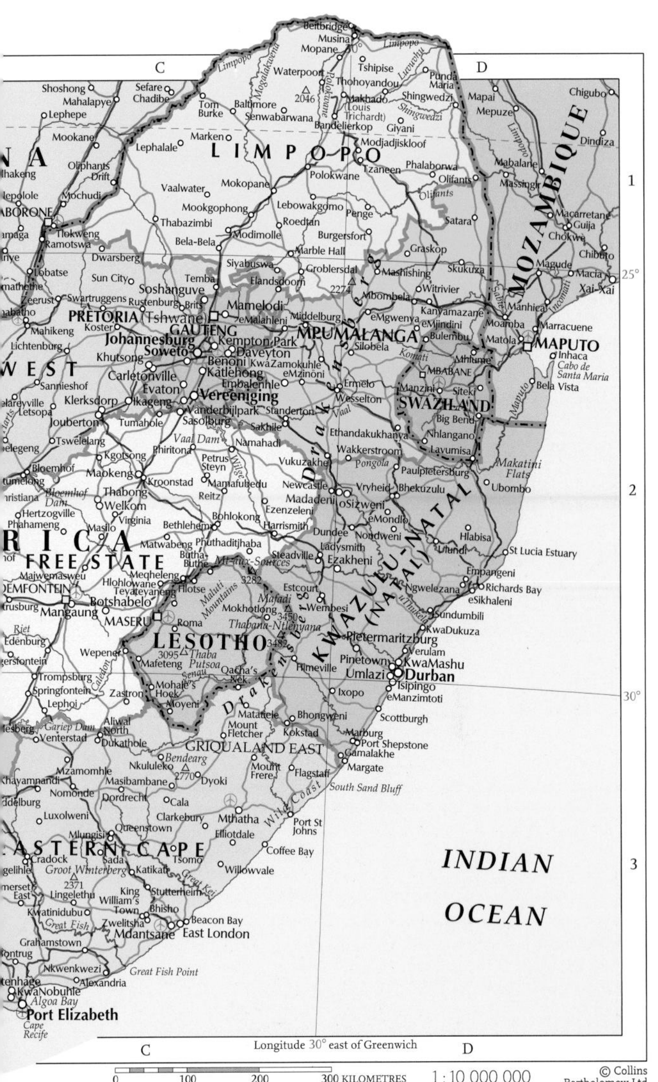

INDIAN OCEAN

LIMPOPO

MOZAMBIQUE

MPUMALANGA

GAUTENG

SWAZILAND

NORTH WEST

FREE STATE

LESOTHO

KWAZULU-NATAL
(NATAL)

EASTERN CAPE

GRIQUALAND EAST

Maputo

MBABANE

MASERU

Johannesburg
Soweto
PRETORIA
Tshwane

Durban

Port Elizabeth

East London

Longitude 30° east of Greenwich

© Collins
Bartholomew Ltd

1 : 10 000 000

0   100   200   300 KILOMETRES

0   100   200 MILES

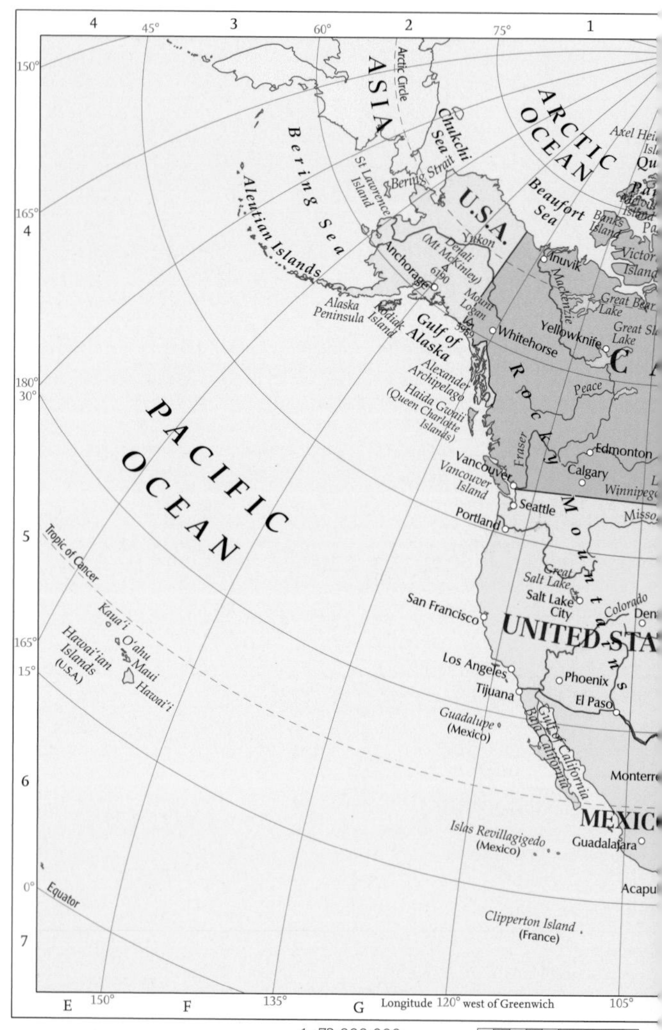

124

1 : 72 000 000 MILES 0 500 1000

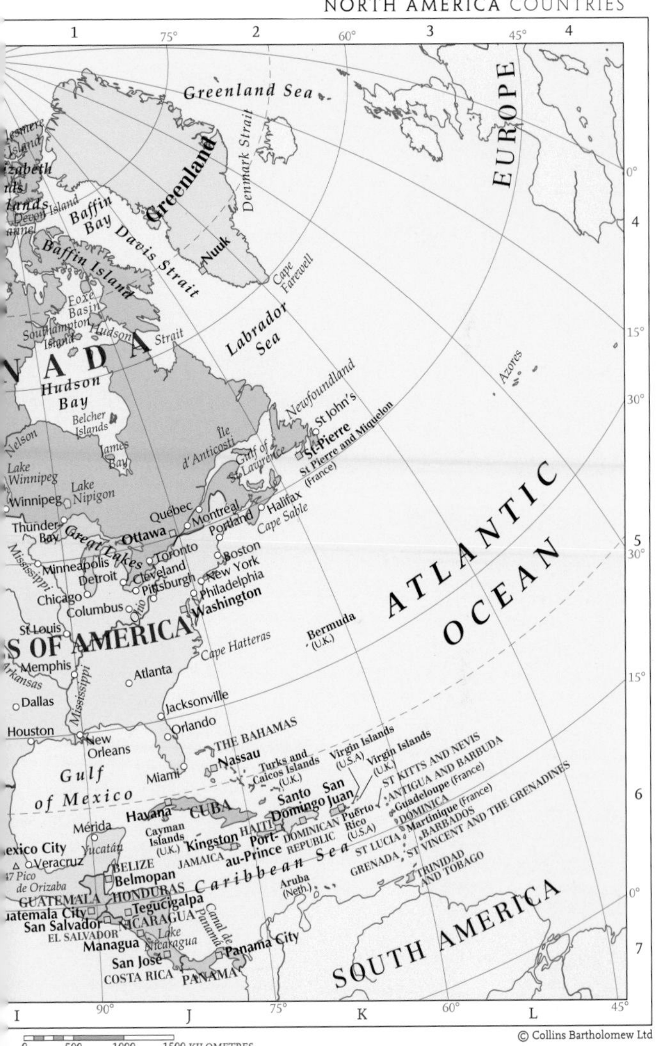

1 75° 2 60° 3 45° 4

Greenland Sea

EUROPE

Ellesmere
Island
Elizabeth
Islands

Devon Island

Baffin
Bay

Greenland

Denmark Strait

Baffin Island

Davis Strait

Nuuk

Cape
Farewell

0°

4

Foxe
Basin

Southampton
Island

Hudson Strait

Labrador
Sea

15°

CANADA

Hudson
Bay

Belcher
Islands

James
Bay

Newfoundland

Île
d'Anticosti

St John's

Azores

30°

Nelson

Lake
Winnipeg

Lake
Nipigon

Winnipeg

Gulf of
St Lawrence

St-Pierre
St Pierre and Miquelon
(France)

Thunder
Bay

Great Lakes

Quebec

Montreal

Portland

Halifax

Cape Sable

5

30°

Minneapolis

Detroit

Cleveland

Ottawa

Toronto

Boston

Mississippi

Chicago

Pittsburgh

New York

Philadelphia

ATLANTIC

St Louis

Columbus

Washington

Memphis

Arkansas

UNITED STATES OF AMERICA

Cape Hatteras

Bermuda
(U.K.)

OCEAN

Dallas

Atlanta

15°

Houston

Jacksonville

New
Orleans

Orlando

Gulf

Miami

THE BAHAMAS

Nassau

Turks and
Caicos Islands
(U.K.)

Virgin Islands
(U.S.A)

Virgin Islands
(U.K.)

ST KITTS AND NEVIS

of Mexico

Havana

CUBA

Santo
Domingo

San
Juan

Puerto
Rico (U.S.A.)

ANTIGUA AND BARBUDA

Guadeloupe (France)

DOMINICA

Mérida

Cayman
Islands
(U.K.)

Kingston

JAMAICA

HAITI

Port-
au-Prince

DOMINICAN
REPUBLIC

Martinique (France)

ST LUCIA

BARBADOS

exico City

Veracruz

Yucatán

BELIZE

Caribbean Sea

Aruba
(Neth.)

GRENADA

ST VINCENT AND THE GRENADINES

Pico
de Orizaba

Belmopan

HONDURAS

TRINIDAD
AND TOBAGO

GUATEMALA

Guatemala City

Tegucigalpa

NICARAGUA

Canal de
Panamá

San Salvador

EL SALVADOR

Managua

Lake
Nicaragua

Panama City

SOUTH AMERICA

San José

COSTA RICA

PANAMA

6

0°

7

45°

I 90° J 75° K 60° L

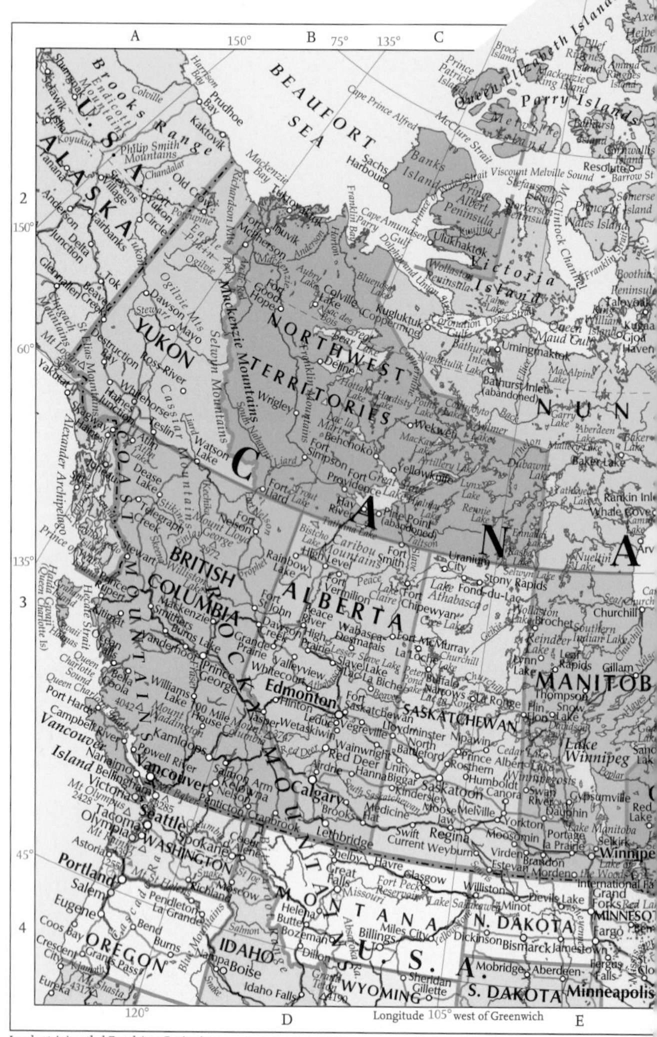

Lambert Azimuthal Equal Area Projection   1 : 30 000 000   MILES 0   200   400   600

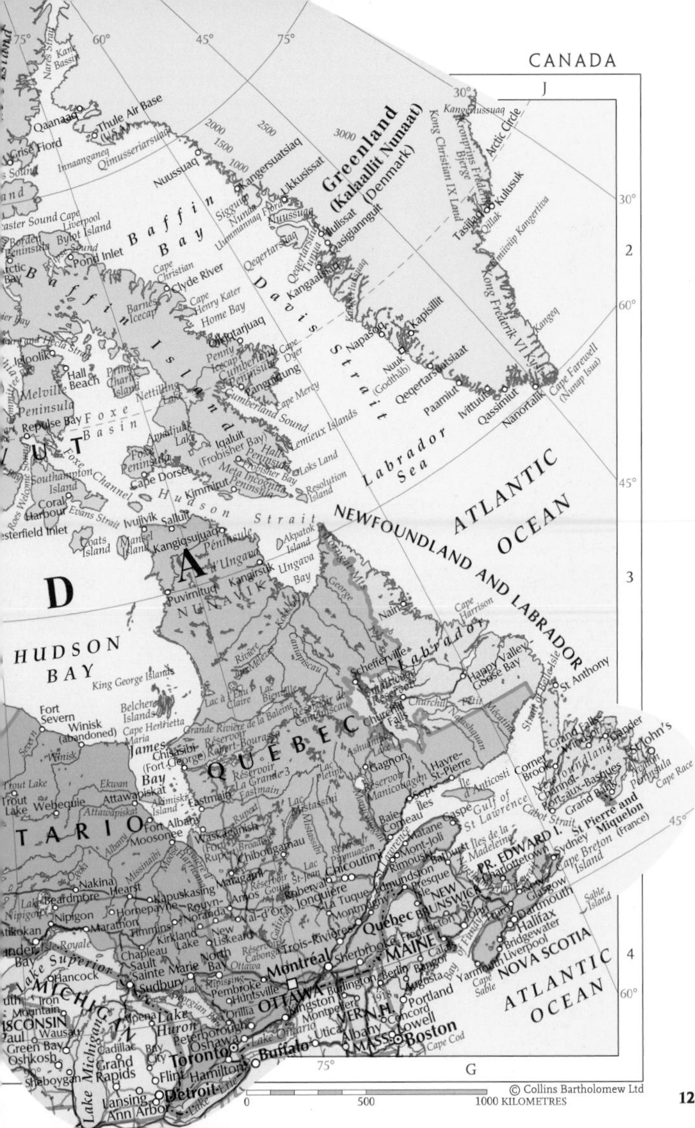

© Collins Bartholomew Ltd

0   500   1000 KILOMETRES

Greenland
(Kalaallit Nunaat)
(Denmark)
Baffin Bay
Baffin Island
Davis Strait
Labrador Sea
ATLANTIC OCEAN
NEWFOUNDLAND AND LABRADOR
HUDSON BAY
Hudson Strait
Foxe Basin
Foxe Channel
QUÉBEC
NUNAVIK
ONTARIO
James Bay
Ungava Bay
Péninsule d'Ungava
Labrador
Gulf of St Lawrence
Cabot Strait
NOVA SCOTIA
NEW BRUNSWICK
PR. EDWARD I.
St Pierre and Miquelon (France)
MAINE
N.H.
MASS.
MICHIGAN
Lake Superior
Lake Huron
Lake Michigan
Lake Ontario
Thule Air Base (U.S.A.)
Qaanaaq
Grise Fiord
Pond Inlet
Clyde River
Qikiqtarjuaq
Pangnirtung
Iqaluit (Frobisher Bay)
Cape Dorset
Kimmirut
Igloolik
Hall Beach
Repulse Bay
Coral Harbour
Ivujivik
Salluit
Kangiqsujuaq
Kangirsuk
Puvirnituq
Nain
Schefferville
Happy Valley-Goose Bay
St Anthony
Churchill Falls
Chisasibi (Fort George)
Gagnon
Sept-Îles
Havre-St-Pierre
Baie-Comeau
Gaspé
Rimouski
Chicoutimi
Roberval
Jonquière
La Tuque
Chibougamau
Matagami
Amos
Rouyn-Noranda
Val-d'Or
Québec
Trois-Rivières
Sherbrooke
Montréal
OTTAWA
Kingston
Toronto
Hamilton
Buffalo
Detroit
Boston
Portland
Bangor
Augusta
Concord
Lowell
Albany
Utica
Burlington
Montpelier
Fredericton
Saint John
Moncton
Charlottetown
Halifax
Dartmouth
Truro
Sydney
Yarmouth
Bridgewater
Liverpool
St John's
Gander
Grand Falls-Windsor
Corner Brook
Channel-Port-aux-Basques
Nuuk (Godthåb)
Paamiut
Qaqortoq
Nanortalik
Kulusuk
Tasiilaq
Sisimiut
Ilulissat
Qasigiannguit
Upernavik
Moosonee
Fort Albany
Attawapiskat
Webequie
Fort Severn
Winisk (abandoned)
Nakina
Hearst
Kapuskasing
Timmins
Kirkland Lake
New Liskeard
North Bay
Sudbury
Sault Sainte Marie
Chapleau
Marathon
Nipigon
Thunder Bay
Atikokan
Pembroke
Huntsville
Orillia
Peterborough
Oshawa
Green Bay
Oshkosh
Sheboygan
Lansing
Ann Arbor
Flint
Saginaw
Grand Rapids
Alpena
Cadillac
Bay City
Hancock
Iron Mountain
Wausau
Cape Farewell (Nunap Isua)
Cape Race
Cape Cod
Cape Sable
Sable Island
Resolution Island
Southampton Island
Coats Island
Mansel Island
Akpatok Island
Belcher Islands
King George Islands
Akimiski Island
Île d'Anticosti
Îles de la Madeleine
Cape Breton Island
Avalon Peninsula
Melville Peninsula
Cumberland Peninsula
Cumberland Sound
Hall Peninsula
Meta Incognita Peninsula
Foxe Peninsula
Réservoir Manicouagan
Réservoir de Caniapiscau
Réservoir Robert-Bourassa
Réservoir La Grande 3
Smallwood Reservoir
Lac Mistassini
Nettilling Lake
Amadjuak Lake
Arctic Circle
Kong Christian IX Land
Kong Frederik VI Kyst
75°
60°
45°
30°
J
G
2
3
4

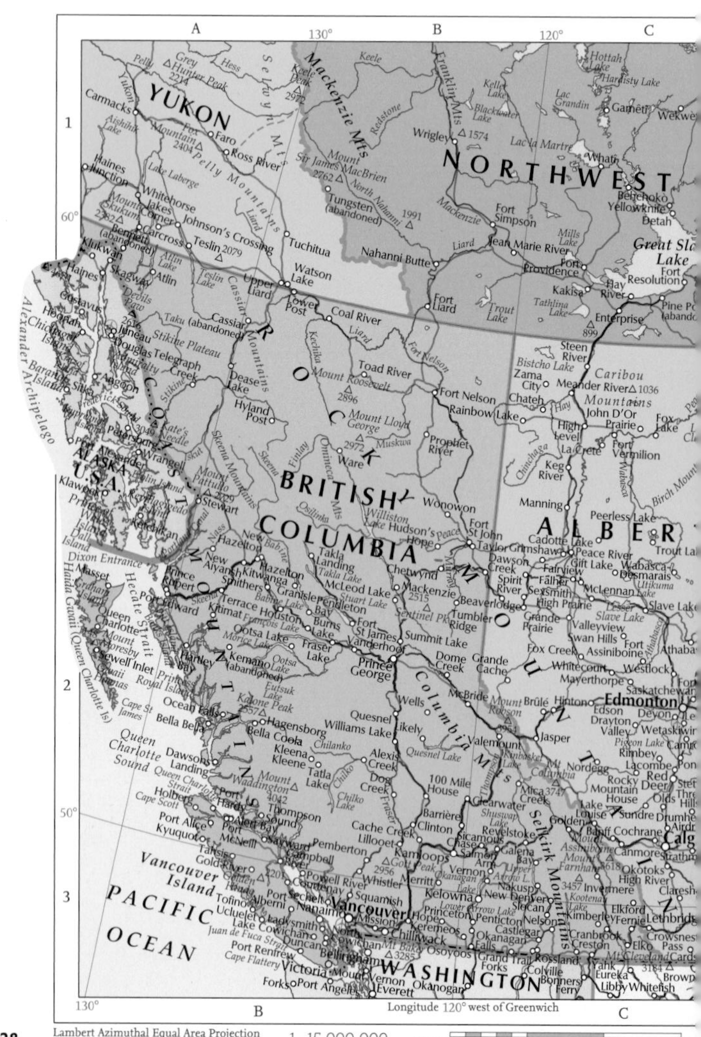

Lambert Azimuthal Equal Area Projection   1 : 15 000 000   MILES 0   100   200   300

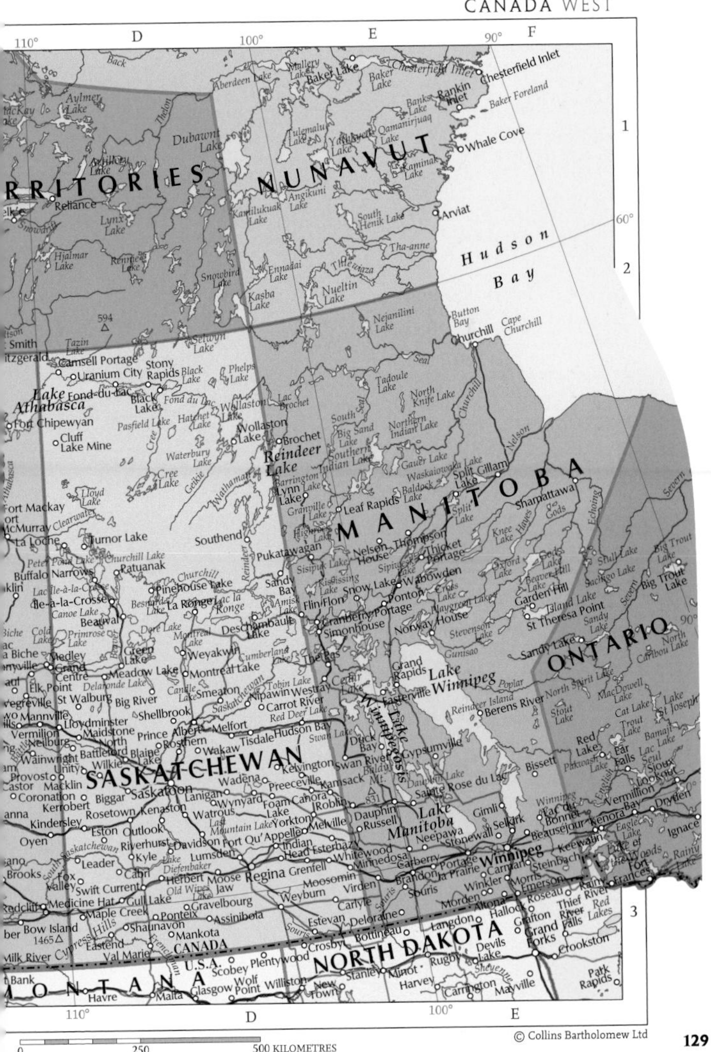
NUNAVUT
MANITOBA
SASKATCHEWAN
ONTARIO
NORTH DAKOTA
MONTANA
Hudson Bay
Chesterfield Inlet
Baker Lake
Rankin Inlet
Whale Cove
Arviat
Churchill
Cape Churchill
Reindeer Lake
Lake Athabasca
Uranium City
Fort Chipewyan
Lake Winnipeg
Lake Manitoba
Lake Winnipegosis
Thompson
Flin Flon
The Pas
Prince Albert
Saskatoon
Regina
Moose Jaw
Swift Current
Medicine Hat
Brandon
Winnipeg
Kenora
Dryden
Lloydminster
North Battleford
Yorkton
Estevan
Minot
Williston
Grand Forks
CANADA
U.S.A.
110°
100°
90°
60°
D
E
F
1
2
3
500 KILOMETRES

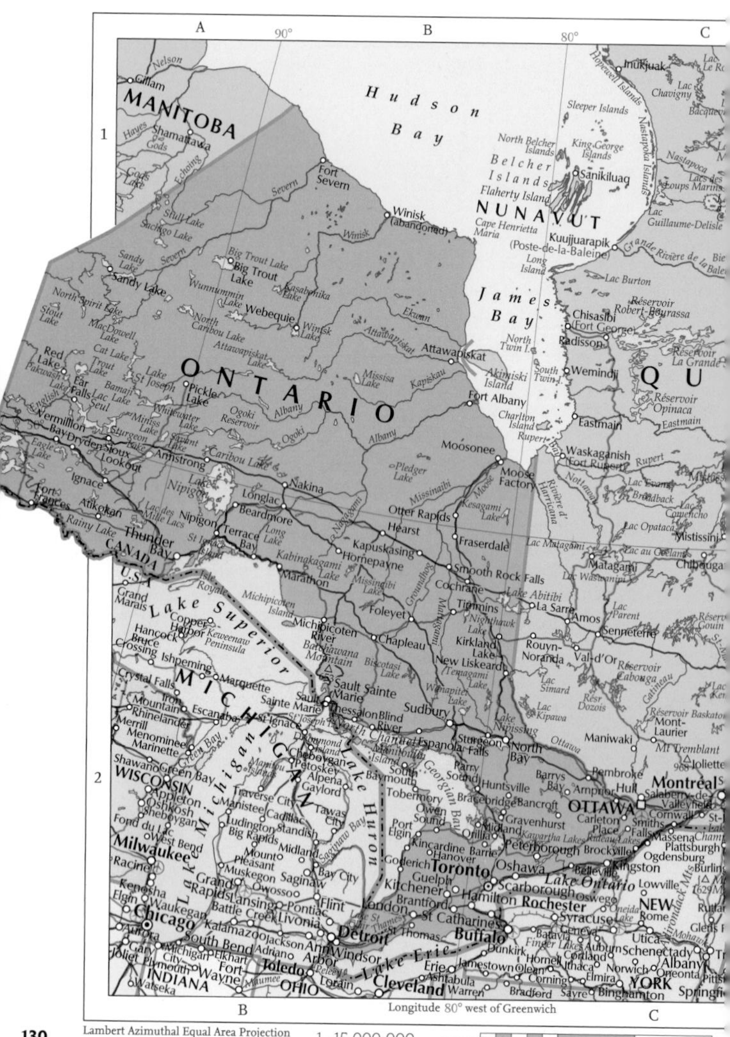

Lambert Azimuthal Equal Area Projection   1 : 15 000 000   MILES 0   100   200   300

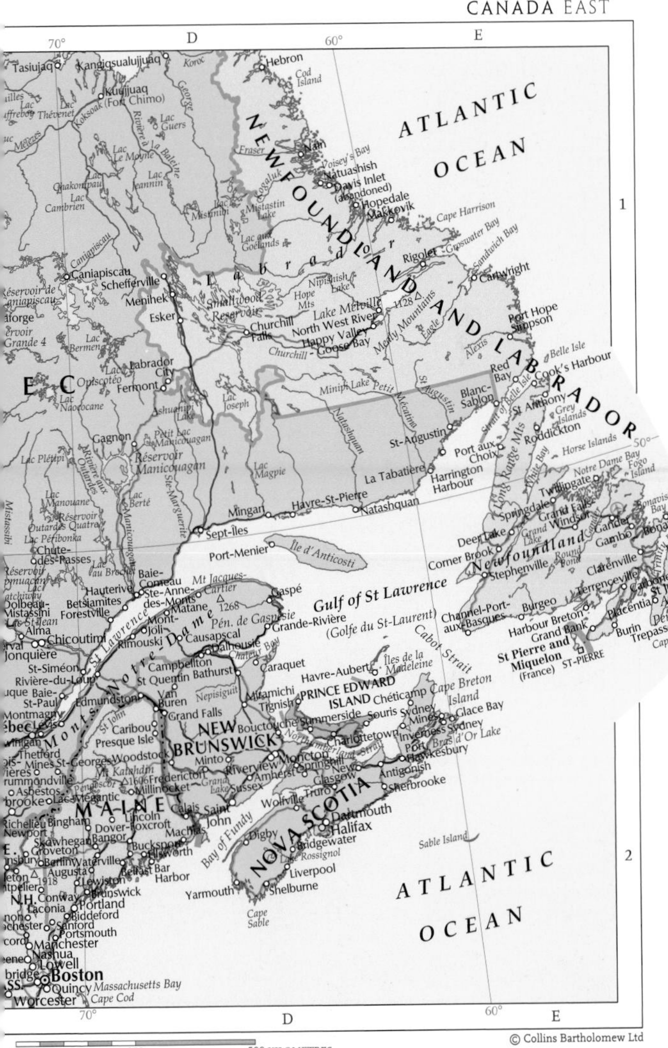

ATLANTIC OCEAN

NEWFOUNDLAND AND LABRADOR

Labrador

Newfoundland

Gulf of St Lawrence
(Golfe du St-Laurent)

St Pierre and Miquelon
(France)

Cabot Strait

PRINCE EDWARD ISLAND

NEW BRUNSWICK

MAINE

NOVA SCOTIA

Bay of Fundy

N.H.

Massachusetts Bay
Cape Cod

ATLANTIC OCEAN

© Collins Bartholomew Ltd

0   250   500 KILOMETRES

131

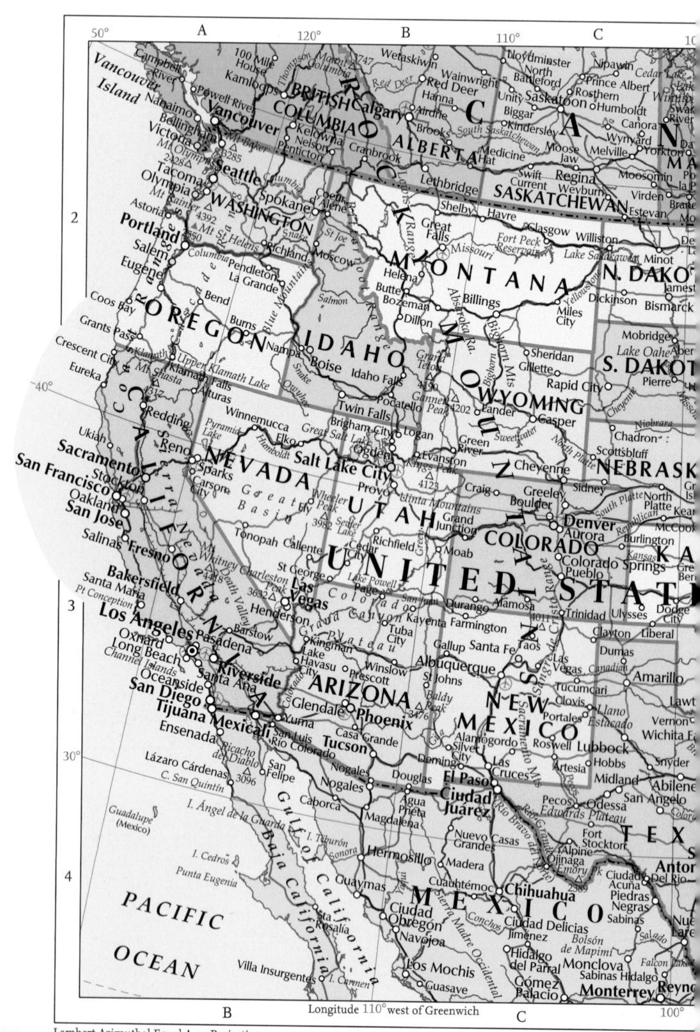

Lambert Azimuthal Equal Area Projection   1 : 25 000 000   MILES 0   250   500

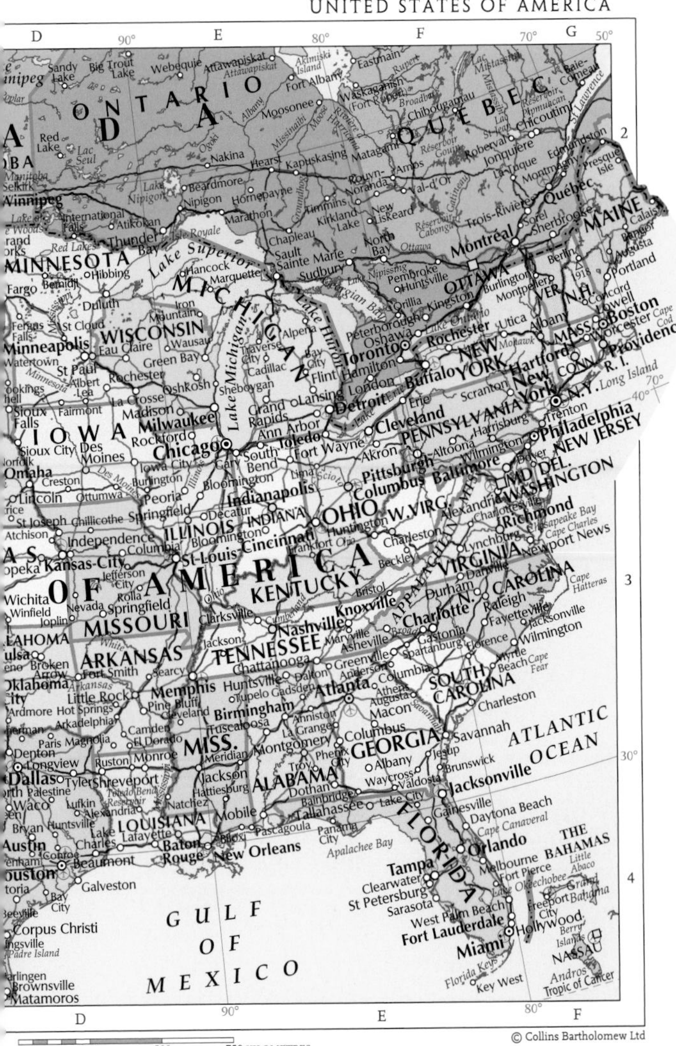
D
E
F
G
90°
80°
70°
50°
2
3
4
40°
30°
ONTARIO
QUÉBEC
CANADA
MINNESOTA
WISCONSIN
MICHIGAN
IOWA
ILLINOIS
INDIANA
OHIO
PENNSYLVANIA
NEW YORK
VER.
N.H.
MAINE
MASS.
CONN.
R.I.
N.Y.
NEW JERSEY
MD.
DEL.
WASHINGTON
W.VIRG.
VIRGINIA
KENTUCKY
TENNESSEE
N. CAROLINA
SOUTH CAROLINA
GEORGIA
ALABAMA
MISS.
LOUISIANA
ARKANSAS
MISSOURI
FLORIDA
UNITED STATES OF AMERICA
APPALACHIAN MTS
ATLANTIC OCEAN
GULF OF MEXICO
THE BAHAMAS
NASSAU
OTTAWA
Lake Superior
Lake Michigan
Lake Huron
Lake Erie
Lake Ontario
Georgian Bay
Lake Nipissing
Lake Nipigon
Lake Okeechobee
Chesapeake Bay
Apalachee Bay
Long Island
Cape Cod
Cape Hatteras
Cape Fear
Cape Charles
Cape Canaveral
Florida Keys
Tropic of Cancer
Moosonee
Attawapiskat
Fort Albany
Eastmain
Chibougamau
Chicoutimi
Thunder Bay
Sault Sainte Marie
Sudbury
North Bay
Montréal
Québec
Toronto
Hamilton
London
Kingston
Trois-Rivières
Sherbrooke
Minneapolis
St Paul
Duluth
Fargo
Milwaukee
Chicago
Detroit
Toledo
Cleveland
Buffalo
Rochester
Albany
Boston
Providence
Hartford
New York
Philadelphia
Baltimore
Pittsburgh
Columbus
Cincinnati
Indianapolis
St Louis
Kansas City
Omaha
Des Moines
Richmond
Norfolk
Newport News
Charlotte
Raleigh
Nashville
Knoxville
Memphis
Atlanta
Birmingham
Montgomery
Jackson
New Orleans
Baton Rouge
Mobile
Tallahassee
Jacksonville
Savannah
Charleston
Orlando
Tampa
St Petersburg
Miami
Fort Lauderdale
Key West
Houston
Dallas
Austin
Galveston
Corpus Christi
Brownsville
Matamoros
Shreveport
Little Rock
Oklahoma City
Tulsa
Wichita
Springfield
Louisville
Lexington

0 250 500 750 KILOMETRES

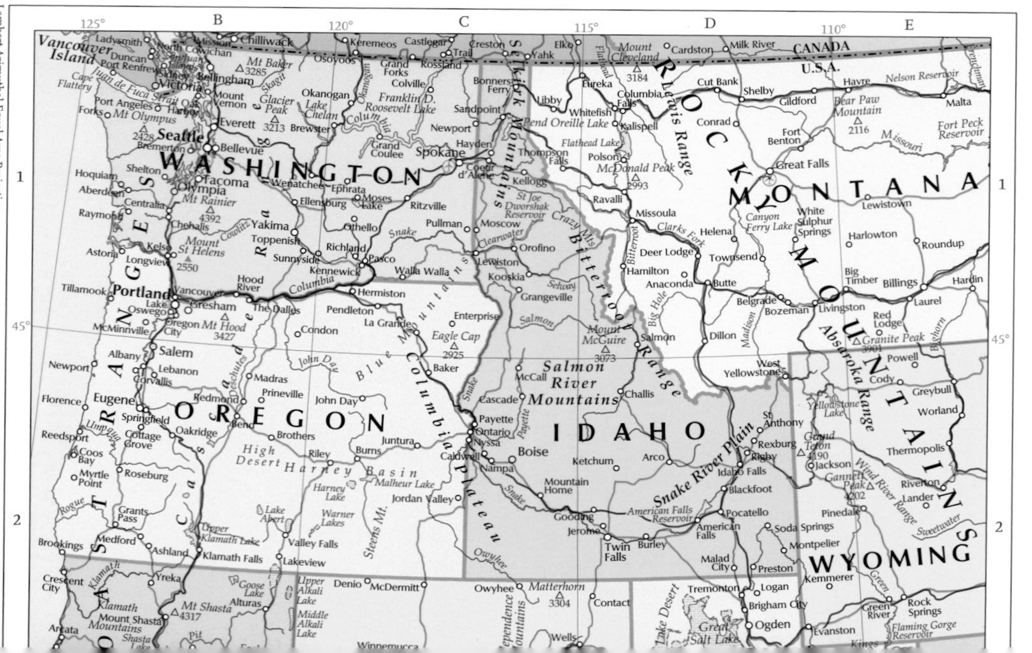

Lambert Azimuthal Equal Area Projection

1 : 11 000 000

MILES 0   100   200

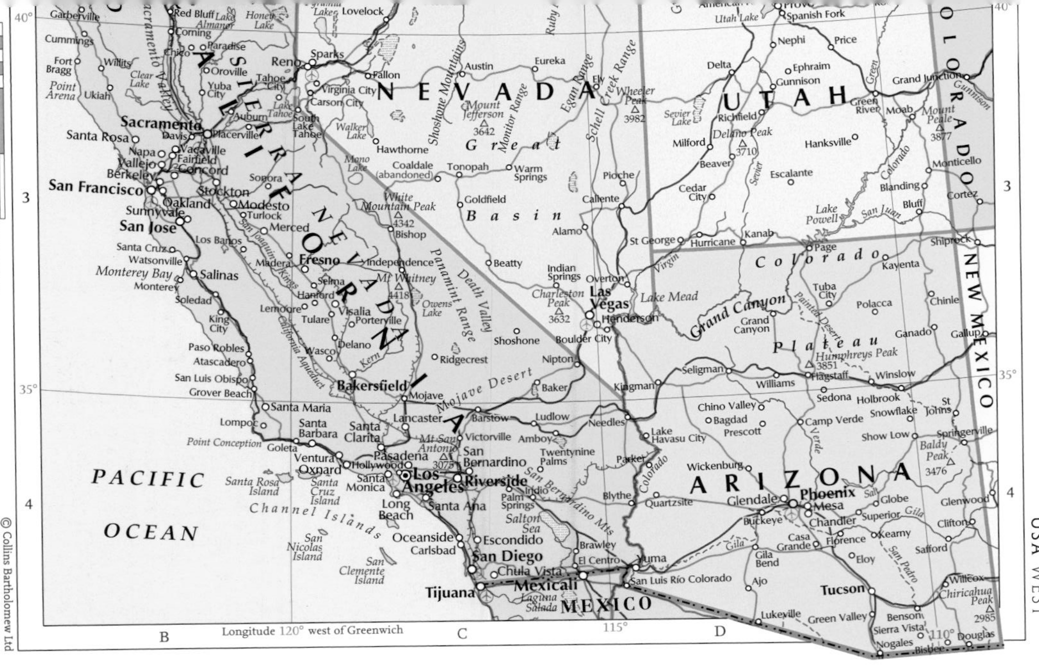

© Collins Bartholomew Ltd

0 100 200 300 KILOMETRES

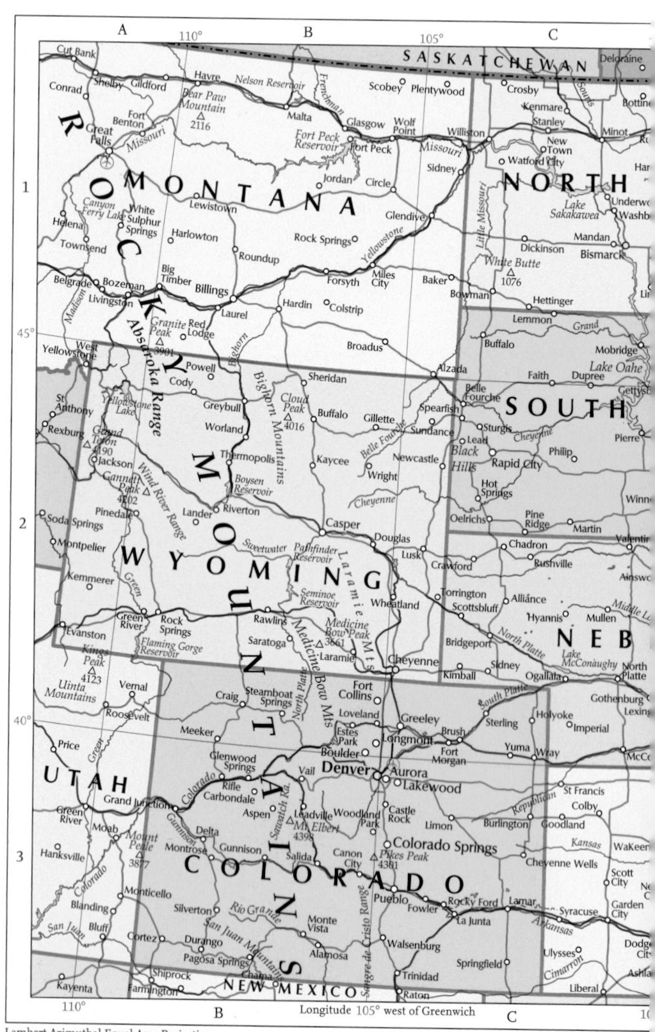

SASKATCHEWAN
MONTANA
NORTH
SOUTH
WYOMING
NEB
UTAH
COLORADO
NEW MEXICO
ROCKY MOUNTAINS
Cut Bank
Shelby
Conrad
Gildford
Havre
Great Falls
Fort Benton
Missouri
Nelson Reservoir
Bear Paw Mountain
2116
Malta
Frenchman
Scobey
Plentywood
Crosby
Kenmare
Stanley
Minot
New Town
Watford City
Glasgow
Wolf Point
Williston
Fort Peck Reservoir
Fort Peck
Sidney
Jordan
Circle
Lewistown
Helena
Canyon Ferry Lake
White Sulphur Springs
Harlowton
Townsend
Glendive
Yellowstone
Rock Springs
Roundup
Lake Sakakawea
Underwood
Washburn
Mandan
Dickinson
Bismarck
White Butte
1076
Little Missouri
Belgrade
Bozeman
Livingston
Madison
Big Timber
Billings
Laurel
Forsyth
Miles City
Baker
Bowman
Hettinger
Hardin
Colstrip
Lemmon
Grand
Granite Peak
3901
Red Lodge
Absaroka Range
Bighorn
Broadus
Buffalo
Mobridge
West Yellowstone
Powell
Cody
Sheridan
Alzada
Faith
Dupree
Lake Oahe
Belle Fourche
St Anthony
Rexburg
Yellowstone Lake
Greybull
Worland
Cloud Peak
4016
Buffalo
Gillette
Spearfish
Sundance
Sturgis
Cheyenne
Pierre
Grand Teton
4190
Jackson
Thermopolis
Bighorn Mountains
Kaycee
Belle Fourche
Newcastle
Lead
Black Hills
Rapid City
Philip
Wright
Gannett Peak
4202
Wind River Range
Boysen Reservoir
Hot Springs
Pinedale
Lander
Riverton
Cheyenne
Soda Springs
Casper
Oelrichs
Pine Ridge
Martin
Montpelier
Sweetwater
Pathfinder Reservoir
Laramie Mts
Douglas
Lusk
Chadron
Rushville
Crawford
Kemmerer
Green
Seminoe Reservoir
Wheatland
Torrington
Scottsbluff
Alliance
Hyannis
Mullen
Green River
Rock Springs
Rawlins
Medicine Bow Peak
3661
Evanston
Flaming Gorge Reservoir
Saratoga
Laramie
Medicine Bow Mts
Cheyenne
Bridgeport
North Platte
Lake McConaughy
Kings Peak
4123
Kimball
Sidney
Ogallala
Uinta Mountains
Vernal
Craig
Steamboat Springs
Fort Collins
South Platte
Gothenburg
Roosevelt
Loveland
Greeley
Sterling
Holyoke
Imperial
Meeker
Estes Park
Longmont
Brush
Fort Morgan
Yuma
Wray
Price
Green
Glenwood Springs
Boulder
Vail
Denver
Aurora
Lakewood
Rifle
Carbondale
Colorado
St Francis
Colby
Grand Junction
Green River
Aspen
Leadville
Mt Elbert
4398
Woodland Park
Castle Rock
Limon
Republican
Burlington
Goodland
Moab
Mount Peale
3877
Delta
Gunnison
Montrose
Sawatch Ra.
Colorado Springs
Kansas
WaKeeney
Hanksville
Salida
Canon City
Pikes Peak
4301
Cheyenne Wells
Scott City
Monticello
Pueblo
Rocky Ford
Lamar
Syracuse
Garden City
Blanding
Silverton
Rio Grande
Fowler
La Junta
Arkansas
San Juan
Bluff
Monte Vista
Sangre de Cristo Range
Cortez
Durango
San Juan Mountains
Alamosa
Walsenburg
Ulysses
Dodge City
Pagosa Springs
Chama
Springfield
Cimarron
Ashland
Kayenta
Shiprock
Farmington
Trinidad
Raton
Liberal
A
B
C
1
2
3
110°
105°
45°
40°
Longitude 105° west of Greenwich
Lambert Azimuthal Equal Area Projection
1 : 11 000 000
MILES 0
100
200

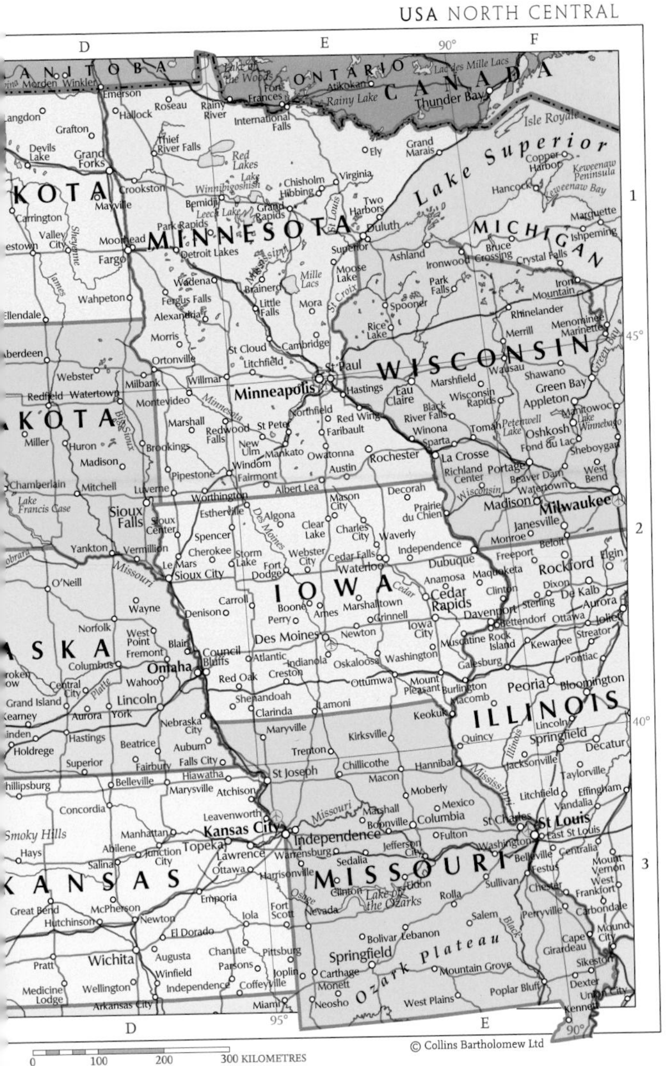

© Collins Bartholomew Ltd

0   100   200   300 KILOMETRES

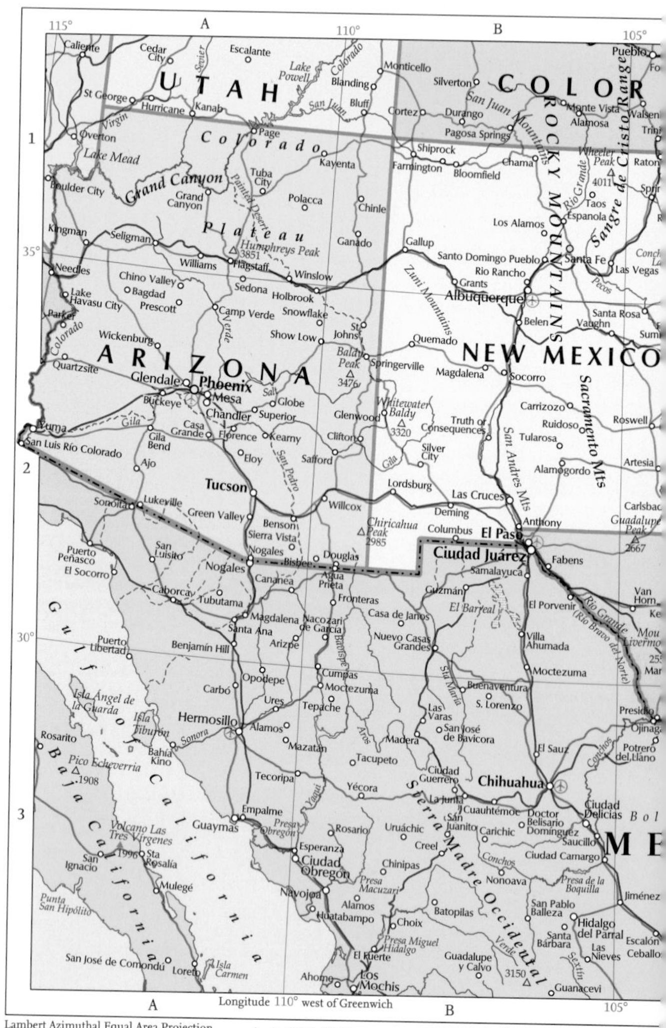

Lambert Azimuthal Equal Area Projection

1 : 11 000 000

MILES 0   100   200

Longitude 110° west of Greenwich

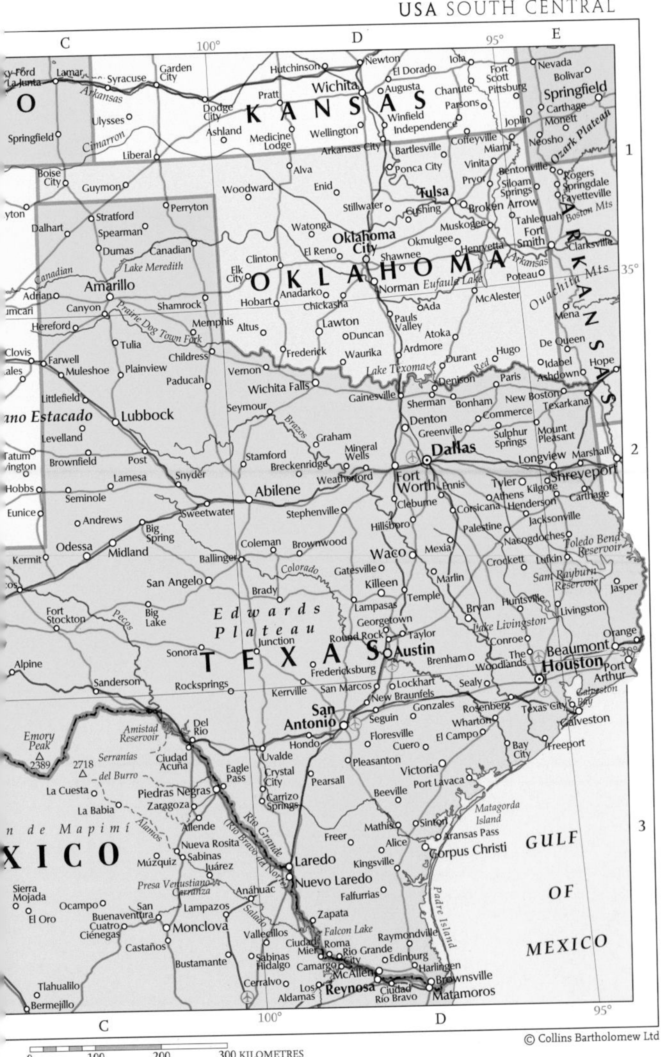

© Collins Bartholomew Ltd

0   100   200   300 KILOMETRES

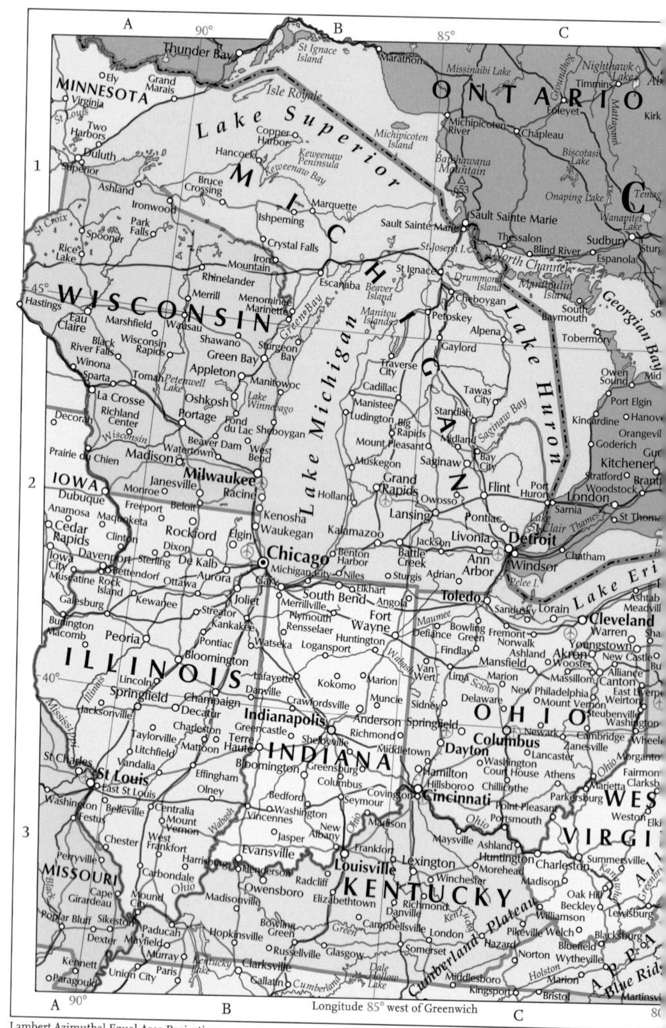

Lambert Azimuthal Equal Area Projection

Longitude 85° west of Greenwich

1 : 11 000 000

MILES 0   100   200

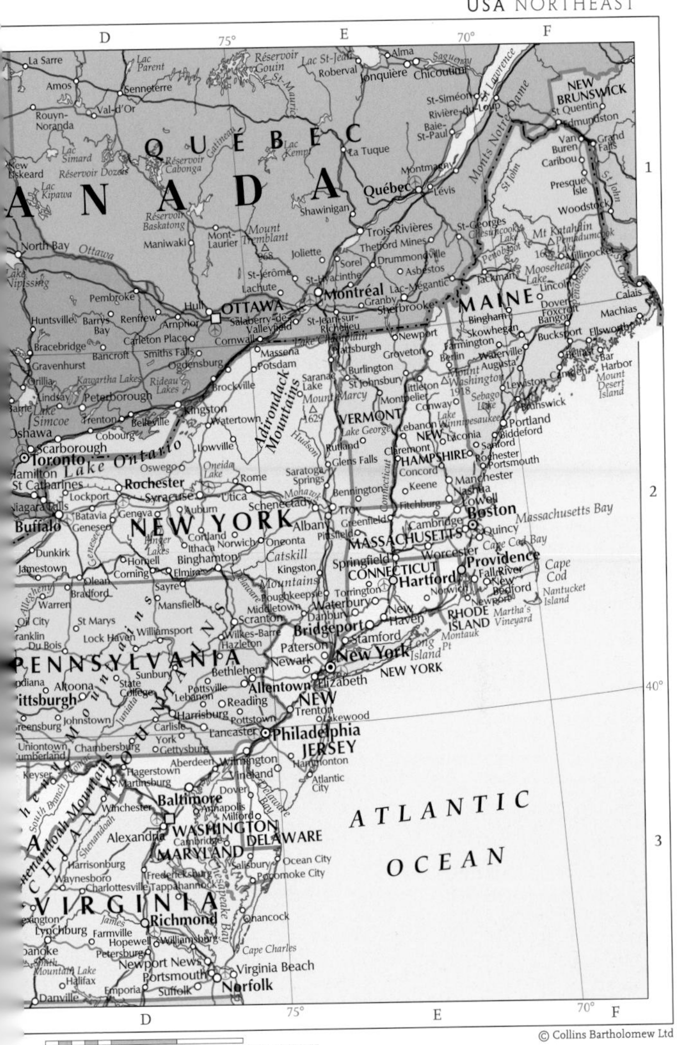
D
75°
E
70°
F
1
2
3
40°
CANADA
QUÉBEC
La Sarre
Lac Parent
Réservoir Gouin
Lac St-Jean
Alma
Roberval
Jonquière
Chicoutimi
Saguenay
St Lawrence
Amos
Senneterre
St-Maurice
NEW BRUNSWICK
St Quentin
Edmundston
Rouyn-Noranda
Val-d'Or
St-Siméon
Rivière-du-Loup
Baie-St-Paul
Monts Notre Dame
Grand Falls
Van Buren
Caribou
Lac Simard
Gatineau
Réservoir Cabonga
Lac Kempt
La Tuque
New Liskeard
Réservoir Dozois
Montmagny
Québec
Lévis
St John
Presque Isle
Lac Kipawa
Shawinigan
Woodstock
Réservoir Baskatong
Mont-Laurier
Mount Tremblant
Trois-Rivières
St-Georges
Mt Katahdin
North Bay
Ottawa
Maniwaki
968
Joliette
Thetford Mines
Drummondville
Asbestos
Lake Nipissing
St-Jérôme
Sorel
St-Hyacinthe
Millinocket
Moosehead Lake
Lachute
Montréal
Lac-Mégantic
Jackman
Lincoln
Penobscot
Calais
Pembroke
Hull
OTTAWA
Granby
Sherbrooke
MAINE
Dover-Foxcroft
Bangor
Machias
Huntsville
Barrys Bay
Renfrew
Arnprior
Salaberry-de-Valleyfield
St-Jean-sur-Richelieu
Bingham
Skowhegan
Bucksport
Ellsworth
Bracebridge
Carleton Place
Cornwall
Lake Champlain
Newport
Farmington
Waterville
Augusta
Bar Harbor
Gravenhurst
Bancroft
Smiths Falls
Massena
Plattsburgh
Groveton
Berlin
Mount Desert Island
Orillia
Kawartha Lakes
Rideau Lakes
Ogdensburg
Potsdam
Burlington
Mount Washington
1918
Lewiston
Camden
Lindsay
Peterborough
Brockville
Saranac Lake
St Johnsbury
Littleton
Brunswick
Barrie
Lake Simcoe
Kingston
Adirondack Mountains
Mount Marcy
1629
Montpelier
Conway
Sebago Lake
Oshawa
Trenton
Belleville
Watertown
VERMONT
Lake Winnipesaukee
Portland
Cobourg
Lake George
Lebanon
Laconia
Biddeford
Scarborough
Toronto
Lake Ontario
Lowville
Hudson
Rutland
NEW HAMPSHIRE
Sanford
Hamilton
Oswego
Oneida Lake
Rome
Glens Falls
Claremont
Rochester
Portsmouth
St Catharines
Rochester
Saratoga Springs
Concord
Manchester
Niagara Falls
Lockport
Syracuse
Utica
Mohawk
Bennington
Keene
Nashua
Lowell
Batavia
Geneva
Auburn
Schenectady
Troy
Fitchburg
Boston
Massachusetts Bay
Buffalo
Geneseo
NEW YORK
Albany
Greenfield
Cambridge
Quincy
Finger Lakes
Cortland
Ithaca
Norwich
Oneonta
Pittsfield
MASSACHUSETTS
Cape Cod Bay
Dunkirk
Genesee
Hornell
Binghamton
Catskill
Worcester
Springfield
Providence
Cape Cod
Jamestown
Corning
Elmira
Kingston
CONNECTICUT
Hartford
Fall River
New Bedford
Olean
Sayre
Catskill Mountains
Torrington
Norwich
Nantucket Island
Allegheny
Bradford
Poughkeepsie
Waterbury
Newport
Warren
Mansfield
Delaware
Middletown
Danbury
New Haven
RHODE ISLAND
Martha's Vineyard
Oil City
St Marys
Scranton
Bridgeport
Franklin
Williamsport
Wilkes-Barre
Stamford
Montauk Pt
Du Bois
Lock Haven
Hazleton
Paterson
Long Island
PENNSYLVANIA
Newark
New York
NEW YORK
Bethlehem
Indiana
Altoona
Sunbury
State College
Pottsville
Allentown
Elizabeth
Pittsburgh
Lebanon
Reading
NEW
Juniata
Harrisburg
Trenton
Greensburg
Johnstown
Pottstown
Lakewood
Carlisle
Lancaster
Philadelphia
York
JERSEY
Uniontown
Chambersburg
Gettysburg
Cumberland
Aberdeen
Wilmington
Hammonton
Keyser
Hagerstown
Vineland
Martinsburg
Atlantic City
APPALACHIAN MOUNTAINS
Allegheny Mountains
South Branch Potomac
Baltimore
Dover
Delaware Bay
Winchester
Annapolis
Milford
Shenandoah Mountain
Shenandoah
WASHINGTON
ATLANTIC OCEAN
Alexandria
Cambridge
DELAWARE
MARYLAND
Salisbury
Ocean City
Harrisonburg
Pocomoke City
Waynesboro
Fredericksburg
Charlottesville
Tappahannock
Chesapeake Bay
VIRGINIA
Lexington
James
Richmond
Chincoteague
Lynchburg
Farmville
Hopewell
Williamsburg
Roanoke
Petersburg
Cape Charles
Smith Mountain Lake
Newport News
Virginia Beach
Halifax
Portsmouth
Emporia
Suffolk
Norfolk
Danville
0
100
200
300 KILOMETRES

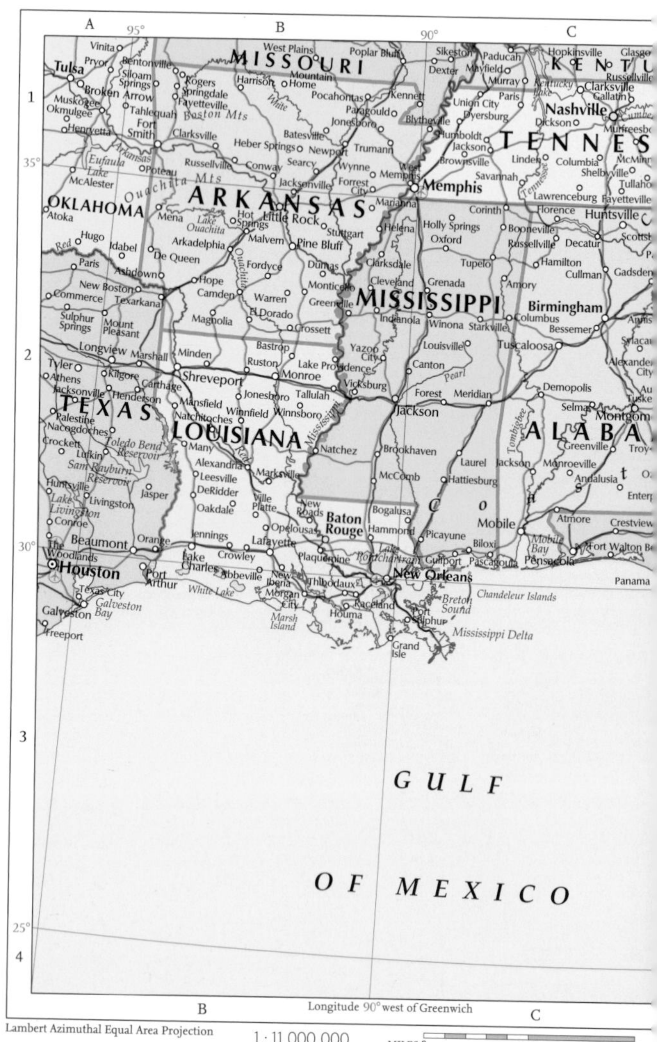

A   95°   B   90°   C

MISSOURI

Vinita
Tulsa Pryor
Siloam Springs
Broken Arrow Rogers Bentonville
Muskogee Springdale West Plains Poplar Bluff
Okmulgee Tahlequah Fayetteville Harrison Mountain Home Sikeston Paducah Hopkinsville Glasgow
Henryetta Fort Smith Boston Mts Pocahontas Dexter Kennett Union City Paris Murray Mayfield Clarksville Russellville
Clarksville White Paragould Blytheville Dyersburg Dickson **Nashville** Murfreesboro
OKLAHOMA Russellville Heber Springs Batesville Jonesboro Trumann Humboldt Jackson Linden Columbia Shelbyville McMinnville
Atoka Conway Newport Memphis Savannah Lawrenceburg Fayetteville Tullahoma
ARKANSAS Jacksonville Wynne Forrest Corinth Florence Huntsville Scotts
Mena Hot Little Marianna Helena Holly Springs Booneville Rossville Decatur
McAlester Springs Rock Stuttgart Oxford **TENNESSEE**
Hugo Arkadelphia Malvern Pine Bluff Clarksdale Tupelo Hamilton Cullman Gadsden
Idabel De Queen Hope Fordyce Dumas Grenada Amory Birmingham
New Boston Camden Warren Monticello Cleveland **MISSISSIPPI** Columbus Bessemer
Commerce Texarkana Magnolia El Dorado Greenville Indianola Winona Starkville Tuscaloosa **ALABAMA**
Longview Minden Crossett Yazoo Louisville Demopolis
Tyler Marshall Bastrop City Canton Selma Montgomery
Athens Kilgore Carthage Ruston Monroe Vicksburg Forest Meridian Greenville Troy
TEXAS Henderson Jonesboro Jackson Demopolis Andalusia
Palestine Mansfield Winnfield Winnsboro Natchez Brookhaven Laurel Jackson Monroeville Enter
Nacogdoches Natchitoches **LOUISIANA** McComb Hattiesburg
Crockett Toledo Bend Many Marksville Bogalusa Atmore Crestview
Lufkin Reservoir Alexandria New Mobile
Huntsville Sam Rayburn Leesville Roads Picayune Pascagoula Fort Walton Be
Conroe Reservoir DeRidder Ville Baton Hammond Biloxi Pensacola
Beaumont Jasper Oakdale Platte Rouge Plaquemine Gulfport Panama
Woodlands Orange Jennings Opelousas Bogalusa
Houston Lake Crowley Lafayette **New Orleans**
Texas City Port Charles Abbeville New Thibodaux
Galveston Arthur White Lake Iberia Morgan Raceland Breton
Galveston Bay Marsh City Houma Sound Chandeleur Islands
Freeport Island Grand Mississippi Delta
Isle

GULF

OF MEXICO

Longitude 90° west of Greenwich
B   C

142   Lambert Azimuthal Equal Area Projection   1 : 11 000 000   MILES 0   100   200

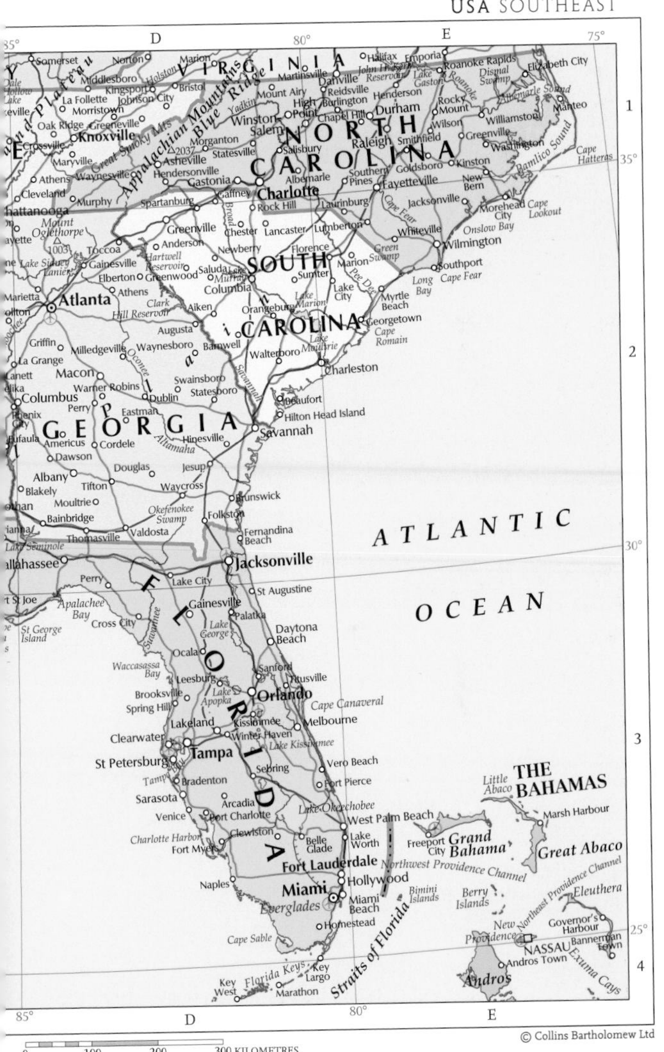

VIRGINIA
NORTH CAROLINA
SOUTH CAROLINA
GEORGIA
FLORIDA
ATLANTIC OCEAN
THE BAHAMAS
Appalachian Mountains
Blue Ridge
Great Smoky Mts
Knoxville
Asheville
Charlotte
Raleigh
Durham
Greensboro
Winston-Salem
Fayetteville
Wilmington
Columbia
Charleston
Savannah
Atlanta
Macon
Columbus
Augusta
Jacksonville
Tallahassee
Orlando
Tampa
St Petersburg
Miami
Fort Lauderdale
Hollywood
West Palm Beach
Cape Canaveral
Cape Hatteras
Pamlico Sound
Albemarle Sound
Straits of Florida
Florida Keys
Key West
Everglades
Lake Okeechobee
Grand Bahama
Great Abaco
Eleuthera
Andros
NASSAU
Northwest Providence Channel
Northeast Providence Channel
Exuma Cays
Marsh Harbour
Freeport City
Bimini Islands
Berry Islands
D
E
80°
75°
85°
35°
30°
25°
1
2
3
4
0
100
200
300 KILOMETRES

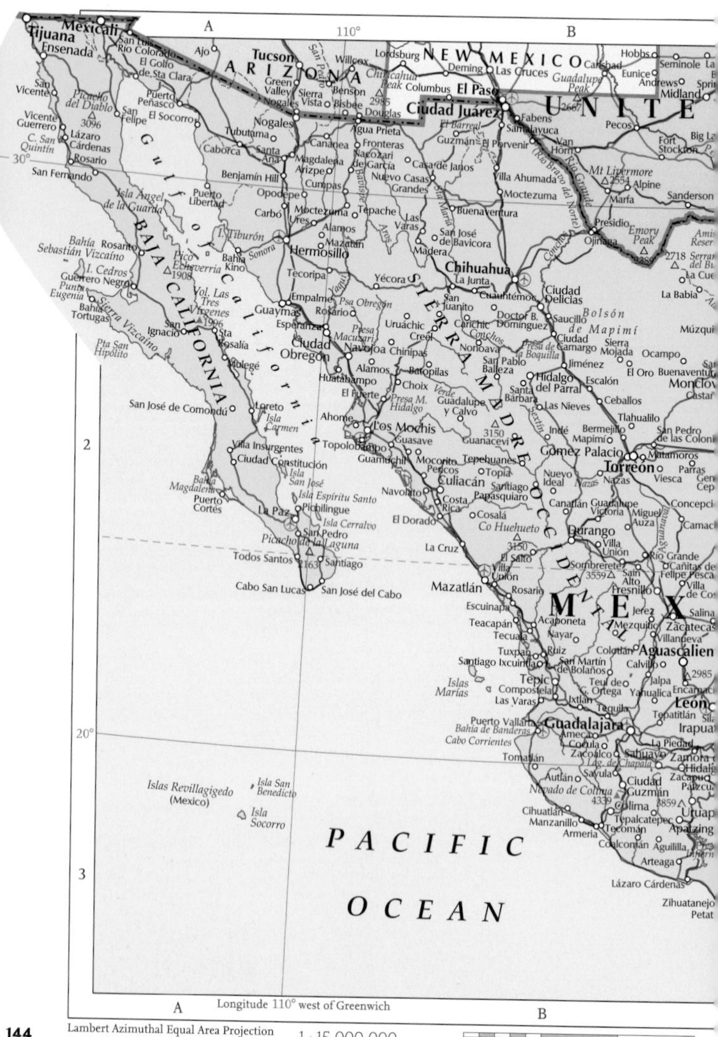

Map of northwestern Mexico and the southwestern United States.

Key place names and features:

United States (ARIZONA, NEW MEXICO): Tucson, Nogales, Green Valley, Sierra Vista, Benson, Willcox, Douglas, Lordsburg, Deming, Las Cruces, El Paso, Columbus, Hobbs, Carlsbad, Eunice, Seminole, Midland, Big Spring, Fort Stockton, Alpine, Marfa, Sanderson, Pecos, Andrews

Mexicali, Tijuana, Ensenada, San Vicente, Vicente Guerrero, Lázaro Cárdenas, C. San Quintín, San Fernando

BAJA CALIFORNIA — Golfo de California

Ciudad Juárez, Agua Prieta, Nuevo Casas Grandes, Chihuahua, Ciudad Delicias, Ciudad Camargo, Hidalgo del Parral, Jiménez, Escalón

Hermosillo, Guaymas, Ciudad Obregón, Navojoa, Los Mochis, Culiacán, Durango, Torreón, Gómez Palacio, Matamoros

SIERRA MADRE

La Paz, Cabo San Lucas, San José del Cabo, Todos Santos, Santiago

Mazatlán, Tepic, Guadalajara, León, Aguascalientes, Zacatecas, Irapuato, La Piedad, Ciudad Guzmán, Colima, Manzanillo, Lázaro Cárdenas, Zihuatanejo

MEXICO

Islas Revillagigedo (Mexico), Isla San Benedicto, Isla Socorro, Islas Marías

PACIFIC OCEAN

110°, 30°, 20°

A · 110° · B

Longitude 110° west of Greenwich

144

Lambert Azimuthal Equal Area Projection

1 : 15 000 000

MILES 0 · 100 · 200 · 300

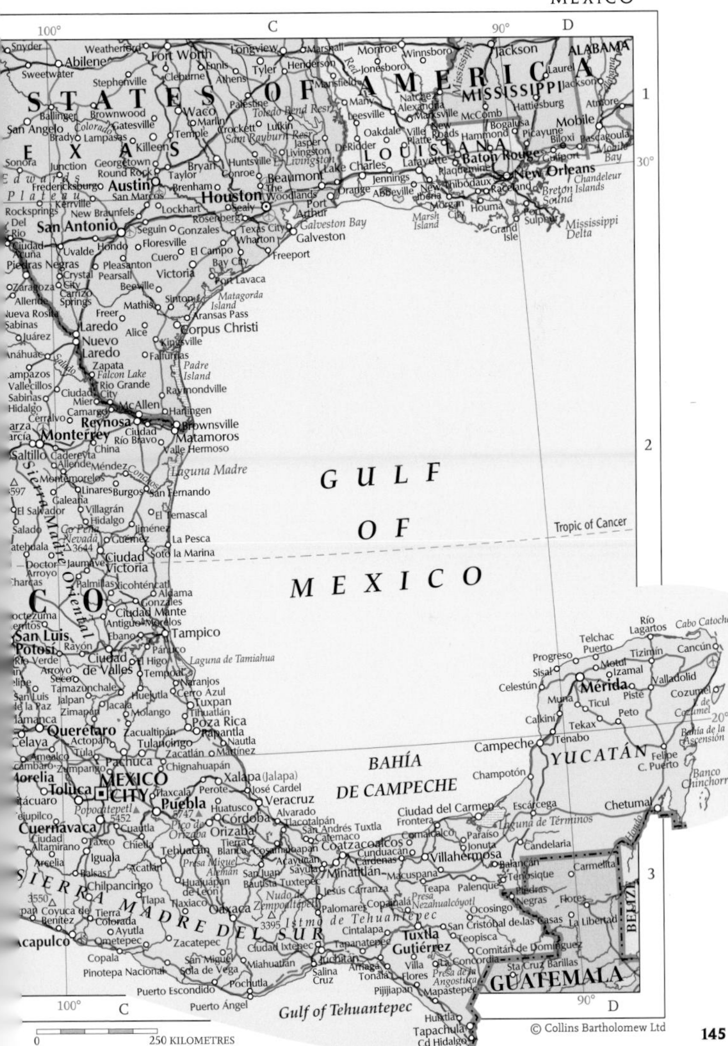

MEXICO

GULF OF MEXICO

GULF OF
MEXICO

BAHÍA
DE CAMPECHE

YUCATÁN

UNITED STATES OF AMERICA

TEXAS

LOUISIANA

MISSISSIPPI

ALABAMA

Tropic of Cancer

Laguna Madre

SIERRA MADRE DEL SUR

Gulf of Tehuantepec

GUATEMALA

BELIZE

© Collins Bartholomew Ltd

145

0   250 KILOMETRES

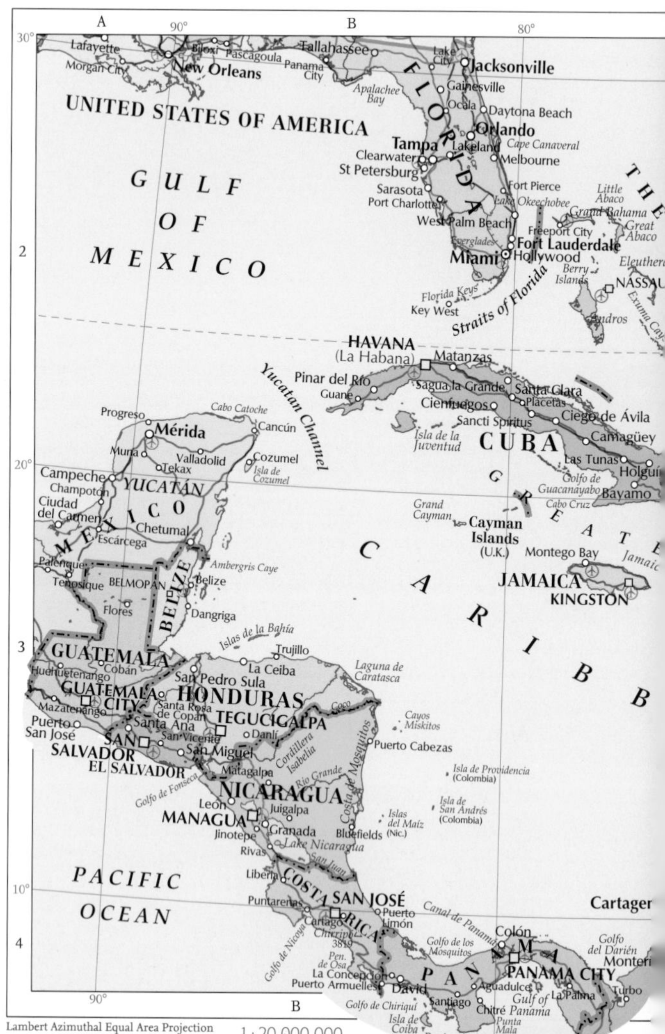

UNITED STATES OF AMERICA

A · 90° · B · 80°

30°

Lafayette · Biloxi · Pascagoula · Tallahassee · Lake City · Jacksonville
Morgan City · New Orleans · Panama City · Gainesville
Apalachee Bay · Ocala · Daytona Beach
FLORIDA · Orlando
Tampa · Lakeland · Cape Canaveral
Clearwater · Melbourne
St Petersburg · Fort Pierce
Sarasota · Lake Okeechobee
Port Charlotte · West Palm Beach · Little Abaco · Grand Bahama
Everglades · Freeport City · Great Abaco
Fort Lauderdale
Miami · Hollywood · Eleuthera
Berry Islands · NASSAU
Florida Keys · Exuma Cays
Key West · Straits of Florida · Andros

GULF

OF

MEXICO

2

HAVANA
(La Habana) · Matanzas
Pinar del Río · Sagua la Grande · Santa Clara
Guane · Cienfuegos · Placetas
Sancti Spíritus · Ciego de Ávila
Progreso · Cabo Catoche · Camagüey
Mérida · Cancún · CUBA
Muna · Valladolid · Isla de la Juventud · Las Tunas
Campeche · Tekax · YUCATÁN · Holguí
Champotón · Cozumel · Golfo de Guacanayabo · Bayamo
Ciudad del Carmen · Isla de Cozumel · Cabo Cruz
Escárcega · Chetumal · GREA
Palenque · Ambergris Caye · Grand Cayman · Cayman Islands (U.K.) · Montego Bay
Tenosique · BELMOPAN · Belize · JAMAICA · Jamaic
Flores · KINGSTON
Dangriga · CARIBB

3

Islas de la Bahía · Trujillo
GUATEMALA · Coban · La Ceiba · Laguna de Caratasca
Huehuetenango · San Pedro Sula · HONDURAS
GUATEMALA CITY · Santa Rosa de Copán · TEGUCIGALPA · Coco
Mazatenango · Santa Ana · San Vicente · Danlí · Cayos Miskitos
Puerto · San José · SAN · San Miguel · Cordillera Isabelia · Puerto Cabezas
SAN SALVADOR · Matagalpa · Río Grande · Isla de Providencia (Colombia)
EL SALVADOR · Golfo de Fonseca · NICARAGUA
León · Juigalpa · Isla de San Andrés (Colombia)
MANAGUA · Granada · Islas del Maíz (Nic.)
Jinotepe · Bluefields
Rivas · Lake Nicaragua

PACIFIC

OCEAN

10°

Liberia · SAN JOSÉ · Cartager
Puntarenas · Puerto Limón
COSTA RICA · Canal de Panamá · Colón · Golfo del Darién · Monter
Golfo de Nicoya · Chiriquí · Golfo de los Mosquitos · Turbo
Pen. de Osa · La Concepción · David · PANAMA · Punta La Palma
Puerto Armuelles · Santiago · Chitré · Panamá · PANAMA CITY
Golfo de Chiriquí · Aguadulce
Isla de Coiba · Punta Mala
Península de Azuero · 80°

4

B

Lambert Azimuthal Equal Area Projection · 1 : 20 000 000

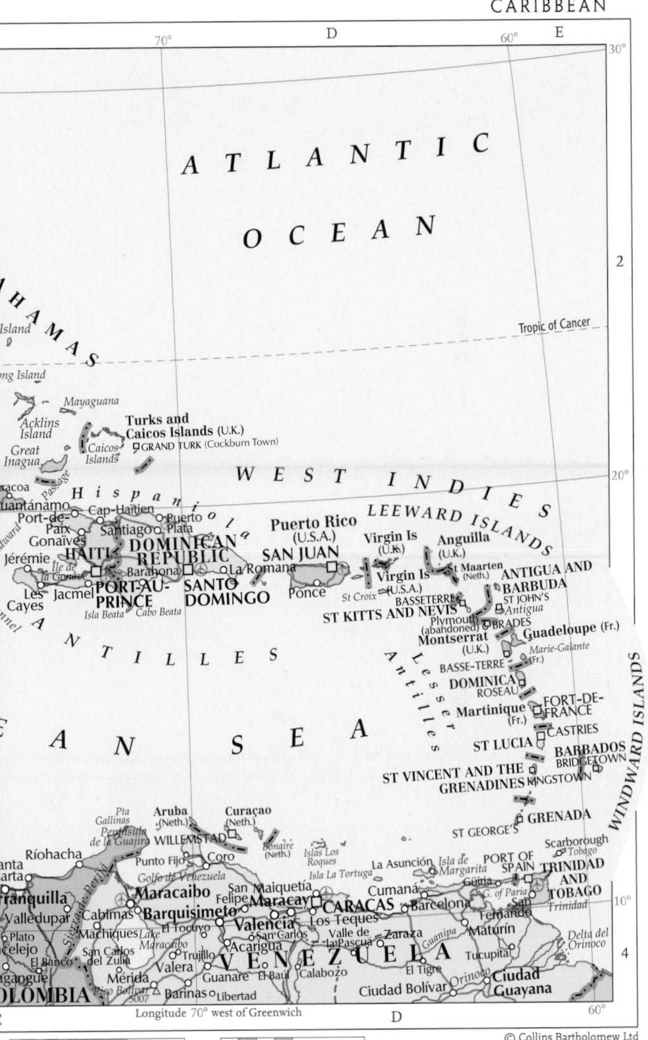

ATLANTIC

OCEAN

Tropic of Cancer

70° D 60° E

30°

2

20°

W E S T I N D I E S

LEEWARD ISLANDS

BAHAMAS

st Island

Long Island

Mayaguana

Acklins
Island

Great
Inagua

Turks and
Caicos Islands (U.K.)
GRAND TURK (Cockburn Town)

Caracoa

Guantánamo

Port-de-
Paix

Gonaïves

Jérémie

Les
Cayes

Jacmel

Île de
la Gonâve

Isla Beata

Cabo Beata

HAITI

PORT-AU-
PRINCE

Cap-Haïtien

Santiago

Barahona

SANTO
DOMINGO

Puerto
Plata

DOMINICAN
REPUBLIC

La Romana

Ponce

Puerto Rico
(U.S.A.)

SAN JUAN

St Croix

Virgin Is
(U.K.)

Virgin Is
(U.S.A.)

BASSETERRE

Anguilla
(U.K.)

St Maarten
(Neth.)

ANTIGUA AND
BARBUDA

ST JOHN'S

Antigua

ST KITTS AND NEVIS

Plymouth
(abandoned)

Montserrat

BRADES

Guadeloupe (Fr.)

Marie-Galante
(Fr.)

BASSE-TERRE

DOMINICA

ROSEAU

FORT-DE-
FRANCE

Martinique
(Fr.)

ST LUCIA

CASTRIES

BARBADOS

BRIDGETOWN

ST VINCENT AND THE
GRENADINES

KINGSTOWN

GRENADA

ST GEORGE'S

Hispaniola

Windward Passage

ANTILLES

EAN SEA

Lesser Antilles

WINDWARD ISLANDS

Aruba
(Neth.)

Curaçao
(Neth.)

Bonaire
(Neth.)

Islas Los
Roques
(Neth.)

Isla La Tortuga

Isla de
Margarita

A Asunción

Tobago

Scarborough

TRINIDAD
AND
TOBAGO

PORT OF
SPAIN

Trinidad

G. of Paria

Pta
Gallinas

Península
de la Guajira

Ríohacha

Santa
Marta

rranquilla

Valledupar

ncelejo

El Banco

Plato

San Carlos
del Zulia

COLOMBIA

Mérida

Valera

Trujillo

Punto Fijo

Coro

WILLEMSTAD

Golfo de Venezuela

Maracaibo

Cabimas

Machiques

Lago de
Maracaibo

San Felipe

El Tocuyo

Barquisimeto

Acarigua

Guanare

Barinas

San
Bolívar
5007

Libertad

Maiquetía

CARACAS

Valencia

Valle de
la Pascua

Zaraza

Calabozo

El Baúl

El Tigre

Los Teques

San Carlos

VENEZUELA

Ciudad Bolívar

Cumaná

Barcelona

San
Fernando

Maturín

Tucupita

Güiria

Delta del
Orinoco

Ciudad
Guayana

Orinoco

Longitude 70° west of Greenwich

D

60°

10°

4

ES 0   100   200   300   0   200   400 KILOMETRES

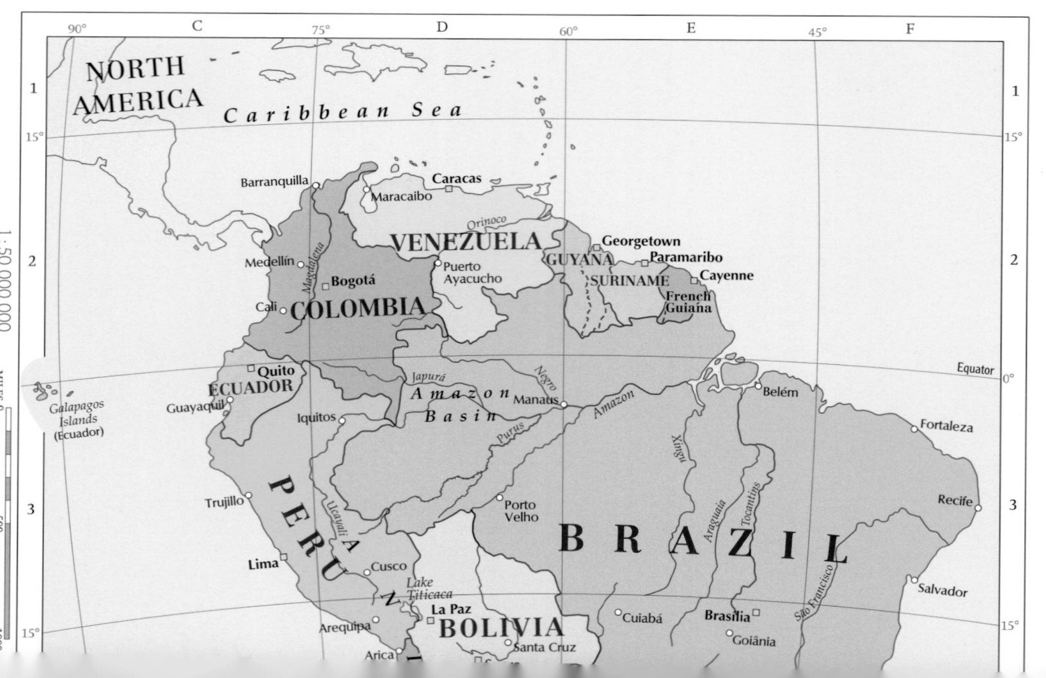
NORTH AMERICA
Caribbean Sea
Equator
BRAZIL
VENEZUELA
COLOMBIA
GUYANA
SURINAME
French Guiana
ECUADOR
PERU
BOLIVIA
Amazon Basin
Caracas
Maracaibo
Barranquilla
Medellín
Bogotá
Cali
Quito
Guayaquil
Georgetown
Paramaribo
Cayenne
Belém
Fortaleza
Recife
Salvador
Brasília
Goiânia
Cuiabá
Manaus
Porto Velho
Santa Cruz
La Paz
Lake Titicaca
Cusco
Arequipa
Arica
Lima
Trujillo
Iquitos
Puerto Ayacucho
Galapagos Islands (Ecuador)
Orinoco
Magdalena
Negro
Japurá
Purus
Amazon
Xingu
Araguaia
Tocantins
São Francisco
Ucayali
MILES 0
500
1000

1 : 50 000 000

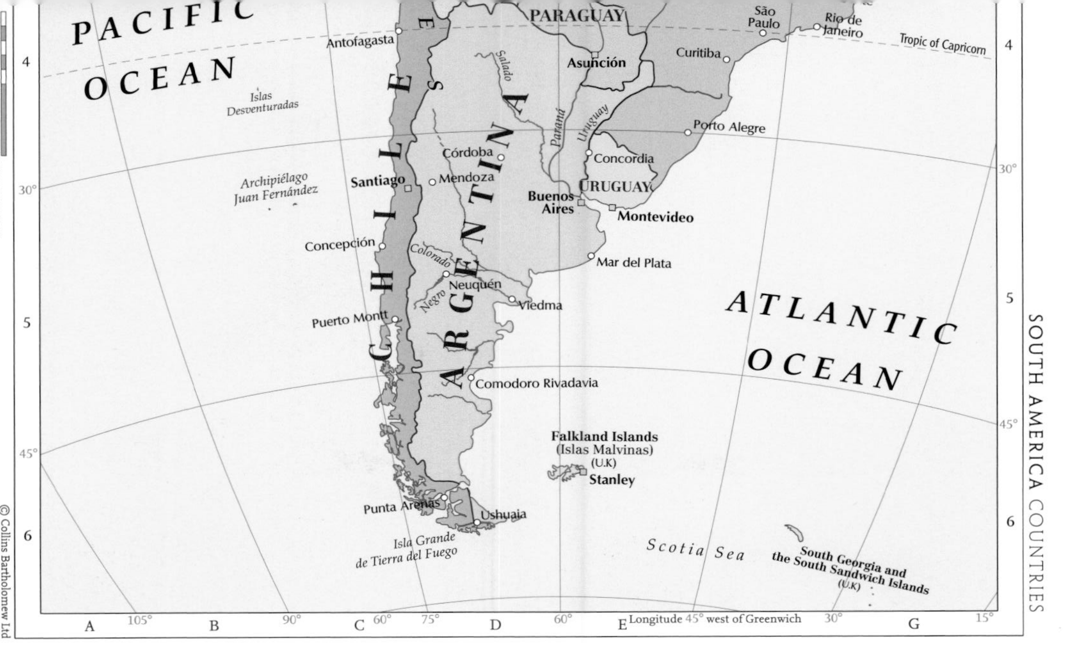

PACIFIC OCEAN

ATLANTIC OCEAN

Tropic of Capricorn

Rio de Janeiro

São Paulo

Curitiba

Porto Alegre

PARAGUAY

Asunción

Pilcomayo

Paraguay

Paraná

Salado

CHILE

ARGENTINA

Concordia

URUGUAY

Montevideo

Mar del Plata

Buenos Aires

Córdoba

Mendoza

Neuquén

Negro

Colorado

Viedma

Comodoro Rivadavia

Falkland Islands
(Islas Malvinas)
(UK)
Stanley

South Georgia and
the South Sandwich Islands
(UK)

Scotia Sea

E Longitude 45° west of Greenwich

Antofagasta

Santiago

Concepción

Puerto Montt

Punta Arenas

Ushuaia

Isla Grande
de Tierra del Fuego

Islas Desventuradas

Archipiélago Juan Fernández

0   500   1000 KILOMETRES

© Collins Bartholomew Ltd   **149**

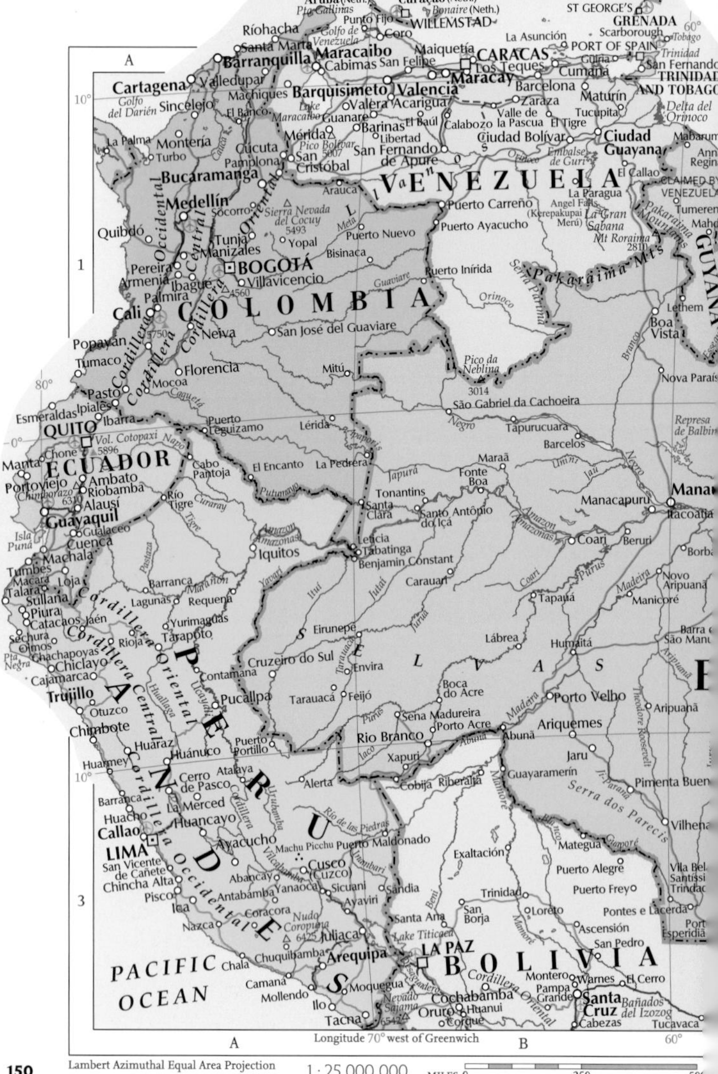

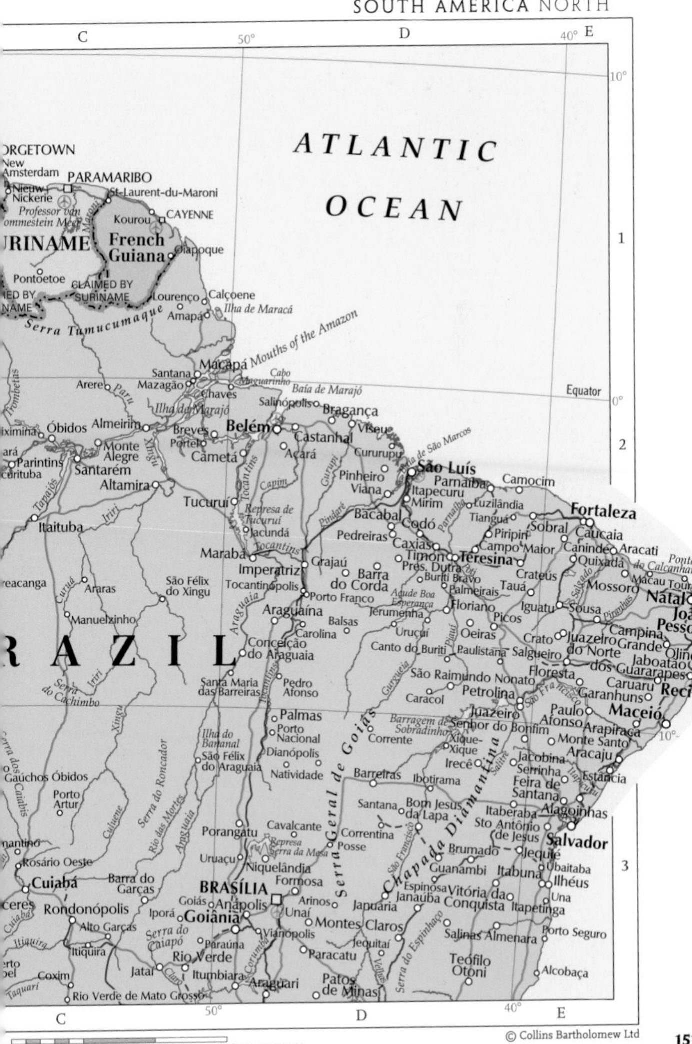

ATLANTIC

OCEAN

ORGETOWN
New
Amsterdam
Nickerie   PARAMARIBO
Nieuw   St-Laurent-du-Maroni
Professor van   Kourou
ommestein Mect   CAYENNE
URINAME   French
Guiana   Oiapoque

Pontoetoe
IED BY   CLAIMED BY
NAME   SURINAME   Lourenço   Calçoene
Serra Tamucumaque   Amapá   Ilha de Maracá

Arere   Mazagão   Macapá   Cabo
Santana   Chaves   Baía de Marajó
mbetas   Paru   Salinópolis   Bragança
iximina   Óbidos   Almeirim   Ilha de Marajó   VBeu
ará   Monte   Breves   Belém   Castanhal
Parintins   Alegre   Portel   Acará   Cururupu   Baía de São Marcos
arituba   Santarém   Câmeta   Pinheiro   São Luís   Camocim
Altamira   Tucuruí   Viana   Parnaíba   Camocim
Itaituba   Represa de   Bacabal   Itapecuru   Tutóia
Tucuruí   Mirim   Odó   Caxias   Sobral   Fortaleza
Marabá   Pedreiras   Timon   Caninde   Aracati
reacanga   Imperatriz   Grajaú   Barra   Pres. Dutra   Teresina   Campo   Quixadá   Ponta do
Araras   Tocantinópolis   do Corda   Burti   Bravo   Maior   Crateús   Calcanhar
São Félix   Porto Franco   Jerumenha   Palmeiras   Tauá   Iguatu   Mossoró
Manuelzinho   do Xingu   Araguaína   Floriano   Oeiras   Picos   Sousa   Natal
RAZIL   Conceição   Balsas   Uruçuí   Canto do Buriti   Crato   Campina   João
do Araguaia   Carolina   São Raimundo Nonato   Juazeiro   Grande   Pessoa
Serra   Santa Maria   Pedro   Paulistana   do Norte   Jaboatão   Olinda
do Cachimbo   das Barreiras   Afonso   Petrolina   Floresta   Caruaru   Recife
Palmas   Paulo   Garanhuns
auchos   Óbidos   Porto   Barragem de   Afonso   Arapiraca   Maceió
Porto   Nacional   Sobradinho   Juazeiro   Monte Santo
do Roncador   Artur   Dianópolis   Xique-   Senhor do Bonfim   Aracaju
rra   Ilha do   Natividade   Xique   Irecê   Jacobina   Estância
dos Gaúchos   Bananal   Barreiras   Ibotirama   Serrinha   Alagoinhas
nantins   Rosário Oeste   Porangatu   Cavalcante   Santana   Bom Jesus   Feira de   Sto Antônio
Cuiabá   Barra do   Uruaçu   Correntina   da Lapa   Santana   de Jesus   Salvador
eres   Rondonópolis   Garças   Formosa   Posse   Guanambi   Brumado   Jequié   Itabuna   Ubaitaba
Alto Garças   Iporá   BRASÍLIA   Goiás   Anápolis   Unaí   Janaúba   Vitória da   Ilhéus
tiquira   Itiquira   Coxim   Goiânia   Vianópolis   Montes Claros   Januária   Conquista   Itapetinga   Una
rto   Rio Verde   Paraúna   Jequitaí   Salinas   Almenara   Porto Seguro
Coxim   Serra do   Itumbiara   Paracatu   Teófilo   Alcobaça
Jataí   Chapão   Araguari   Patos   Otoni
Rio Verde de Mato Grosso   Rio Verde   de Minas

© Collins Bartholomew Ltd

0   250   500   750 KILOMETRES

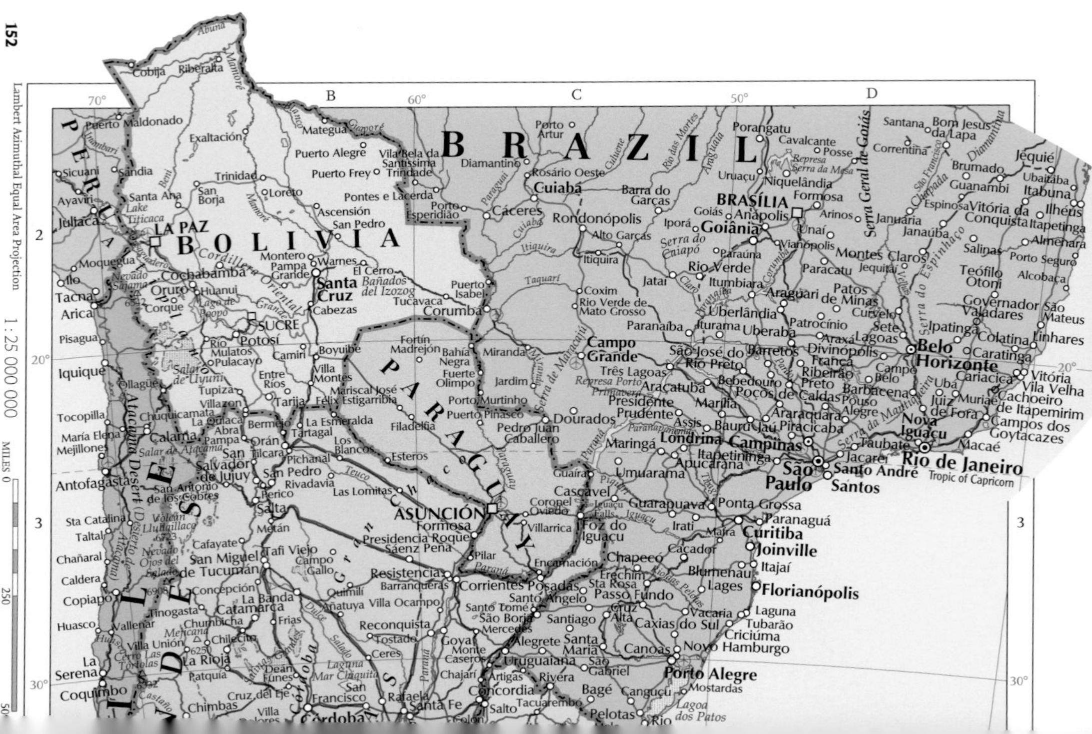

Lambert Azimuthal Equal Area Projection

1 : 25 000 000

MILES 0   250   50

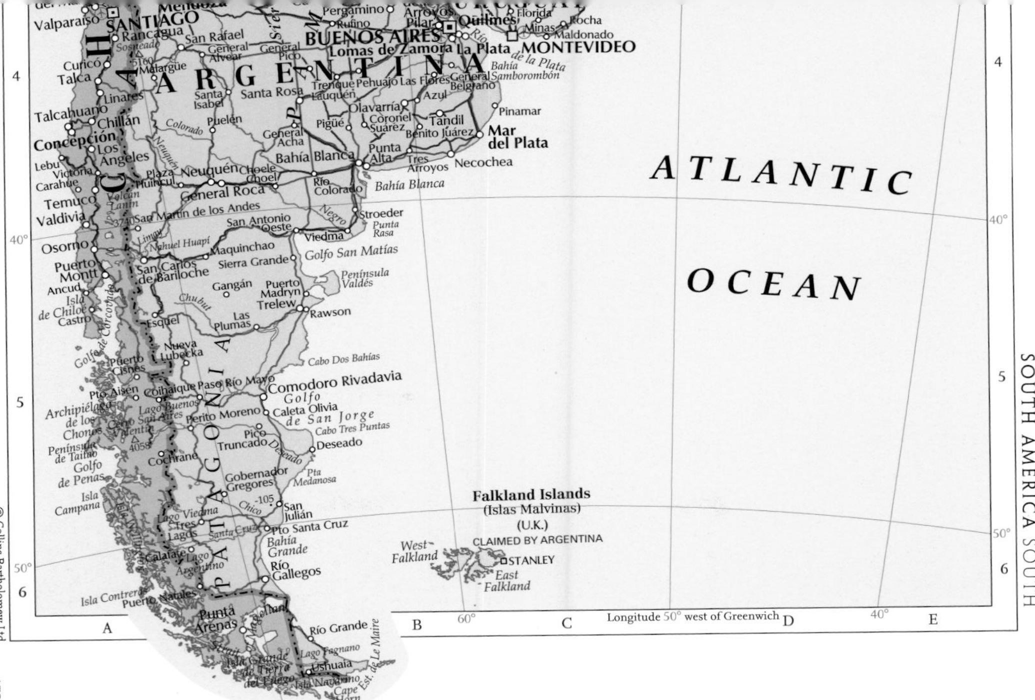

ATLANTIC

OCEAN

Falkland Islands
(Islas Malvinas)
(U.K.)
CLAIMED BY ARGENTINA

STANLEY

West
Falkland

East
Falkland

Longitude 50° west of Greenwich

Valparaíso
SANTIAGO
Rancagua San Rafael
Curicó
Sagrada
Talca
Talcahuano Chillán
Concepción Los Ángeles
Lebu
Victoria
Carahue Temuco
Valdivia Lanín
Osorno
Puerto
Montt Ancud
Isla de Chiloé

Mendoza
Pergamino
Junín
San Luis
Santa Rosa
General
Acha
Neuquén
Plaza Huincul
San Martín de los Andes
Lago Nahuel Huapí San Carlos
de Bariloche
Esquel

BUENOS AIRES
Quilmes
La Plata
Chascomús
Dolores

Pigüe
Coronel
Suárez Tandil
Olavarría Azul
Benito Juárez
Tres
Arroyos Necochea
Bahía Blanca
Colorado
Punta
Alta

MONTEVIDEO

Mar
del Plata

Pinamar

Bahía
Blanca

Patagones Roca
Viedma
Golfo San Matías
Península
Valdés

ARGENTINA

A R G E N T I N A

General Roca
Andes
Catriel
General
Conesa
San Antonio
Oeste

Sierra Grande

Gangán Chubut

Puerto
Madryn
Rawson
Trelew

Cabo Dos Bahías

Las
Plumas

Comodoro Rivadavia
Golfo
Pico
Salamanca San
Jorge
Caleta Olivia Cabo Tres Puntas

Nueva
Lubecka
Paso de
Indios

Sarmiento
Colhué
Huapí
Perito Moreno
Deseado

Río Mayo

PATAGONIA

P A T A G O N I A

Puerto
Aisén
Coihaique

Lago
Buenos Aires
Pico
Truncado
–105
Puerto
Deseado

Cochrane

Archipiélago
de los
Chonos

Lago
Cardiel
Gobernador
Gregores

San
Julián

Bahía
Grande

Golfo
de Penas

Península
de
Taitao

Lago
Viedma
Río Chico
Lago
Argentino

Puerto
Santa Cruz
Río
Santa Cruz

Isla
Campana

Puerto
Natales

Río
Gallegos

Isla
Contreras

Punta
Arenas

Río Grande

Est. de La Maire

Ushuaia

Isla de Chiloé

© Collins Bartholomew Ltd

0   250   500   750 KILOMETRES

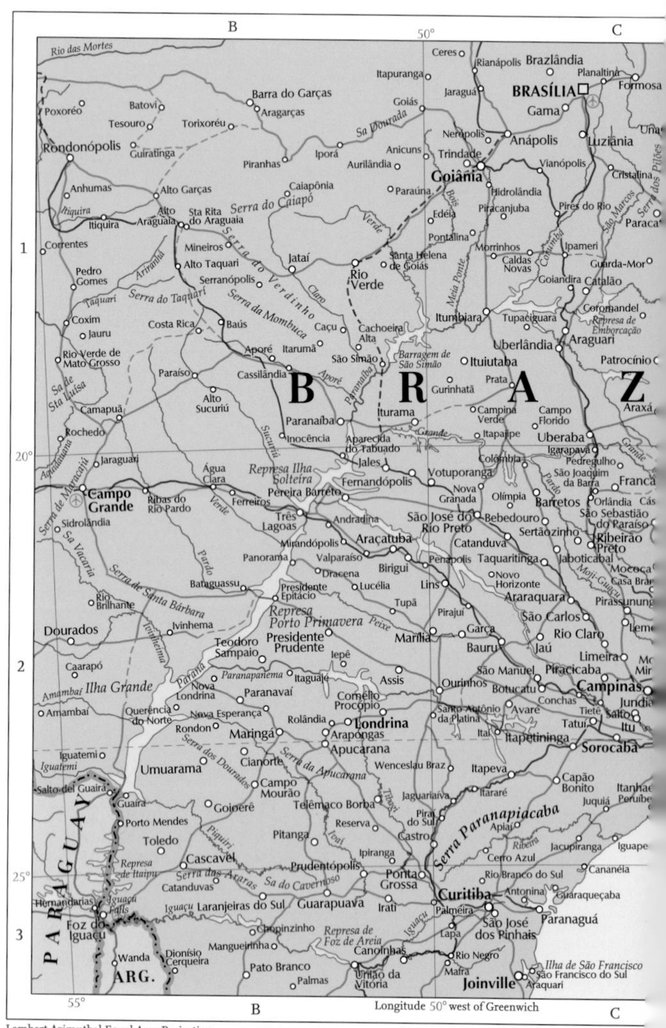

Map of central-southern Brazil

Rio das Mortes

B — 50° — C

Poxoréo · Batoví · Barra do Garças · Jaraguá · Planaltina · Brazlândia · Ceres · Rianápolis · Formosa

BRASÍLIA · Gama · Uña

Tesouro · Torixoréu · Aragarças · Sa Bonsuca · Goiás · Itapuranga · Nemópolis · Anápolis · Luziânia · Cristalina

Rondonópolis · Guiratinga · Piranhas · Iporá · Anicuns · Trindade · Vianópolis · Cristalina · Sa dos Pilões

Anhumas · Alto Garças · Caiapônia · Aurilândia · Paraúna · **Goiânia** · Hidrolândia · Pires do Rio · Paraca

Itiquira · Alto Araguaia · Sta Rita do Araguaia · Serra do Caiapó · Verde · Edéia · Piracanjuba · Guarda-Mor

Corrientes · Mineiros · Santa Helena de Goiás · Pontalina · Morrinhos · Caldas Novas · Goiandira · Catalão · Comandel

Pedro Gomes · Alto Taquari · Jataí · Serra do Verdinho · Rio Verde · Itumbiara · Tupaciguara · Araguari · Represa de Emborcação · Patrocínio

Coxim · Costa Rica · Serra da Mombuca · Caçu · Cachoeira Alta · Barragem de São Simão · Uberlândia

Jauru · Baús · Aporé · Itarumã · São Simão · Ituiutaba · **B R A Z**

Rio Verde de Mato Grosso · Paraíso · Cassilândia · Gurinhatã · Araxá

Camapuã · Alto Sucuriú · Paranaíba · Iturama · Campina Verde · Campo Florido

Rochedo · Inocência · Aparecida do Tabuado · Itapagipe · Uberaba · Igarapava

20° · Jaraguari · Água Clara · Represa Ilha Solteira · Jales · Votuporanga · Colômbia · Pedregulho · São Joaquim da Barra · Franca

Campo Grande · Ribas do Rio Pardo · Pereira Barreto · Fernandópolis · Nova Granada · Olímpia · Barretos · Orlândia · Cás · Sebastião do Paraíso

Sidrolândia · Ferreiros · Três Lagoas · Andradina · São José do Rio Preto · Bebedouro · Sertãozinho · Ribeirão Preto

Mirandópolis · Araçatuba · Catanduva · Taquaritinga · Jaboticabal · Mococa · Casa Bra

Panorama · Valparaíso · Birigui · Penápolis · Novo Horizonte · Araraquara · Pirassununga

Baiaguassu · Dracena · Lucélia · Tupã · Lins · São Carlos · Rio Claro · Leme

Presidente Epitácio · Represa Porto Primavera · Pexe · Pirajuí · Garça · Jaú · Limeira · Mo

Dourados · Ivinhema · Teodoro Sampaio · Marília · Bauru · São Manuel · Piracicaba · Mir

2 · Presidente Prudente · Iepê · Assis · Ourinhos · Botucatu · Conchas · **Campinas**

Caarapó · Ilha Grande · Paranapanema · Itaguaí · Sto Antônio da Platina · Avaré · Tietê · Salto · Jundiaí

Amambaí · Nova Londrina · Paranavaí · Cornélio Procópio · **Londrina** · Tatuí · Itu · Sorocaba

Iguatemi · Querência do Norte · Nova Esperança · Rolândia · Arapongas · Itapetininga

Salto del Guaíra · Rondon · Maringá · Apucarana · Wenceslau Braz · Itapeva · Capão Bonito · Itanhaé

Guaíra · Cianorte · Serra da Apucarana · Jaguariaíva · Itararé · Juquiá · Perube

Umuarama · Goioerê · Campo Mourão · Telêmaco Borba · Reserva · Rio do Sul · Apiaí · Jacupiranga · Iguape

Toledo · Pitanga · Ipiranga · Castro · Serra Paranapiacaba · Cerro Azul

25° · Cascavel · Serra das Araras · Prudentópolis · Ponta Grossa · Rio Branco do Sul · Cananéia

Catanduvas · Sa do Cavernoso · Irati · **Curitiba** · Antonina · Guaraqueçaba

Hernandárias · Foz do Iguaçu · Iguaçu Falls · Laranjeiras do Sul · Guarapuava · Palmeira · **São José dos Pinhais** · Paranaguá

Wanda · Dionísio Cerqueira · Chopinzinho · Represa de Foz de Areia · Canoinhas · Lapa · Rio Negro · Ilha de São Francisco

ARG. · Pato Branco · União da Vitória · Mafra · **Joinville** · São Francisco do Sul

Mangueirinha · Palmas · Araquari

3 · **P A R A G U A Y**

55° · B · Longitude 50° west of Greenwich · C

154 · Lambert Azimuthal Equal Area Projection · 1 : 10 000 000 · MILES 0 · 100 · 200

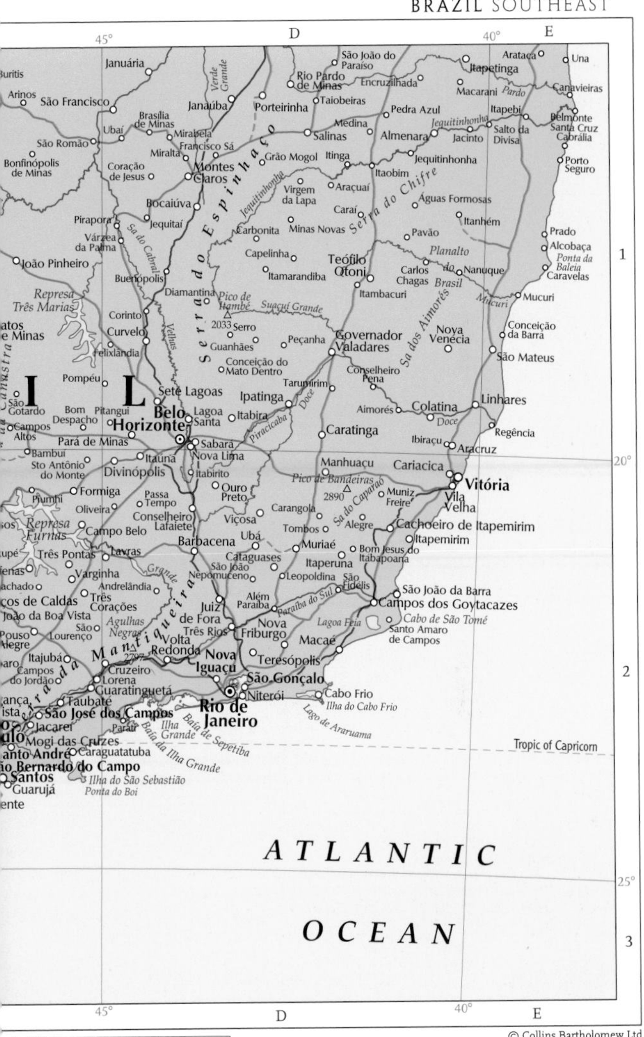

D
E
45°
40°
Januária
Janaúba
Porteirinha
São João do Paraíso
Rio Pardo de Minas
Encruzilhada
Taiobeiras
Arataca
Una
Itapetinga
Macarani
Canavieiras
São Francisco
Arinos
Brasília de Minas
Mirabela
Ubaí
São Romão
Francisco Sá
Miralta
Montes Claros
Bonfinópolis de Minas
Coração de Jesus
Pedra Azul
Medina
Salinas
Almenara
Jacinto
Salto da Divisa
Itapebi
Belmonte
Santa Cruz Cabrália
Porto Seguro
Grão Mogol
Itinga
Jequitinhonha
Itaobim
Virgem da Lapa
Araçuaí
Serra do Chifre
Águas Formosas
Bocaiúva
Pirapora
Jequitaí
Várzea da Palma
Carbonita
Minas Novas
Caraí
Itanhém
Pavão
Prado
Alcobaça
Ponta da Baleia
Caravelas
Capelinha
Teófilo Otoni
Carlos Chagas
Planalto do Brasil
Nanuque
João Pinheiro
Buenópolis
Itamarandiba
Itambacuri
Mucuri
Diamantina
Pico de Itambé
Suaçuí Grande
Represa Três Marias
Corinto
Curvelo
2033
Serro
Guanhães
Peçanha
Governador Valadares
Nova Venécia
Conceição da Barra
São Mateus
Felixlândia
Conceição do Mato Dentro
Conselheiro Pena
Pompéu
Sete Lagoas
Ipatinga
Tarumirim
Linhares
São Gotardo
Bom Despacho
Pitangui
Belo Horizonte
Lagoa Santa
Itabira
Aimorés
Colatina
Doce
Regência
Campos Altos
Pará de Minas
Sabará
Caratinga
Ibiraçu
Aracruz
Bambuí
Itaúna
Nova Lima
Manhuaçu
Cariacica
20°
Sto Antônio do Monte
Divinópolis
Itabirito
Pico de Bandeiras
2890
Vitória
Vila Velha
Muniz Freire
Piumhi
Formiga
Ouro Preto
Passa Tempo
Carangola
Oliveira
Conselheiro Lafaiete
Viçosa
Tombos
Alegre
Cachoeiro de Itapemirim
Represa Furnas
Campo Belo
Itapemirim
Barbacena
Ubá
Muriaé
Bom Jesus do Itabapoana
Três Pontas
Lavras
Cataguases
Itaperuna
Varginha
São João Nepomuceno
Leopoldina
São Fidélis
Andrelândia
Três Corações
Além Paraíba
São João da Barra
Campos dos Goytacazes
Juiz de Fora
Nova Friburgo
Lagoa Feia
Cabo de São Tomé
Santo Amaro de Campos
Agulhas Negras
São Lourenço
Três Rios
Macaé
Pouso Alegre
Volta Redonda
Nova Iguaçu
Itajubá
2797
Teresópolis
2
Campos do Jordão
Cruzeiro
Lorena
São Gonçalo
Guaratinguetá
Niterói
Cabo Frio
Ilha do Cabo Frio
Taubaté
Rio de Janeiro
São José dos Campos
Lago de Araruama
Jacareí
Paraty
Ilha Grande
Baía de Sepetiba
Mogi das Cruzes
Caraguatatuba
Tropic of Capricorn
Baía da Ilha Grande
Santo André
São Bernardo do Campo
Santos
Ilha do São Sebastião
Guarujá
Ponta do Boi
ATLANTIC
OCEAN
25°
1
3
300 KILOMETRES
0
100
200

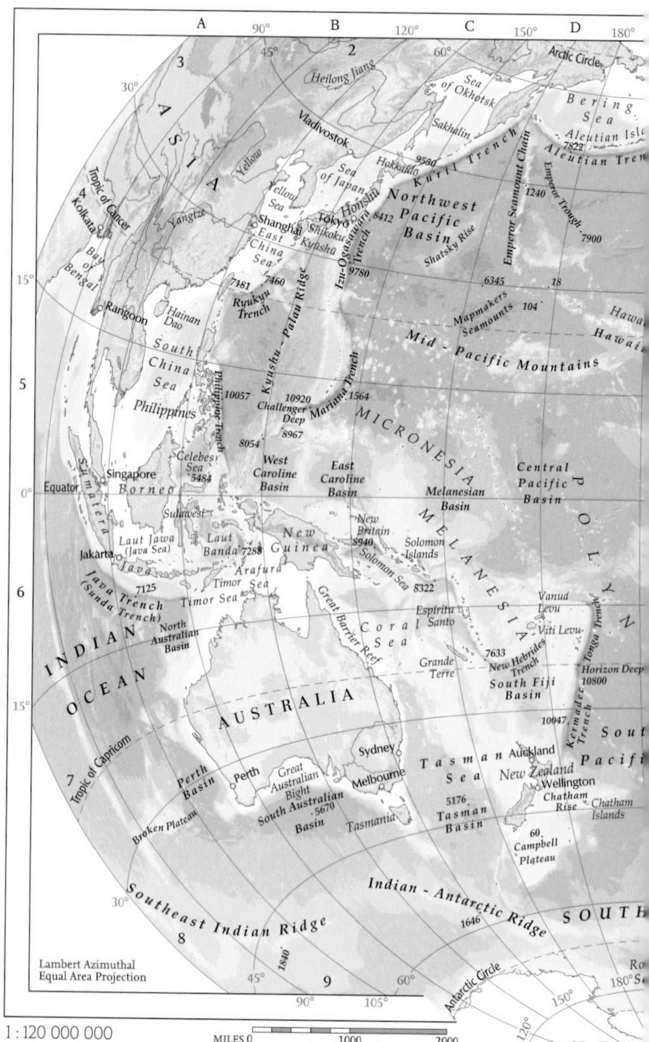

A 90° B 120° C 150° D 180°

3 45° 2

30°

A S I A

Tropic of Cancer

Kolkata

Bay of Bengal

Rangoon

Hainan Dao

South China Sea

Singapore

Borneo

Sumatra

Equator

Jakarta

Java

Laut Jawa (Java Sea)

Sulawesi

Laut Banda

Timor Sea

7125

Java Trench (Sunda Trench)

North Australian Basin

INDIAN

OCEAN

Tropic of Capricorn

AUSTRALIA

Perth Basin

Perth

Great Australian Bight

South Australian Basin

5670

Broken Plateau

Melbourne

Sydney

Tasmania

Tasman Sea

Southeast Indian Ridge

1840

30°

Heilong Jiang

Vladivostok

Yellow

Yangtze

Yellow Sea

Shanghai

East China Sea

Sea of Japan

Hokkaido

3550

Kuril Trench

Honshu

Tokyo

Shikoku

Kyushu

8412

Izu-Ogasawara Trench

9780

Northwest Pacific Basin

Shatsky Rise

1240

Emperor Seamount Chain

Emperor Trough

7822

Aleutian Isl.

Aleutian Trench

Bering Sea

Arctic Circle

7900

6345

18

Mapmakers Seamounts 104

Hawa

Hawai

Mid - Pacific Mountains

7181

7460

Ryukyu Trench

Kyushu - Palau Ridge

10057

Philippine Trench

Philippines

10920

Challenger Deep

8967

Mariana Trench

1564

8054

West Caroline Basin

East Caroline Basin

M I C R O N E S I A

Central Pacific Basin

P O L

Celebes Sea

5484

New Guinea

New Britain

8940

Solomon Sea

8322

Solomon Islands

Melanesian Basin

M E L A N E S I A

7288

Arafura Sea

Great Barrier Reef

Coral Sea

Espiritu Santo

Grande Terre

7633

South New Hebrides Trench Fiji Basin

Vanua Levu

Viti Levu

Kermadec Trench

Horizon Deep 10800

10047

South

Pacific

Tasman Sea

5176

Auckland

New Zealand

Wellington

Chatham Rise

Chatham Islands

Tasman Basin

60

Campbell Plateau

Indian - Antarctic Ridge

1646

S O U T H

Antarctic Circle

A N T A

Lambert Azimuthal Equal Area Projection

45° 9 60°

90° 105°

120°

150°

180°

Ro

MILES 0   1000   2000

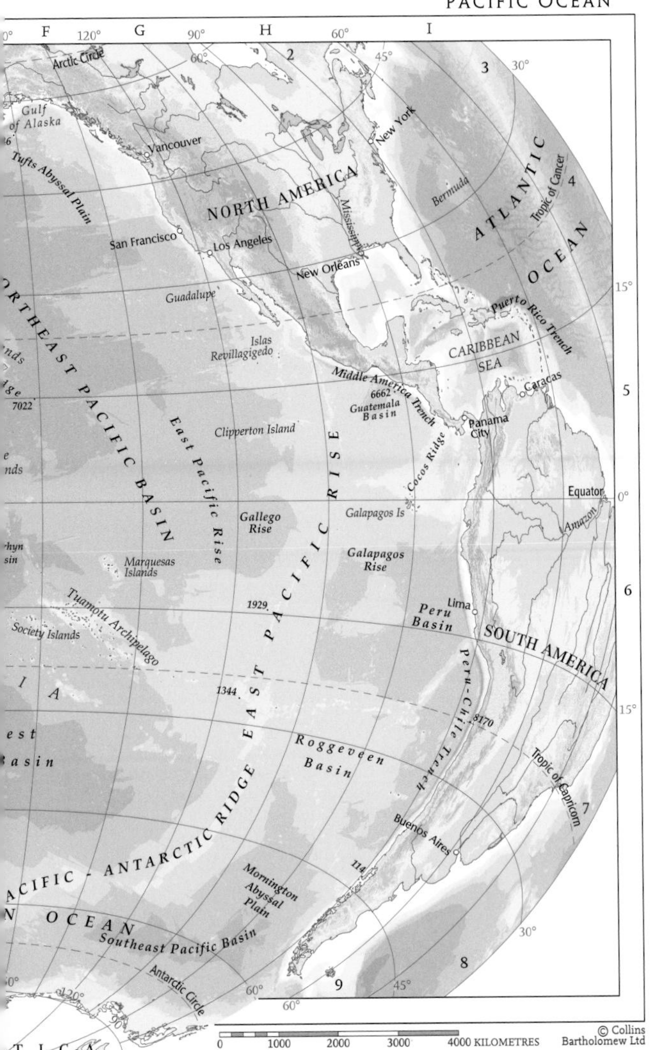

Arctic Circle
Gulf of Alaska
Tufts Abyssal Plain
Vancouver
NORTH AMERICA
San Francisco
Los Angeles
New York
Bermuda
ATLANTIC
Tropic of Cancer
Guadalupe
New Orleans
OCEAN
Islas Revillagigedo
CARIBBEAN SEA
Puerto Rico Trench
Middle America Trench
SOUTHEAST PACIFIC BASIN
7022
East Pacific Rise
Clipperton Island
6662
Guatemala Basin
Cocos Ridge
Panama City
Caracas
Gallego Rise
Galapagos Is
Equator
Amazon
Galapagos Rise
Marquesas Islands
1929
Lima
Peru Basin
SOUTH AMERICA
Tuamotu Archipelago
Society Islands
1344
Peru-Chile Trench
5170
Roggeveen Basin
Tropic of Capricorn
Buenos Aires
114
Mornington Abyssal Plain
PACIFIC - ANTARCTIC RIDGE
EAST PACIFIC RISE
OCEAN
Southeast Pacific Basin
Antarctic Circle
TICA

© Collins Bartholomew Ltd

0   1000   2000   3000   4000 KILOMETRES

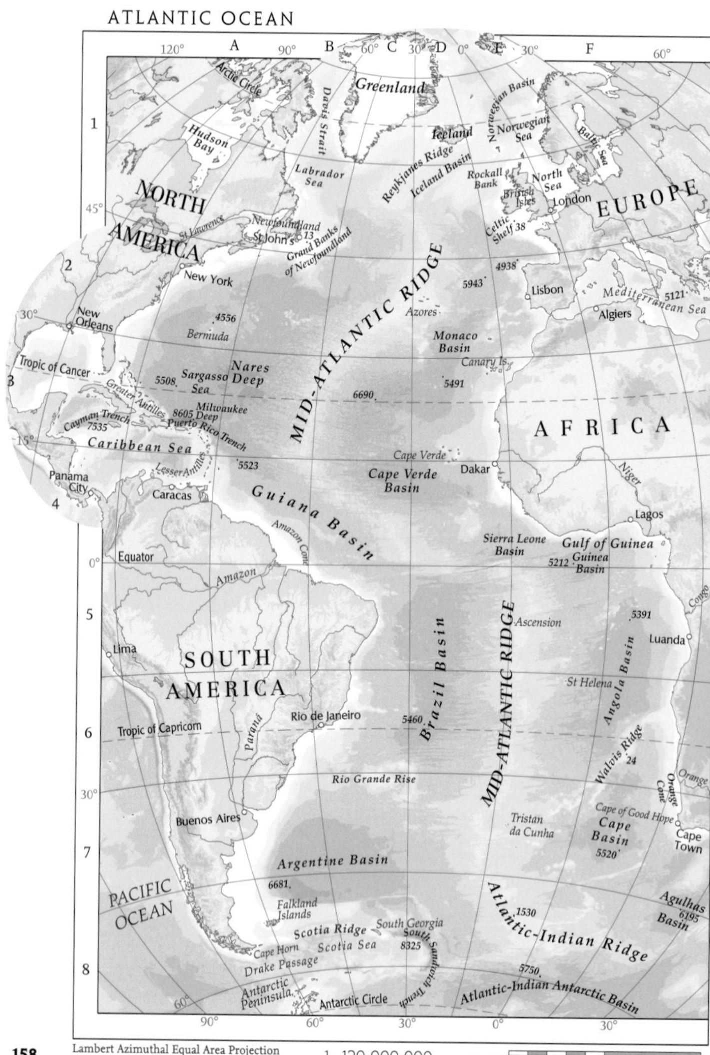

Lambert Azimuthal Equal Area Projection

1 : 120 000 000

MILES 0   1000   2000

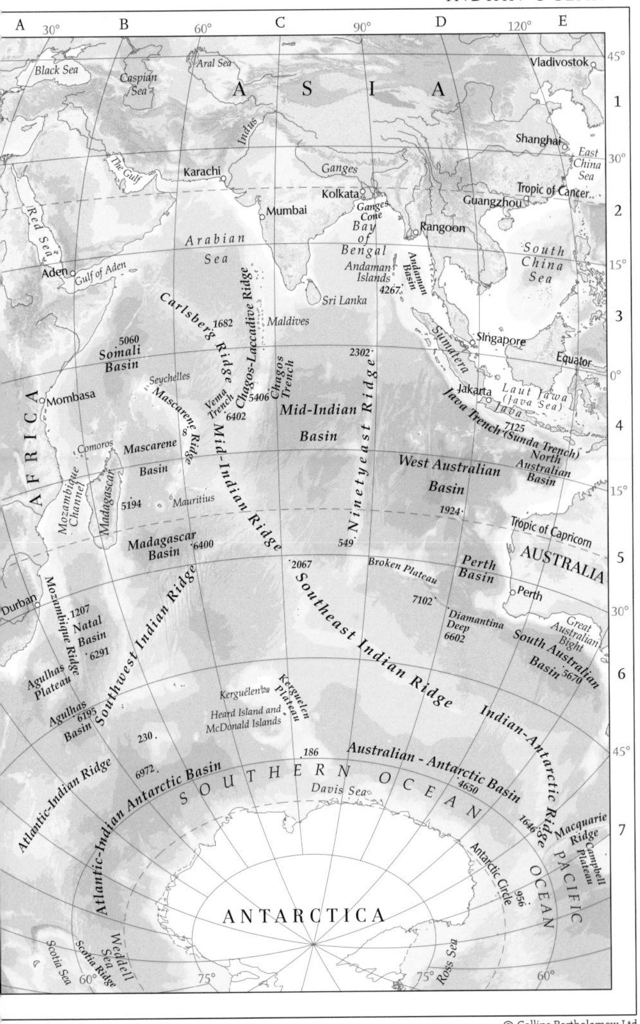

A 30° B 60° C 90° D 120° E

Black Sea
Caspian Sea
Aral Sea
Vladivostok

A S I A

The Gulf
Karachi
Indus
Ganges
Shanghai
East China Sea

Red Sea
Aden
Gulf of Aden
Mumbai
Kolkata
Ganges Cone
Bay of Bengal
Rangoon
Guangzhou
Tropic of Cancer

Arabian Sea
Andaman Islands
Andaman Basin
4267
South China Sea

Somali Basin
5060
Carlsberg Ridge
1682
Chagos-Laccadive Ridge
Maldives
Sri Lanka
2302
Sumatra
Singapore
Equator

Mombasa
Seychelles
Mascarene Ridge
Vema Trench
6402
Ganges Trench
5406
Mid-Indian Basin
Ninetyeast Ridge
Jakarta
Java
Laut Jawa (Java Sea)
Java Trench (Sunda Trench)
7125
North Australian Basin

AFRICA
Comoros
Mascarene Basin
Mid-Indian Ridge
5194
Mauritius
Tropic of Capricorn

Mozambique Channel
Madagascar
Madagascar Basin
6400
2067
549
Broken Plateau
West Australian Basin
1924
Perth Basin
AUSTRALIA
Perth

Durban
Agulhas Plateau
1207
Natal Basin
6291
Southwest Indian Ridge
7102
Diamantina Deep
6602
South Australian Basin
5670
Great Australian Bight

Agulhas Basin
6195
Atlantic-Indian Ridge
Southwest Indian Ridge
Kerguélen Plateau
Heard Island and McDonald Islands
230
186
Australian - Antarctic Basin
4650
Indian-Antarctic Ridge
1664
Macquarie Ridge

6972
S O U T H E R N   O C E A N
Davis Sea
956
Antarctic Circle
Ross Sea
PACIFIC OCEAN
Campbell Plateau

Scotia Sea
Weddell Sea
ANTARCTICA
75°
60°

0 1000 2000 3000 4000 KILOMETRES

© Collins Bartholomew Ltd

159

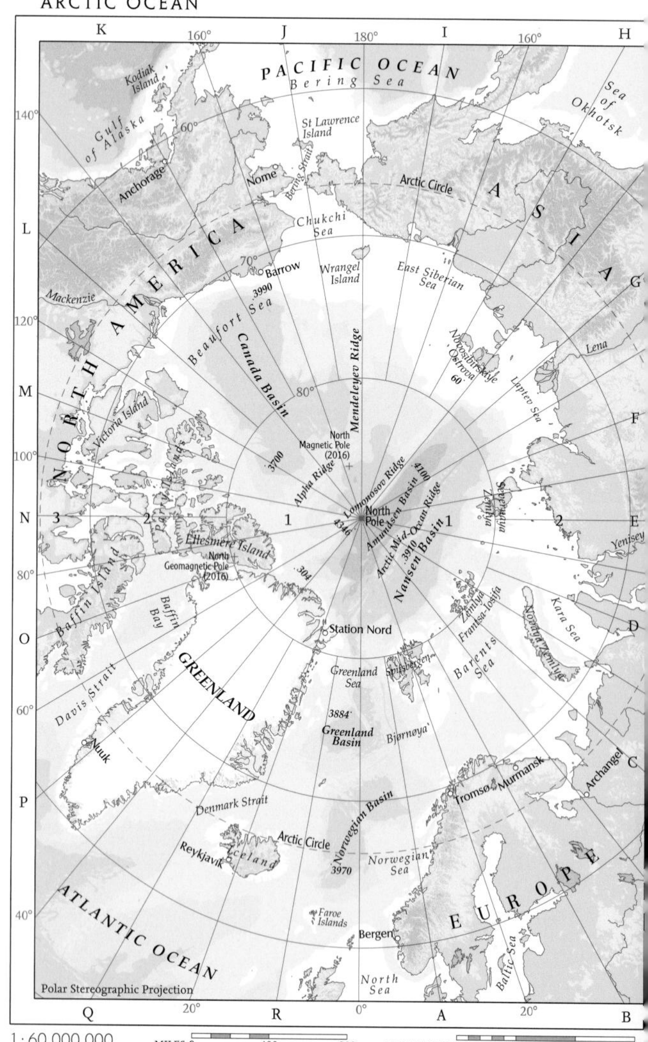

1 : 60 000 000

MILES 0   400   800   KILOMETRES 0   500   1000   1500

INTRODUCTION TO THE INDEX

The index includes all names shown on the maps in the Atlas of the World. Names are referenced by page number and by a grid reference. The grid reference correlates to the alphanumeric values which appear within each map frame. Each entry also includes the country or geographical area in which the feature is located. Entries relating to names appearing on insets are indicated by a small box symbol: □, followed by a grid reference if the inset has its own alphanumeric values.

Name forms are as they appear on the maps, with additional alternative names or name forms included as cross-references which refer the user to the entry for the map form of the name. Names beginning with Mc or Mac are alphabetized exactly as they appear. The terms Saint, Sainte, etc., are abbreviated to St, Ste, etc., but alphabetized as if in the full form.

Names of physical features beginning with generic, geographical terms are permuted – the descriptive term is placed after the main part of the name. For example, Lake Superior is indexed as Superior, Lake; Mount Everest as Everest, Mount. This policy is applied to all languages.

Entries, other than those for towns and cities, include a descriptor indicating the type of geographical feature. Descriptors are not included where the type of feature is implicit in the name itself.

Administrative divisions are included to differentiate entries of the same name and feature type within the one country. In such cases, duplicate names are alphabetized in order of administrative division. Additional qualifiers are also included for names within selected geographical areas.

INDEX ABBREVIATIONS

admin. div.	administrative division	Fin.	Finland	Phil.	Philippines
		for.	forest	plat.	plateau
Afgh.	Afghanistan	Fr.	French	P.N.G.	Papua New
Alg.	Algeria	g.	gulf		Guinea
Arg.	Argentina	Ger.	Germany	Pol.	Poland
Austr.	Australia	Guat.	Guatemala	Port.	Portugal
aut. reg.	autonomous region	h.	hill	prov.	province
Azer.	Azerbaijan	hd	head	r.	river
b.	bay	Hond.	Honduras	reg.	region
Bangl.	Bangladesh	imp. l.	impermanent lake	resr.	reservoir
Bol.	Bolivia	Indon.	Indonesia	S.	South
Bos. & Herz.	Bosnia and	i.	island	str.	strait
	Herzegovina	is	Islands	Switz.	Switzerland
Bulg.	Bulgaria	isth.	isthmus	Tajik.	Tajikistan
c.	cape	Kazakh.	Kazakhstan	Tanz.	Tanzania
Can.	Canada	Kyrg.	Kyrgyzstan	terr.	territory
C.A.R.	Central African	lag.	lagoon	Thai.	Thailand
	Republic	Lith.	Lithuania	Trin. and Tob.	Trinidad and
chan.	channel	Lux.	Luxembourg		Tobago
Col.	Colombia	Madag.	Madagascar	Turkm.	Turkmenistan
Czech Rep.	Czech Republic	Maur.	Mauritania	U.A.E.	United Arab
Dem. Rep.	Democratic	Mex.	Mexico		Emirates
Congo	Republic of the	Moz.	Mozambique	U.K.	United Kingdom
	Congo	mt.	mountain	Ukr.	Ukraine
depr.	depression	mun.	municipality	Uru.	Uruguay
des.	desert	N.	North	U.S.A.	United States of
disp. terr.	disputed territory	Neth.	Netherlands		America
Dom. Rep.	Dominican	Nic.	Nicaragua	Uzbek.	Uzbekistan
	Republic	N.Z.	New Zealand	val.	valley
esc.	escarpment	Pak.	Pakistan	Venez.	Venezuela
est.	estuary	Para.	Paraguay	vol.	volcano
Eth.	Ethiopia	pen.	peninsula		

1

128 B2 100 Mile House Can.

A

93 E4 **Aabenraa** Denmark
100 C2 **Aachen** Ger.
93 E4 **Aalborg** Denmark
100 B2 **Aalst** Belgium
93 F4 **Aarhus** Denmark
100 B2 **Aarschot** Belgium
68 C2 **Aba** China
115 C4 **Aba** Nigeria
81 C2 **Ābādān** Iran
81 D2 **Ābādeh** Iran
114 B1 **Abadla** Alg.
115 C4 **Abakaliki** Nigeria
83 H3 **Abakan** Russia
150 A3 **Abancay** Peru
81 D2 **Abarkūh** Iran
66 D2 **Abashiri** Japan
117 B4 **Abaya, Lake** Eth.
Ābay Wenz *r.* Eth./Sudan *see* **Blue Nile**
82 G3 **Abaza** Russia
108 A2 **Abbasanta** *Sardinia* Italy
104 C1 **Abbeville** France
142 B3 **Abbeville** U.S.A.
55 O2 **Abbot Ice Shelf** Antarctica
74 B1 **Abbottabad** Pak.
115 E3 **Abéché** Chad
114 B4 **Abengourou** Côte d'Ivoire
114 C4 **Abeokuta** Nigeria
99 A2 **Aberaeron** U.K.
96 C2 **Aberchirder** U.K.
99 B3 **Aberdare** U.K.
99 A2 **Aberdaron** U.K.
122 B3 **Aberdeen** S. Africa
96 C2 **Aberdeen** U.K.
141 D3 **Aberdeen** *MD* U.S.A.
137 D1 **Aberdeen** *SD* U.S.A.
134 B1 **Aberdeen** *WA* U.S.A.
129 E1 **Aberdeen Lake** Can.
134 B2 **Abert, Lake** U.S.A.
99 A2 **Aberystwyth** U.K.
86 F2 **Abez'** Russia
78 B3 **Abhā** Saudi Arabia
Abiad, Bahr el *r.* Sudan/Uganda *see* **White Nile**
114 B4 **Abidjan** Côte d'Ivoire
137 D3 **Abilene** *KS* U.S.A.
139 D2 **Abilene** *TX* U.S.A.
99 C3 **Abingdon** U.K.
91 D3 **Abinsk** Russia
130 B2 **Abitibi, Lake** Can.
81 C1 **Abkhazia** *disp. terr.* Georgia
Åbo Fin. *see* **Turku**
74 B1 **Abohar** India
114 C4 **Abomey** Benin
60 A1 **Abongabong, Gunung** *mt.* Indon.
118 B2 **Abong Mbang** Cameroon
64 A2 **Aborlan** Phil.
115 D3 **Abou Déia** Chad
106 B2 **Abrantes** Port.
152 B3 **Abra Pampa** Arg.
136 A2 **Absaroka Range** *mts* U.S.A.
81 C1 **Abşeron Yarımadası** *pen.* Azer.
78 B3 **Abū 'Arīsh** Saudi Arabia
79 C2 **Abu Dhabi** U.A.E.
116 B3 **Abu Hamed** Sudan
115 C4 **Abuja** Nigeria
81 C2 **Abū Kamāl** Syria
152 B1 **Abunã** *r.* Bol./Brazil
150 B2 **Abunã** Brazil
74 B2 **Abu Road** India
116 B2 **Abū Sunbul** Egypt
117 A3 **Abu Zabad** Sudan
Abū Ẓaby U.A.E. *see* **Abu Dhabi**
117 A4 **Abyei** Sudan
145 B2 **Acambaro** Mex.
106 B1 **A Cañiza** Spain
144 B2 **Acaponeta** Mex.
145 C3 **Acapulco** Mex.
151 D2 **Acará** Brazil
150 B1 **Acarigua** Venez.
145 C3 **Acatlán** Mex.
145 C3 **Acayucán** Mex.
114 B4 **Accra** Ghana
98 B2 **Accrington** U.K.
74 B2 **Achalpur** India
97 A2 **Achill Island** Ireland
101 D1 **Achim** Ger.
96 B2 **Achnasheen** U.K.
91 D2 **Achuyevo** Russia
111 C3 **Acıpayam** Turkey
109 C3 **Acireale** *Sicily* Italy
147 C2 **Acklins Island** Bahamas
153 B4 **Aconcagua, Cerro** *mt.* Arg.
106 B1 **A Coruña** Spain
108 A2 **Acqui Terme** Italy
103 D2 **Ács** Hungary
145 C2 **Actopán** Mex.
139 D2 **Ada** U.S.A.
79 C2 **Adam** Oman
49 J4 **Adamstown** Pitcairn Is
'Adan Yemen *see* **Aden**
80 B2 **Adana** Turkey
111 D2 **Adapazarı** Turkey
Adapazari Turkey *see* **Adapazarı**
108 A1 **Adda** *r.* Italy
78 B2 **Ad Dafīnah** Saudi Arabia
78 B2 **Ad Dahnā'** *des.* Saudi Arabia
78 B2 **Ad Dahnā'** *des.* Saudi Arabia
Ad Dammām Saudi Arabia *see* **Dammam**
78 A2 **Ad Dār al Ḥamrā'** Saudi Arabia
78 B3 **Ad Darb** Saudi Arabia
78 B2 **Ad Dawādimī** Saudi Arabia
Ad Dawḥah Qatar *see* **Doha**
78 B2 **Ad Dilam** Saudi Arabia
116 C2 **Ad Dir'īyah** Saudi Arabia
117 B4 **Addis Ababa** Eth.
81 C2 **Ad Dīwānīyah** Iraq
52 A2 **Adelaide** Austr.
50 C1 **Adelaide River** Austr.
101 D2 **Adelebsen** Ger.
55 J2 **Adélie Land** Antarctica
78 B3 **Aden** Yemen
117 C3 **Aden, Gulf of** Somalia/Yemen
100 C2 **Adenau** Ger.
79 C2 **Adh Dhayd** U.A.E.
59 C3 **Adi** *i.* Indon.
78 A3 **Ādī Ārk'ay** Eth.
116 B3 **Ādīgrat** Eth.
75 B3 **Adilabad** India
141 E2 **Adirondack Mountains** U.S.A.
Ādīs Ābeba Eth. *see* **Addis Ababa**
117 B4 **Ādīs Alem** Eth.
110 C1 **Adjud** Romania
50 B1 **Admiralty Gulf** Austr.
128 A2 **Admiralty Island** U.S.A.
104 B3 **Adour** *r.* France
106 C2 **Adra** Spain
114 B2 **Adrar** Alg.
140 C2 **Adrian** *MI* U.S.A.
139 C1 **Adrian** *TX* U.S.A.
108 B2 **Adriatic Sea** Europe
116 B3 **Ādwa** Eth.
83 K2 **Adycha** *r.* Russia
91 D3 **Adygeysk** Russia
114 B4 **Adzopé** Côte d'Ivoire
111 B3 **Aegean Sea** Greece/Turkey
101 D1 **Aerzen** Ger.
106 B1 **A Estrada** Spain
116 B3 **Afabet** Eritrea
76 C3 **Afghanistan** *country* Asia
78 B2 **'Afīf** Saudi Arabia
80 B2 **Afyon** Turkey
115 C3 **Agadez** Niger
114 B1 **Agadir** Morocco
74 B2 **Agar** India
75 D2 **Agartala** India
81 C1 **Ağdam** Azer.
81 C2 **Ağdam (abandoned)** Azer.
105 C3 **Agde** France
104 C3 **Agen** France
122 A2 **Aggeneys** S. Africa
111 C3 **Agia Varvara** Greece
111 B3 **Agios Dimitrios** Greece
111 C3 **Agios Efstratios** *i.* Greece
111 C3 **Agios Nikolaos** Greece
110 B1 **Agnita** Romania
75 B2 **Agra** India
81 C2 **Ağrı** Turkey
Ağrı Dağı *mt.* Turkey *see* **Ararat, Mount**
108 B3 **Agrigento** *Sicily* Italy
111 B3 **Agrinio** Greece
109 B2 **Agropoli** Italy
154 B2 **Água Clara** Brazil
146 B4 **Aguadulce** Panama
144 B2 **Aguanaval** *r.* Mex.
144 B1 **Agua Prieta** Mex.
144 B2 **Aguascalientes** Mex.
155 D1 **Águas Formosas** Brazil
106 B1 **Águeda** Port.
106 C1 **Aguilar de Campoo** Spain
107 C2 **Águilas** Spain
144 B3 **Aguililla** Mex.
122 B3 **Agulhas, Cape** S. Africa
155 D2 **Agulhas Negras** *mt.* Brazil
111 C2 **Ağva** Turkey
115 C2 **Ahaggar** *plat.* Alg.
115 C2 **Ahaggar, Tassili oua-n-** *plat.* Alg.
81 C2 **Ahar** Iran
100 C1 **Ahaus** Ger.
81 C2 **Ahlat** Turkey
100 C2 **Ahlen** Ger.
74 B2 **Ahmadabad** India
73 B3 **Ahmadnagar** India
74 B2 **Ahmadpur East** Pak.
74 B1 **Ahmadpur Sial** Pak.
Ahmedabad India *see* **Ahmadabad**
Ahmednagar India *see* **Ahmadnagar**
144 B2 **Ahome** Mex.
79 C2 **Ahram** Iran
101 E1 **Ahrensburg** Ger.
104 C2 **Ahun** France
81 C2 **Ahvāz** Iran
122 A2 **Ai-Ais** Namibia
80 B2 **Aigialousa** Cyprus
111 B3 **Aigio** Greece
143 D2 **Aiken** U.S.A.
97 B1 **Ailt an Chorráin** Ireland
155 D1 **Aimorés** Brazil
155 D1 **Aimorés, Serra dos** *hills* Brazil
114 B2 **'Aïn Ben Tili** Maur.
107 D2 **Aïn Defla** Alg.
114 B1 **Aïn Sefra** Alg.
136 D2 **Ainsworth** U.S.A.
Aintab Turkey *see* **Gaziantep**
107 D2 **Aïn Taya** Alg.
107 D2 **Aïn Tédélès** Alg.

115 C3 Aïr, Massif de l' mts Niger
60 A1 Airbangis Indon.
128 C2 Airdrie Can.
104 B3 Aire-sur-l'Adour France
101 E3 Aisch r. Ger.
128 C2 Aishihik Lake Can.
100 A3 Aisne r. France
59 D3 Aitape P.N.G.
110 B1 Aiud Romania
105 D3 Aix-en-Provence France
105 D2 Aix-les-Bains France
62 A1 Aizawl India
88 C2 Aizkraukle Latvia
67 C3 Aizuwakamatsu Japan
105 D3 Ajaccio Corsica France
115 E1 Ajdābiyā Libya
115 E1 Ajjer, Tassili n' plat. Alg.
74 B2 Ajmer India
138 A2 Ajo U.S.A.
77 D2 Akadyr Kazakh.
87 E3 Akbulak Russia
80 B2 Akçakale Turkey
80 B2 Akdağmadeni Turkey
88 A2 Åkersberga Sweden
118 C2 Aketi Dem. Rep. Congo
87 D4 Akhalkalaki Georgia
79 C2 Akhdar, Jabal mts Oman
111 C3 Akhisar Turkey
87 D4 Akhtubinsk Russia
130 B1 Akimiski Island Can.
66 D3 Akita Japan
114 A3 Akjoujt Maur.
77 D1 Akkol' Kazakh.
88 B2 Akmenrags pt Latvia
Akmola Kazakh. see Astana
117 B4 Akobo South Sudan
74 B2 Akola India
78 A3 Akordat Eritrea
127 G2 Akpatok Island Can.
92 □A3 Akranes Iceland
140 C2 Akron U.S.A.
75 B1 Aksai Chin terr. Asia
80 B2 Aksaray Turkey
76 B1 Aksay Kazakh.
91 D2 Aksay Russia
80 B2 Akşehir Turkey
76 C2 Akshiganak Kazakh.
77 E2 Aksu China
78 A3 Āksum Eth.
76 B1 Aktau Kazakh.
76 B1 Aktobe Kazakh.
77 D2 Aktogay Kazakh.
88 C3 Aktsyabrski Belarus
115 C4 Akure Nigeria
92 □B2 Akureyri Iceland
62 A1 Akyab Myanmar
142 C2 Alabama r. U.S.A.
142 C2 Alabama state U.S.A.
111 C3 Alaçatı Turkey
81 C1 Alagir Russia
151 E3 Alagoinhas Brazil
107 C1 Alagón Spain
79 B2 Al Aḥmadī Kuwait
77 E2 Alakol', Ozero salt l. Kazakh.
92 J2 Alakurtti Russia
78 B3 Al 'Alayyah Saudi Arabia
80 A2 Al 'Āmirīyah Egypt
135 C3 Alamo U.S.A.
138 B2 Alamogordo U.S.A.
144 A2 Alamos Sonora Mex.
144 A2 Alamos Sonora Mex.
144 B2 Alamos r. Mex.
136 B3 Alamosa U.S.A.
93 G3 Åland Islands Fin.
78 B4 Alappuzha India
80 B3 Al 'Aqabah Jordan

78 B2 Al 'Aqīq Saudi Arabia
107 C2 Alarcón, Embalse de resr Spain
80 B2 Al 'Arīsh Egypt
78 B2 Al Arţāwīyah Saudi Arabia
61 C2 Alas Indon.
111 C3 Alaşehir Turkey
128 A2 Alaska state U.S.A.
124 F3 Alaska, Gulf of U.S.A.
81 C2 Älät Azer.
87 D3 Alatyr' Russia
150 A2 Alausí Ecuador
93 H3 Alavus Fin.
52 B2 Alawoona Austr.
108 A2 Alba Italy
107 C2 Albacete Spain
110 B1 Alba Iulia Romania
109 C2 Albania country Europe
50 A3 Albany Austr.
130 B1 Albany r. Can.
143 D2 Albany GA U.S.A.
141 E2 Albany NY U.S.A.
134 B2 Albany OR U.S.A.
Al Başrah Iraq see Basra
51 D1 Albatross Bay Austr.
116 A2 Al Bawīţī Egypt
115 E1 Al Bayḍā' Libya
78 B3 Al Bayḍā' Yemen
143 D1 Albemarle U.S.A.
143 E1 Albemarle Sound sea chan. U.S.A.
108 A2 Albenga Italy
51 C2 Alberga watercourse Austr.
119 D2 Albert, Lake Dem. Rep. Congo/Uganda
128 C2 Alberta prov. Can.
100 B2 Albert Kanaal canal Belgium
137 E2 Albert Lea U.S.A.
104 C3 Albi France
78 A2 Al Bi'r Saudi Arabia
78 B3 Al Birk Saudi Arabia
78 B2 Al Biyāḍh reg. Saudi Arabia
106 B2 Alboran Sea Europe
Alborz, Reshteh-ye mts Iran see Elburz Mountains
106 B2 Albufeira Port.
138 B1 Albuquerque U.S.A.
79 C2 Al Buraymī Oman
53 C3 Albury Austr.
106 B2 Alcácer do Sal Port.
106 C1 Alcalá de Henares Spain
106 C2 Alcalá la Real Spain
108 B3 Alcamo Sicily Italy
107 C1 Alcañiz Spain
106 B2 Alcántara Spain
106 C2 Alcaraz Spain
106 C2 Alcaraz, Sierra de mts Spain
106 C2 Alcaudete Spain
106 C2 Alcázar de San Juan Spain
91 D2 Alchevs'k Ukr.
155 E1 Alcobaça Brazil
106 C2 Alcoy-Alcoi Spain
107 D2 Alcúdia Spain
145 C2 Aldama Mex.
83 J3 Aldan Russia
83 J2 Aldan r. Russia
95 C4 Alderney i. Channel Is
114 A3 Aleg Maur.
155 D2 Alegre Brazil
152 C3 Alegrete Brazil
83 K3 Aleksandrovsk-Sakhalinskiy Russia
91 D1 Alekseyevka Belgorodskaya Oblast' Russia
91 D1 Alekseyevka Belgorodskaya Oblast' Russia
89 E3 Aleksin Russia
109 D2 Aleksinac Serbia

118 B3 Alèmbé Gabon
155 D2 Além Paraíba Brazil
93 F3 Ålen Norway
104 C2 Alençon France
80 B2 Aleppo Syria
150 A3 Alerta Peru
128 B2 Alert Bay Can.
105 C3 Alès France
110 B1 Aleşd Romania
108 A2 Alessandria Italy
93 E3 Ålesund Norway
124 C3 Aleutian Islands U.S.A.
83 L3 Alevina, Mys c. Russia
128 A2 Alexander Archipelago is U.S.A.
122 A2 Alexander Bay S. Africa
142 C2 Alexander City U.S.A.
55 O2 Alexander Island Antarctica
53 C3 Alexandra Austr.
54 A3 Alexandra N.Z.
111 B2 Alexandreia Greece
Alexandretta Turkey see İskenderun
116 A1 Alexandria Egypt
110 C2 Alexandria Romania
123 C3 Alexandria S. Africa
142 B2 Alexandria LA U.S.A.
137 D1 Alexandria MN U.S.A.
141 D3 Alexandria VA U.S.A.
52 A3 Alexandrina, Lake Austr.
111 C2 Alexandroupoli Greece
131 E1 Alexis r. Can.
128 B2 Alexis Creek Can.
77 E1 Aleysk Russia
107 C1 Alfaro Spain
81 C3 Al Fāw Iraq
101 D2 Alfeld (Leine) Ger.
155 C2 Alfenas Brazil
Al Fujayrah U.A.E. see Fujairah
Al Furāt r. Iraq/Syria see Euphrates
106 B2 Algeciras Spain
107 C2 Algemesí Spain
78 A3 Algena Eritrea
Alger Alg. see Algiers
114 C2 Algeria country Africa
Al Ghardaqah Egypt see Hurghada
79 C3 Al Ghaydah Yemen
108 A2 Alghero Sardinia Italy
79 B2 Al Ghwaybiyah Saudi Arabia
115 C1 Algiers Alg.
123 C3 Algoa Bay S. Africa
137 E2 Algona U.S.A.
106 C1 Algorta Spain
81 C2 Al Ḥadīthah Iraq
79 C2 Al Hajar al Gharbī mts Oman
107 C2 Alhama de Murcia Spain
78 B2 Al Ḥammām Egypt
78 B2 Al Ḥanākīyah Saudi Arabia
81 C2 Al Ḥasakah Syria
81 C2 Al Ḥayy Iraq
79 C3 Al Ḥibāk des. Saudi Arabia
78 B2 Al Ḥillah Saudi Arabia
78 B3 Al Ḥinnāh Saudi Arabia
Al Ḥudaydah Yemen see Hodeidah
79 B2 Al Hufūf Saudi Arabia
115 D2 Al Ḥulayq al Kabīr hills Libya
79 C2 'Alīābād Iran
111 C3 Aliağa Turkey
111 B2 Aliakmonas r. Greece
107 C2 Alicante Spain
139 D3 Alice U.S.A.
109 C3 Alice, Punta pt Italy
51 C2 Alice Springs Austr.
75 B2 Aligarh India

Alīgūdarz

81 C2 Alīgūdarz Iran
69 E1 Alihe China
118 B3 Alima r. Congo
111 C3 Aliova r. Turkey
117 C3 Ali Sabieh Djibouti
Al Iskandarīyah Egypt see
Alexandria
116 B1 Al Ismā'īlīyah Egypt
123 E3 Aliwal North S. Africa
115 E2 Al Jaghbūb Libya
78 B2 Al Jahrah Kuwait
115 D1 Al Jawsh Libya
106 B2 Aljezur Port.
Al Jīzah Egypt see Giza
79 B2 Al Jubayl Saudi Arabia
79 C2 Al Jumaylīyah Qatar
78 B2 Al Junaynah Saudi Arabia
106 B2 Aljustrel Port.
78 B2 Al Kahfah Saudi Arabia
79 C2 Al Kāmil Oman
80 B2 Al Karak Jordan
79 C2 Al Khābūrah Oman
78 B2 Al Khamāsīn Saudi Arabia
116 B1 Al Khārijah Egypt
79 C2 Al Khaṣab Oman
78 B3 Al Khawkhah Yemen
79 C2 Al Khawr Qatar
115 D1 Al Khums Libya
79 B2 Al Khunn Saudi Arabia
79 C2 Al Kir'ānah Qatar
100 B1 Alkmaar Neth.
115 E2 Al Kufrah Libya
81 C2 Al Kūt Iraq
Al Kuwayt Kuwait see Kuwait
Al Lādhiqīyah Syria see
Latakia
75 C2 Allahabad India
83 K2 Allakh-Yun' Russia
141 D2 Allegheny r. U.S.A.
140 C3 Allegheny Mountains U.S.A.
97 B1 Allen, Lough l. Ireland
145 B2 Allende Coahuila Mex.
145 B2 Allende Nuevo León Mex.
141 D2 Allentown U.S.A.
101 D1 Aller r. Ger.
136 C2 Alliance NE U.S.A.
140 C2 Alliance OH U.S.A.
78 B2 Al Līth Saudi Arabia
96 C2 Alloa U.K.
131 C2 Alma Can.
Alma-Ata Kazakh. see Almaty
106 B2 Almada Port.
106 C2 Almadén Spain
Al Madīnah Saudi Arabia see
Medina
116 B1 Al Mafraq Jordan
78 B3 Al Maḥwīt Yemen
78 B2 Al Majma'ah Saudi Arabia
135 B2 Almanor, Lake U.S.A.
107 C2 Almansa Spain
80 B2 Al Manṣūrah Egypt
79 C2 Al Mariyyah U.A.E.
115 E1 Al Marj Libya
77 D2 Almaty Kazakh.
106 C1 Almazán Spain
151 E2 Almeirim Brazil
100 C1 Almelo Neth.
155 D1 Almenara Brazil
106 B1 Almendra, Embalse de resr
Spain
106 B2 Almendralejo Spain
106 C2 Almería Spain
106 C2 Almería, Golfo de b. Spain
87 E3 Al'met'yevsk Russia
78 B2 Al Mindak Saudi Arabia
116 B1 Al Minyā Egypt
79 B2 Al Mish'āb Saudi Arabia
106 B2 Almodôvar Port.

106 B2 Almonte Spain
75 B2 Almora India
79 B2 Al Mubarrez Saudi Arabia
79 C2 Al Muḍaibī Oman
80 B3 Al Mudawwarah Jordan
Al Mukallā Yemen see
Mukalla
Al Mukhā Yemen see Mocha
106 C2 Almuñécar Spain
78 A2 Al Muwaylih Saudi Arabia
111 B3 Almyros Greece
96 B2 Alness U.K.
98 C1 Alnwick U.K.
49 G4 Alofi Niue
111 B3 Alonnisos i. Greece
59 C3 Alor i. Indon.
59 C3 Alor, Kepulauan is Indon.
60 B1 Alor Star Malaysia
Alost Belgium see Aalst
86 C2 Alozero (abandoned) Russia
140 C1 Alpena U.S.A.
139 C2 Alpine U.S.A.
105 D2 Alps mts Europe
79 B3 Al Qa'āmīyāt reg.
Saudi Arabia
115 D1 Al Qaddāhīyah Libya
Al Qāhirah Egypt see Cairo
78 B2 Al Qā'īyah Saudi Arabia
81 C2 Al Qāmishlī Syria
80 B2 Al Qaryatayn Syria
79 B3 Al Qaṭn Yemen
106 B2 Alqueva, Barragem de resr
Port.
80 B2 Al Qunayṭirah (abandoned)
Syria
78 B3 Al Qunfidhah Saudi Arabia
116 B2 Al Quṣayr Egypt
78 B2 Al Quwayīyah Saudi Arabia
101 D2 Alsfeld Ger.
92 H1 Alta Norway
92 H1 Altaelva r. Norway
68 B1 Altai Mountains Asia
143 D2 Altamaha r. U.S.A.
151 C2 Altamira Brazil
109 C2 Altamura Italy
68 B1 Altay China
68 C1 Altay Mongolia
105 D2 Altdorf Switz.
107 C2 Altea Spain
101 F2 Altenburg Ger.
100 C2 Altenkirchen (Westerwald)
Ger.
111 C3 Altınoluk Turkey
111 D3 Altıntaş Turkey
152 B2 Altiplano plain Bol.
154 B1 Alto Araguaia Brazil
154 B1 Alto Garças Brazil
121 C2 Alto Molócuè Moz.
129 E3 Altona Can.
141 D2 Altoona U.S.A.
154 B1 Alto Sucuriú Brazil
154 B1 Alto Taquari Brazil
102 C2 Altötting Ger.
68 B2 Altun Shan mts China
134 B2 Alturas U.S.A.
139 D2 Altus U.S.A.
88 C2 Alūksne Latvia
78 A2 Al 'Ulā Saudi Arabia
115 D1 Al 'Uqaylah Libya
Al Uqsur Egypt see Luxor
91 C3 Alushta Ukr.
139 D1 Alva U.S.A.
145 C3 Alvarado Mex.
93 F3 Älvdalen Sweden
93 F3 Älvdalen val. Sweden
92 H2 Älvsbyn Sweden
78 A2 Al Wajh Saudi Arabia
74 B2 Alwar India

81 C2 Al Widyān plat. Iraq/
Saudi Arabia
Alxa Youqi China see
Ehen Hudag
Alxa Zuoqi China see
Bayan Hot
51 C2 Alyangula Austr.
88 B3 Alytus Lith.
136 C1 Alzada U.S.A.
100 D3 Alzey Ger.
50 C2 Amadeus, Lake imp. l. Austr.
81 C2 'Amādīyah/Amēdî Iraq
127 G2 Amadjuak Lake Can.
106 B2 Amadora Port.
78 B2 Amā'ir Saudi Arabia
93 F4 Åmål Sweden
111 B3 Amaliada Greece
59 D3 Amamapare Indon.
154 A2 Amambaí Brazil
154 B2 Amambaí r. Brazil
69 E3 Amami-Ō-shima i. Japan
69 E3 Amami-shotō is Japan
77 C1 Amankel'dy Kazakh.
109 C3 Amantea Italy
151 C1 Amapá Brazil
106 B2 Amareleja Port.
139 C1 Amarillo U.S.A.
108 B2 Amaro, Monte mt. Italy
80 B1 Amasya Turkey
150 C1 Amazon r. S. America
151 D1 Amazon, Mouths of the Brazil
Amazonas r. S. America see
Amazon
74 B1 Ambala India
121 □D2 Ambalavao Madag.
121 □D2 Ambanja Madag.
150 A2 Ambato Ecuador
121 □D2 Ambato Boeny Madag.
121 □D2 Ambato Finandrahana Madag.
121 □D2 Ambatolampy Madag.
121 □D2 Ambatondrazaka Madag.
101 E3 Amberg Ger.
146 B3 Ambergris Caye i. Belize
75 C2 Ambikapur India
121 □D2 Ambilobe Madag.
98 B1 Ambleside U.K.
121 □D3 Amboasary Madag.
121 □D2 Ambohimahasoa Madag.
59 C3 Ambon Indon.
59 C3 Ambon i. Indon.
121 □D3 Ambositra Madag.
121 □D3 Ambovombe Madag.
135 C4 Amboy U.S.A.
120 A1 Ambriz Angola
145 B2 Amealco Mex.
144 B2 Ameca Mex.
100 B1 Ameland i. Neth.
134 D2 American Falls U.S.A.
134 D2 American Falls Reservoir
U.S.A.
135 D2 American Fork U.S.A.
49 G3 American Samoa terr.
S. Pacific Ocean
143 D2 Americus U.S.A.
100 B1 Amersfoort Neth.
55 F2 Amery Ice Shelf Antarctica
137 E2 Ames U.S.A.
111 B3 Amfissa Greece
83 J2 Amga Russia
66 C1 Amgu Russia
115 C2 Amguid Alg.
83 K3 Amgun' r. Russia
131 D2 Amherst Can.
104 C2 Amiens France
73 B3 Amindivi Islands India
122 A1 Aminuis Namibia
74 A2 Amir Chah Pak.
129 D2 Amisk Lake Can.

139 C3 **Amistad Reservoir** Mex./U.S.A.
98 A2 **Amlwch** U.K.
80 B2 **'Ammān** Jordan
81 D2 **Āmol** Iran
111 C3 **Amorgos** *i.* Greece
142 C2 **Amory** U.S.A.
130 C2 **Amos** Can.
Amoy China *see* **Xiamen**
155 C2 **Amparo** Brazil
107 D1 **Amposta** Spain
75 B2 **Amravati** India
74 B1 **Amritsar** India
100 B1 **Amstelveen** Neth.
100 B1 **Amsterdam** Neth.
103 C2 **Amstetten** Austria
115 E3 **Am Timan** Chad
76 B2 **Amudar'ya** *r.* Asia
126 E1 **Amund Ringnes Island** Can.
126 C2 **Amundsen Gulf** Can.
55 M2 **Amundsen Sea** Antarctica
61 C2 **Amuntai** Indon.
Amur *r.* China *see* **Heilong Jiang**
78 A3 **'Amur, Wadi** *watercourse* Sudan
83 I2 **Anabar** *r.* Russia
83 I2 **Anabarskiy Zaliv** *b.* Russia
134 D1 **Anaconda** U.S.A.
139 D1 **Anadarko** U.S.A.
80 B1 **Anadolu Dağları** *mts* Turkey
83 M2 **Anadyr'** *r.* Russia
81 C2 **'Ānah** Iraq
145 B2 **Anáhuac** Mex.
60 B1 **Anambas, Kepulauan** *is* Indon.
137 E2 **Anamosa** U.S.A.
80 B2 **Anamur** Turkey
67 B4 **Anan** Japan
73 B3 **Anantapur** India
74 B1 **Anantnag** India
90 B2 **Anan'yiv** Ukr.
91 D3 **Anapa** Russia
154 C1 **Anápolis** Brazil
152 B3 **Añatuya** Arg.
97 B1 **An Baile Thiar** Ireland
65 B2 **Anbyon** N. Korea
104 B2 **Ancenis** France
108 B2 **Ancona** Italy
153 A5 **Ancud** Chile
Anda China *see* **Daqing**
93 E3 **Åndalsnes** Norway
142 C2 **Andalusia** U.S.A.
73 D3 **Andaman Islands** India
63 A2 **Andaman Sea** Indian Ocean
121 □D2 **Andapa** Madag.
100 B2 **Andelst** Neth.
92 G2 **Andenes** Norway
100 B2 **Andenne** Belgium
100 B2 **Anderlecht** Belgium
126 C2 **Anderson** *r.* Can.
126 B2 **Anderson** *AK* U.S.A.
140 B2 **Anderson** *IN* U.S.A.
143 D2 **Anderson** *SC* U.S.A.
148 D3 **Andes** *mts* S. America
77 D2 **Andijon** Uzbek.
121 □D2 **Andilamena** Madag.
121 □D2 **Andilanatoby** Madag.
121 □D2 **Andoany** Madag.
Andong China *see* **Dandong**
65 B2 **Andong** S. Korea
107 D1 **Andorra** *country* Europe
107 D1 **Andorra la Vella** Andorra
99 C3 **Andover** U.K.
154 B2 **Andradina** Brazil
89 D2 **Andreapol'** Russia
155 D2 **Andrelândia** Brazil
139 C2 **Andrews** U.S.A.
109 C2 **Andria** Italy
121 □D3 **Androka** Madag.
146 C2 **Andros** *i.* Bahamas
111 B3 **Andros** *i.* Greece
143 E4 **Andros Town** Bahamas
73 B3 **Andrott** *i.* India
90 B1 **Andrushivka** Ukr.
92 G2 **Andselv** Norway
106 C2 **Andújar** Spain
120 A2 **Andulo** Angola
114 C3 **Anéfis** Mali
115 D3 **Aney** Niger
83 H3 **Angara** *r.* Russia
69 C1 **Angarsk** Russia
93 G3 **Ånge** Sweden
144 A2 **Ángel de la Guarda, Isla** *i.* Mex.
64 B1 **Angeles** Phil.
150 B1 **Angel Falls** Venez.
93 F4 **Ängelholm** Sweden
92 G3 **Ångermanälven** *r.* Sweden
104 B2 **Angers** France
129 E1 **Angikuni Lake** Can.
98 A2 **Anglesey** *i.* U.K.
121 C2 **Angoche** Moz.
120 A2 **Angola** *country* Africa
140 C2 **Angola** U.S.A.
128 A2 **Angoon** U.S.A.
104 C2 **Angoulême** France
77 D2 **Angren** Uzbek.
147 D3 **Anguilla** *terr.* West Indies
75 C2 **Angul** India
93 F4 **Anholt** *i.* Denmark
71 B3 **Anhua** China
70 B2 **Anhui** *prov.* China
154 B1 **Anhumas** Brazil
154 C1 **Anicuns** Brazil
66 D1 **Aniva, Mys** *c.* Russia
65 B2 **Anjŭ** N. Korea
70 A2 **Ankang** China
80 B2 **Ankara** Turkey
70 B2 **Anlu** China
91 E1 **Anna** Russia
115 C1 **Annaba** Alg.
101 F2 **Annaberg-Buchholz** Ger.
78 B3 **An Nābiyah** Yemen
78 B2 **An Nafūd** *des.* Saudi Arabia
81 C2 **An Najaf** Iraq
96 C3 **Annan** U.K.
141 D3 **Annapolis** U.S.A.
75 C2 **Annapurna I** *mt.* Nepal
140 C2 **Ann Arbor** U.S.A.
150 C1 **Anna Regina** Guyana
81 C2 **An Nāşirīyah** Iraq
105 D2 **Annecy** France
78 B3 **An Nimāş** Saudi Arabia
62 B1 **Anning** China
142 C2 **Anniston** U.S.A.
105 C2 **Annonay** France
79 B2 **An Nu'ayrīyah** Saudi Arabia
121 □D2 **Anorontany, Tanjona** *hd* Madag.
71 B3 **Anpu** China
70 B2 **Anqing** China
102 C2 **Ansbach** Ger.
70 C1 **Anshan** China
71 A3 **Anshun** China
114 C3 **Ansongo** Mali
150 A3 **Antabamba** Peru
80 B2 **Antakya** Turkey
121 □E2 **Antalaha** Madag.
80 B2 **Antalya** Turkey
80 B2 **Antalya Körfezi** *g.* Turkey
121 □D2 **Antananarivo** Madag.
55 Q3 **Antarctic Peninsula** Antarctica
96 B2 **An Teallach** *mt.* U.K.
106 C2 **Antequera** Spain
138 B2 **Anthony** U.S.A.
114 B2 **Anti-Atlas** *mts* Morocco
105 D3 **Antibes** France
131 D2 **Anticosti, Île d'** *i.* Can.
131 D2 **Antigonish** Can.
147 D3 **Antigua** *i.* Antigua
147 D3 **Antigua and Barbuda** *country* West Indies
145 C2 **Antiguo-Morelos** Mex.
111 B3 **Antikythira** *i.* Greece
Antioch Turkey *see* **Antakya**
49 F6 **Antipodes Islands** N.Z.
152 A3 **Antofagasta** Chile
154 C3 **Antonina** Brazil
97 C1 **Antrim** U.K.
97 C1 **Antrim Hills** U.K.
121 □D2 **Antsalova** Madag.
121 □D2 **Antsirabe** Madag.
121 □D2 **Antsirañana** Madag.
121 □D2 **Antsohihy** Madag.
Antwerp Belgium *see* **Antwerpen**
100 B2 **Antwerpen** Belgium
An Uaimh Ireland *see* **Navan**
74 B2 **Anupgarh** India
73 C4 **Anuradhapura** Sri Lanka
51 C3 **Anxious Bay** Austr.
70 B2 **Anyang** China
65 B2 **Anyang** S. Korea
108 B2 **Anzio** Italy
Aomen China *see* **Macao**
66 D2 **Aomori** Japan
54 B2 **Aoraki/Mount Cook** *mt.* N.Z.
108 A1 **Aosta** Italy
143 D3 **Apalachee Bay** U.S.A.
150 B2 **Apaporis** *r.* Col.
154 B2 **Aparecida do Tabuado** Brazil
64 B1 **Aparri** Phil.
86 C2 **Apatity** Russia
144 B3 **Apatzingán** Mex.
100 B1 **Apeldoorn** Neth.
100 C1 **Apen** Ger.
108 A2 **Apennines** *mts* Italy
49 G3 **Apia** Samoa
154 C2 **Apiaí** Brazil
64 B2 **Apo, Mount** *vol.* Phil.
101 E2 **Apolda** Ger.
52 B3 **Apollo Bay** Austr.
143 D3 **Apopka, Lake** U.S.A.
154 B1 **Aporé** Brazil
154 B1 **Aporé** *r.* Brazil
80 B2 **Apostolos Andreas, Cape** Cyprus
133 E3 **Appalachian Mountains** U.S.A.
53 D2 **Appin** Austr.
140 B2 **Appleton** U.S.A.
108 B2 **Aprilia** Italy
91 D3 **Apsheronsk** Russia
154 B2 **Apucarana** Brazil
154 B2 **Apucarana, Serra da** *hills* Brazil
78 A2 **Aqaba, Gulf of** Asia
154 A1 **Aquidauana** *r.* Brazil
117 A4 **Arab, Bahr el** *watercourse* Sudan
Arabian Gulf Asia *see* **The Gulf**
78 B2 **Arabian Peninsula** Asia
56 G5 **Arabian Sea** Indian Ocean
151 E3 **Aracaju** Brazil
151 E2 **Aracati** Brazil
154 B2 **Araçatuba** Brazil
155 D1 **Aracruz** Brazil
155 D1 **Araçuaí** Brazil
110 B1 **Arad** Romania
115 E3 **Arada** Chad
156 C6 **Arafura Sea** Austr./Indon.
154 B1 **Aragarças** Brazil
107 C1 **Aragón** *r.* Spain
151 D2 **Araguaia** *r.* Brazil
151 D2 **Araguaína** Brazil

154 C1 **Araguari** Brazil
115 C2 **Arak** Alg.
81 C2 **Ārāk** Iran
62 A1 **Arakan Yoma** *mts* Myanmar
81 C1 **Arak's** *r.* Armenia
76 B2 **Aralkum** *des.* Uzbek.
76 C2 **Aral Sea** *salt l.* Kazakh./Uzbek.
76 C2 **Aral'sk** Kazakh.
Aral'skoye More *salt l.* Kazakh./Uzbek. *see* **Aral Sea**
106 C1 **Aranda de Duero** Spain
109 D2 **Aranđelovac** Serbia
97 B2 **Aran Islands** Ireland
106 C1 **Aranjuez** Spain
122 A1 **Aranos** Namibia
139 D3 **Aransas Pass** U.S.A.
67 B4 **Arao** Japan
114 B3 **Araouane** Mali
151 E2 **Arapiraca** Brazil
154 B2 **Arapongas** Brazil
154 C3 **Araquari** Brazil
78 B1 **'Ar'ar** Saudi Arabia
154 C2 **Araraquara** Brazil
151 C2 **Araras** Brazil
154 B3 **Araras, Serra das** *mts* Brazil
52 B3 **Ararat** Austr.
81 C2 **Ararat, Mount** Turkey
155 D2 **Araruama, Lago de** *lag.* Brazil
155 E1 **Arataca** Brazil
Aratürük China *see* **Yiwu**
150 A1 **Arauca** Col.
154 C1 **Araxá** Brazil
81 C2 **Arbīl/Hewlêr** Iraq
96 C2 **Arbroath** U.K.
74 A2 **Arbū-ye Shamālī, Dasht-e** *des.* Afgh.
104 B3 **Arcachon** France
143 D3 **Arcadia** U.S.A.
134 B2 **Arcata** U.S.A.
145 B3 **Arcelia** Mex.
86 D2 **Archangel** Russia
51 D1 **Archer** *r.* Austr.
134 D2 **Arco** U.S.A.
106 B2 **Arcos de la Frontera** Spain
127 F2 **Arctic Bay** Can.
160 **Arctic Ocean**
126 C2 **Arctic Red** *r.* Can.
81 C2 **Ardabīl** Iran
81 C1 **Ardahan** Turkey
93 E3 **Årdalstangen** Norway
100 B3 **Ardennes** *plat.* Belgium
81 D2 **Ardestān** Iran
53 C2 **Ardlethan** Austr.
139 D2 **Ardmore** U.S.A.
96 A2 **Ardnamurchan, Point of** U.K.
52 A2 **Ardrossan** Austr.
96 B3 **Ardrossan** U.K.
135 B3 **Arena, Point** U.S.A.
93 E4 **Arendal** Norway
101 E1 **Arendsee (Altmark)** Ger.
150 A3 **Arequipa** Peru
151 C2 **Arere** Brazil
106 C1 **Arévalo** Spain
108 B2 **Arezzo** Italy
104 B2 **Argentan** France
153 B4 **Argentina** *country* S. America
153 A6 **Argentino, Lago** *l.* Arg.
104 C2 **Argenton-sur-Creuse** France
110 C2 **Argeș** *r.* Romania
74 A1 **Arghandāb Rōd** *r.* Afgh.
111 B3 **Argolikos Kolpos** *b.* Greece
111 B3 **Argos** Greece
111 B3 **Argostoli** Greece
107 C1 **Arguis** Spain
69 E1 **Argun'** *r.* China/Russia
50 B1 **Argyle, Lake** Austr.
122 A2 **Ariamsvlei** Namibia
152 A2 **Arica** Chile
96 A2 **Arinagour** U.K.
155 C1 **Arinos** Brazil
150 C3 **Aripuanã** Brazil
150 B2 **Aripuanã** *r.* Brazil
150 B2 **Ariquemes** Brazil
154 B1 **Ariranhá** *r.* Brazil
138 A2 **Arizona** *state* U.S.A.
144 A1 **Arizpe** Mex.
78 B2 **'Arjah** Saudi Arabia
92 G2 **Arjeplog** Sweden
142 B2 **Arkadelphia** U.S.A.
77 C1 **Arkalyk** Kazakh.
142 B2 **Arkansas** *r.* U.S.A.
142 B1 **Arkansas** *state* U.S.A.
137 D3 **Arkansas City** U.S.A.
Arkhangel'sk Russia *see* **Archangel**
97 C2 **Arklow** Ireland
102 C1 **Arkona, Kap** *c.* Ger.
82 G1 **Arkticheskogo Instituta, Ostrova** *is* Russia
105 C3 **Arles** France
115 C3 **Arlit** Niger
100 B3 **Arlon** Belgium
97 C1 **Armagh** U.K.
116 B2 **Armant** Egypt
81 C1 **Armavir** Russia
81 C1 **Armenia** *country* Asia
150 A1 **Armenia** Col.
144 B3 **Armeria** Mex.
53 D2 **Armidale** Austr.
130 B1 **Armstrong** Can.
80 B1 **Armyans'k** Ukr.
80 B2 **Arnauti, Cape** Cyprus
100 B2 **Arnhem** Neth.
51 C1 **Arnhem, Cape** Austr.
51 C1 **Arnhem Bay** Austr.
51 C1 **Arnhem Land** *reg.* Austr.
108 B2 **Arno** *r.* Italy
52 A2 **Arno Bay** Austr.
130 C2 **Arnprior** Can.
100 D2 **Arnsberg** Ger.
101 E2 **Arnstadt** Ger.
122 A2 **Aroab** Namibia
101 D2 **Arolsen** Ger.
78 A3 **Aroma** Sudan
108 A1 **Arona** Italy
144 B2 **Aros** *r.* Mex.
75 C2 **Arrah** India
81 C2 **Ar Ramādī** Iraq
96 B3 **Arran** *i.* U.K.
97 B1 **Arranmore Island** Ireland
80 B2 **Ar Raqqah** Syria
104 C1 **Arras** France
78 B2 **Ar Rass** Saudi Arabia
145 C3 **Arriagá** Mex.
79 C2 **Ar Rimāl** *reg.* Saudi Arabia
Ar Riyāḍ Saudi Arabia *see* **Riyadh**
145 C2 **Arroyo Seco** Mex.
79 C2 **Ar Rustāq** Oman
80 C2 **Ar Ruṭbah** Iraq
81 D3 **Arsanajān** Iran
66 B2 **Arsen'yev** Russia
111 B3 **Arta** Greece
144 B3 **Arteaga** Mex.
66 B2 **Artem** Russia
91 D2 **Artemivs'k** Ukr.
104 C2 **Artenay** France
138 C2 **Artesia** U.S.A.
51 E2 **Arthur Point** Austr.
54 B2 **Arthur's Pass** N.Z.
152 C4 **Artigas** Uru.
129 D1 **Artillery Lake** Can.
90 B2 **Artsyz** Ukr.
77 D3 **Artux** China
81 C1 **Artvin** Turkey
59 C3 **Aru, Kepulauan** *is* Indon.
119 D2 **Arua** Uganda
147 D3 **Aruba** *terr.* West Indies
119 D3 **Arusha** Tanz.
69 C1 **Arvayheer** Mongolia
129 E1 **Arviat** Can.
92 G2 **Arvidsjaur** Sweden
93 F4 **Arvika** Sweden
87 D3 **Arzamas** Russia
107 C2 **Arzew** Alg.
100 C2 **Arzfeld** Ger.
101 F2 **Aš** Czech Rep.
115 C4 **Asaba** Nigeria
74 B1 **Asadābād** Afgh.
66 D2 **Asahi-dake** *vol.* Japan
66 D2 **Asahikawa** Japan
78 B3 **Āsalē** *l.* Eth.
75 C2 **Asansol** India
131 C2 **Asbestos** Can.
109 C2 **Ascea** Italy
152 B2 **Ascensión** Bol.
113 D7 **Ascension** *i.* S. Atlantic Ocean
145 D3 **Ascensión, Bahía de la** *b.* Mex.
101 D3 **Aschaffenburg** Ger.
100 C2 **Ascheberg** Ger.
101 E2 **Aschersleben** Ger.
108 B2 **Ascoli Piceno** Italy
92 G3 **Åsele** Sweden
110 B2 **Asenovgrad** Bulg.
76 B3 **Aşgabat** Turkm.
50 A2 **Ashburton** *watercourse* Austr.
54 B2 **Ashburton** N.Z.
142 B2 **Ashdown** U.S.A.
143 D1 **Asheville** U.S.A.
53 D1 **Ashford** Austr.
99 D3 **Ashford** U.K.
Ashgabat Turkm. *see* **Aşgabat**
98 C1 **Ashington** U.K.
67 B4 **Ashizuri-misaki** *pt* Japan
136 D3 **Ashland** *KS* U.S.A.
140 C3 **Ashland** *KY* U.S.A.
140 C2 **Ashland** *OH* U.S.A.
134 B2 **Ashland** *OR* U.S.A.
140 A1 **Ashland** *WI* U.S.A.
88 C3 **Ashmyany** Belarus
78 B3 **Ash Sharawrah** Saudi Arabia
Ash Shāriqah U.A.E. *see* **Sharjah**
81 C2 **Ash Sharqāṭ** Iraq
81 C2 **Ash Shaṭrah** Iraq
79 B3 **Ash Shiḥr** Yemen
79 C2 **Ash Şhināş** Oman
78 B2 **Ash Shu'bah** Saudi Arabia
78 B2 **Ash Shumlūl** Saudi Arabia
140 C2 **Ashtabula** U.S.A.
131 D1 **Ashuanipi Lake** Can.
106 B2 **Asilah** Morocco
108 A2 **Asinara, Golfo dell'** *b. Sardinia* Italy
82 G3 **Asino** Russia
88 C3 **Asipovichy** Belarus
78 B2 **'Asīr** *reg.* Saudi Arabia
93 F4 **Askim** Norway
68 C1 **Askiz** Russia
116 B3 **Asmara** Eritrea
93 F4 **Åsnen** *l.* Sweden
116 B2 **Asoteriba, Jebel** *mt.* Sudan
103 D2 **Aspang-Markt** Austria
136 B3 **Aspen** U.S.A.
54 A2 **Aspiring, Mount** N.Z.
117 C3 **Assab** Eritrea
Aş Şaḥrā' al Gharbīyah *des.* Egypt *see* **Western Desert**
Aş Şaḥrā' ash Sharqīyah *des.* Egypt *see* **Eastern Desert**
78 B2 **As Salamiyah** Saudi Arabia
81 C2 **As Samāwah** Iraq
79 C2 **Aş Şanām** *reg.* Saudi Arabia

115	E2	As Sarīr reg. Libya
100	C1	Assen Neth.
100	B2	Assesse Belgium
115	D1	As Sidrah Libya
129	D3	Assiniboia Can.
128	C2	Assiniboine, Mount Can.
154	B2	Assis Brazil
78	B2	Aş Şubayḩīyah Kuwait
81	C2	As Sulaymānīyah/Slēmānī Iraq
78	B2	As Sulaymī Saudi Arabia
78	B2	As Sulayyil Saudi Arabia
78	B2	As Sūq Saudi Arabia
80	B2	As Suwaydā' Syria
79	C2	As Suwayq Oman
		As Suways see Suez
111	B3	Astakos Greece
77	D1	Astana Kazakh.
81	C2	Āstārā Iran
108	A2	Asti Italy
74	B1	Astor Pak.
106	B1	Astorga Spain
134	B1	Astoria U.S.A.
87	D4	Astrakhan' Russia
88	C3	Astravyets Belarus
111	C3	Astypalaia i. Greece
152	C3	Asunción Para.
116	B2	Aswān Egypt
116	B2	Asyūṭ Egypt
		Atacama, Desierto de des. Chile see Atacama Desert
152	B3	Atacama, Salar de salt flat Chile
152	B3	Atacama Desert Chile
114	C4	Atakpamé Togo
111	B3	Atalanti Greece
150	A3	Atalaya Peru
77	C3	Atamyrat Turkm.
78	B3	'Ataq Yemen
114	A2	Atâr Maur.
135	B3	Atascadero U.S.A.
77	D2	Atasu Kazakh.
111	C3	Atavyros mt. Greece
116	B3	Atbara Sudan
116	B3	Atbara r. Sudan
77	C1	Atbasar Kazakh.
137	D3	Atchison U.S.A.
108	B2	Aterno r. Italy
108	B3	Atessa Italy
100	A2	Ath Belgium
128	C2	Athabasca Can.
129	C2	Athabasca r. Can.
129	D2	Athabasca, Lake Can.
111	B3	Athens Greece
143	D2	Athens GA U.S.A.
140	C3	Athens OH U.S.A.
143	D1	Athens TN U.S.A.
139	D2	Athens TX U.S.A.
		Athina Greece see Athens
97	C2	Athlone Ireland
111	B3	Athos mt. Greece
97	C2	Athy Ireland
115	D3	Ati Chad
130	A2	Atikokan Can.
87	D3	Atkarsk Russia
143	D2	Atlanta U.S.A.
137	D2	Atlantic U.S.A.
141	E3	Atlantic City U.S.A.
158		Atlantic Ocean
122	A3	Atlantis S. Africa
114	B1	Atlas Mountains Africa
114	C1	Atlas Saharien mts Alg.
128	A2	Atlin Can.
128	A2	Atlin Lake Can.
142	C2	Atmore U.S.A.
139	D2	Atoka U.S.A.
75	C2	Atrai r. India
78	B2	Aṭ Ṭā'if Saudi Arabia

63	B2	Attapu Laos
130	B1	Attawapiskat Can.
130	B1	Attawapiskat r. Can.
130	B1	Attawapiskat Lake Can.
100	C2	Attendorn Ger.
78	B3	Aṭ Ṭur Egypt
78	B3	At Turbah Yemen
76	B2	Atyrau Kazakh.
105	C3	Aubenas France
126	C2	Aubry Lake Can.
142	C2	Auburn AL U.S.A.
135	B3	Auburn CA U.S.A.
137	D2	Auburn NE U.S.A.
141	D2	Auburn NY U.S.A.
104	C2	Aubusson France
104	C3	Auch France
54	B1	Auckland N.Z.
48	F6	Auckland Islands is N.Z.
101	F2	Aue Ger.
102	C2	Augsburg Ger.
109	C3	Augusta Sicily Italy
143	D2	Augusta GA U.S.A.
137	D3	Augusta KS U.S.A.
141	F2	Augusta ME U.S.A.
50	A2	Augustus, Mount Austr.
100	A2	Aulnoye-Aymeries France
62	A2	Aunglan Myanmar
122	B2	Auob watercourse Namibia/S. Africa
74	B3	Aurangabad India
100	C1	Aurich Ger.
154	B1	Aurilândia Brazil
104	C3	Aurillac France
136	C3	Aurora CO U.S.A.
140	B2	Aurora IL U.S.A.
137	D2	Aurora NE U.S.A.
122	A2	Aus Namibia
137	E2	Austin MN U.S.A.
135	C3	Austin NV U.S.A.
139	D2	Austin TX U.S.A.
50	B2	Australia country Oceania
159	D7	Australian-Antarctic Basin Southern Ocean
53	C3	Australian Capital Territory admin. div. Austr.
102	C2	Austria country Europe
144	B3	Autlán Mex.
105	C2	Autun France
105	C2	Auxerre France
105	D2	Auxonne France
105	C2	Avallon France
131	E2	Avalon Peninsula Can.
154	C2	Avaré Brazil
49	H4	Avarua Cook Is
91	D2	Avdiyivka Ukr.
106	B1	Aveiro Port.
109	B2	Avellino Italy
100	A2	Avesnes-sur-Helpe France
93	G3	Avesta Sweden
108	B2	Avezzano Italy
96	C2	Aviemore U.K.
109	C2	Avigliano Italy
105	C3	Avignon France
106	C1	Ávila Spain
106	B1	Avilés Spain
109	C3	Avola Sicily Italy
99	B2	Avon r. England U.K.
99	C3	Avon r. England U.K.
104	B2	Avranches France
54	B1	Awanui N.Z.
117	C4	Āwash Eth.
117	C3	Āwash r. Eth.
115	D2	Awbārī Libya
115	D2	Awbārī, Idhān des. Libya
117	C4	Aw Dheegle Somalia
96	B2	Awe, Loch l. U.K.
117	A4	Aweil South Sudan
126	E1	Axel Heiberg Island Can.

114	B4	Axim Ghana
150	A3	Ayacucho Peru
77	E2	Ayagoz Kazakh.
68	B2	Ayakkum Hu salt l. China
106	B2	Ayamonte Spain
83	K3	Ayan Russia
150	A3	Ayaviri Peru
76	A2	Aybas Kazakh.
91	D2	Aydar r. Ukr.
77	C2	Aydarko'l ko'li l. Uzbek.
111	C3	Aydın Turkey
99	C3	Aylesbury U.K.
106	C1	Ayllón Spain
129	D1	Aylmer Lake Can.
117	B4	Ayod South Sudan
83	M2	Ayon, Ostrov i. Russia
114	B3	'Ayoûn el 'Atroûs Maur.
51	D1	Ayr Austr.
96	B3	Ayr U.K.
98	A1	Ayre, Point of Isle of Man
76	C2	Ayteke Bi Kazakh.
110	C2	Aytos Bulg.
145	C3	Ayutla Mex.
63	B2	Ayutthaya Thai.
111	C3	Ayvacık Turkey
111	C3	Ayvalık Turkey
		Azania reg. Somalia see Jubaland
114	C3	Azaouagh, Vallée de watercourse Mali/Niger
77	C2	Azat, Gory h. Kazakh.
114	B3	Azawad reg. Mali
		Azbine mts Niger see Aïr, Massif de l'
81	C1	Azerbaijan country Asia
77	C1	Azhibeksor, Ozero salt l. Kazakh.
86	D2	Azopol'ye Russia
84	D5	Azores aut. reg. Port.
91	D2	Azov Russia
91	D2	Azov, Sea of Russia/Ukr.
		Azraq, Bahr el r. Eth./Sudan see Blue Nile
106	B2	Azuaga Spain
146	B4	Azuero, Península de pen. Panama
153	C4	Azul Arg.
80	B2	Az Zaqāzīq Egypt
80	B2	Az Zarqā' Jordan
78	B3	Az Zaydiyah Yemen
114	C2	Azzel Matti, Sebkha salt pan Alg.
78	B2	Az Zilfī Saudi Arabia
78	B3	Az Zuqur i. Yemen

B

63	B2	Ba, Sông r. Vietnam
117	C4	Baardheere Somalia
77	C3	Bābā, Kōh-e mts Afgh.
110	C2	Babadag Romania
111	C2	Babaeski Turkey
117	C3	Bāb al Mandab str. Africa/Asia
61	C2	Babana Indon.
59	C3	Babar i. Indon.
119	D3	Babati Tanz.
89	E2	Babayevo Russia
128	B2	Babine r. Can.
128	B2	Babine Lake Can.
59	C3	Babo Indon.
81	D2	Bābol Iran
122	A3	Baboon Point S. Africa
88	C3	Babruysk Belarus
64	B1	Babuyan i. Phil.
64	B1	Babuyan Islands Phil.

151 D2 **Bacabal** Brazil
59 C3 **Bacan** i. Indon.
110 C1 **Bacău** Romania
52 B3 **Bacchus Marsh** Austr.
77 D3 **Bachu** China
129 E1 **Back** r. Can.
109 C1 **Bačka Palanka** Serbia
63 B3 **Bac Liêu** Vietnam
64 B1 **Bacolod** Phil.
130 C1 **Bacqueville, Lac** l. Can.
106 B2 **Badajoz** Spain
62 A1 **Badarpur** India
101 F1 **Bad Belzig** Ger.
101 E2 **Bad Berka** Ger.
101 D2 **Bad Berleburg** Ger.
101 E1 **Bad Bevensen** Ger.
100 C2 **Bad Ems** Ger.
103 D2 **Baden** Austria
102 B2 **Baden-Baden** Ger.
101 E2 **Bad Harzburg** Ger.
101 D2 **Bad Hersfeld** Ger.
102 C2 **Bad Hofgastein** Austria
101 D2 **Bad Homburg vor der Höhe** Ger.
74 A2 **Badin** Pak.
Bādiyat ash Shām des. Asia see **Syrian Desert**
101 E2 **Bad Kissingen** Ger.
100 C3 **Bad Kreuznach** Ger.
101 E2 **Bad Lauterberg im Harz** Ger.
101 D2 **Bad Lippspringe** Ger.
101 D3 **Bad Mergentheim** Ger.
101 D2 **Bad Nauheim** Ger.
100 C2 **Bad Neuenahr-Ahrweiler** Ger.
101 E2 **Bad Neustadt an der Saale** Ger.
101 E1 **Bad Oldesloe** Ger.
101 D2 **Bad Pyrmont** Ger.
78 A2 **Badr Ḥunayn** Saudi Arabia
101 D1 **Bad Salzuflen** Ger.
101 E2 **Bad Salzungen** Ger.
102 C1 **Bad Schwartau** Ger.
101 E1 **Bad Segeberg** Ger.
100 C3 **Bad Sobernheim** Ger.
101 E2 **Bad Staffelstein** Ger.
73 C4 **Badulla** Sri Lanka
100 D1 **Bad Zwischenahn** Ger.
65 A2 **Baengnyeong-do** i. S. Korea
106 C2 **Baeza** Spain
114 A3 **Bafatá** Guinea-Bissau
160 O2 **Baffin Bay** sea Can./Greenland
127 G2 **Baffin Island** Can.
118 B2 **Bafia** Cameroon
114 A3 **Bafing** r. Africa
114 A3 **Bafoulabé** Mali
118 B2 **Bafoussam** Cameroon
76 B3 **Bāfq** Iran
80 B1 **Bafra** Turkey
79 C2 **Bāft** Iran
119 C2 **Bafwasende** Dem. Rep. Congo
119 D3 **Bagamoyo** Tanz.
60 B1 **Bagan Datuk** Malaysia
120 B2 **Bagani** Namibia
60 B1 **Bagansiapiapi** Indon.
138 A2 **Bagdad** U.S.A.
152 C4 **Bagé** Brazil
97 C2 **Bagenalstown** Ireland
81 C2 **Baghdād** Iraq
77 C3 **Baghlān** Afgh.
104 C3 **Bagnères-de-Luchon** France
Bago Myanmar see **Pegu**
88 B3 **Bagrationovsk** Russia
Bagrax China see **Bohu**
64 B1 **Baguio** Phil.
115 C3 **Bagzane, Monts** mts Niger
Bahamas country West Indies see **The Bahamas**
146 C2 **Bahamas, The** country West Indies
75 C2 **Baharampur** India
Bahariya Oasis Egypt see **Bahrīyah, Wāḥāt al**
76 B3 **Baharly** Turkm.
60 B1 **Bahau** Malaysia
72 B2 **Bahawalnagar** Pak.
74 B2 **Bahawalpur** Pak.
153 B4 **Bahía Blanca** Arg.
144 A2 **Bahía Kino** Mex.
152 C3 **Bahía Negra** Para.
144 A2 **Bahía Tortugas** Mex.
117 B3 **Bahir Dar** Eth.
75 C2 **Bahraich** India
79 C2 **Bahrain** country Asia
116 A2 **Bahrīyah, Wāḥāt al** oasis Egypt
79 D2 **Bāhū Kālāt** Iran
110 B1 **Baia Mare** Romania
69 E1 **Baicheng** China
131 D2 **Baie-Comeau** Can.
Baie-du-Poste. Can. see **Mistissini**
131 C2 **Baie-St-Paul** Can.
65 B1 **Baihe** China
69 D1 **Baikal, Lake** Russia
110 B2 **Băileşti** Romania
68 C2 **Baima** China
143 D2 **Bainbridge** U.S.A.
48 F2 **Bairiki** Kiribati
Bairin Youqi China see **Daban**
53 C3 **Bairnsdale** Austr.
71 A3 **Baise** China
65 B1 **Baishan** Jilin China
65 B1 **Baishan** Jilin China
65 B1 **Baitou Shan** mt. China/N. Korea
70 A2 **Baiyin** China
116 B3 **Baiyuda Desert** Sudan
103 D2 **Baja** Hungary
144 A1 **Baja California** pen. Mex.
78 B3 **Bājil** Yemen
114 A3 **Bakel** Senegal
135 C3 **Baker** CA U.S.A.
136 C1 **Baker** MT U.S.A.
134 C2 **Baker** OR U.S.A.
134 B1 **Baker, Mount** vol. U.S.A.
129 E1 **Baker Foreland** hd Can.
49 G2 **Baker Island** N. Pacific Ocean
129 E1 **Baker Lake** Can.
129 E1 **Baker Lake** Can.
135 C3 **Bakersfield** U.S.A.
91 C3 **Bakhchysaray** Ukr.
91 C1 **Bakhmach** Ukr.
Bākhtarān Iran see **Kermānshāh**
Bakı Azer. see **Baku**
111 C2 **Bakırköy** Turkey
92 □C2 **Bakkaflói** b. Iceland
118 C2 **Bakouma** C.A.R.
81 C1 **Baku** Azer.
64 A2 **Balabac** Phil.
64 A2 **Balabac** i. Phil.
64 A2 **Balabac Strait** Malaysia/Phil.
61 C2 **Balaiberkuak** Indon.
52 A2 **Balaklava** Austr.
91 C3 **Balaklava** Ukr.
91 D2 **Balakliya** Ukr.
87 D3 **Balakovo** Russia
76 C3 **Bālā Murghāb** Afgh.
145 C3 **Balancán** Mex.
111 C3 **Balan Dağı** h. Turkey
64 B1 **Balanga** Phil.
75 C2 **Balangir** India
79 C2 **Bālā Shahr** Iran
87 D3 **Balashov** Russia
103 D2 **Balaton, Lake** Hungary
103 D2 **Balatonboglár** Hungary
150 C2 **Balbina, Represa de** resr Brazil
97 C2 **Balbriggan** Ireland
52 A2 **Balcanoona** Austr.
110 C2 **Balchik** Bulg.
54 A3 **Balclutha** N.Z.
129 E2 **Baldock Lake** Can.
129 D2 **Baldy Mountain** h. Can.
138 B2 **Baldy Peak** U.S.A.
Baleares, Islas is Spain see **Balearic Islands**
107 D2 **Balearic Islands** Spain
107 D1 **Balearic Sea** Spain
155 E1 **Baleia, Ponta da** pt Brazil
130 C1 **Baleine, Grande Rivière de la** r. Can.
131 D1 **Baleine, Rivière à la** r. Can.
75 C2 **Baleshwar** India
108 A2 **Balestrieri, Punta** mt. Sardinia Italy
61 C2 **Bali** i. Indon.
115 D4 **Bali** Nigeria
60 A1 **Balige** Indon.
75 C2 **Baligurha** India
111 C3 **Balıkesir** Turkey
61 C2 **Balikpapan** Indon.
59 D3 **Balimo** P.N.G.
102 B2 **Balingen** Ger.
Bali Sea Indon. see **Laut Bali**
76 B3 **Balkanabat** Turkm.
110 B2 **Balkan Mountains** Bulgaria/Serbia
77 D2 **Balkash** Kazakh.
77 D2 **Balkhash, Lake** Kazakh.
Balkhash, Ozero l. Kazakh. see **Balkhash, Lake**
50 B3 **Balladonia** Austr.
97 B2 **Ballaghaderreen** Ireland
92 G2 **Ballangen** Norway
52 B3 **Ballarat** Austr.
50 B2 **Ballard, Lake** imp. l. Austr.
96 C2 **Ballater** U.K.
114 B3 **Ballé** Mali
53 D1 **Ballina** Austr.
97 B1 **Ballina** Ireland
97 B2 **Ballinasloe** Ireland
139 D2 **Ballinger** U.S.A.
97 B2 **Ballinrobe** Ireland
97 B1 **Ballycastle** Ireland
97 C1 **Ballycastle** U.K.
97 D1 **Ballyclare** U.K.
97 C1 **Ballymena** U.K.
97 C1 **Ballymoney** U.K.
97 D1 **Ballynahinch** U.K.
97 B1 **Ballyshannon** Ireland
95 B2 **Ballyvoy** U.K.
51 D2 **Balonne** r. Austr.
74 B2 **Balotra** India
75 C2 **Balrampur** India
52 B2 **Balranald** Austr.
110 B2 **Balş** Romania
151 D2 **Balsas** Brazil
145 C3 **Balsas** Mex.
90 B2 **Balta** Ukr.
90 B2 **Bălţi** Moldova
93 G4 **Baltic Sea** g. Europe
80 B2 **Balṭīm** Egypt
123 C1 **Baltimore** S. Africa
141 D3 **Baltimore** U.S.A.
88 A3 **Baltiysk** Russia
62 A1 **Balu** India
88 C2 **Balvi** Latvia
77 D2 **Balykchy** Kyrg.
87 E4 **Balykshi** Kazakh.
79 C2 **Bam** Iran
51 D1 **Bamaga** Austr.
130 A1 **Bamaji Lake** Can.

114	B3	Bamako Mali
118	C2	Bambari C.A.R.
101	E3	Bamberg Ger.
119	C2	Bambili Dem. Rep. Congo
119	C2	Bambouti C.A.R.
155	C2	Bambuí Brazil
118	B2	Bamenda Cameroon
74	A1	Bāmyān Afgh.
119	C2	Banalia Dem. Rep. Congo
151	C3	Bananal, Ilha do i. Brazil
74	B2	Banas r. India
111	C3	Banaz Turkey
62	B2	Ban Ban Laos
97	C1	Banbridge U.K.
99	C2	Banbury U.K.
130	C2	Bancroft Can.
119	C2	Banda Dem. Rep. Congo
75	C2	Banda India
59	C3	Banda, Kepulauan is Indon.
59	C3	Banda, Laut sea Indon.
60	A1	Banda Aceh Indon.
79	C2	Bandar-e 'Abbās Iran
81	C2	Bandar-e Anzalī Iran
79	C2	Bandar-e Būshehr Iran
79	C2	Bandar-e Chārak Iran
81	C2	Bandar-e Emām Khomeynī Iran
81	D3	Bandar-e Ganāveh Iran
79	C2	Bandar-e Jāsk Iran
79	C2	Bandar-e Kangān Iran
79	C2	Bandar-e Lengeh Iran
79	C2	Bandar-e Moqām Iran
60	B2	Bandar Lampung Indon.
61	C1	Bandar Seri Begawan Brunei Banda Sea Indon. see Banda, Laut
155	D2	Bandeiras, Pico de mt. Brazil
123	C1	Bandelierkop S. Africa
144	B2	Banderas, Bahía de b. Mex.
114	B3	Bandiagara Mali
111	C3	Bandırma Turkey
97	B3	Bandon Ireland
118	B3	Bandundu Dem. Rep. Congo
61	B2	Bandung Indon.
128	C2	Banff Can.
96	C2	Banff U.K.
114	B3	Banfora Burkina Faso
64	B2	Banga Phil.
		Bangalore India see Bengaluru
118	C2	Bangassou C.A.R.
59	C3	Banggai Indon.
59	C3	Banggai, Kepulauan is Indon.
61	C1	Banggi i. Sabah Malaysia
60	B2	Bangka i. Indon.
61	C2	Bangkalan Indon.
60	B1	Bangkinang Indon.
60	B2	Bangko Indon.
63	B2	Bangkok Thai.
75	C2	Bangladesh country Asia
63	B2	Ba Ngoi Vietnam
97	D1	Bangor Northern Ireland U.K.
98	A2	Bangor Wales U.K.
141	F2	Bangor U.S.A.
63	A2	Bang Saphan Yai Thai.
64	B1	Bangued Phil.
118	B2	Bangui C.A.R.
121	B2	Bangweulu, Lake Zambia
62	B2	Ban Huai Khon Thai.
116	B2	Banī Suwayf Egypt
115	D1	Banī Walīd Libya
80	B2	Bāniyās Syria
109	C2	Banja Luka Bos. & Herz.
61	C2	Banjarmasin Indon.
114	A3	Banjul Gambia
128	A2	Banks Island B.C. Can.
126	C2	Banks Island N.W.T. Can.
48	F3	Banks Islands Vanuatu
129	E1	Banks Lake Can.

54	B2	Banks Peninsula N.Z.
75	C2	Bankura India
62	A1	Banmauk Myanmar
62	B2	Ban Mouang Laos
97	C1	Bann r. U.K.
62	B2	Ban Napè Laos
63	A3	Ban Na San Thai.
143	E4	Bannerman Town Bahamas
74	B1	Bannu Pak.
74	B2	Banswara India
63	A3	Ban Tha Kham Thai.
62	A2	Ban Tha Song Yang Thai.
63	B2	Ban Tôp Laos
97	B3	Bantry Ireland
97	B3	Bantry Bay Ireland
118	B2	Banyo Cameroon
107	D1	Banyoles Spain
61	C2	Banyuwangi Indon.
		Bao'an China see Shenzhen
69	D2	Baochang China
70	B2	Baoding China
70	A2	Baoji China
62	B1	Bao Lac Vietnam
63	B2	Bao Lôc Vietnam
66	B1	Baoqing China
62	A1	Baoshan China
70	B1	Baotou China
81	C2	Ba'qūbah Iraq
109	C2	Bar Montenegro
117	C4	Baraawe Somalia
147	C2	Baracoa Cuba
53	C2	Baradine Austr.
147	C3	Barahona Dom. Rep.
116	B3	Baraka watercourse Eritrea/Sudan
61	C1	Baram r. Sarawak Malaysia
74	B1	Baramulla India
88	C3	Baranavichy Belarus
78	A2	Baranis Egypt
90	B1	Baranivka Ukr.
128	A2	Baranof Island U.S.A.
59	C3	Barat Daya, Kepulauan is Indon.
155	D2	Barbacena Brazil
147	E3	Barbados country West Indies
107	D1	Barbastro Spain
104	B2	Barbezieux-St-Hilaire France
51	D2	Barcaldine Austr.
107	D1	Barcelona Spain
150	B1	Barcelona Venez.
105	D3	Barcelonnette France
150	B2	Barcelos Brazil
		Barcoo Creek watercourse Austr. see Cooper Creek
103	D2	Barcs Hungary
75	C2	Barddhaman India
103	E2	Bardejov Slovakia
79	C2	Bardsīr Iran
75	B2	Bareilly India
160	B2	Barents Sea Arctic Ocean
78	A3	Barentu Eritrea
75	C2	Barh India
141	F2	Bar Harbor U.S.A.
109	C2	Bari Italy
74	B1	Barī Kōt Afgh.
150	A1	Barinas Venez.
75	C2	Baripada India
75	D2	Barisal Bangl.
60	B2	Barisan, Pegunungan mts Indon.
61	C2	Barito r. Indon.
79	C2	Barkā Oman
88	C2	Barkava Latvia
51	C1	Barkly Tableland reg. Austr.
68	C2	Barkol China
110	C1	Bârlad Romania
105	D2	Bar-le-Duc France
50	A2	Barlee, Lake l. Austr.

109	C2	Barletta Italy
53	C2	Barmedman Austr.
74	B2	Barmer India
99	A2	Barmouth U.K.
101	D1	Barmstedt Ger.
98	C1	Barnard Castle U.K.
53	B2	Barnato Austr.
82	G3	Barnaul Russia
127	G2	Barnes Icecap Can.
100	B1	Barneveld Neth.
98	C2	Barnsley U.K.
99	A3	Barnstaple U.K.
99	A3	Barnstaple Bay U.K.
143	D2	Barnwell U.S.A.
		Baroda India see Vadodara
150	B1	Barquisimeto Venez.
96	A2	Barra i. U.K.
53	D2	Barraba Austr.
151	D3	Barra do Corda Brazil
154	B1	Barra do Garças Brazil
150	C2	Barra do São Manuel Brazil
150	A3	Barranca Lima Peru
150	A2	Barranca Loreto Peru
152	C3	Barranqueras Arg.
150	A1	Barranquilla Col.
151	D3	Barreiras Brazil
154	C2	Barretos Brazil
130	C2	Barrie Can.
128	B2	Barrière Can.
52	B2	Barrier Range hills Austr.
53	D2	Barrington, Mount Austr.
129	D2	Barrington Lake Can.
53	C1	Barringun Austr.
97	C2	Barrow r. Ireland
51	C2	Barrow Creek Austr.
98	B1	Barrow-in-Furness U.K.
50	A2	Barrow Island Austr.
126	E2	Barrow Strait Can.
99	B3	Barry U.K.
130	C2	Barrys Bay Can.
74	B2	Barsalpur India
135	C4	Barstow U.S.A.
105	C2	Bar-sur-Aube France
80	B1	Bartın Turkey
51	D1	Bartle Frere, Mount Austr.
139	D1	Bartlesville U.S.A.
103	E1	Bartoszyce Pol.
61	C2	Barung i. Indon.
69	D1	Baruun-Urt Mongolia
91	D2	Barvinkove Ukr.
53	C2	Barwon r. Austr.
88	C3	Barysaw Belarus
110	C2	Basarabi Romania
105	D2	Basel Switz.
91	C2	Bashtanka Ukr.
64	B2	Basilan i. Phil.
99	D3	Basildon U.K.
99	C3	Basingstoke U.K.
81	C2	Başkale Turkey
130	C2	Baskatong, Réservoir resr Can.
		Basle Switz. see Basel
118	C2	Basoko Dem. Rep. Congo
81	C2	Basra Iraq
128	C2	Bassano Can.
114	C4	Bassar Togo
63	A2	Bassein Myanmar
147	D3	Basse-Terre Guadeloupe
147	D3	Basseterre St Kitts and Nevis
114	B3	Bassikounou Maur.
51	D3	Bass Strait Austr.
79	C2	Bastak Iran
101	E2	Bastheim Ger.
75	C2	Basti India
105	D3	Bastia Corsica France
100	B2	Bastogne Belgium
142	B2	Bastrop U.S.A.
		Basuo China see Dongfang

118 A2 Bata Equat. Guinea
83 J2 Batagay Russia
154 B2 Bataguassu Brazil
106 D2 Batalha Port.
60 B1 Batam Indon.
71 C3 Batan i. Phil.
118 D2 Batangafo C.A.R.
64 B1 Batangas Phil.
60 B1 Batanghari r. Indon.
71 C3 Batan Islands Phil.
141 D2 Batavia U.S.A.
91 D2 Bataysk Russia
130 B2 Batchawana Mountain h. Can.
50 C1 Batchelor Austr.
63 C3 Bâtdâmbâng Cambodia
53 D3 Batemans Bay Austr.
142 B1 Batesville U.S.A.
89 D2 Batetskiy Russia
99 B3 Bath U.K.
74 B1 Bathinda India
53 C2 Bathurst Austr.
131 D2 Bathurst Can.
126 D2 Bathurst Inlet Can.
126 D2 Bathurst Inlet (abandoned) Can.
50 C1 Bathurst Island Austr.
126 E1 Bathurst Island Can.
78 B1 Bāṭin, Wādī al watercourse Asia
81 C2 Batman Turkey
115 C1 Batna Alg.
142 B2 Baton Rouge U.S.A.
144 B2 Batopilas Mex.
118 B2 Batouri Cameroon
154 B1 Batovi Brazil
92 I1 Båtsfjord Norway
73 C4 Batticaloa Sri Lanka
109 B2 Battipaglia Italy
129 D2 Battle r. Can.
140 B2 Battle Creek U.S.A.
135 C2 Battle Mountain U.S.A.
74 B1 Battura Glacier Pak.
117 B4 Batu mt. Eth.
60 A2 Batu, Pulau-pulau is Indon.
81 C1 Bat'umi Georgia
60 B1 Batu Pahat Malaysia
59 C3 Baubau Indon.
115 C3 Bauchi Nigeria
104 B2 Baugé-en-Anjou France
105 D2 Baume-les-Dames France
154 C2 Bauru Brazil
154 B1 Baús Brazil
88 B2 Bauska Latvia
102 C1 Bautzen Ger.
144 B2 Bavispe r. Mex.
87 E3 Bavly Russia
62 A1 Bawdwin Myanmar
61 C2 Bawean i. Indon.
114 B3 Bawku Ghana
146 C2 Bayamo Cuba
Bayan Gol China see Dengkou
68 C1 Bayanhongor Mongolia
70 A2 Bayan Hot China
70 A1 Bayannur China
69 D2 Bayan Shutu China
69 D1 Bayan-Uul Mongolia
64 B2 Bayawan Phil.
80 C1 Bayburt Turkey
140 C2 Bay City MI U.S.A.
139 D3 Bay City TX U.S.A.
86 F2 Baydaratskaya Guba Russia
117 C4 Baydhabo Somalia
81 C2 Bayjī Iraq
Baykal, Ozero l. Russia see Baikal, Lake
83 I3 Baykal'skiy Khrebet mts Russia

76 C2 Baykonyr Kazakh.
87 E3 Baymak Russia
64 B1 Bayombong Phil.
104 B3 Bayonne France
111 C3 Bayramiç Turkey
101 E3 Bayreuth Ger.
78 B3 Bayt al Faqīh Yemen
106 C2 Baza Spain
106 C2 Baza, Sierra de mts Spain
74 A1 Bāzārak Afgh.
76 A2 Bazardyuzyu, Gora mt. Azer./ Russia
104 B3 Bazas France
74 A2 Bazdar Pak.
70 A2 Bazhong China
79 D2 Bazmān Iran
79 D2 Bazmān, Kūh-e mt. Iran
121 □D2 Bé, Nosy i. Madag.
99 D3 Beachy Head U.K.
123 C3 Beacon Bay S. Africa
50 B1 Beagle Gulf Austr.
121 □D2 Beal an Mhuirthead Ireland
97 B1 Beal an Mhuirthead Ireland
130 B2 Beardmore Can.
Bear Island Arctic Ocean see Bjørnøya
134 E1 Bear Paw Mountain U.S.A.
147 C3 Beata, Cabo c. Dom. Rep.
147 C3 Beata, Isla i. Dom. Rep.
137 D2 Beatrice U.S.A.
135 C3 Beatty U.S.A.
53 D1 Beaudesert Austr.
52 B3 Beaufort Austr.
61 C1 Beaufort Sabah Malaysia
143 D2 Beaufort U.S.A.
160 L2 Beaufort Sea Can./U.S.A.
122 B3 Beaufort West S. Africa
96 B2 Beauly r. U.K.
100 B2 Beaumont Belgium
54 A3 Beaumont N.Z.
139 E2 Beaumont U.S.A.
105 C2 Beaune France
100 B2 Beauraing Belgium
129 E2 Beauséjour Can.
104 C2 Beauvais France
129 D2 Beauval Can.
129 D2 Beaver r. Can.
135 D3 Beaver U.S.A.
126 B2 Beaver Creek Can.
140 B2 Beaver Dam U.S.A.
129 E2 Beaver Hill Lake Can.
140 B1 Beaver Island U.S.A.
128 C2 Beaverlodge Can.
74 B2 Beawar India
154 C2 Bebedouro Brazil
101 D2 Bebra Ger.
106 B1 Becerreá Spain
114 B1 Béchar Alg.
140 C3 Beckley U.S.A.
117 B4 Bedele Eth.
99 C2 Bedford U.K.
140 B3 Bedford U.S.A.
100 C1 Bedum Neth.
53 D2 Beecroft Peninsula Austr.
101 F1 Beelitz Ger.
53 D1 Beenleigh Austr.
80 B2 Beersheba Israel
139 D3 Beeville U.S.A.
53 C3 Bega Austr.
107 D1 Begur, Cap de c. Spain
128 C1 Behchokò Can.
81 D2 Behshahr Iran
69 E1 Bei'an China
71 A3 Beihai China
70 B2 Beijing China
100 C1 Beilen Neth.
96 A2 Beinn Mhòr h. U.K.
121 C2 Beira Moz.

80 B2 Beirut Lebanon
123 C1 Beitbridge Zimbabwe
106 B2 Beja Port.
115 C1 Bejaïa Alg.
106 B1 Béjar Spain
74 A2 Beji r. Pak.
103 E2 Békés Hungary
103 E2 Békéscsaba Hungary
121 □D3 Bekily Madag.
74 A2 Bela Pak.
123 C1 Bela-Bela S. Africa
118 B2 Bélabo Cameroon
109 D2 Bela Crkva Serbia
61 C1 Belaga Sarawak Malaysia
73 B3 Belagavi India
88 C3 Belarus country Europe
Belau country N. Pacific Ocean see Palau
123 D2 Bela Vista Moz.
60 A1 Belawan Indon.
83 M2 Belaya r. Russia
103 D1 Bełchatów Pol.
127 G2 Belcher Islands Can.
117 C4 Beledweyne Somalia
151 D2 Belém Brazil
138 B2 Belen U.S.A.
89 E3 Belev Russia
97 D1 Belfast U.K.
141 F2 Belfast U.S.A.
105 D2 Belfort France
Belgaum India see Belagavi
100 B2 Belgium country Europe
91 D1 Belgorod Russia
109 D2 Belgrade Serbia
134 D1 Belgrade U.S.A.
60 B2 Belinyu Indon.
61 B2 Belitung i. Indon.
146 B3 Belize Belize
146 B3 Belize country Central America
83 K1 Bel'kovskiy, Ostrov i. Russia
128 B2 Bella Bella Can.
104 C2 Bellac France
128 B2 Bella Coola Can.
53 C1 Bellata Austr.
136 C2 Belle Fourche U.S.A.
136 C2 Belle Fourche r. U.S.A.
143 D3 Belle Glade U.S.A.
104 B2 Belle-Île i. France
131 E1 Belle Isle i. Can.
131 E1 Belle Isle, Strait of Can.
130 C2 Belleville Can.
140 B3 Belleville IL U.S.A.
137 D3 Belleville KS U.S.A.
134 B1 Bellevue U.S.A.
134 B1 Bellingham U.S.A.
55 O2 Bellingshausen Sea Antarctica
105 D2 Bellinzona Switz.
108 B1 Belluno Italy
122 A3 Bellville S. Africa
155 E1 Belmonte Brazil
146 B3 Belmopan Belize
69 E1 Belogorsk Russia
121 □D3 Beloha Madag.
155 D1 Belo Horizonte Brazil
140 B2 Beloit U.S.A.
86 C2 Belomorsk Russia
91 D3 Belorechensk Russia
87 E3 Beloretsk Russia
Belorussia country Europe see Belarus
86 F2 Beloyarskiy Russia
86 C2 Beloye, Ozero l. Russia
Beloye More sea Russia see White Sea
86 C2 Belozersk Russia
77 E2 Belukha, Gora mt. Kazakh./ Russia
86 D2 Belush'ye Russia

89 D2 Belyy Russia
82 F2 Belyy, Ostrov i. Russia
137 E1 Bemidji U.S.A.
118 C3 Bena Dibele Dem. Rep. Congo
53 C3 Benalla Austr.
106 B1 Benavente Spain
96 A2 Benbecula i. U.K.
134 B2 Bend U.S.A.
123 C3 Bendearg mt. S. Africa
90 B2 Bender Moldova
52 B3 Bendigo Austr.
121 C2 Bene Moz.
102 C2 Benešov Czech Rep.
109 B2 Benevento Italy
159 C2 Bengal, Bay of sea
         Indian Ocean
73 B3 Bengaluru India
70 B2 Bengbu China
115 E1 Benghazi Libya
61 B1 Bengkayang Indon.
60 B2 Bengkulu Indon.
120 A2 Benguela Angola
96 B1 Ben Hope h. U.K.
152 B2 Beni r. Bol.
119 C2 Beni Dem. Rep. Congo
114 B1 Beni Abbès Alg.
107 C2 Benidorm Spain
114 B1 Beni Mellal Morocco
114 C3 Benin country Africa
114 C4 Benin, Bight of g. Africa
115 C4 Benin City Nigeria
153 C4 Benito Juárez Arg.
150 B2 Benjamin Constant Brazil
144 A1 Benjamín Hill Mex.
59 C3 Benjina Indon.
96 B2 Ben Lawers mt. U.K.
96 B2 Ben Lomond h. U.K.
96 C2 Ben Macdui mt. U.K.
96 A2 Ben More h. U.K.
54 B2 Benmore, Lake N.Z.
96 B1 Ben More Assynt h. U.K.
128 A2 Bennett (abandoned) Can.
83 K1 Bennetta, Ostrov i. Russia
96 B2 Ben Nevis mt. U.K.
141 E2 Bennington U.S.A.
123 C2 Benoni S. Africa
101 D3 Bensheim Ger.
138 A2 Benson U.S.A.
58 C3 Benteng Indon.
140 B2 Benton Harbor U.S.A.
142 B1 Bentonville U.S.A.
115 C4 Benue r. Nigeria
97 B1 Benwee Head Ireland
96 B2 Ben Wyvis mt. U.K.
70 C1 Benxi China
       Beograd Serbia see Belgrade
75 C2 Beohari India
67 B4 Beppu Japan
109 C2 Berane Montenegro
109 C2 Berat Albania
59 C3 Berau, Teluk b. Indon.
116 B3 Berber Sudan
117 C3 Berbera Somalia
118 B2 Berbérati C.A.R.
104 C1 Berck France
91 D2 Berdyans'k Ukr.
90 B2 Berdychiv Ukr.
90 A2 Berehove Ukr.
59 D3 Bereina P.N.G.
76 B3 Bereket Turkm.
129 E2 Berens River Can.
90 B1 Berezhany Ukr.
90 C2 Berezivka Ukr.
90 B1 Berezne Ukr.
86 D2 Bereznik Russia
86 E3 Berezniki Russia
86 F2 Berezovo Russia
107 D1 Berga Spain

111 C3 Bergama Turkey
108 A1 Bergamo Italy
93 E3 Bergen Norway
102 C1 Bergen Ger.
100 B2 Bergen auf Rügen Ger.
100 B2 Bergen op Zoom Neth.
104 C3 Bergerac France
100 C2 Bergheim Ger.
100 C2 Bergisch Gladbach Ger.
122 A1 Bergland Namibia
92 H2 Bergsviken Sweden
83 M3 Beringa, Ostrov i. Russia
100 B2 Beringen Belgium
124 C3 Bering Sea N. Pacific Ocean
83 N2 Bering Strait Russia/U.S.A.
135 B3 Berkeley U.S.A.
100 B1 Berkhout Neth.
55 Q2 Berkner Sub-glacial Island
         Antarctica
110 B2 Berkovitsa Bulg.
92 I1 Berlevåg Norway
101 F1 Berlin Ger.
141 E2 Berlin U.S.A.
101 E2 Berlingerode Ger.
53 D3 Bermagui Austr.
144 B2 Bermejillo Mex.
152 B3 Bermejo Bol.
131 D1 Bermen, Lac l. Can.
125 K4 Bermuda terr.
         N. Atlantic Ocean
105 D2 Bern Switz.
101 E2 Bernburg (Saale) Ger.
127 F2 Bernier Bay Can.
100 C3 Bernkastel-Kues Ger.
121 □D3 Beroroha Madag.
52 B2 Berri Austr.
107 D2 Berrouaghia Alg.
146 C2 Berry Islands Bahamas
100 C1 Bersenbrück Ger.
90 B2 Bershad' Ukr.
131 D1 Berté, Lac l. Can.
118 B2 Bertoua Cameroon
150 B2 Beruri Brazil
98 B1 Berwick-upon-Tweed U.K.
91 C2 Beryslav Ukr.
121 □D2 Besalampy Madag.
105 D2 Besançon France
129 D2 Besnard Lake Can.
142 C2 Bessemer U.S.A.
76 B2 Besshoky, Gora h. Kazakh.
106 B1 Betanzos Spain
118 B2 Bétaré Oya Cameroon
122 A2 Bethanie Namibia
123 C2 Bethlehem S. Africa
141 D2 Bethlehem U.S.A.
121 □D3 Betioky Madag.
77 D3 Betpakdala plain Kazakh.
121 □D3 Betroka Madag.
131 D2 Betsiamites Can.
121 □D2 Betsiboka r. Madag.
137 E2 Bettendorf U.S.A.
75 C2 Bettiah India
75 B2 Betul India
75 B2 Betwa r. India
98 B1 Betws-y-coed U.K.
98 C2 Beverley U.K.
101 D2 Beverungen Ger.
100 B1 Beverwijk Neth.
99 D3 Bexhill U.K.
111 C3 Beykoz Turkey
114 B4 Beyla Guinea
76 B2 Beyneu Kazakh.
80 B1 Beypazarı Turkey
       Beyrouth Lebanon see Beirut
80 B2 Beyşehir Turkey
80 B2 Beyşehir Gölü l. Turkey
91 D2 Beysug r. Russia
88 C2 Bezhanitsy Russia

89 E2 Bezhetsk Russia
105 C3 Béziers France
75 C2 Bhadrak India
73 B3 Bhadravati India
75 C2 Bhagalpur India
74 A2 Bhairi Hol mt. Pak.
74 B1 Bhakkar Pak.
62 A1 Bhamo Myanmar
75 C2 Bhanjanagar India
74 B2 Bharatpur India
74 B2 Bharuch India
74 B2 Bhavnagar India
75 C3 Bhawanipatna India
123 D2 Bhekuzulu S. Africa
74 B2 Bhilwara India
73 B3 Bhima r. India
123 C3 Bhisho S. Africa
74 B2 Bhiwani India
123 C3 Bhongweni S. Africa
74 B2 Bhopal India
75 C2 Bhubaneshwar India
74 A2 Bhuj India
74 B2 Bhusawal India
75 D2 Bhutan country Asia
59 D3 Biak Indon.
59 D3 Biak i. Indon.
103 E1 Biała Podlaska Pol.
103 D1 Białogard Pol.
103 E1 Białystok Pol.
109 C3 Bianco Italy
104 B3 Biarritz France
105 D2 Biasca Switz.
66 D2 Bibai Japan
120 A2 Bibala Angola
102 B2 Biberach an der Riß Ger.
115 C4 Bida Nigeria
141 E2 Biddeford U.S.A.
99 A3 Bideford U.K.
101 D2 Biedenkopf Ger.
105 D2 Biel/Bienne Switz.
101 D1 Bielefeld Ger.
108 A1 Biella Italy
103 D2 Bielsko-Biała Pol.
63 B2 Biên Hoa Vietnam
130 C1 Bienville, Lac l. Can.
100 B3 Bièvre Belgium
118 B3 Bifoun Gabon
111 C2 Biga Turkey
123 D2 Big Bend Swaziland
129 D2 Biggar Can.
96 C3 Biggar U.K.
134 D1 Big Hole r. U.S.A.
136 B1 Bighorn r. U.S.A.
136 B2 Bighorn Mountains U.S.A.
139 C2 Big Lake U.S.A.
140 B2 Big Rapids U.S.A.
129 D2 Big River Can.
129 E2 Big Sand Lake Can.
137 D2 Big Sioux r. U.S.A.
139 C2 Big Spring U.S.A.
134 E1 Big Timber U.S.A.
130 B1 Big Trout Lake Can.
130 A1 Big Trout Lake Can.
109 C2 Bihać Bos. & Herz.
75 C2 Bihar Sharif India
110 B1 Bihor, Vârful mt. Romania
114 A3 Bijagós, Arquipélago dos is
         Guinea-Bissau
81 C2 Bijār Iran
109 C2 Bijeljina Bos. & Herz.
109 C2 Bijelo Polje Montenegro
71 A3 Bijie China
74 B2 Bikaner India
69 E1 Bikin Russia
118 B3 Bikoro Dem. Rep. Congo
79 C2 Bilād Banī Bū 'Alī Oman
75 C2 Bilaspur India
90 C2 Bila Tserkva Ukr.

63 A2 **Bilauktaung Range** *mts* Myanmar/Thai.
106 C1 **Bilbao** Spain
111 C2 **Bilecik** Turkey
103 E1 **Biłgoraj** Pol.
90 C2 **Bilhorod-Dnistrovs'kyy** Ukr.
119 C2 **Bili** Dem. Rep. Congo
83 M2 **Bilibino** Russia
134 E1 **Billings** U.S.A.
99 B3 **Bill of Portland** *hd* U.K.
115 D3 **Bilma** Niger
115 D3 **Bilma, Grand Erg de** *des.* Niger
51 E2 **Biloela** Austr.
91 C2 **Bilohirs'k** Ukr.
90 B1 **Bilohir"ya** Ukr.
91 C1 **Bilopillya** Ukr.
91 D2 **Bilovods'k** Ukr.
142 C2 **Biloxi** U.S.A.
51 C2 **Bilpa Morea Claypan** *salt flat* Austr.
101 E2 **Bilshausen** Ger.
115 E3 **Biltine** Chad
90 C2 **Bilyayivka** Ukr.
143 E3 **Bimini Islands** Bahamas
75 B2 **Bina-Etawa** India
59 C3 **Binaija, Gunung** *mt.* Indon.
118 B3 **Bindu** Dem. Rep. Congo
121 C2 **Bindura** Zimbabwe
107 D1 **Binéfar** Spain
53 D1 **Bingara** Austr.
100 C3 **Bingen am Rhein** Ger.
141 F1 **Bingham** U.S.A.
141 D2 **Binghamton** U.S.A.
81 C2 **Bingöl** Turkey
60 A1 **Binjai** Indon.
60 B2 **Bintuhan** Indon.
61 C1 **Bintulu** *Sarawak* Malaysia
70 B2 **Binzhou** China
109 C2 **Biograd na Moru** Croatia
118 A2 **Bioko** *i.* Equat. Guinea
118 C1 **Birao** C.A.R.
75 C2 **Biratnagar** Nepal
128 C2 **Birch Mountains** Can.
51 C2 **Birdsville** Austr.
80 B2 **Birecik** Turkey
60 A1 **Bireun** Indon.
75 C2 **Birganj** Nepal
154 B2 **Birigui** Brazil
76 B3 **Bīrjand** Iran
98 B2 **Birkenhead** U.K.
114 B2 **Bir Lahlou** Western Sahara
99 C2 **Birmingham** U.K.
142 C2 **Birmingham** U.S.A.
114 A2 **Bîr Mogreïn** Maur.
115 C3 **Birnin-Kebbi** Nigeria
115 C3 **Birnin Konni** Niger
69 E1 **Birobidzhan** Russia
97 C2 **Birr** Ireland
96 C1 **Birsay** U.K.
78 A2 **Bi'r Shalatayn** Egypt
91 D1 **Biryuch** Russia
88 B2 **Biržai** Lith.
75 B2 **Bisalpur** India
138 B2 **Bisbee** U.S.A.
104 A2 **Biscay, Bay of** *sea* France/Spain
102 C2 **Bischofshofen** Austria
130 B2 **Biscotasi Lake** Can.
77 D2 **Bishkek** Kyrg.
135 C3 **Bishop** U.S.A.
69 E1 **Bishui** China
150 B1 **Bisinaca** Col.
115 C1 **Biskra** Alg.
136 C1 **Bismarck** U.S.A.
48 D3 **Bismarck Sea** *sea* P.N.G.
107 D2 **Bissa, Djebel** *mt.* Alg.
114 A3 **Bissau** Guinea-Bissau
129 E2 **Bissett** Can.
128 C2 **Bistcho Lake** Can.
110 B1 **Bistrița** Romania
110 C1 **Bistrița** *r.* Romania
100 C3 **Bitburg** Ger.
105 D2 **Bitche** France
115 D3 **Bitkine** Chad
109 D2 **Bitola** Macedonia
109 C2 **Bitonto** Italy
122 A3 **Bitterfontein** S. Africa
134 D1 **Bitterroot** *r.* U.S.A.
134 C1 **Bitterroot Range** *mts* U.S.A.
91 D1 **Bityug** *r.* Russia
115 D3 **Biu** Nigeria
67 C3 **Biwa-ko** *l.* Japan
68 B1 **Biysk** Russia
115 C1 **Bizerte** Tunisia
92 □A2 **Bjargtangar** *hd* Iceland
92 G3 **Bjästa** Sweden
92 G2 **Bjerkvik** Norway
82 C2 **Bjørnøya** *i.* Arctic Ocean
114 B3 **Bla** Mali
137 E3 **Black** *r.* U.S.A.
51 D2 **Blackall** Austr.
98 B2 **Blackburn** U.K.
134 D2 **Blackfoot** U.S.A.
102 B2 **Black Forest** *mts* Ger.
136 C2 **Black Hills** U.S.A.
96 B2 **Black Isle** *pen.* U.K.
129 D2 **Black Lake** Can.
129 D2 **Black Lake** Can.
99 B3 **Black Mountains** *hills* U.K.
98 B2 **Blackpool** U.K.
62 B1 **Black River** *r.* Vietnam
140 A2 **Black River Falls** U.S.A.
140 C3 **Blacksburg** U.S.A.
80 B1 **Black Sea** Asia/Europe
97 A1 **Blacksod Bay** Ireland
97 C2 **Blackstairs Mountains** *hills* Ireland
114 B4 **Black Volta** *r.* Africa
97 C2 **Blackwater** *r.* Ireland
128 B1 **Blackwater Lake** Can.
50 A3 **Blackwood** *r.* Austr.
76 A2 **Blagodarnyy** Russia
110 B2 **Blagoevgrad** Bulg.
69 E1 **Blagoveshchensk** Russia
129 D2 **Blaine Lake** Can.
137 D2 **Blair** U.S.A.
96 C2 **Blair Atholl** U.K.
96 C2 **Blairgowrie** U.K.
143 D2 **Blakely** U.S.A.
105 D2 **Blanc, Mont** *mt.* France/Italy
153 B4 **Blanca, Bahía** *b.* Arg.
52 A1 **Blanche, Lake** *imp. l.* Austr.
152 B2 **Blanco** *r.* Bol.
131 E1 **Blanc-Sablon** Can.
92 □A2 **Blanda** *r.* Iceland
99 B3 **Blandford Forum** U.K.
135 E3 **Blanding** U.S.A.
107 D1 **Blanes** Spain
60 A1 **Blangkejeren** Indon.
100 C2 **Blankenheim** Ger.
100 C2 **Blankenrath** Ger.
103 D2 **Blansko** Czech Rep.
121 C2 **Blantyre** Malawi
98 C1 **Blaydon** U.K.
53 C2 **Blayney** Austr.
54 B2 **Blenheim** N.Z.
115 C1 **Blida** Alg.
130 B2 **Blind River** Can.
123 C2 **Bloemfontein** S. Africa
123 C2 **Bloemhof** S. Africa
123 C2 **Bloemhof Dam** S. Africa
92 □A2 **Blönduós** Iceland
97 B1 **Bloody Foreland** *pt* Ireland
138 B1 **Bloomfield** U.S.A.
140 B2 **Bloomington** *IL* U.S.A.
140 B3 **Bloomington** *IN* U.S.A.
140 C3 **Bluefield** U.S.A.
146 B3 **Bluefields** Nic.
53 C2 **Blue Mountains** Austr.
134 C1 **Blue Mountains** U.S.A.
116 B3 **Blue Nile** *r.* Eth./Sudan
126 D2 **Bluenose Lake** Can.
140 C3 **Blue Ridge** *mts* U.S.A.
97 B1 **Blue Stack Mountains** *hills* Ireland
54 A3 **Bluff** N.Z.
135 E3 **Bluff** U.S.A.
152 D3 **Blumenau** Brazil
52 A2 **Blyth** Austr.
98 C1 **Blyth** U.K.
135 D4 **Blythe** U.S.A.
142 C1 **Blytheville** U.S.A.
114 A4 **Bo** Sierra Leone
64 B1 **Boac** Phil.
151 D2 **Boa Esperança, Açude** *resr* Brazil
150 B1 **Boa Vista** Brazil
53 C2 **Bobadah** Austr.
71 B3 **Bobai** China
121 □D2 **Bobaomby, Tanjona** *c.* Madag.
114 B3 **Bobo-Dioulasso** Burkina Faso
89 F3 **Bobrov** Russia
91 C1 **Bobrovytsya** Ukr.
91 C2 **Bobrynets'** Ukr.
121 □D3 **Boby** *mt.* Madag.
150 B2 **Boca do Acre** Brazil
155 D1 **Bocaiúva** Brazil
118 B2 **Bocaranga** C.A.R.
103 E2 **Bochnia** Pol.
100 C2 **Bocholt** Ger.
100 C2 **Bochum** Ger.
101 E1 **Bockenem** Ger.
118 B2 **Boda** C.A.R.
83 I3 **Bodaybo** Russia
115 D3 **Bodélé** *reg.* Chad
92 H2 **Boden** Sweden
99 A3 **Bodmin** U.K.
99 A3 **Bodmin Moor** *moorland* U.K.
92 F2 **Bodø** Norway
111 C3 **Bodrum** Turkey
118 C3 **Boende** Dem. Rep. Congo
142 C2 **Bogalusa** U.S.A.
114 B3 **Bogandé** Burkina Faso
68 B2 **Bogda Shan** *mts* China
53 D1 **Boggabilla** Austr.
97 B2 **Boggeragh Mountains** *hills* Ireland
100 B3 **Bogny-sur-Meuse** France
97 C2 **Bog of Allen** *reg.* Ireland
53 C3 **Bogong, Mount** Austr.
60 B2 **Bogor** Indon.
89 E3 **Bogoroditsk** Russia
150 A1 **Bogotá** Col.
82 G3 **Bogotol** Russia
83 H3 **Boguchany** Russia
91 E2 **Boguchar** Russia
70 B2 **Bo Hai** *g.* China
100 A3 **Bohain-en-Vermandois** France
70 B2 **Bohai Wan** *b.* China
123 C2 **Bohlokong** S. Africa
102 C2 **Böhmer Wald** *mts* Ger.
91 D1 **Bohodukhiv** Ukr.
64 B2 **Bohol** *i.* Phil.
64 B2 **Bohol Sea** Phil.
68 B2 **Bohu** China
155 C2 **Boi, Ponta do** *pt* Brazil
154 B1 **Bois** *r.* Brazil
126 C2 **Bois, Lac des** *l.* Can.
134 C2 **Boise** U.S.A.
139 C1 **Boise City** U.S.A.
123 C2 **Boitumelong** S. Africa

101	E1	Boizenburg/Elbe Ger.
76	B3	Bojnürd Iran
118	B3	Bokatola Dem. Rep. Congo
114	A3	Boké Guinea
118	C3	Bokele Dem. Rep. Congo
93	E4	Boknafjorden *sea chan.* Norway
115	D3	Bokoro Chad
63	A2	Bokpyin Myanmar
89	D2	Boksitogorsk Russia
122	B2	Bokspits Botswana
114	A3	Bolama Guinea-Bissau
63	B2	Bolaven, Phouphiang *plat.* Laos
104	C2	Bolbec France
77	E2	Bole China
118	B3	Boleko Dem. Rep. Congo
114	B3	Bolgatanga Ghana
90	B2	Bolhrad Ukr.
66	B1	Boli China
62	B2	Bolikhamxai Laos
110	C2	Bolintin-Vale Romania
137	E3	Bolívar U.S.A.
150	A1	Bolívar, Pico *mt.* Venez.
152	B2	Bolivia *country* South America
89	E3	Bolkhov Russia
105	C3	Bollène France
93	G3	Bollnäs Sweden
53	C1	Bollon Austr.
101	E2	Bollstedt Ger.
93	F4	Bolmen *l.* Sweden
118	B3	Bolobo Dem. Rep. Congo
108	B2	Bologna Italy
89	D2	Bologovo Russia
89	D2	Bologoye Russia
118	B3	Bolomba Dem. Rep. Congo
83	H1	Bol'shevik, Ostrov *i.* Russia
86	E2	Bol'shezemel'skaya Tundra *lowland* Russia
66	B2	Bol'shoy Kamen' Russia
		Bol'shoy Kavkaz *mts* Asia/Europe *see* Caucasus
83	K2	Bol'shoy Lyakhovskiy, Ostrov *i.* Russia
100	B1	Bolsward Neth.
98	B2	Bolton U.K.
80	B1	Bolu Turkey
92	□A2	Bolungarvík Iceland
108	B1	Bolzano Italy
118	B3	Boma Dem. Rep. Congo
53	D2	Bomaderry Austr.
53	C3	Bombala Austr.
		Bombay India *see* Mumbai
155	C1	Bom Despacho Brazil
75	D2	Bomdila India
151	D3	Bom Jesus da Lapa Brazil
155	D2	Bom Jesus do Itabapoana Brazil
115	D1	Bon, Cap *c.* Tunisia
147	D3	Bonaire *i.* West Indies
147	D3	Bonaire *mun.* West Indies
50	B1	Bonaparte Archipelago *is* Austr.
131	E2	Bonavista Can.
131	E2	Bonavista Bay Can.
118	C3	Bondo Dem. Rep. Congo
114	B4	Bondoukou Côte d'Ivoire
58	C3	Bonerate, Kepulauan *is* Indon.
155	C1	Bonfinópolis de Minas Brazil
117	B4	Bonga Eth.
75	D2	Bongaigaon India
118	C3	Bongandanga Dem. Rep. Congo
118	C2	Bongo, Massif des *mts* C.A.R.
115	D3	Bongor Chad
114	B4	Bongouanou Côte d'Ivoire
63	B2	Bông Sơn Vietnam
139	D2	Bonham U.S.A.

105	D3	Bonifacio *Corsica* France
108	A2	Bonifacio, Strait of France/Italy
100	C2	Bonn Ger.
134	E2	Bonners Ferry U.S.A.
50	A3	Bonnie Rock Austr.
129	C2	Bonnyville Can.
108	A2	Bonorva *Sardinia* Italy
61	C1	Bontang Indon.
64	B1	Bontoc Phil.
58	B3	Bontosunggu Indon.
123	C3	Bontrug S. Africa
52	B2	Booligal Austr.
53	C1	Boomi Austr.
53	D1	Boonah Austr.
137	E2	Boone U.S.A.
142	C2	Booneville U.S.A.
137	E3	Boonville U.S.A.
53	C2	Boorowa Austr.
126	F2	Boothia, Gulf of Can.
126	E2	Boothia Peninsula Can.
100	C2	Boppard Ger.
144	B2	Boquilla, Presa de la *resr* Mex.
109	D2	Bor Serbia
117	B4	Bor South Sudan
80	B2	Bor Turkey
121	□E2	Boraha, Nosy *i.* Madag.
76	B2	Borankul Kazakh.
93	F4	Borås Sweden
79	C2	Borāzjān Iran
150	C2	Borba Brazil
104	B3	Bordeaux France
127	F3	Borden Peninsula Can.
52	B3	Bordertown Austr.
107	D2	Bordj Bou Arréridj Alg.
107	D2	Bordj Bounaama Alg.
115	C1	Bordj Messaouda Alg.
115	C2	Bordj Omer Driss Alg.
94	B1	Borðoy *i.* Faroe Is
92	□A3	Borgarnes Iceland
108	A1	Borgosesia Italy
87	D3	Borisoglebsk Russia
89	E2	Borisoglebskiy Russia
91	D1	Borisovka Russia
100	C2	Borken Ger.
92	G2	Borkenes Norway
100	C1	Borkum Ger.
100	C1	Borkum *i.* Ger.
93	G3	Borlänge Sweden
101	F2	Borna Ger.
61	C1	Borneo *i.* Asia
93	F4	Bornholm *i.* Denmark
111	C3	Bornova Turkey
89	D2	Borodyanka Ukr.
89	E2	Borovichi Russia
89	E2	Borovsk Russia
107	C2	Borriana Spain
51	C1	Borroloola Austr.
110	B1	Borşa Romania
90	B2	Borshchiv Ukr.
69	D1	Borshchovochnyy Khrebet *mts* Russia
101	E1	Börßum Ger.
		Bortala China *see* Bole
81	C2	Borüjerd Iran
65	B2	Boryeong S. Korea
90	A2	Boryslav Ukr.
90	C1	Boryspil' Ukr.
91	C1	Borzna Ukr.
69	D1	Borzya Russia
109	C1	Bosanska Dubica Bos. & Herz.
109	C1	Bosanska Gradiška Bos. & Herz.
109	C2	Bosanska Krupa Bos. & Herz.
109	C1	Bosanski Novi Bos. & Herz.
109	C2	Bosansko Grahovo Bos. & Herz.
123	C2	Boshof S. Africa

109	C2	Bosnia and Herzegovina *country* Europe
118	B3	Bosobolo Dem. Rep. Congo
110	C2	Bosporus *str.* Turkey
118	B2	Bossangoa C.A.R.
118	B2	Bossembélé C.A.R.
68	A2	Bosten Hu *l.* China
98	C2	Boston U.K.
141	E2	Boston U.S.A.
142	B1	Boston Mountains U.S.A.
53	D2	Botany Bay Austr.
110	B2	Botev *mt.* Bulg.
93	G3	Bothnia, Gulf of Fin./Sweden
110	C1	Botoşani Romania
123	C2	Botshabelo S. Africa
120	B3	Botswana *country* Africa
109	C3	Botte Donato, Monte *mt.* Italy
136	C1	Bottineau U.S.A.
100	C2	Bottrop Ger.
154	C2	Botucatu Brazil
114	B4	Bouaké Côte d'Ivoire
118	B2	Bouar C.A.R.
114	B1	Bou Arfa Morocco
131	D2	Bouctouche Can.
107	C2	Bougaa Alg.
114	B3	Bougouni Mali
100	B3	Bouillon Belgium
107	C2	Bouira Alg.
114	A2	Boujdour Western Sahara
136	C2	Boulder U.S.A.
135	D3	Boulder City U.S.A.
51	C2	Boulia Austr.
104	C2	Boulogne-Billancourt France
99	D3	Boulogne-sur-Mer France
118	B3	Boumango Gabon
118	B2	Boumba *r.* Cameroon
107	D2	Boumerdès Alg.
114	B4	Bouna Côte d'Ivoire
114	B4	Boundiali Côte d'Ivoire
134	D2	Bountiful U.S.A.
114	B3	Bourem Mali
104	C2	Bourganeuf France
105	D2	Bourg-en-Bresse France
104	C2	Bourges France
53	C2	Bourke Austr.
99	C3	Bournemouth U.K.
115	C1	Bou Saâda Alg.
115	D3	Bousso Chad
114	A3	Boutilimit Maur.
129	C3	Bow *r.* Can.
51	D2	Bowen Austr.
129	C3	Bow Island Can.
140	B3	Bowling Green KY U.S.A.
140	C2	Bowling Green OH U.S.A.
136	C1	Bowman U.S.A.
101	D3	Boxberg Ger.
100	B2	Boxtel Neth.
80	B1	Boyabat Turkey
		Boyang China *see* Poyang
97	C2	Boyle Ireland
97	C2	Boyne *r.* Ireland
136	B2	Boysen Reservoir U.S.A.
152	B3	Boyuíbe Bol.
111	C3	Bozburun Turkey
111	C3	Bozcaada *i.* Turkey
111	C3	Bozdağ *mt.* Turkey
111	C3	Boz Dağları *mts* Turkey
111	C3	Bozdoğan Turkey
134	D1	Bozeman U.S.A.
118	B2	Bozoum C.A.R.
111	D3	Bozüyük Turkey
109	C2	Brač *i.* Croatia
130	C2	Bracebridge Can.
99	C3	Bracknell U.K.
109	C2	Bradano *r.* Italy
143	D3	Bradenton U.S.A.

Bradford

98	C2	**Bradford** U.K.
141	D2	**Bradford** U.S.A.
139	D2	**Brady** U.S.A.
96	C2	**Braemar** U.K.
106	B1	**Braga** Port.
151	D2	**Bragança** Brazil
106	B1	**Bragança** Port.
155	C2	**Bragança Paulista** Brazil
89	D3	**Brahin** Belarus
73	C3	**Brahmapur** India
62	A1	**Brahmaputra** r. China/India
110	C1	**Brăila** Romania
137	E1	**Brainerd** U.S.A.
99	D3	**Braintree** U.K.
100	B2	**Braives** Belgium
101	D1	**Brake (Unterweser)** Ger.
100	D1	**Bramsche** Ger.
150	B2	**Branco** r. Brazil
101	F1	**Brandenburg an der Havel** Ger.
129	E3	**Brandon** Can.
97	A2	**Brandon Mountain** h. Ireland
122	B3	**Brandvlei** S. Africa
103	D1	**Braniewo** Pol.
130	B2	**Brantford** Can.
131	D2	**Bras d'Or Lake** Can.
155	D2	**Brasil, Planalto do** plat. Brazil
154	C1	**Brasília** Brazil
155	C1	**Brasília de Minas** Brazil
88	C2	**Braslaw** Belarus
110	C1	**Braşov** Romania
103	D2	**Bratislava** Slovakia
83	H3	**Bratsk** Russia
102	C2	**Braunau am Inn** Austria
101	E1	**Braunschweig** Ger.
92	□A2	**Bráarhtindt** Iceland
		Bravo del Norte, Rio r. Mex./U.S.A. see **Rio Grande**
135	C4	**Brawley** U.S.A.
97	C2	**Bray** Ireland
150	C2	**Brazil** country S. America
154	C1	**Brazlândia** Brazil
139	D3	**Brazos** r. U.S.A.
118	B3	**Brazzaville** Congo
109	C2	**Brčko** Bos. & Herz.
96	C2	**Brechin** U.K.
100	B2	**Brecht** Belgium
139	D2	**Breckenridge** U.S.A.
103	D2	**Břeclav** Czech Rep.
99	B3	**Brecon** U.K.
99	B3	**Brecon Beacons** reg. U.K.
100	B2	**Breda** Neth.
122	B3	**Bredasdorp** S. Africa
102	B2	**Bregenz** Austria
92	H1	**Breivikbotn** Norway
92	E3	**Brekstad** Norway
101	D1	**Bremen** Ger.
101	D1	**Bremerhaven** Ger.
134	B1	**Bremerton** U.S.A.
101	D1	**Bremervörde** Ger.
139	D2	**Brenham** U.S.A.
108	B1	**Brennero** Italy
102	C2	**Brenner Pass** Austria/Italy
99	D3	**Brentwood** U.K.
108	B1	**Brescia** Italy
108	B1	**Bressanone** Italy
96	□	**Bressay** i. U.K.
104	B2	**Bressuire** France
88	B3	**Brest** Belarus
104	B2	**Brest** France
142	C3	**Breton Sound** b. U.S.A.
151	C2	**Breves** Brazil
53	C1	**Brewarrina** Austr.
134	C1	**Brewster** U.S.A.
89	E2	**Breytovo** Russia
109	C1	**Brezovo Polje** plain Croatia
118	C2	**Bria** C.A.R.
105	D3	**Briançon** France
90	B2	**Briceni** Moldova
99	B3	**Bridgend** U.K.
141	E2	**Bridgeport** CT U.S.A.
136	C2	**Bridgeport** NE U.S.A.
147	E3	**Bridgetown** Barbados
131	D2	**Bridgewater** Can.
99	B3	**Bridgwater** U.K.
98	C1	**Bridlington** U.K.
98	C1	**Bridlington Bay** U.K.
105	D2	**Brig** Switz.
134	D2	**Brigham City** U.S.A.
54	B3	**Brighton** N.Z.
99	C3	**Brighton** U.K.
105	D3	**Brignoles** France
101	D2	**Brilon** Ger.
109	C2	**Brindisi** Italy
91	D2	**Brin'kovskaya** Russia
53	D1	**Brisbane** Austr.
99	B3	**Bristol** U.K.
143	D1	**Bristol** U.S.A.
99	A3	**Bristol Channel** est. U.K.
128	B2	**British Columbia** prov. Can.
56	G7	**British Indian Ocean Territory** terr. Indian Ocean
95	B2	**British Isles** is Europe
123	C2	**Brits** S. Africa
122	B3	**Britstown** S. Africa
104	C2	**Brive-la-Gaillarde** France
106	C1	**Briviesca** Spain
103	D2	**Brno** Czech Rep.
143	D2	**Broad** r. U.S.A.
130	C1	**Broadback** r. Can.
96	C3	**Broad Law** h. U.K.
136	B1	**Broadus** U.S.A.
129	D2	**Brochet** Can.
129	D2	**Brochet, Lac** l. Can.
131	D2	**Brochet, Lac au** l. Can.
101	E1	**Bröckel** Ger.
126	D1	**Brock Island** Can.
130	C2	**Brockville** Can.
127	F2	**Brodeur Peninsula** Can.
96	B3	**Brodick** U.K.
103	D1	**Brodnica** Pol.
90	B1	**Brody** Ukr.
139	D1	**Broken Arrow** U.S.A.
137	D2	**Broken Bow** U.S.A.
52	B2	**Broken Hill** Austr.
99	B2	**Bromsgrove** U.K.
93	E4	**Brønderslev** Denmark
92	F2	**Brønnøysund** Norway
64	A2	**Brooke's Point** Phil.
142	B2	**Brookhaven** U.S.A.
134	B2	**Brookings** OR U.S.A.
137	D2	**Brookings** SD U.S.A.
129	C2	**Brooks** Can.
126	B2	**Brooks Range** mts U.S.A.
143	D3	**Brooksville** U.S.A.
96	B2	**Broom, Loch** inlet U.K.
50	B1	**Broome** Austr.
134	B2	**Brothers** U.S.A.
		Broughton Island Can. see **Qikiqtarjuaq**
90	C1	**Brovary** Ukr.
139	C2	**Brownfield** U.S.A.
128	C3	**Browning** U.S.A.
142	C1	**Brownsville** TN U.S.A.
139	D3	**Brownsville** TX U.S.A.
139	D2	**Brownwood** U.S.A.
92	□A2	**Brú** Iceland
104	C2	**Bruay-la-Buissière** France
140	B1	**Bruce Crossing** U.S.A.
103	D2	**Bruck an der Mur** Austria
		Bruges Belgium see **Brugge**
100	A2	**Brugge** Belgium
128	C2	**Brûlé** Can.
151	D3	**Brumado** Brazil
93	F3	**Brumunddal** Norway
61	C1	**Brunei** country Asia
102	C2	**Brunico** Italy
101	D1	**Brunsbüttel** Ger.
143	D2	**Brunswick** GA U.S.A.
141	F2	**Brunswick** ME U.S.A.
53	D1	**Brunswick Heads** Austr.
136	C2	**Brush** U.S.A.
		Brussel Belgium see **Brussels**
100	B2	**Brussels** Belgium
		Bruxelles Belgium see **Brussels**
139	D2	**Bryan** U.S.A.
89	D3	**Bryansk** Russia
91	D2	**Bryukhovetskaya** Russia
103	D1	**Brzeg** Pol.
114	A3	**Buba** Guinea-Bissau
80	B2	**Bucak** Turkey
150	A1	**Bucaramanga** Col.
53	C3	**Buchan** Austr.
114	A4	**Buchanan** Liberia
110	C2	**Bucharest** Romania
65	B2	**Bucheon** S. Korea
101	D1	**Buchholz in der Nordheide** Ger.
110	C1	**Bucin, Pasul** pass Romania
101	D1	**Bückeburg** Ger.
138	A2	**Buckeye** U.S.A.
141	A3	**Buckhaven** U.K.
96	C2	**Buckie** U.K.
51	C1	**Buckingham Bay** Austr.
51	D2	**Buckland Tableland** reg. Austr.
52	A2	**Buckleboo** Austr.
141	F2	**Bucksport** U.S.A.
		Bucureşti Romania see **Bucharest**
89	D3	**Buda-Kashalyova** Belarus
103	D2	**Budapest** Hungary
75	B2	**Budaun** India
108	A2	**Buddusò** Sardinia Italy
83	M3	**Bude** U.K.
87	D4	**Budennovsk** Russia
89	D2	**Budogoshch'** Russia
108	A2	**Budoni** Sardinia Italy
118	A2	**Buea** Cameroon
144	B2	**Buenaventura** Mex.
106	C1	**Buendía, Embalse de** resr Spain
155	D1	**Buenópolis** Brazil
153	C4	**Buenos Aires** Arg.
153	A5	**Buenos Aires, Lago** l. Arg./Chile
141	D2	**Buffalo** NY U.S.A.
136	C1	**Buffalo** SD U.S.A.
136	B2	**Buffalo** WY U.S.A.
129	D2	**Buffalo Narrows** Can.
122	A2	**Buffels** watercourse S. Africa
110	C2	**Buftea** Romania
103	E1	**Bug** r. Pol.
61	C2	**Bugel, Tanjung** pt Indon.
109	C2	**Bugojno** Bos. & Herz.
64	A2	**Bugsuk** i. Phil.
87	E3	**Buguruslan** Russia
110	C1	**Buhuşi** Romania
99	B2	**Builth Wells** U.K.
120	A3	**Buitepos** Namibia
109	D2	**Bujanovac** Serbia
119	C3	**Bujumbura** Burundi
69	D1	**Bukachacha** Russia
119	C3	**Bukavu** Dem. Rep. Congo
60	B2	**Bukittinggi** Indon.
119	D3	**Bukoba** Tanz.
53	D2	**Bulahdelah** Austr.
121	B3	**Bulawayo** Zimbabwe
111	C3	**Buldan** Turkey
123	D2	**Bulembu** Swaziland
69	C1	**Bulgan** Mongolia
110	C2	**Bulgaria** country Europe
54	B2	**Buller** r. N.Z.

52	B1	**Bulloo Downs** Austr.
122	A1	**Büllsport** Namibia
58	C3	**Bulukumba** Indon.
118	B3	**Bulungu** Dem. Rep. Congo
118	C2	**Bumba** Dem. Rep. Congo
62	A1	**Bumhkang** Myanmar
118	B3	**Buna** Dem. Rep. Congo
50	A3	**Bunbury** Austr.
97	C1	**Buncrana** Ireland
119	D3	**Bunda** Tanz.
51	E2	**Bundaberg** Austr.
53	D2	**Bundarra** Austr.
74	B2	**Bundi** India
97	B1	**Bundoran** Ireland
53	C3	**Bungendore** Austr.
67	B4	**Bungo-suidō** sea chan. Japan
119	C3	**Bunia** Dem. Rep. Congo
118	C3	**Bunianga** Dem. Rep. Congo
63	B2	**Buôn Ma Thuột** Vietnam
119	D3	**Bura** Kenya
78	B3	**Buraydah** Saudi Arabia
100	D2	**Burbach** Ger.
117	C4	**Burco** Somalia
100	B1	**Burdaard** Neth.
80	B2	**Burdur** Turkey
117	B3	**Burē** Eth.
99	D2	**Bure** r. U.K.
74	B1	**Burewala** Pak.
101	E1	**Burg** Ger.
110	C2	**Burgas** Bulg.
101	E1	**Burgdorf** Niedersachsen Ger.
101	E1	**Burgdorf** Niedersachsen Ger.
131	E2	**Burgeo** Can.
123	D1	**Burgersfort** S. Africa
100	A2	**Burgh-Haamstede** Neth.
145	C2	**Burgos** Mex.
106	C1	**Burgos** Spain
111	C3	**Burhaniye** Turkey
74	B2	**Burhanpur** India
101	D1	**Burhave (Butjadingen)** Ger.
131	E2	**Burin** Can.
151	D2	**Buriti Bravo** Brazil
155	C1	**Buritis** Brazil
51	C1	**Burketown** Austr.
114	B3	**Burkina Faso** country Africa
134	D2	**Burley** U.S.A.
136	C3	**Burlington** CO U.S.A.
137	E2	**Burlington** IA U.S.A.
143	E1	**Burlington** NC U.S.A.
141	E2	**Burlington** VT U.S.A.
		Burma country Asia see
		Myanmar
134	B2	**Burney** U.S.A.
51	D4	**Burnie** Austr.
98	B2	**Burnley** U.K.
134	C2	**Burns** U.S.A.
128	B2	**Burns Lake** Can.
77	E2	**Burqin** China
52	A2	**Burra** Austr.
109	D2	**Burrel** Albania
97	B2	**Burren** reg. Ireland
53	C2	**Burrendong, Lake** resr Austr.
53	C2	**Burren Junction** Austr.
		Burriana Spain see Borriana
53	C2	**Burrinjuck Reservoir** Austr.
144	B2	**Burro, Serranías del** mts Mex.
111	C2	**Bursa** Turkey
116	B2	**Bür Safājah** Egypt
		Bür Sa'īd Egypt see **Port Said**
130	C1	**Burton, Lac** l. Can.
99	C2	**Burton upon Trent** U.K.
59	C3	**Buru** i. Indon.
119	C3	**Burundi** country Africa
119	C3	**Bururi** Burundi
91	C1	**Buryn'** Ukr.
99	D2	**Bury St Edmunds** U.K.
108	B3	**Busambra, Rocca** mt. Sicily
		Italy

65	B2	**Busan** S. Korea
118	C3	**Busanga** Dem. Rep. Congo
119	D3	**Bushenyi** Uganda
118	C2	**Businga** Dem. Rep. Congo
50	A3	**Busselton** Austr.
139	C3	**Bustamante** Mex.
118	C2	**Buta** Dem. Rep. Congo
119	C3	**Butare** Rwanda
123	C2	**Butha-Buthe** Lesotho
140	D2	**Butler** U.S.A.
59	C3	**Buton** i. Indon.
134	D1	**Butte** U.S.A.
60	B1	**Butterworth** Malaysia
96	A1	**Butt of Lewis** hd U.K.
129	E2	**Button Bay** Can.
64	B2	**Butuan** Phil.
91	E1	**Buturlinovka** Russia
75	C2	**Butwal** Nepal
101	D2	**Butzbach** Ger.
117	C4	**Buulobarde** Somalia
117	C5	**Buur Gaabo** Somalia
117	C4	**Buurhabaka** Somalia
76	C3	**Buxoro** Uzbek.
101	D1	**Buxtehude** Ger.
89	F2	**Buy** Russia
87	D4	**Buynaksk** Russia
69	D1	**Buyr Nuur** l. Mongolia
111	C3	**Büyükmenderes** r. Turkey
110	B1	**Buzău** Romania
121	C2	**Búzi** Moz.
87	E3	**Buzuluk** Russia
110	C2	**Byala** Bulg.
88	C3	**Byalynichy** Belarus
88	D3	**Byarezina** r. Belarus
88	B3	**Byaroza** Belarus
103	D1	**Bydgoszcz** Pol.
88	C3	**Byerazino** Belarus
88	C2	**Byeshankovichy** Belarus
89	D3	**Bykhaw** Belarus
127	F2	**Bylot Island** Can.
53	C2	**Byrock** Austr.
53	D1	**Byron Bay** Austr.
83	J2	**Bytantay** r. Russia
103	D1	**Bytom** Pol.
103	D1	**Bytów** Pol.

C

154	B2	**Caarapó** Brazil
64	B1	**Cabanatuan** Phil.
117	C3	**Cabdul Qaadir** Somalia
106	B2	**Cabeza del Buey** Spain
152	B2	**Cabezas** Bol.
150	A1	**Cabimas** Venez.
120	A1	**Cabinda** Angola
118	B3	**Cabinda** prov. Angola
155	D2	**Cabo Frio** Brazil
155	D2	**Cabo Frio, Ilha do** i. Brazil
130	C2	**Cabonga, Réservoir** resr Can.
51	E2	**Caboolture** Austr.
150	A2	**Cabo Pantoja** Peru
144	A1	**Caborca** Mex.
144	B2	**Cabo San Lucas** Mex.
131	D2	**Cabot Strait** Can.
155	D1	**Cabral, Serra do** mts Brazil
107	D2	**Cabrera, Illa de** i. Spain
106	B1	**Cabrera, Sierra de la** mts
		Spain
129	D2	**Cabri** Can.
107	C2	**Cabriel** r. Spain
152	C3	**Caçador** Brazil
109	D2	**Čačak** Serbia
108	A2	**Caccia, Capo** c. Sardinia Italy
151	C3	**Cáceres** Brazil
106	B2	**Cáceres** Spain
128	B2	**Cache Creek** Can.

114	A3	**Cacheu** Guinea-Bissau
151	C2	**Cachimbo, Serra do** hills
		Brazil
154	B1	**Cachoeira Alta** Brazil
155	D2	**Cachoeiro de Itapemirim**
		Brazil
114	A3	**Cacine** Guinea-Bissau
120	A2	**Cacolo** Angola
154	B1	**Caçu** Brazil
103	D2	**Čadca** Slovakia
101	D1	**Cadenberge** Ger.
145	B2	**Cadereyta** Mex.
140	B2	**Cadillac** U.S.A.
106	B2	**Cádiz** Spain
106	B2	**Cádiz, Golfo de** g. Spain
128	C2	**Cadotte Lake** Can.
104	B2	**Caen** France
98	A2	**Caernarfon** U.K.
98	A2	**Caernarfon Bay** U.K.
152	B3	**Cafayate** Arg.
64	B2	**Cagayan de Oro** Phil.
64	A2	**Cagayan de Tawi-Tawi** i.
		Phil.
108	B2	**Cagli** Italy
108	A3	**Cagliari** Sardinia Italy
108	A3	**Cagliari, Golfo di** b. Sardinia
		Italy
76	B2	**Çagyl** Turkm.
97	B3	**Caha Mountains** hills Ireland
97	A3	**Cahermore** Ireland
97	C2	**Cahir** Ireland
97	A3	**Cahirsiveen** Ireland
97	C2	**Cahore Point** Ireland
104	C3	**Cahors** France
90	B2	**Cahul** Moldova
121	C2	**Caia** Moz.
151	C3	**Caiabis, Serra dos** hills Brazil
120	B2	**Caianda** Angola
154	B1	**Caiapó, Serra de** mts Brazil
154	B1	**Caiapônia** Brazil
147	C2	**Caicos Islands**
		Turks and Caicos Is
96	C2	**Cairngorm Mountains** U.K.
98	A1	**Cairnryan** U.K.
51	D1	**Cairns** Austr.
116	B1	**Cairo** Egypt
98	C2	**Caistor** U.K.
120	A2	**Caiundo** Angola
150	A2	**Cajamarca** Peru
109	C2	**Čakovec** Croatia
123	C3	**Cala** S. Africa
150	B1	**Calabozo** Venez.
110	B2	**Calafat** Romania
153	A6	**Calafate** Arg.
107	C1	**Calahorra** Spain
104	C1	**Calais** France
141	F1	**Calais** U.S.A.
152	B3	**Calama** Chile
64	A1	**Calamian Group** is Phil.
107	C2	**Calamocha** Spain
120	A1	**Calandula** Angola
60	A1	**Calang** Indon.
64	B1	**Calapan** Phil.
110	C2	**Călărași** Romania
107	C1	**Calatayud** Spain
64	B1	**Calayan** i. Phil.
64	B1	**Calbayog** Phil.
151	E2	**Calcanhar, Ponta do** pt
		Brazil
151	C1	**Calçoene** Brazil
		Calcutta India see **Kolkata**
106	B2	**Caldas da Rainha** Port.
154	C1	**Caldas Novas** Brazil
152	A3	**Caldera** Chile
134	C2	**Caldwell** U.S.A.
123	C3	**Caledon** r. Lesotho/S. Africa
122	A3	**Caledon** S. Africa
153	B5	**Caleta Olivia** Arg.

98 A1 Calf of Man i. Isle of Man
128 C2 Calgary Can.
150 A1 Cali Col.
Calicut India see Kozhikode
135 D3 Caliente U.S.A.
135 B2 California state U.S.A.
144 A1 California, Gulf of Mex.
135 B3 California Aqueduct canal U.S.A.
122 B3 Calitzdorp S. Africa
145 E2 Calkini Mex.
52 B1 Callabonna, Lake imp. l. Austr.
96 B2 Callander U.K.
150 A3 Callao Peru
108 B3 Caltagirone Sicily Italy
108 B3 Caltanissetta Sicily Italy
120 A2 Caluquembe Angola
117 D3 Caluula Somalia
105 D3 Calvi Corsica France
107 D2 Calvià Spain
144 B2 Calvillo Mex.
122 A3 Calvinia S. Africa
109 C2 Calvo, Monte mt. Italy
144 B2 Camacho Mex.
120 A2 Camacupa Angola
146 C2 Camagüey Cuba
150 A3 Camaná Peru
154 B1 Camapuã Brazil
145 C2 Camargo Mex.
63 B3 Ca Mau Vietnam
63 B3 Ca Mau, Mui c. Vietnam
63 B2 Cambodia country Asia
99 A3 Camborne U.K.
105 C1 Cambrai France
99 B2 Cambrian Mountains hills U.K.
54 C1 Cambridge N.Z.
99 D2 Cambridge U.K.
141 E2 Cambridge MA U.S.A.
141 D3 Cambridge MD U.S.A.
137 E1 Cambridge MN U.S.A.
140 C2 Cambridge OH U.S.A.
131 D1 Cambrien, Lac l. Can.
53 D2 Camden Austr.
142 B2 Camden AR U.S.A.
141 F2 Camden ME U.S.A.
118 B2 Cameroon country Africa
118 B2 Cameroon Highlands slope Cameroon/Nigeria
151 D2 Cametá Brazil
64 B1 Camiguin i. Phil.
152 B3 Camiri Bol.
151 D2 Camocim Brazil
51 C1 Camooweal Austr.
63 A3 Camorta i. India
153 A5 Campana, Isla i. Chile
122 B3 Campbell S. Africa
54 B2 Campbell, Cape N.Z.
48 F6 Campbell Island N.Z.
128 B2 Campbell River Can.
140 B3 Campbellsville U.S.A.
131 D2 Campbellton Can.
96 B3 Campbeltown U.K.
145 C3 Campeche Mex.
145 C3 Campeche, Bahía de g. Mex.
52 B3 Camperdown Austr.
110 C1 Câmpina Romania
151 E2 Campina Grande Brazil
154 C2 Campinas Brazil
154 C1 Campina Verde Brazil
108 B2 Campobasso Italy
155 C2 Campo Belo Brazil
154 C1 Campo Florido Brazil
152 B3 Campo Gallo Arg.
154 B2 Campo Grande Brazil
151 D2 Campo Maior Brazil
106 B2 Campo Maior Port.
154 B2 Campo Mourão Brazil

155 C1 Campos Altos Brazil
155 C2 Campos do Jordão Brazil
155 C2 Campos dos Goytacazes Brazil
110 C1 Câmpulung Romania
138 A2 Camp Verde U.S.A.
128 C2 Camrose Can.
129 D2 Camsell Portage Can.
111 C2 Çan Turkey
126 F2 Canada country N. America
139 C1 Canadian U.S.A.
139 D1 Canadian r. U.S.A.
111 C2 Çanakkale Turkey
144 A1 Cananea Mex.
154 C2 Cananéia Brazil
Canarias, Islas terr. N. Atlantic Ocean see Canary Islands
114 A2 Canary Islands terr. N. Atlantic Ocean
155 C1 Canastra, Serra da mts Brazil
144 B2 Canatlán Mex.
143 D3 Canaveral, Cape U.S.A.
155 E1 Canavieiras Brazil
53 C3 Canberra Austr.
145 D2 Cancún Mex.
111 C3 Çandarlı Turkey
145 D2 Candelaria Mex.
129 D2 Candle Lake Can.
120 A2 Cangamba Angola
106 B1 Cangas del Narcea Spain
152 C4 Canguçu Brazil
70 B2 Cangzhou China
131 D1 Caniapiscau Can.
131 D1 Caniapiscau r. Can.
131 C1 Caniapiscau, Réservoir de resr Can.
108 B3 Canicattì Sicily Italy
151 E2 Canindé Brazil
144 B2 Cañitas de Felipe Pescador Mex.
80 B1 Çankırı Turkey
128 C2 Canmore Can.
96 A2 Canna i. U.K.
Cannanore India see Kannur
105 D3 Cannes France
99 B2 Cannock U.K.
53 C3 Cann River Austr.
152 C3 Canoas Brazil
129 D2 Canoe Lake Can.
154 B3 Canoinhas Brazil
136 B3 Canon City U.S.A.
129 D2 Canora Can.
53 C2 Canowindra Austr.
106 C1 Cantábrica, Cordillera mts Spain
106 B1 Cantábrico, Mar sea Spain
99 D3 Canterbury U.K.
54 B2 Canterbury Bight b. N.Z.
54 B2 Canterbury Plains N.Z.
63 B2 Cân Thơ Vietnam
151 D2 Canto do Buriti Brazil
Canton China see Guangzhou
142 C2 Canton MS U.S.A.
140 C2 Canton OH U.S.A.
139 C1 Canyon U.S.A.
136 B2 Canyon Ferry Lake U.S.A.
62 B1 Cao Bằng Vietnam
154 C2 Capão Bonito Brazil
155 D2 Caparaó, Serra do mts Brazil
51 D4 Cape Barren Island Austr.
52 A3 Cape Borda Austr.
131 D2 Cape Breton Island Can.
114 B4 Cape Coast Ghana
141 E2 Cape Cod Bay U.S.A.
127 F2 Cape Dorset Can.
143 E2 Cape Fear r. U.S.A.
137 F3 Cape Girardeau U.S.A.

155 D1 Capelinha Brazil
100 B2 Capelle aan den IJssel Neth.
120 A1 Capenda-Camulemba Angola
122 A3 Cape Town S. Africa
114 C5 Cape Verde country Africa
51 D1 Cape York Peninsula Austr.
147 C3 Cap-Haïtien Haiti
151 D2 Capim r. Brazil
58 D1 Capitol Hill N. Mariana Is
109 C2 Čapljina Bos. & Herz.
109 B3 Capo d'Orlando Sicily Italy
108 A2 Capraia, Isola di i. Italy
108 A2 Caprara, Punta pt Sardinia Italy
108 B2 Capri, Isola di i. Italy
120 B2 Caprivi Strip reg. Namibia
150 B2 Caquetá r. Col.
110 C2 Caracal Romania
150 B1 Caracas Venez.
151 D2 Caracol Brazil
155 C2 Caraguatatuba Brazil
153 A4 Carahue Chile
155 D1 Caraí Brazil
155 D2 Carangola Brazil
110 B1 Caransebeș Romania
131 D2 Caraquet Can.
146 B3 Caratasca, Laguna de lag. Hond.
155 D1 Caratinga Brazil
150 B2 Carauari Brazil
155 E1 Caravaca de la Cruz Spain
155 E1 Caravelas Brazil
129 E3 Carberry Can.
144 A2 Carbó Mex.
107 C2 Carbon, Cap c. Alg.
153 B5 Carbón, Laguna del l. Arg.
108 A3 Carbonara, Capo c. Sardinia Italy
136 B3 Carbondale CO U.S.A.
140 B3 Carbondale IL U.S.A.
131 E2 Carbonear Can.
107 C2 Carboneras Spain
107 C2 Carcaixent Spain
104 C3 Carcassonne France
128 A1 Carcross Can.
145 C3 Cárdenas Mex.
99 B3 Cardiff U.K.
99 A2 Cardigan U.K.
99 A2 Cardigan Bay U.K.
128 C3 Cardston Can.
110 B1 Carei Romania
104 B2 Carentan France
52 B2 Carey, Lake imp. l. Austr.
155 D2 Cariacica Brazil
Caribbean Sea N. Atlantic Ocean
141 F1 Caribou U.S.A.
130 B1 Caribou r. Can.
128 C2 Caribou Lake Can.
128 B2 Caribou Mountains Can.
100 B3 Carichic Mex.
104 B2 Carignan France
53 C2 Carinda Austr.
107 C1 Cariñena Spain
130 C2 Carleton Place Can.
123 C2 Carletonville S. Africa
97 C1 Carlingford Lough inlet Ireland/U.K.
98 B1 Carlisle U.K.
141 D2 Carlisle U.S.A.
155 D1 Carlos Chagas Brazil
97 C2 Carlow Ireland
135 C4 Carlsbad CA U.S.A.
138 C2 Carlsbad NM U.S.A.
129 D3 Carlyle Can.
128 A1 Carmacks Can.
129 E3 Carman Can.
99 A3 Carmarthen U.K.
99 A3 Carmarthen Bay U.K.

104 C3 **Carmaux** France
145 C3 **Carmelita** Guat.
144 A2 **Carmen, Isla** i. Mex.
104 B2 **Carnac** France
122 B3 **Carnarvon** S. Africa
97 C1 **Carndonagh** Ireland
50 B2 **Carnegie, Lake** imp. l. Austr.
96 B2 **Carn Eige** mt. U.K.
63 A3 **Car Nicobar** i. India
118 B2 **Carnot** C.A.R.
52 A2 **Carnot, Cape** Austr.
97 C2 **Carnsore Point** Ireland
151 D2 **Carolina** Brazil
48 D2 **Caroline Islands**
N. Pacific Ocean
103 D2 **Carpathian Mountains**
Europe
Carpaţii Meridionali
mts Romania see
Transylvanian Alps
51 C1 **Carpentaria, Gulf of** Austr.
105 D3 **Carpentras** France
97 B3 **Carrantuohill** mt. Ireland
105 E3 **Carrara** Italy
97 C2 **Carrickmacross** Ireland
97 B2 **Carrick-on-Shannon** Ireland
97 C2 **Carrick-on-Suir** Ireland
137 D1 **Carrington** U.S.A.
139 D3 **Carrizo Springs** U.S.A.
138 B2 **Carrizozo** U.S.A.
137 E2 **Carroll** U.S.A.
143 C2 **Carrollton** U.S.A.
129 D2 **Carrot River** Can.
135 C3 **Carson City** U.S.A.
150 A1 **Cartagena** Col.
107 C2 **Cartagena** Spain
146 B4 **Cartago** Costa Rica
137 E3 **Carthage** MO U.S.A.
139 E2 **Carthage** TX U.S.A.
151 E1 **Cartwright** Can.
151 D2 **Caruaru** Brazil
114 B1 **Casablanca** Morocco
154 C2 **Casa Branca** Brazil
144 B1 **Casa de Janos** Mex.
138 A2 **Casa Grande** U.S.A.
108 A1 **Casale Monferrato** Italy
109 C2 **Casarano** Italy
134 C2 **Cascade** U.S.A.
134 B2 **Cascade Range** mts Can./
U.S.A.
106 B2 **Cascais** Port.
154 B2 **Cascavel** Brazil
108 B2 **Caserta** Italy
97 C2 **Cashel** Ireland
53 D1 **Casino** Austr.
107 C1 **Caspe** Spain
136 B2 **Casper** U.S.A.
76 A2 **Caspian Lowland** Kazakh./
Russia
81 C1 **Caspian Sea** Asia/Europe
154 C2 **Cássia** Brazil
128 B2 **Cassiar (abandoned)** Can.
128 A2 **Cassiar Mountains** Can.
154 B1 **Cassilândia** Brazil
108 B2 **Cassino** Italy
96 B2 **Cassley** r. U.K.
151 D2 **Castanhal** Brazil
152 B4 **Castaño** r. Arg.
144 B2 **Castaños** Mex.
104 C3 **Casteljaloux** France
107 C2 **Castellón de la Plana** Spain
106 B2 **Castelo Branco** Port.
108 B3 **Castelvetrano** Sicily Italy
52 B3 **Casterton** Austr.
108 B2 **Castiglione della Pescaia**
Italy
97 B2 **Castlebar** Ireland
96 A2 **Castlebay** U.K.

97 C1 **Castleblayney** Ireland
97 C1 **Castlederg** U.K.
96 C3 **Castle Douglas** U.K.
128 C3 **Castlegar** Can.
97 B2 **Castleisland** Ireland
52 B3 **Castlemaine** Austr.
97 B2 **Castlerea** Ireland
53 C2 **Castlereagh** r. Austr.
136 C3 **Castle Rock** U.S.A.
129 C2 **Castor** Can.
104 C3 **Castres** France
100 B1 **Castricum** Neth.
147 D3 **Castries** St Lucia
154 C2 **Castro** Brazil
153 A5 **Castro** Chile
106 B2 **Castro Verde** Port.
109 C3 **Castrovillari** Italy
150 A2 **Catacaos** Peru
155 D2 **Cataguases** Brazil
154 B3 **Catalão** Brazil
152 B3 **Catamarca** Arg.
64 B1 **Catanduanes** i. Phil.
154 C2 **Catanduva** Brazil
154 B3 **Catanduvas** Brazil
109 C3 **Catania** Sicily Italy
109 C3 **Catanzaro** Italy
64 B1 **Catarman** Phil.
145 C2 **Catemaco** Mex.
147 C2 **Cat Island** Bahamas
130 A1 **Cat Lake** Can.
145 D2 **Catoche, Cabo** c. Mex.
141 E2 **Catskill Mountains** U.S.A.
64 B1 **Cauayan** Phil.
150 A1 **Cauca** r. Col.
151 E2 **Caucaia** Brazil
81 C1 **Caucasus** mts Asia/Europe
100 A2 **Caudry** France
109 C3 **Caulonia** Italy
131 D2 **Causapscal** Can.
105 D3 **Cavaillon** France
151 D3 **Cavalcante** Brazil
97 C2 **Cavan** Ireland
154 B3 **Cavernoso, Serra do** mts
Brazil
151 D2 **Caxias** Brazil
152 C3 **Caxias do Sul** Brazil
120 A1 **Caxito** Angola
151 C1 **Cayenne** Fr. Guiana
146 B3 **Cayman Islands** terr.
West Indies
117 C4 **Caynaba** Somalia
120 B2 **Cazombo** Angola
144 B2 **Ceballos** Mex.
64 B1 **Cebu** Phil.
64 B1 **Cebu** i. Phil.
108 B2 **Cecina** Italy
137 F2 **Cedar** r. U.S.A.
135 D3 **Cedar City** U.S.A.
137 E2 **Cedar Falls** U.S.A.
129 D2 **Cedar Lake** Can.
137 E2 **Cedar Rapids** U.S.A.
144 A2 **Cedros, Isla** i. Mex.
51 C3 **Ceduna** Austr.
117 C4 **Ceeldheere** Somalia
117 C3 **Ceerigaabo** Somalia
108 B3 **Cefalù** Sicily Italy
145 B2 **Celaya** Mex.
58 C3 **Celebes** i. Indon.
156 B5 **Celebes Sea** Indon./Phil.
145 C2 **Celestún** Mex.
101 E1 **Celle** Ger.
95 B3 **Celtic Sea** Ireland/U.K.
59 D3 **Cenderawasih, Teluk** b.
Indon.
150 A1 **Central, Cordillera** mts Col.
150 A3 **Central, Cordillera** mts Peru
64 B1 **Central, Cordillera** mts Phil.

118 C2 **Central African Republic**
country Africa
74 A2 **Central Brahui Range** mts
Pak.
137 D3 **Central City** U.S.A.
140 B3 **Centralia** IL U.S.A.
134 B1 **Centralia** WA U.S.A.
74 A2 **Central Makran Range** mts
Pak.
59 D3 **Central Range** mts P.N.G.
89 E3 **Central Russian Upland** hills
Russia
83 I2 **Central Siberian Plateau**
Russia
111 B3 **Cephalonia** i. Greece
Ceram Sea Indon. see
Seram, Laut
152 B3 **Ceres** Arg.
154 C1 **Ceres** Brazil
122 A3 **Ceres** S. Africa
104 C3 **Céret** France
106 C2 **Cerezo de Abajo** Spain
109 C2 **Cerignola** Italy
110 C2 **Cernavodă** Romania
145 C2 **Cerralvo** Mex.
144 B2 **Cerralvo, Isla** i. Mex.
145 B2 **Cerritos** Mex.
154 C2 **Cerro Azul** Brazil
145 C2 **Cerro Azul** Mex.
150 A3 **Cerro de Pasco** Peru
105 D3 **Cervione** Corsica France
106 B1 **Cervo** Spain
108 B2 **Cesena** Italy
88 C2 **Cēsis** Latvia
102 C2 **České Budějovice** Czech Rep.
101 F3 **Český les** mts Czech Rep.
111 C3 **Çeşme** Turkey
53 D2 **Cessnock** Austr.
109 C2 **Cetinje** Montenegro
109 C3 **Cetraro** Italy
106 B2 **Ceuta** N. Africa
105 C3 **Cévennes** mts France
Ceylon country Asia see
Sri Lanka
79 D2 **Chābahār** Iran
150 A2 **Chachapoyas** Peru
89 D3 **Chachersk** Belarus
63 B2 **Chachoengsao** Thai.
115 D3 **Chad** country Africa
115 D3 **Chad, Lake** Africa
69 C1 **Chadaasan** Mongolia
68 C1 **Chadan** Russia
123 C1 **Chadibe** Botswana
136 C2 **Chadron** U.S.A.
65 B2 **Chaeryŏng** N. Korea
74 A2 **Chagai** Pak.
77 C3 **Chaghcharān** Afgh.
89 E2 **Chagoda** Russia
75 C2 **Chaibasa** India
63 B2 **Chainat** Thai.
63 B2 **Chaiyaphum** Thai.
152 C4 **Chajarí** Arg.
131 D1 **Chakonipau, Lac** l. Can.
150 A3 **Chala** Peru
131 D2 **Chaleur Bay** inlet Can.
74 B2 **Chalisgaon** India
111 C3 **Chalki** i. Greece
111 B3 **Chalkida** Greece
104 B2 **Challans** France
134 D2 **Challis** U.S.A.
105 C2 **Châlons-en-Champagne**
France
105 C2 **Chalon-sur-Saône** France
121 C2 **Chama** Zambia
138 B1 **Chama** U.S.A.
74 A2 **Chaman** Pak.
74 B1 **Chamba** India
74 B2 **Chambal** r. India

Chamberlain

137 D2	Chamberlain U.S.A.
141 D3	Chambersburg U.S.A.
105 D2	Chambéry France
121 C2	Chambeshi Zambia
140 B2	Champaign U.S.A.
141 E2	Champlain, Lake Can./U.S.A.
145 C3	Champotón Mex.
152 A3	Chañaral Chile
126 B2	Chandalar r. U.S.A.
142 C3	Chandeleur Islands U.S.A.
74 B1	Chandigarh India
138 A2	Chandler U.S.A.
75 B3	Chandrapur India
63 B2	Chang, Ko i. Thai.
	Chang'an China see Rong'an
121 C3	Changane r. Moz.
121 C2	Changara Moz.
65 B1	Changbai China
65 B1	Changbai Shan mts China/N. Korea
69 E2	Changchun China
71 B3	Changde China
65 B2	Ch'angdo N. Korea
	Chang Jiang r. China see Yangtze
65 B3	Changjin N. Korea
65 B1	Changjin-gang r. N. Korea
71 B3	Changsha China
71 B3	Changting China
65 B2	Changwon S. Korea
70 B2	Changyuan China
70 B2	Changzhi China
70 B2	Changzhou China
111 B3	Chania Greece
95 C4	Channel Islands English Chan.
135 C4	Channel Islands U.S.A.
131 E2	Channel-Port-aux-Basques Can.
63 B2	Chanthaburi Thai.
104 C2	Chantilly France
137 D3	Chanute U.S.A.
82 G3	Chany, Ozero salt l. Russia
71 B3	Chaoyang Guangdong China
	Chaoyang China see Huinan
71 B3	Chaozhou China
144 B2	Chapala, Laguna de l. Mex.
76 B1	Chapayevo Kazakh.
152 C3	Chapecó Brazil
143 E1	Chapel Hill U.S.A.
130 B2	Chapleau Can.
89 E3	Chaplygin Russia
91 C2	Chaplynka Ukr.
75 C2	Chapra India
145 B2	Charcas Mex.
99 B3	Chard U.K.
104 B2	Charente r. France
77 C3	Chārīkār Afgh.
86 E2	Charkayuvom Russia
100 B2	Charleroi Belgium
141 D3	Charles, Cape U.S.A.
137 E2	Charles City U.S.A.
140 B3	Charleston IL U.S.A.
143 E2	Charleston SC U.S.A.
140 C3	Charleston WV U.S.A.
135 C3	Charleston Peak U.S.A.
51 D2	Charleville Austr.
105 C2	Charleville-Mézières France
143 D1	Charlotte U.S.A.
143 D3	Charlotte Harbor b. U.S.A.
141 D3	Charlottesville U.S.A.
131 D2	Charlottetown Can.
52 B3	Charlton Austr.
130 C1	Charlton Island Can.
51 D2	Charters Towers Austr.
104 C2	Chartres France
128 C2	Chase Can.
88 C3	Chashniki Belarus

54 A3	Chaslands Mistake c. N.Z.
65 B1	Chasŏng N. Korea
104 B2	Chassiron, Pointe de pt France
104 B2	Châteaubriant France
104 C2	Château-du-Loir France
104 C2	Châteaudun France
104 B2	Châteaulin France
105 D3	Châteauneuf-les-Martigues France
104 C2	Châteauneuf-sur-Loire France
105 C2	Château-Thierry France
128 C2	Chateh Can.
100 B2	Châtelet Belgium
104 C2	Châtellerault France
140 C2	Chatham Can.
49 G5	Chatham Islands is N.Z.
105 C2	Châtillon-sur-Seine France
143 D2	Chattahoochee r. U.S.A.
143 C1	Chattanooga U.S.A.
63 B2	Châu Đốc Vietnam
62 A1	Chauk Myanmar
105 D2	Chaumont France
105 C2	Chauny France
151 D2	Chaves Brazil
106 B1	Chaves Port.
130 C1	Chavigny, Lac l. Can.
89 D3	Chavusy Belarus
89 E2	Chayevo Russia
86 E3	Chaykovskiy Russia
102 C1	Cheb Czech Rep.
87 D3	Cheboksary Russia
140 C1	Cheboygan U.S.A.
114 B2	Chegga Maur.
134 B1	Chehalis U.S.A.
	Cheju S. Korea see Jeju
	Cheju-do i. S. Korea see Jeju-do
89 E2	Chekhov Russia
134 B1	Chelan, Lake U.S.A.
103 E1	Chełm Pol.
99 D3	Chelmer r. U.K.
103 D1	Chełmno Pol.
99 D3	Chelmsford U.K.
99 B3	Cheltenham U.K.
87 F2	Chelyabinsk Russia
101 F2	Chemnitz Ger.
114 B2	Chenachane Alg.
70 B1	Chengde China
70 A2	Chengdu China
71 A3	Chengguan China
71 B4	Chengmai China
	Chengshou China see Yingshan
70 A2	Chengxian China
	Chengyang China see Juxian
73 C3	Chennai India
71 B3	Chenzhou China
65 B2	Cheonan S. Korea
65 B2	Cheorwon S. Korea
99 C3	Chepstow U.K.
104 B2	Cherbourg-Octeville France
89 E3	Cheremisinovo Russia
69 C1	Cheremkhovo Russia
89 E2	Cherepovets Russia
91 C2	Cherkasy Ukr.
87 D4	Cherkessk Russia
91 C1	Chernihiv Ukr.
91 C2	Chernihivka Ukr.
90 B2	Chernivtsi Ukr.
90 B1	Chernyakhiv Ukr.
88 B3	Chernyakhovsk Russia
89 E3	Chernyanka Russia
83 I2	Chernyshevskiy Russia
137 D2	Cherokee U.S.A.
83 K2	Cherskogo, Khrebet mts Russia

91 E2	Chertkovo Russia
90 A1	Chervonohrad Ukr.
88 C3	Chervyen' Belarus
89 D3	Cherykaw Belarus
141 D3	Chesapeake Bay U.S.A.
86 D2	Cheshskaya Guba b. Russia
98 B2	Chester U.K.
140 B3	Chester IL U.S.A.
143 D2	Chester SC U.S.A.
98 C2	Chesterfield U.K.
129 E1	Chesterfield Inlet Can.
129 E1	Chesterfield Inlet Can.
141 F1	Chesuncook Lake U.S.A.
131 D2	Chéticamp Can.
145 D3	Chetumal Mex.
98 B1	Chetwynd Can.
98 B1	Cheviot Hills U.K.
136 C2	Cheyenne U.S.A.
136 C2	Cheyenne r. U.S.A.
136 C3	Cheyenne Wells U.S.A.
75 B2	Chhatarpur India
62 A2	Chiang Dao Thai.
62 A2	Chiang Mai Thai.
62 A2	Chiang Rai Thai.
108 A1	Chiavenna Italy
70 B3	Chibi China
121 C3	Chiboma Moz.
130 C2	Chibougamau Can.
123 D1	Chibuto Moz.
75 D1	Chibuzhang Co l. China
140 B2	Chicago U.S.A.
128 A2	Chichagof Island U.S.A.
99 C3	Chichester U.K.
50 A2	Chichester Range mts Austr.
139 D1	Chickasha U.S.A.
150 A2	Chiclayo Peru
153 B5	Chico r. Arg.
135 B3	Chico U.S.A.
131 C2	Chicoutimi Can.
108 B2	Chieti Italy
145 C3	Chietla Mex.
69 D2	Chifeng China
155 D1	Chifre, Serra do mts Brazil
145 C3	Chignahuapán Mex.
121 C3	Chigubo Moz.
74 A1	Chihil Abdālān, Kōh-e mts Afgh.
144 B2	Chihuahua Mex.
88 C2	Chikhachevo Russia
67 C3	Chikuma-gawa r. Japan
128 B2	Chilanko r. Can.
74 B1	Chilas Pak.
139 C2	Childress U.S.A.
153 A4	Chile country S. America
152 B3	Chilecito Arg.
75 C3	Chilika Lake India
121 B2	Chililabombwe Zambia
128 B2	Chilko r. Can.
128 B2	Chilko Lake Can.
153 A4	Chillán Chile
137 E3	Chillicothe MO U.S.A.
140 C3	Chillicothe OH U.S.A.
128 B3	Chilliwack Can.
153 A5	Chiloé, Isla de i. Chile
145 C3	Chilpancingo Mex.
53 C3	Chiltern Austr.
71 C3	Chilung Taiwan
119 D3	Chimala Tanz.
152 B4	Chimbas Arg.
150 A2	Chimborazo mt. Ecuador
150 A2	Chimbote Peru
87 E4	Chimboy Uzbek.
	Chimkent Kazakh. see Shymkent
121 C2	Chimoio Moz.
77 C2	Chimtargha, Qullai mt. Tajik.
68 C2	China country Asia
145 C2	China Mex.

150 A3 **Chincha Alta** Peru
128 C2 **Chinchaga** *r.* Can.
145 D3 **Chinchorro, Banco** Mex.
121 C2 **Chinde** Moz.
65 B3 **Chindo** S. Korea
68 C2 **Chindu** China
62 A1 **Chindwin** *r.* Myanmar
65 B2 **Chinghwa** N. Korea
120 B2 **Chingola** Zambia
120 A2 **Chinguar** Angola
121 C2 **Chinhoyi** Zimbabwe
74 B1 **Chiniot** Pak.
144 B2 **Chinipas** Mex.
65 B2 **Chinju** S. Korea
118 C2 **Chinko** *r.* C.A.R.
138 B1 **Chinle** U.S.A.
67 C3 **Chino** Japan
104 C2 **Chinon** France
138 A2 **Chino Valley** U.S.A.
121 C2 **Chinsali** Zambia
108 B1 **Chioggia** Italy
111 C3 **Chios** Greece
111 C3 **Chios** *i.* Greece
121 C2 **Chipata** Zambia
120 A2 **Chipindo** Angola
121 C3 **Chipinge** Zimbabwe
73 B3 **Chiplun** India
99 B3 **Chippenham** U.K.
99 C3 **Chipping Norton** U.K.
77 C2 **Chirchiq** Uzbek.
121 C3 **Chiredzi** Zimbabwe
138 B2 **Chiricahua Peak** U.S.A.
146 B4 **Chiriquí, Golfo de** *b.* Panama
146 B4 **Chirripó** *mt.* Costa Rica
121 B2 **Chirundu** Zimbabwe
130 C1 **Chisasibi** Can.
137 E1 **Chisholm** U.S.A.
90 B2 **Chișinău** Moldova
87 E3 **Chistopol'** Russia
69 D1 **Chita** Russia
120 A2 **Chitado** Angola
121 C2 **Chitambo** Zambia
120 B1 **Chitato** Angola
121 C1 **Chitipa** Malawi
73 B3 **Chitradurga** India
74 B1 **Chitral** Pak.
146 B4 **Chitré** Panama
75 D2 **Chittagong** Bangl.
74 B2 **Chittaurgarh** India
121 C2 **Chitungwiza** Zimbabwe
120 B2 **Chiume** Angola
121 C2 **Chivhu** Zimbabwe
70 B2 **Chizhou** China
114 C1 **Chlef** Alg.
107 D2 **Chlef, Oued** *r.* Alg.
153 B4 **Choele Choel** Arg.
144 B2 **Choix** Mex.
103 D1 **Chojnice** Pol.
117 B3 **Ch'ok'ē Mountains** Eth.
117 B3 **Ch'ok'ē Terara** *mt.* Eth.
83 K2 **Chokurdakh** Russia
121 C3 **Chókwè** Moz.
104 B2 **Cholet** France
102 C1 **Chomutov** Czech Rep.
83 I2 **Chona** *r.* Russia
150 A2 **Chone** Ecuador
65 B1 **Ch'ŏngjin** N. Korea
65 B2 **Chŏngju** N. Korea
65 B2 **Chŏngp'yŏng** N. Korea
70 A3 **Chongqing** China
70 A2 **Chongqing** *mun.* China
71 A3 **Chongzuo** China
Chŏnju S. Korea *see* **Jeonju**
153 A5 **Chonos, Archipiélago de los** *is* Chile
154 B3 **Chopinzinho** Brazil
111 B3 **Chora Sfakion** Greece
98 B2 **Chorley** U.K.
91 C2 **Chornomors'ke** Ukr.
90 B2 **Chortkiv** Ukr.
65 B1 **Ch'osan** N. Korea
67 D3 **Chōshi** Japan
103 D1 **Choszczno** Pol.
114 A2 **Choûm** Maur.
69 D1 **Choybalsan** Mongolia
69 D1 **Choyr** Mongolia
54 B2 **Christchurch** N.Z.
99 C3 **Christchurch** U.K.
127 G2 **Christian, Cape** Can.
123 C2 **Christiana** S. Africa
54 A2 **Christina, Mount** N.Z.
153 B5 **Chubut** *r.* Arg.
90 B1 **Chudniv** Ukr.
89 D2 **Chudovo** Russia
Chudskoye Ozero *l.* Estonia/Russia *see* **Peipus, Lake**
126 B2 **Chugach Mountains** U.S.A.
67 B4 **Chūgoku-sanchi** *mts* Japan
66 B2 **Chuguyevka** Russia
91 D2 **Chuhuyiv** Ukr.
160 J3 **Chukchi Sea** Russia/U.S.A.
83 N2 **Chukotskiy Poluostrov** *pen.* Russia
135 C4 **Chula Vista** U.S.A.
82 G3 **Chulym** Russia
152 B3 **Chumbicha** Arg.
83 K3 **Chumikan** Russia
63 A2 **Chumphon** Thai.
65 B2 **Chuncheon** S. Korea
Chungking China *see* **Chongqing**
83 H2 **Chunya** *r.* Russia
150 A3 **Chuquibamba** Peru
152 B3 **Chuquicamata** Chile
105 D2 **Chur** Switz.
62 A1 **Churachandpur** India
129 E2 **Churchill** Can.
129 E2 **Churchill** *r. Man.* Can.
131 D1 **Churchill** *r. Nfld. and Lab.* Can.
129 E2 **Churchill, Cape** Can.
131 D1 **Churchill Falls** Can.
129 D2 **Churchill Lake** Can.
74 B2 **Churu** India
131 C2 **Chute-des-Passes** Can.
62 B1 **Chuxiong** China
90 B2 **Ciadîr-Lunga** Moldova
61 B2 **Ciamis** Indon.
60 B2 **Cianjur** Indon.
154 B2 **Cianorte** Brazil
103 E1 **Ciechanów** Pol.
146 C2 **Ciego de Ávila** Cuba
146 B2 **Cienfuegos** Cuba
107 C2 **Cieza** Spain
106 C2 **Cigüela** *r.* Spain
80 B2 **Cihanbeyli** Turkey
144 B3 **Cihuatlán** Mex.
106 C2 **Cíjara, Embalse de** *resr* Spain
61 B2 **Cilacap** Indon.
139 C1 **Cimarron** *r.* U.S.A.
90 B2 **Cimișlia** Moldova
108 B2 **Cimone, Monte** *mt.* Italy
140 C3 **Cincinnati** U.S.A.
111 C3 **Çine** Turkey
100 B2 **Ciney** Belgium
145 C3 **Cintalapa** Mex.
71 B3 **Ciping** China
126 B2 **Circle** *AK* U.S.A.
136 B1 **Circle** *MT* U.S.A.
58 B3 **Cirebon** Indon.
99 C3 **Cirencester** U.K.
108 A1 **Ciriè** Italy
109 C3 **Cirò Marina** Italy
109 C2 **Čitluk** Bos. & Herz.
122 A3 **Citrusdal** S. Africa
145 B2 **Ciudad Acuña** Mex.
145 B3 **Ciudad Altamirano** Mex.
150 B1 **Ciudad Bolívar** Venez.
144 B2 **Ciudad Camargo** Mex.
144 A2 **Ciudad Constitución** Mex.
145 C3 **Ciudad del Carmen** Mex.
144 B2 **Ciudad Delicias** Mex.
145 C2 **Ciudad de Valles** Mex.
150 B1 **Ciudad Guayana** Venez.
138 B3 **Ciudad Guerrero** Mex.
144 B3 **Ciudad Guzmán** Mex.
145 C3 **Ciudad Hidalgo** Mex.
145 C3 **Ciudad Ixtepec** Mex.
144 B1 **Ciudad Juárez** Mex.
145 C2 **Ciudad Mante** Mex.
145 C2 **Ciudad Mier** Mex.
144 B2 **Ciudad Obregón** Mex.
106 C2 **Ciudad Real** Spain
145 C2 **Ciudad Río Bravo** Mex.
106 B1 **Ciudad Rodrigo** Spain
145 C2 **Ciudad Victoria** Mex.
107 D1 **Ciutadella** Spain
111 C3 **Cıvan Dağ** *mt.* Turkey
108 B1 **Cividale del Friuli** Italy
108 B2 **Civitanova Marche** Italy
108 B2 **Civitavecchia** Italy
104 C2 **Civray** France
111 C3 **Çivril** Turkey
70 C2 **Cixi** China
99 D3 **Clacton-on-Sea** U.K.
128 C2 **Claire, Lake** Can.
105 C2 **Clamecy** France
122 A3 **Clanwilliam** S. Africa
52 A2 **Clare** Austr.
97 A2 **Clare Island** Ireland
141 E2 **Claremont** U.S.A.
97 B2 **Claremorris** Ireland
54 B2 **Clarence** N.Z.
131 E2 **Clarenville** Can.
128 C2 **Claresholm** Can.
137 D2 **Clarinda** U.S.A.
123 C3 **Clarkebury** S. Africa
143 D2 **Clark Hill Reservoir** U.S.A.
140 C3 **Clarksburg** U.S.A.
142 B2 **Clarksdale** U.S.A.
134 C1 **Clarks Fork** *r.* U.S.A.
142 B1 **Clarksville** *AR* U.S.A.
142 C1 **Clarksville** *TN* U.S.A.
154 B1 **Claro** *r.* Brazil
139 C1 **Clayton** U.S.A.
97 B3 **Clear, Cape** Ireland
137 E2 **Clear Lake** U.S.A.
135 B3 **Clear Lake** U.S.A.
128 C2 **Clearwater** Can.
129 C2 **Clearwater** *r.* Can.
143 D3 **Clearwater** U.S.A.
134 C1 **Clearwater** *r.* U.S.A.
139 D2 **Cleburne** U.S.A.
51 D2 **Clermont** Austr.
105 C2 **Clermont-Ferrand** France
52 A2 **Cleve** Austr.
142 B2 **Cleveland** *MS* U.S.A.
140 C2 **Cleveland** *OH* U.S.A.
143 D1 **Cleveland** *TN* U.S.A.
134 D1 **Cleveland, Mount** U.S.A.
143 D3 **Clewiston** U.S.A.
97 A2 **Clifden** Ireland
53 D1 **Clifton** Austr.
138 B2 **Clifton** U.S.A.
128 B2 **Clinton** Can.
137 E2 **Clinton** *IA* U.S.A.
137 E3 **Clinton** *MO* U.S.A.
139 D1 **Clinton** *OK* U.S.A.
96 A2 **Clisham** *h.* U.K.
98 B2 **Clitheroe** U.K.
97 B3 **Clonakilty** Ireland
51 D2 **Cloncurry** Austr.
97 C1 **Clones** Ireland
97 C2 **Clonmel** Ireland

Cloppenburg

100 D1 Cloppenburg Ger.
50 A2 Cloud Break Austr.
136 B2 Cloud Peak U.S.A.
139 C2 Clovis U.S.A.
129 D2 Cluff Lake Mine Can.
110 B1 Cluj-Napoca Romania
51 C2 Cluny Austr.
105 D2 Cluses France
54 A3 Clutha r. N.Z.
96 B3 Clyde r. U.K.
96 B3 Clyde, Firth of est. U.K.
96 B3 Clydebank U.K.
127 G2 Clyde River Can.
144 B3 Coalcomán Mex.
135 C3 Coaldale (abandoned) U.S.A.
128 B2 Coal River Can.
150 B2 Coari Brazil
150 B2 Coari r. Brazil
142 B2 Coastal Plain U.S.A.
128 B2 Coast Mountains Can.
135 B2 Coast Ranges mts U.S.A.
96 B3 Coatbridge U.K.
127 F2 Coats Island Can.
55 R2 Coats Land reg. Antarctica
145 C3 Coatzacoalcos Mex.
146 A3 Cobán Guat.
53 C2 Cobar Austr.
97 B3 Cobh Ireland
152 B2 Cobija Bol.
141 D2 Cobourg U.S.A.
50 C1 Cobourg Peninsula Austr.
53 C3 Cobram Austr.
101 E2 Coburg Ger.
152 B2 Cochabamba Bol.
100 C2 Cochem Ger.
Cochin India see Kochi
128 C2 Cochrane Alta Can.
130 B2 Cochrane Ont. Can.
153 A5 Cochrane Chile
52 B2 Cockburn Austr.
Cockburn Town Turks and Caicos Is see Grand Turk
98 B1 Cockermouth U.K.
122 B1 Cockscomb mt. S. Africa
146 B3 Coco r. Hond./Nic.
144 B2 Cocula Mex.
150 A1 Cocuy, Sierra Nevada del mt. Col.
141 E2 Cod, Cape U.S.A.
108 B2 Codigoro Italy
131 D1 Cod Island Can.
151 D2 Codó Brazil
136 B2 Cody U.S.A.
51 D1 Coen Austr.
100 C2 Coesfeld Ger.
134 C1 Coeur d'Alene U.S.A.
123 D3 Coffee Bay S. Africa
137 D3 Coffeyville U.S.A.
53 D2 Coffs Harbour Austr.
104 B2 Cognac France
118 A2 Cogo Equat. Guinea
52 B3 Cohuna Austr.
146 B4 Coiba, Isla de i. Panama
153 A5 Coihaique Chile
73 B3 Coimbatore India
106 B1 Coimbra Port.
52 B3 Colac Austr.
155 D1 Colatina Brazil
136 C3 Colby U.S.A.
99 D3 Colchester U.K.
129 C2 Cold Lake Can.
96 C3 Coldstream U.K.
139 D2 Coleman U.S.A.
52 B3 Coleraine Austr.
97 C1 Coleraine U.K.
123 C3 Colesberg S. Africa
144 B3 Colima Mex.
144 B3 Colima, Nevado de vol. Mex.

96 A2 Coll i. U.K.
53 C1 Collarenebri Austr.
50 B1 Collier Bay Austr.
54 B2 Collingwood N.Z.
97 B1 Collooney Ireland
105 D2 Colmar France
100 C2 Cologne Ger.
154 C2 Colômbia Brazil
150 A1 Colombia country S. America
73 B4 Colombo Sri Lanka
104 C3 Colomiers France
152 C4 Colón Arg.
146 C4 Colón Panama
109 C3 Colonna, Capo c. Italy
96 A2 Colonsay i. U.K.
153 B4 Colorado r. Arg.
138 A2 Colorado r. Mex./U.S.A.
139 D3 Colorado r. U.S.A.
136 B3 Colorado state U.S.A.
135 E3 Colorado Plateau U.S.A.
136 C3 Colorado Springs U.S.A.
144 B2 Colotlán Mex.
136 B1 Colstrip U.S.A.
137 E3 Columbia MO U.S.A.
143 D2 Columbia SC U.S.A.
142 C1 Columbia TN U.S.A.
134 B1 Columbia r. U.S.A.
128 C2 Columbia, Mount Can.
134 D1 Columbia Falls U.S.A.
128 B2 Columbia Mountains Can.
134 C1 Columbia Plateau U.S.A.
143 D2 Columbus GA U.S.A.
140 B3 Columbus IN U.S.A.
142 C2 Columbus MS U.S.A.
137 D2 Columbus NE U.S.A.
138 B2 Columbus NM U.S.A.
140 C3 Columbus OH U.S.A.
134 C1 Colville U.S.A.
126 A2 Colville r. U.S.A.
126 C2 Colville Lake Can.
98 B2 Colwyn Bay U.K.
108 B2 Comacchio Italy
145 C3 Comalcalco Mex.
110 C1 Comăneşti Romania
130 C1 Comencho, Lac l. Can.
97 C2 Comeragh Mountains hills Ireland
75 D2 Comilla Bangl.
108 A2 Comino, Capo c. Sardinia Italy
145 C3 Comitán de Domínguez Mex.
104 C2 Commentry France
139 D2 Commerce U.S.A.
127 F2 Committee Bay Can.
108 A1 Como Italy
153 B5 Comodoro Rivadavia Arg.
121 D2 Comoros country Africa
104 C2 Compiègne France
144 B2 Compostela Mex.
90 B2 Comrat Moldova
114 A4 Conakry Guinea
155 E1 Conceição da Barra Brazil
151 D2 Conceição do Araguaia Brazil
155 D1 Conceição do Mato Dentro Brazil
152 B3 Concepción Arg.
153 A4 Concepción Chile
144 B2 Concepción Mex.
135 B4 Conception, Point U.S.A.
154 C2 Conchas Brazil
138 C1 Conchas Lake U.S.A.
144 B2 Conchos r. Chihuahua Mex.
145 C2 Conchos r. Nuevo León/Tamaulipas Mex.
135 B3 Concord CA U.S.A.
141 E2 Concord NH U.S.A.
152 C4 Concordia Arg.
122 A2 Concordia S. Africa

137 D3 Concordia U.S.A.
53 C2 Condobolin Austr.
104 C3 Condom France
134 B1 Condon U.S.A.
108 B1 Conegliano Italy
104 C2 Confolens France
75 C2 Congdü China
118 B3 Congo country Africa
118 B3 Congo r. Congo/Dem. Rep. Congo
118 C3 Congo, Democratic Republic of the country Africa
129 C2 Conklin Can.
97 B1 Conn, Lough l. Ireland
97 B2 Connacht reg. Ireland
141 E2 Connecticut r. U.S.A.
141 E2 Connecticut state U.S.A.
97 B2 Connemara reg. Ireland
134 D1 Conrad U.S.A.
139 D2 Conroe U.S.A.
155 D2 Conselheiro Lafaiete Brazil
155 D1 Conselheiro Pena Brazil
98 C1 Consett U.K.
63 B3 Côn Sơn, Đao i. Vietnam
110 C2 Constanţa Romania
106 B2 Constantina Spain
115 C1 Constantine Alg.
134 D2 Contact U.S.A.
150 A2 Contamana Peru
153 A6 Contreras, Isla i. Chile
126 D2 Contwoyto Lake Can.
142 B1 Conway AR U.S.A.
141 E2 Conway NH U.S.A.
51 C2 Coober Pedy Austr.
Cook, Mount mt. N.Z. see Aoraki/Mount Cook
143 C1 Cookeville U.S.A.
49 H4 Cook Islands S. Pacific Ocean
131 E1 Cook's Harbour Can.
97 C1 Cookstown U.K.
54 B2 Cook Strait N.Z.
53 D1 Cooktown Austr.
53 C2 Coolabah Austr.
53 C2 Coolamon Austr.
50 B3 Coolgardie Austr.
53 C2 Coolangatta Austr.
53 C3 Cooma Austr.
52 B2 Coombah Austr.
53 C2 Coonabarabran Austr.
52 A3 Coonalpyn Austr.
53 C2 Coonamble Austr.
52 A1 Cooper Creek watercourse Austr.
134 B2 Coos Bay U.S.A.
122 B2 Cootamundra Austr.
145 C3 Copainalá Mex.
145 C3 Copala Mex.
93 F4 Copenhagen Denmark
109 C2 Copertino Italy
152 A3 Copiapó Chile
140 B1 Copper Harbor U.S.A.
Coppermine Can. see Kugluktuk
126 D2 Coppermine r. Can.
122 B2 Copperton S. Africa
152 A3 Coquimbo Chile
110 B2 Corabia Romania
155 D1 Coração de Jesus Brazil
150 A3 Coracora Peru
53 D1 Coraki Austr.
50 A2 Coral Bay Austr.
127 F2 Coral Harbour Can.
156 D7 Coral Sea S. Pacific Ocean
48 D4 Coral Sea Islands Territory Austr.
52 B3 Corangamite, Lake Austr.
99 C2 Corby U.K.

153 A5 Corcovado, Golfo de *sea chan.* Chile
143 D2 Cordele U.S.A.
64 B1 Cordilleras Range *mts* Phil.
152 B4 Córdoba Arg.
145 C3 Córdoba Mex.
106 C2 Córdoba Spain
153 B4 Córdoba, Sierras de *mts* Arg.
111 A3 Corfu *i.* Greece
106 B2 Coria Spain
111 B3 Corinth Greece
142 C2 Corinth U.S.A.
155 D1 Corinto Brazil
97 B3 Cork Ireland
111 C2 Çorlu Turkey
154 B2 Cornélio Procópio Brazil
131 E2 Corner Brook Can.
53 C3 Corner Inlet *b.* Austr.
135 B3 Corning CA U.S.A.
141 D2 Corning NY U.S.A.
Corn Islands *is* Nic. *see* Maíz, Islas del
108 B2 Corno Grande *mt.* Italy
130 C2 Cornwall Can.
126 E1 Cornwallis Island Can.
150 B1 Coro Venez.
154 C1 Coromandel Brazil
54 C1 Coromandel Peninsula N.Z.
129 C2 Coronation Can.
126 D2 Coronation Gulf Can.
152 C3 Coronel Oviedo Para.
153 B4 Coronel Suárez Arg.
150 A3 Coropuna, Nudo *mt.* Peru
109 D2 Çorovodë Albania
139 D3 Corpus Christi U.S.A.
152 B2 Corque Bol.
151 D3 Corrente Brazil
154 B1 Correntes Brazil
151 B3 Correntina Brazil
97 B3 Corrib, Lough *l.* Ireland
152 C3 Corrientes Arg.
144 B2 Corrientes, Cabo *c.* Mex.
53 C3 Corryong Austr.
Corse *i.* France *see* Corsica
105 D3 Corse, Cap *c.* Corsica France
105 D3 Corsica *i.* France
139 D2 Corsicana U.S.A.
105 D3 Corte Corsica France
106 B2 Cortegana Spain
136 B3 Cortez U.S.A.
108 D1 Cortina d'Ampezzo Italy
141 D2 Cortland U.S.A.
108 D2 Cortona Italy
106 B2 Coruche Port.
Çoruh Turkey *see* Artvin
80 B1 Çorum Turkey
152 C2 Corumbá Brazil
154 C1 Corumbá *r.* Brazil
134 B2 Corvallis U.S.A.
98 B2 Corwen U.K.
144 B2 Cosalá Mex.
145 C3 Cosamaloapan Mex.
109 C3 Cosenza Italy
105 C2 Cosne-Cours-sur-Loire France
107 C2 Costa Blanca *coastal area* Spain
107 D1 Costa Brava *coastal area* Spain
106 B2 Costa del Sol *coastal area* Spain
154 B1 Costa Rica Brazil
146 B3 Costa Rica *country* Central America
144 B2 Costa Rica Mex.
110 B2 Coșteşti Romania
64 B2 Cotabato Phil.
114 B4 Côte d'Ivoire *country* Africa

150 A2 Cotopaxi, Volcán *vol.* Ecuador
99 B3 Cotswold Hills U.K.
134 B2 Cottage Grove U.S.A.
102 C1 Cottbus Ger.
104 B2 Coubre, Pointe de la *pt* France
52 A3 Couedic, Cape du Austr.
137 D2 Council Bluffs U.S.A.
88 B2 Courland Lagoon *b.* Lith./ Russia
128 B3 Courtenay Can.
104 B2 Coutances France
104 B2 Coutras France
100 B2 Couvin Belgium
99 C2 Coventry U.K.
140 C3 Covington U.S.A.
50 B3 Cowan, Lake *imp. l.* Austr.
96 C2 Cowdenbeath U.K.
53 C3 Cowes U.K.
134 B1 Cowlitz *r.* U.S.A.
53 C2 Cowra Austr.
154 B1 Coxim Brazil
75 D2 Cox's Bazar Bangl.
145 B3 Coyuca de Benítez Mex.
75 C1 Cozhê China
145 D2 Cozumel Mex.
145 D2 Cozumel, Isla de *i.* Mex.
123 C3 Cradock S. Africa
136 B2 Craig U.S.A.
102 C2 Crailsheim Ger.
110 B2 Craiova Romania
129 D2 Cranberry Portage Can.
53 C3 Cranbourne Austr.
128 C3 Cranbrook Can.
151 D2 Crateús Brazil
151 E2 Crato Brazil
136 C2 Crawford U.S.A.
140 B2 Crawfordsville U.S.A.
99 C3 Crawley U.K.
134 D1 Crazy Mountains U.S.A.
129 C2 Cree *r.* Can.
144 B2 Creel Mex.
129 D2 Cree Lake Can.
100 B1 Creil Neth.
105 D2 Crema Italy
108 B1 Cremona Italy
108 B2 Cres *i.* Croatia
134 B2 Crescent City U.S.A.
128 C3 Creston Can.
137 E2 Creston U.S.A.
142 C2 Crestview U.S.A.
111 B3 Crete *i.* Greece
107 D1 Creus, Cap de *c.* Spain
98 B2 Crewe U.K.
96 B2 Crianlarich U.K.
152 D3 Criciúma Brazil
96 C2 Crieff U.K.
108 B1 Crikvenica Croatia
91 C2 Crimea *disp. terr.* Ukr.
91 C2 Crimea *pen.* Ukr.
101 F2 Crimmitschau Ger.
154 C1 Cristina Brazil
101 E1 Crivitz Ger.
Crna Gora *country* Europe *see* Montenegro
109 D2 Crni Vrh *h.* Serbia
109 C1 Črnomelj Slovenia
97 B2 Croagh Patrick *h.* Ireland
109 C1 Croatia *country* Europe
61 C1 Crocker, Banjaran *mts* Malaysia
139 D2 Crockett U.S.A.
59 C3 Croker Island Austr.
99 D2 Cromer U.K.
54 A3 Cromwell N.Z.
137 D1 Crookston U.S.A.
53 C2 Crookwell Austr.
136 C1 Crosby U.S.A.

143 D3 Cross City U.S.A.
142 B2 Crossett U.S.A.
98 B1 Cross Fell *h.* U.K.
129 E2 Cross Lake Can.
143 C1 Crossville U.S.A.
109 C3 Crotone Italy
99 D3 Crowborough U.K.
142 B2 Crowley U.S.A.
128 C3 Crowsnest Pass Can.
146 C3 Cruz, Cabo *c.* Cuba
152 C3 Cruz Alta Brazil
152 B4 Cruz del Eje Arg.
155 D2 Cruzeiro Brazil
150 A2 Cruzeiro do Sul Brazil
52 A2 Crystal Brook Austr.
139 D3 Crystal City U.S.A.
140 B1 Crystal Falls U.S.A.
103 E2 Csongrád Hungary
120 B2 Cuando *r.* Angola/Zambia
120 A2 Cuangar Angola
118 B3 Cuango *r.* Angola/Dem. Rep. Congo
120 A1 Cuanza *r.* Angola
139 C3 Cuatro Ciénegas Mex.
144 B2 Cuauhtémoc Mex.
145 C3 Cuautla Mex.
146 B2 Cuba *country* West Indies
120 A2 Cubal Angola
120 B2 Cubango *r.* Angola/Namibia
150 A1 Cúcuta Col.
73 B3 Cuddalore India
Cuddapah. India *see* Kadapa
106 C1 Cuéllar Spain
120 A2 Cuemba Angola
150 A2 Cuenca Ecuador
107 C1 Cuenca Spain
107 C1 Cuenca, Serranía de *mts* Spain
145 C3 Cuernavaca Mex.
139 D3 Cuero U.S.A.
151 C3 Cuiabá Brazil
151 C3 Cuiabá *r.* Brazil
96 A2 Cuillin Sound *sea chan.* U.K.
120 A1 Cuilo Angola
120 A2 Cuito *r.* Angola
120 A2 Cuito Cuanavale Angola
60 B1 Cukai Malaysia
53 C3 Culcairn Austr.
53 C1 Culgoa *r.* Austr.
144 B2 Culiacán Mex.
64 A1 Culion *i.* Phil.
107 C2 Cullera Spain
142 C2 Cullman U.S.A.
97 C1 Cullybackey U.K.
151 C3 Culuene *r.* Brazil
150 B1 Cumaná Venez.
141 D3 Cumberland U.S.A.
140 C3 Cumberland *r.* U.S.A.
129 D2 Cumberland Lake Can.
127 G2 Cumberland Peninsula Can.
142 C1 Cumberland Plateau U.S.A.
127 G2 Cumberland Sound *sea chan.* Can.
96 C3 Cumbernauld U.K.
135 B3 Cummings U.S.A.
96 B3 Cumnock U.K.
144 B1 Cumpas Mex.
145 C3 Cunduacán Mex.
120 A2 Cunene *r.* Angola
108 A2 Cuneo Italy
53 C1 Cunnamulla Austr.
108 A1 Cuorgnè Italy
96 C2 Cupar U.K.
147 D3 Curaçao *i.* West Indies
147 D3 Curaçao *terr.* West Indies
150 A2 Curaray *r.* Ecuador
153 A4 Curicó Chile
154 C3 Curitiba Brazil

Curnamona

52 A2 **Curnamona** Austr.
51 D3 **Currie** Austr.
51 E2 **Curtis Island** Austr.
151 C2 **Curuá** *r.* Brazil
60 B2 **Curup** Indon.
151 D2 **Cururupu** Brazil
155 B2 **Curvelo** Brazil
150 A3 **Cusco** Peru
139 D1 **Cushing** U.S.A.
134 D1 **Cut Bank** U.S.A.
75 C2 **Cuttack** India
101 D1 **Cuxhaven** Ger.
64 B1 **Cuyo Islands** Phil.
Cuzco Peru *see* **Cusco**
119 D3 **Cyangugu** Rwanda
111 B3 **Cyclades** *is* Greece
129 C3 **Cypress Hills** Can.
80 B2 **Cyprus** *country* Asia
102 C2 **Czech Republic** *country*
Europe
103 D1 **Czersk** Pol.
103 D1 **Częstochowa** Pol.

D

Đa, Sông *r.* Vietnam *see*
Black River
69 B1 **Daban** China
114 A3 **Dabola** Guinea
Dacca Bangl. *see* **Dhaka**
102 C2 **Dachau** Ger.
117 C4 **Dadaab** Kenya
74 A2 **Dadu** Pak.
65 B2 **Daegu** S. Korea
65 B2 **Daejeon** S. Korea
65 B3 **Daejeong** S. Korea
64 B1 **Daet** Phil.
114 A3 **Dagana** Senegal
64 B1 **Dagupan** Phil.
74 A1 **Dahanu** India
69 D2 **Da Hinggan Ling** *mts* China
116 C3 **Dahlak Archipelago** *is* Eritrea
100 C2 **Dahlem** Ger.
78 B3 **Dahm, Ramlat** *des.* Saudi
Arabia /Yemen
60 B2 **Daik** Indon.
106 C2 **Daimiel** Spain
97 A2 **Daingean Uí Chúis** Ireland
51 C2 **Dajarra** Austr.
114 A3 **Dakar** Senegal
116 A2 **Dākhilah, Wāḥāt ad** *oasis*
Egypt
114 A2 **Dakhla** Western Sahara
Dakhla Oasis Egypt *see*
Dākhilah, Wāḥāt ad
63 A3 **Dakoank** India
88 C3 **Dakol'ka** *r.* Belarus
Đakovica Kosovo *see* **Gjakovë**
109 C1 **Đakovo** Croatia
120 B2 **Dala** Angola
68 C2 **Dalain Hob** China
93 G3 **Dalälven** *r.* Sweden
111 C3 **Dalaman** Turkey
111 C3 **Dalaman** *r.* Turkey
69 C2 **Dalandzadgad** Mongolia
63 B2 **Đa Lat** Vietnam
74 A2 **Dalbandin** Pak.
96 C3 **Dalbeattie** U.K.
51 E2 **Dalby** Austr.
143 C1 **Dale Hollow Lake** U.S.A.
53 C3 **Dalgety** Austr.
139 C1 **Dalhart** U.S.A.
131 D2 **Dalhousie** Can.
62 B1 **Dali** China
70 C2 **Dalian** China
96 C3 **Dalkeith** U.K.

139 D2 **Dallas** U.S.A.
128 A2 **Dall Island** U.S.A.
109 C2 **Dalmatia** *reg.* Bos. & Herz./
Croatia
66 C2 **Dal'negorsk** Russia
66 B1 **Dal'nerechensk** Russia
114 B4 **Daloa** Côte d'Ivoire
51 D2 **Dalrymple, Mount** Austr.
92 ☐A3 **Dalsmynni** Iceland
143 D2 **Dalton** U.S.A.
75 C2 **Daltonganj** India
60 B1 **Daludalu** Indon.
92 ☐B2 **Dalvík** Iceland
50 C1 **Daly** *r.* Austr.
51 C1 **Daly Waters** Austr.
74 B2 **Daman** India
80 B2 **Damanhûr** Egypt
59 C3 **Damar** *i.* Indon.
80 B2 **Damascus** Syria
115 D3 **Damaturu** Nigeria
76 B3 **Damāvand** Iran
81 D2 **Damāvand, Qolleh-ye** *mt.* Iran
81 D2 **Dāmghān** Iran
79 C2 **Dammam** Saudi Arabia
101 D1 **Damme** Ger.
75 B2 **Damoh** India
114 B4 **Damongo** Ghana
59 C3 **Dampir, Selat** *sea chan.* Indon.
75 C2 **Damqoq Zangbo** *r.* China
117 C3 **Danakil** *reg.* Africa
114 B4 **Danané** Côte d'Ivoire
63 B2 **Đa Nẵng** Vietnam
141 E2 **Danbury** U.S.A.
65 A1 **Dandong** China
146 B3 **Dangriga** Belize
70 B2 **Dangshan** China
89 F2 **Danilov** Russia
89 E2 **Danilovskaya Vozvyshennost'**
hills Russia
70 B2 **Danjiangkou** China
89 E3 **Dankov** Russia
146 B3 **Danlí** Hond.
101 E1 **Dannenberg (Elbe)** Ger.
54 C2 **Dannevirke** N.Z.
62 B2 **Dan Sai** Thai.
Dantu China *see* **Zhenjiang**
110 C1 **Danube** *r.* Europe
110 C1 **Danube Delta** Romania/Ukr.
140 B2 **Danville** *IL* U.S.A.
140 C3 **Danville** *KY* U.S.A.
141 D3 **Danville** *VA* U.S.A.
71 B3 **Danzhou** China
71 B3 **Daoxian** China
114 C3 **Dapaong** Togo
64 B2 **Dapitan** Phil.
68 C2 **Da Qaidam** China
69 E1 **Daqing** China
80 B2 **Dar'ā** Syria
79 C2 **Dārāb** Iran
81 D2 **Dārān** Iran
75 C2 **Darbhanga** India
119 D3 **Dar es Salaam** Tanz.
117 A3 **Darfur** *reg.* Sudan
74 B1 **Dargai** Pak.
54 B1 **Dargaville** N.Z.
53 C3 **Dargo** Austr.
69 D1 **Darhan** Mongolia
150 A1 **Darién, Golfo del** *g.* Col.
75 C2 **Darjiling** India
52 B2 **Darling** *r.* Austr.
53 C1 **Darling Downs** *hills* Austr.
50 A3 **Darling Range** *hills* Austr.
98 C1 **Darlington** U.K.
53 C2 **Darlington Point** Austr.
103 D1 **Darłowo** Pol.
101 D3 **Darmstadt** Ger.
115 E1 **Darnah** Libya
52 B2 **Darnick** Austr.
107 C1 **Daroca** Spain

99 D3 **Dartford** U.K.
99 A3 **Dartmoor** *hills* U.K.
131 D2 **Dartmouth** Can.
99 B3 **Dartmouth** U.K.
59 D3 **Daru** P.N.G.
50 C1 **Darwin** Austr.
74 A1 **Dasht** *r.* Pak.
76 B2 **Daşoguz** Turkm.
61 C1 **Datadian** Indon.
111 C3 **Datça** Turkey
70 B1 **Datong** China
64 B2 **Datu Piang** Phil.
74 B1 **Daud Khel** Pak.
88 B2 **Daugava** *r.* Latvia
88 C2 **Daugavpils** Latvia
100 C2 **Daun** Ger.
129 D2 **Dauphin** Can.
129 E2 **Dauphin Lake** Can.
73 B3 **Davangere** India
64 B2 **Davao** Phil.
64 B2 **Davao Gulf** Phil.
137 F2 **Davenport** U.S.A.
99 C2 **Daventry** U.K.
123 C2 **Daveyton** S. Africa
146 B4 **David** Panama
129 D2 **Davidson** Can.
126 E3 **Davidson Lake** Can.
135 B3 **Davis** U.S.A.
131 D1 **Davis Inlet (abandoned)**
Can.
159 F3 **Davis Sea** Antarctica
160 P3 **Davis Strait** Can./Greenland
105 D2 **Davos** Switz.
Dawei Myanmar *see* **Tavoy**
78 A2 **Dawmat al Jandal**
Saudi Arabia
79 C3 **Dawqah** Oman
126 B2 **Dawson** Can.
143 D2 **Dawson** U.S.A.
128 B2 **Dawson Creek** Can.
128 B2 **Dawsons Landing** Can.
68 C2 **Dawu** China
Dawukou China *see*
Shizuishan
104 B3 **Dax** France
68 C2 **Da Xueshan** *mts* China
80 C2 **Dayr az Zawr** Syria
140 C3 **Dayton** U.S.A.
143 D3 **Daytona Beach** U.S.A.
70 A2 **Dazhou** China
122 B3 **De Aar** S. Africa
80 B2 **Dead Sea** *salt l.* Asia
71 B3 **De'an** China
152 B4 **Deán Funes** Arg.
128 B2 **Dease Lake** Can.
126 D2 **Dease Strait** Can.
135 C3 **Death Valley** *depr.* U.S.A.
104 C2 **Deauville** France
61 C1 **Debak** Sarawak Malaysia
109 D2 **Debar** Macedonia
103 E2 **Debrecen** Hungary
117 B3 **Debre Markos** Eth.
117 B3 **Debre Tabor** Eth.
117 B4 **Debre Zeyit** Eth.
142 C2 **Decatur** *AL* U.S.A.
140 B3 **Decatur** *IL* U.S.A.
73 B3 **Deccan** *plat.* India
102 C1 **Děčín** Czech Rep.
137 E2 **Decorah** U.S.A.
88 C2 **Dedovichi** Russia
121 C2 **Dedza** Malawi
98 B2 **Dee** *r. England/Wales* U.K.
96 C2 **Dee** *r. Scotland* U.K.
53 D1 **Deepwater** Austr.
131 E2 **Deer Lake** Can.
134 D1 **Deer Lodge** U.S.A.
140 C3 **Defiance** U.S.A.
68 C2 **Dêgê** China

117 C4 Degeh Bur Eth.
102 C2 Deggendorf Ger.
91 E2 Degtevo Russia
75 B1 Dehra Dun India
75 C2 Dehri India
69 E2 Dehui China
100 A2 Deinze Belgium
110 B1 Dej Romania
140 B2 De Kalb U.S.A.
78 A3 Dekemhare Eritrea
118 C3 Dekese Dem. Rep. Congo
135 C3 Delano U.S.A.
135 D3 Delano Peak U.S.A.
123 C2 Delareyville S. Africa
129 D2 Delaronde Lake Can.
140 C2 Delaware U.S.A.
141 D3 Delaware r. U.S.A.
141 D3 Delaware state U.S.A.
141 D3 Delaware Bay U.S.A.
53 C3 Delegate Austr.
105 D2 Delémont Switz.
100 B1 Delft Neth.
100 C1 Delfzijl Neth.
121 D2 Delgado, Cabo c. Moz.
69 C1 Delgerhaan Mongolia
75 B2 Delhi India
60 B2 Deli i. Indon.
126 C2 Déline Can.
101 F2 Delitzsch Ger.
107 D2 Dellys Alg.
101 D1 Delmenhorst Ger.
109 B1 Delnice Croatia
83 L1 De-Longa, Ostrova is Russia
129 D2 Deloraine Can.
111 B3 Delphi tourist site Greece
139 C3 Del Rio U.S.A.
136 B3 Delta CO U.S.A.
135 D3 Delta UT U.S.A.
126 B2 Delta Junction U.S.A.
109 D3 Delvinë Albania
106 C1 Demanda, Sierra de la mts Spain
118 C3 Demba Dem. Rep. Congo
117 B4 Dembi Dolo Eth.
89 D2 Demidov Russia
138 B2 Deming U.S.A.
111 C3 Demirci Turkey
110 C2 Demirköy Turkey
102 C1 Demmin Ger.
142 C2 Demopolis U.S.A.
60 B2 Dempo, Gunung vol. Indon.
89 D2 Demyansk Russia
122 B3 De Naawte S. Africa
124 E2 Denali U.S.A.
100 B1 Den Burg Neth.
100 B2 Dendermonde Belgium
70 A1 Dengkou China
70 B2 Dengzhou China
Den Haag Neth. see The Hague
50 B2 Denham Austr.
100 B1 Den Helder Neth.
52 B3 Deniliquin Austr.
134 C2 Denio U.S.A.
137 D2 Denison IA U.S.A.
139 D2 Denison TX U.S.A.
111 C3 Denizli Turkey
53 D2 Denman Austr.
50 A3 Denmark Austr.
94 country Europe
160 Q3 Denmark Strait Greenland/ Iceland
61 C2 Denpasar Indon.
139 D2 Denton U.S.A.
50 A3 D'Entrecasteaux, Point Austr.
136 B3 Denver U.S.A.
75 C2 Deogarh Odisha India
74 B2 Deogarh Rajasthan India

75 C2 Deoghar India
83 K2 Deputatskiy Russia
68 C3 Dêqên China
142 B2 De Queen U.S.A.
74 A2 Dera Bugti Pak.
74 B1 Dera Ghazi Khan Pak.
74 B1 Dera Ismail Khan Pak.
87 D4 Derbent Russia
50 B1 Derby Austr.
99 C2 Derby U.K.
99 D2 Dereham U.K.
97 B2 Derg, Lough l. Ireland
91 D1 Derhachi Ukr.
142 B2 DeRidder U.S.A.
91 D2 Derkul r. Russia/Ukr.
Derry U.K. see Londonderry
75 C1 Dêrub China
116 B3 Derudeb Sudan
122 B3 De Rust S. Africa
109 C2 Derventa Bos. & Herz.
98 C2 Derwent r. U.K.
98 B1 Derwent Water l. U.K.
77 C1 Derzhavinsk Kazakh.
152 B3 Desaguadero r. Bol.
129 D2 Deschambault Lake Can.
134 B1 Deschutes r. U.S.A.
117 B3 Desē Eth.
153 B5 Deseado Arg.
153 B5 Deseado r. Arg.
137 E2 Des Moines U.S.A.
137 E2 Des Moines r. U.S.A.
91 C1 Desna r. Russia/Ukr.
89 D3 Desnogorsk Russia
101 F2 Dessau-Roßlau Ger.
126 B2 Destruction Bay Can.
149 C3 Desventuradas, Islas is S. Pacific Ocean
128 C1 Detah Can.
101 D2 Detmold Ger.
140 C2 Detroit U.S.A.
137 D1 Detroit Lakes U.S.A.
100 B2 Deurne Neth.
110 B1 Deva Romania
100 C1 Deventer Neth.
96 C2 Deveron r. U.K.
103 D2 Devét skal h. Czech Rep.
137 D1 Devils Lake U.S.A.
128 A2 Devils Paw mt. U.S.A.
99 C3 Devizes U.K.
74 B2 Devli India
110 C2 Devnya Bulg.
128 C2 Devon Can.
51 D4 Devonport Austr.
74 B2 Dewas India
137 F3 Dexter U.S.A.
70 A2 Deyang China
59 D3 Deyong, Tanjung pt Indon.
81 C2 Dezful Iran
70 B2 Dezhou China
79 C2 Dhahran Saudi Arabia
75 D2 Dhaka Bangl.
78 B3 Dhamār Yemen
75 C2 Dhamtari India
74 B2 Dhanbad India
74 B2 Dhankuta Nepal
62 A1 Dharmanagar India
75 C2 Dharmjaygarh India
73 B3 Dharwad India
74 B2 Dhasa India
78 B3 Dhubāb Yemen
74 B2 Dhule India
144 A1 Diablo, Picacho del mt. Mex.
155 D1 Diamantina watercourse Austr.
151 D3 Diamantina Brazil
155 D1 Diamantina, Chapada plat. Brazil
151 C3 Diamantino Brazil

71 B3 Dianbai China
151 D3 Dianópolis Brazil
114 B4 Dianra Côte d'Ivoire
114 C3 Diapaga Burkina Faso
79 C2 Dibā al Ḥiṣn U.A.E.
118 C3 Dibaya Dem. Rep. Congo
62 A1 Dibrugarh India
136 C1 Dickinson U.S.A.
142 C1 Dickson U.S.A.
Dicle r. Turkey see Tigris
105 D3 Die France
129 D2 Diefenbaker, Lake Can.
114 B3 Diéma Mali
62 B2 Diên Châu Vietnam
101 D1 Diepholz Ger.
104 C2 Dieppe France
115 D3 Diffa Niger
131 D2 Digby Can.
105 D3 Digne-les-Bains France
105 C2 Digoin France
64 B2 Digos Phil.
59 D3 Digul r. Indon.
Dihang r. China/India see Brahmaputra
Dihang r. China/India see Yarlung Zangbo
105 D2 Dijon France
117 C3 Dikhil Djibouti
111 C3 Dikili Turkey
100 A2 Diksmuide Belgium
115 D3 Dikwa Nigeria
117 B4 Dila Eth.
74 A1 Dilārām Afgh.
59 C3 Dili East Timor
81 D2 Dīlīngham U.S.A.
134 D1 Dillon U.S.A.
118 C4 Dilolo Dem. Rep. Congo
72 D2 Dimapur India
Dimashq Syria see Damascus
52 B3 Dimboola Austr.
110 C2 Dimitrovgrad Bulg.
87 D3 Dimitrovgrad Russia
64 B1 Dinagat i. Phil.
104 C2 Dinan France
100 B2 Dinant Belgium
111 D3 Dinar Turkey
81 C2 Dīnār, Kūh-e mt. Iran
73 B3 Dindigul India
123 D1 Dindiza Moz.
101 E2 Dingelstädt Ger.
75 C2 Dinggyê China
Dingle Bay Ireland see Daingean Uí Chúis
97 A2 Dingle Bay Ireland
96 C2 Dingwall U.K.
70 A2 Dingxi China
154 B3 Dionísio Cerqueira Brazil
114 A3 Diourbel Senegal
74 A2 Dir Pak.
51 D1 Direction, Cape Austr.
117 C4 Dirê Dawa Eth.
120 B2 Dirico Angola
115 D1 Dirj Libya
50 A2 Dirk Hartog Island Austr.
53 C1 Dirranbandi Austr.
78 B3 Dirs Saudi Arabia
50 B2 Disappointment, Lake imp. l. Austr.
52 B3 Discovery Bay Austr.
143 E1 Dismal Swamp U.S.A.
99 D2 Diss U.K.
108 B3 Dittaino r. Sicily Italy
74 B2 Diu India
155 D2 Divinópolis Brazil
87 D4 Divnoye Russia
114 B4 Divo Côte d'Ivoire
80 B2 Divriği Turkey

140	B2	Dixon U.S.A.
128	A2	Dixon Entrance sea chan. Can./U.S.A.
80	C2	Diyarbakır Turkey
74	A2	Diz Pak.
115	D2	Djado Niger
115	D2	Djado, Plateau du Niger
118	B3	Djambala Congo
115	C2	Djanet Alg.
115	C1	Djelfa Alg.
119	C2	Djéma C.A.R.
114	B3	Djenné Mali
118	B2	Djibloho Equat. Guinea
114	B3	Djibo Burkina Faso
117	C3	Djibouti country Africa
117	C3	Djibouti Djibouti
114	C4	Djougou Benin
92	C3	Djúpivogur Iceland
91	E1	Dmitriyevka Russia
89	D3	Dmitriyev-L'govskiy Russia
89	E2	Dmitrov Russia
		Dnepr r. Ukr. see Dnieper
91	C2	Dnieper r. Ukr.
90	B2	Dniester r. Ukr.
		Dnipro r. Ukr. see Dnieper
91	C2	Dniprodzerzhyns'k Ukr.
91	D2	Dnipropetrovs'k Ukr.
91	C2	Dniprorudne Ukr.
		Dnister r. Ukr. see Dniester
88	C2	Dno Russia
115	D4	Doba Chad
88	B2	Dobele Latvia
101	F2	Döbeln Ger.
59	C3	Doberai, Jazirah pen. Indon.
59	C3	Dobo Indon.
109	C2	Doboj Bos. & Herz.
110	C2	Dobrich Bulg.
89	F3	Dobrinka Russia
89	E3	Dobroye Russia
89	D3	Dobrush Belarus
155	E1	Doce r. Brazil
145	B2	Doctor Arroyo Mex.
144	B2	Doctor Belisario Domínguez Mex.
111	C3	Dodecanese is Greece
		Dodekanisos is Greece see Dodecanese
136	C3	Dodge City U.S.A.
119	D3	Dodoma Tanz.
100	C2	Doetinchem Neth.
59	C3	Dofa Indon.
75	C1	Dogai Coring salt l. China
128	B2	Dog Creek Can.
67	B3	Dōgo i. Japan
115	C3	Dogondoutchi Niger
81	C2	Doğubeyazıt Turkey
79	C2	Doha Qatar
62	A2	Doi Saket Thai.
100	B1	Dokkum Neth.
88	C3	Dokshytsy Belarus
91	D2	Dokuchayevs'k Ukr.
131	C2	Dolbeau-Mistassini Can.
104	B2	Dol-de-Bretagne France
105	D2	Dole France
99	B2	Dolgellau U.K.
89	E3	Dolgorukovo Russia
89	E3	Dolgoye Russia
		Dolisie Congo see Loubomo
59	D3	Dolok, Pulau i. Indon.
108	B1	Dolomites mts Italy
117	C4	Dolo Odo Eth.
126	D2	Dolphin and Union Strait Can.
90	A2	Dolyna Ukr.
102	C2	Domažlice Czech Rep.
93	E3	Dombås Norway
103	D2	Dombóvár Hungary
128	B2	Dome Creek Can.

147	D3	Dominica country West Indies
147	C3	Dominican Republic country West Indies
89	E2	Domodedovo Russia
111	B3	Domokos Greece
61	C2	Dompu Indon.
89	E3	Don r. Russia
96	C2	Don r. U.K.
97	D1	Donaghadee U.K.
52	B3	Donald Austr.
		Donau r. Austria/Ger. see Danube
102	C2	Donauwörth Ger.
106	B2	Don Benito Spain
98	C2	Doncaster U.K.
120	A1	Dondo Angola
121	C2	Dondo Moz.
73	C4	Dondra Head Sri Lanka
97	B1	Donegal Ireland
97	B1	Donegal Bay Ireland
91	D2	Donets'k Ukr.
91	D2	Donets'ky Kryazh hills Russia/Ukr.
50	A2	Dongara Austr.
62	B1	Dongchuan China
65	B2	Dongducheon S. Korea
71	A4	Dongfang China
66	B1	Dongfanghong China
58	B3	Donggala Indon.
65	A2	Donggang China
71	B3	Dongguan China
62	B2	Đông Ha Vietnam
65	B2	Donghae S. Korea
		Dong Hai sea N. Pacific Ocean see East China Sea
62	B2	Đông Hơi Vietnam
118	B2	Dongou Congo
71	B3	Dongshan China
70	B2	Dongsheng China
70	C2	Dongtai China
71	B3	Dongting Hu l. China
		Dong Ujimqin Qi China see Uliastai
70	B2	Dongying China
63	B2	Don Kêv Cambodia
54	B1	Donnellys Crossing N.Z.
99	B3	Dorchester U.K.
122	A1	Dordabis Namibia
104	B2	Dordogne r. France
100	B2	Dordrecht Neth.
123	C3	Dordrecht S. Africa
129	D2	Doré Lake Can.
101	D1	Dorfmark Ger.
68	C1	Dörgön Nuur salt l. Mongolia
114	B3	Dori Burkina Faso
122	A3	Doring r. S. Africa
96	B2	Dornoch U.K.
96	B2	Dornoch Firth est. U.K.
89	D3	Dorogobuzh Russia
90	B2	Dorohoi Romania
92	G3	Dorotea Sweden
50	A2	Dorre Island Austr.
53	D2	Dorrigo Austr.
100	C2	Dortmund Ger.
100	C2	Dortmund-Ems-Kanal canal Ger.
153	B5	Dos Bahías, Cabo c. Arg.
77	C3	Dōshī Afgh.
101	F1	Dosse r. Ger.
114	C3	Dosso Niger
143	C2	Dothan U.S.A.
101	D1	Dötlingen Ger.
105	C1	Douai France
118	A2	Douala Cameroon
104	B2	Douarnenez France
114	B3	Douentza Mali
98	A1	Douglas Isle of Man
122	B2	Douglas S. Africa

128	A2	Douglas AK U.S.A.
138	B2	Douglas AZ U.S.A.
143	D2	Douglas GA U.S.A.
136	B2	Douglas WY U.S.A.
71	C3	Douliu Taiwan
104	C1	Doullens France
154	B1	Dourada, Serra hills Brazil
154	B2	Dourados Brazil
154	B2	Dourados, Serra dos hills Brazil
106	B1	Douro r. Port.
99	D3	Dover U.K.
141	D3	Dover U.S.A.
95	D3	Dover, Strait of France/U.K.
141	F1	Dover-Foxcroft U.S.A.
79	C2	Dowlatābād Būshehr Iran
79	C2	Dowlatābād Kermān Iran
97	D1	Downpatrick U.K.
67	B3	Dōzen is Japan
130	C2	Dozois, Réservoir resr Can.
114	B2	Drâa, Hamada du plat. Alg.
53	A3	Dracena Brazil
100	C1	Drachten Neth.
110	B2	Drăgăneşti-Olt Romania
110	B2	Drăgăşani Romania
88	C3	Drahichyn Belarus
123	C2	Drakensberg mts Lesotho/S. Africa
123	C2	Drakensberg mts S. Africa
158	B8	Drake Passage S. Atlantic Ocean
111	B2	Drama Greece
93	F4	Drammen Norway
109	C1	Drava r. Europe
128	C2	Drayton Valley Can.
102	C1	Dresden Ger.
104	C2	Dreux France
100	B1	Driemond Neth.
109	C2	Drina r. Bos. & Herz./Serbia
109	C2	Drniš Croatia
110	B2	Drobeta-Turnu Severin Romania
101	D1	Drochtersen Ger.
97	C2	Drogheda Ireland
90	A2	Drohobych Ukr.
97	C1	Dromore U.K.
74	B1	Drosh Pak.
53	C3	Drouin Austr.
128	C2	Drumheller Can.
140	C1	Drummond Island U.S.A.
131	C2	Drummondville Can.
88	B3	Druskininkai Lith.
91	D2	Druzhkivka Ukr.
88	D2	Druzhnaya Gorka Russia
130	A2	Dryden Can.
50	B1	Drysdale r. Austr.
78	A2	Dubā Saudi Arabia
79	C2	Dubai U.A.E.
129	D1	Dubawnt Lake Can.
		Dubayy U.A.E. see Dubai
78	A2	Dubbagh, Jabal ad mt. Saudi Arabia
53	C2	Dubbo Austr.
97	C2	Dublin Ireland
143	D2	Dublin U.S.A.
90	B1	Dubno Ukr.
141	D2	Du Bois U.S.A.
114	A4	Dubréka Guinea
109	C2	Dubrovnik Croatia
90	B1	Dubrovytsya Ukr.
89	D3	Dubrowna Belarus
137	E2	Dubuque U.S.A.
129	D2	Duck Bay Can.
101	E2	Duderstadt Ger.
82	G2	Dudinka Russia
99	B2	Dudley U.K.
106	B1	Duero r. Spain

131 C1 Duffreboy, Lac l. Can.
96 C2 Dufftown U.K.
109 C2 Dugi Rat Croatia
100 C2 Duisburg Ger.
123 C3 Dukathole S. Africa
79 C3 Dukhān Qatar
89 D2 Dukhovshchina Russia
Dukou China see Panzhihua
88 C2 Dūkštas Lith.
68 C2 Dulan China
152 B4 Dulce r. Arg.
100 C2 Dülmen Ger.
110 C2 Dulovo Bulg.
137 E1 Duluth U.S.A.
64 B2 Dumaguete Phil.
60 B1 Dumai Indon.
64 B1 Dumaran i. Phil.
142 B2 Dumas AR U.S.A.
139 C2 Dumas TX U.S.A.
96 B3 Dumbarton U.K.
103 D2 Ďumbier mt. Slovakia
96 C3 Dumfries U.K.
55 J3 Dumont d'Urville Sea
Antarctica
116 B1 Dumyāṭ Egypt
Duna r. Hungary see Danube
Dunaj r. Slovakia see Danube
103 D2 Dunakeszi Hungary
Dunărea r. Romania see
Danube
103 D2 Dunaújváros Hungary
Dunav r. Serbia see Danube
90 B2 Dunayivtsi Ukr.
96 C2 Dunbar U.K.
96 C1 Dunbeath U.K.
128 B3 Duncan Can.
139 D2 Duncan U.S.A.
96 C1 Duncansby Head U.K.
97 C1 Dundalk Ireland
97 C2 Dundalk Bay Ireland
123 D2 Dundee S. Africa
96 C2 Dundee U.K.
97 D1 Dundrum Bay U.K.
54 B3 Dunedin N.Z.
96 C2 Dunedin U.K.
96 C2 Dunfermline U.K.
97 C1 Dungannon U.K.
74 B2 Dungarpur India
97 C2 Dungarvan Ireland
99 D3 Dungeness hd U.K.
97 C1 Dungiven U.K.
53 D2 Dungog Austr.
119 C2 Dungu Dem. Rep. Congo
60 B1 Dungun Malaysia
116 B2 Dungunab Sudan
69 E2 Dunhua China
68 C2 Dunhuang China
Dunkerque France see
Dunkirk
104 C1 Dunkirk France
141 D2 Dunkirk U.S.A.
97 C2 Dún Laoghaire Ireland
97 C1 Dunmurry U.K.
96 C1 Dunnet Head U.K.
96 C3 Duns U.K.
99 C3 Dunstable U.K.
110 B2 Dupnitsa Bulg.
136 C1 Dupree U.S.A.
50 B1 Durack r. Austr.
144 B2 Durango Mex.
106 C1 Durango Spain
136 B3 Durango U.S.A.
139 D2 Durant U.S.A.
153 C4 Durazno Uru.
123 D2 Durban S. Africa
104 C3 Durban-Corbières France
122 A3 Durbanville S. Africa
100 B2 Durbuy Belgium
100 C2 Düren Ger.

73 C2 Durg India
98 C1 Durham U.K.
143 E1 Durham U.S.A.
60 B1 Duri Indon.
109 C2 Durmitor mt. Montenegro
96 B1 Durness U.K.
109 C2 Durrës Albania
97 A3 Dursey Island Ireland
111 C3 Dursunbey Turkey
59 D3 D'Urville, Tanjung pt Indon.
54 B2 D'Urville Island N.Z.
71 A3 Dushan China
77 C3 Dushanbe Tajik.
100 C2 Düsseldorf Ger.
71 A3 Duyun China
91 D2 Dvorichna Ukr.
74 A2 Dwarka India
123 C1 Dwarsberg S. Africa
134 C1 Dworshak Reservoir U.S.A.
89 D3 Dyat'kovo Russia
96 C2 Dyce U.K.
127 G2 Dyer, Cape Can.
142 C1 Dyersburg U.S.A.
103 D2 Dyje r. Austria/Czech Rep.
103 D1 Dylewska Góra h. Pol.
91 D1 Dymytrov Ukr.
123 C3 Dyoki S. Africa
87 E3 Dyurtyuli Russia
69 D2 Dzamïn Üüd Mongolia
120 D2 Dzaoudzi Mayotte
Dzhambul Kazakh. see Taraz
91 C2 Dzhankoy Ukr.
91 D3 Dzhubga Russia
83 K3 Dzhugdzhur, Khrebet mts
Russia
103 E1 Działdowo Pol.
69 D1 Dzuunmod Mongolia
88 C3 Dzyarzhynsk Belarus
88 C3 Dzyatlavichy Belarus

E

131 E1 Eagle r. Can.
134 C1 Eagle Cap mt. U.S.A.
130 A2 Eagle Lake Can.
134 B2 Eagle Lake U.S.A.
139 C3 Eagle Pass U.S.A.
126 B2 Eagle Plain Can.
130 A1 Ear Falls Can.
99 D3 Eastbourne U.K.
156 C3 East China Sea
N. Pacific Ocean
54 B1 East Coast Bays N.Z.
129 D3 Eastend Can.
123 C3 Eastern Cape prov. S. Africa
116 B2 Eastern Desert Egypt
73 B3 Eastern Ghats mts India
129 E2 Easterville Can.
153 C6 East Falkland i. Falkland Is
100 C1 East Frisian Islands Ger.
96 B3 East Kilbride U.K.
99 C3 Eastleigh U.K.
140 C2 East Liverpool U.S.A.
123 C3 East London S. Africa
130 C1 Eastmain Can.
130 C1 Eastmain r. Can.
143 D2 Eastman U.S.A.
140 A3 East St Louis U.S.A.
East Sea N. Pacific Ocean see
Japan, Sea of
83 K2 East Siberian Sea Russia
59 C3 East Timor country Asia
140 A2 Eau Claire U.S.A.
130 C1 Eau Claire, Lac à l' l. Can.
59 D2 Eauripik atoll Micronesia
145 C2 Ébano Mex.

99 B3 Ebbw Vale U.K.
102 C1 Eberswalde Ger.
109 C2 Eboli Italy
118 B2 Ebolowa Cameroon
107 C1 Ebro r. Spain
144 A2 Echeverría, Pico mt. Mex.
67 C3 Echizen Japan
129 E2 Echoing r. Can.
100 C3 Echternach Lux.
52 B3 Echuca Austr.
106 B2 Écija Spain
102 B1 Eckernförde Ger.
127 F2 Eclipse Sound sea chan.
Can.
150 A2 Ecuador country S. America
116 C3 Ed Eritrea
117 A3 Ed Da'ein Sudan
117 B3 Ed Damazin Sudan
116 B3 Ed Damer Sudan
116 B3 Ed Debba Sudan
116 B3 Ed Dueim Sudan
51 D4 Eddystone Point Austr.
100 B1 Ede Neth.
118 B2 Edéa Cameroon
154 C1 Edéia Brazil
53 C3 Eden Austr.
98 B1 Eden r. U.K.
123 C2 Edenburg S. Africa
97 C2 Edenderry Ireland
52 B3 Edenhope Austr.
111 B2 Edessa Greece
82 C1 Edgeøya i. Svalbard
139 D3 Edinburg U.S.A.
96 C3 Edinburgh U.K.
110 C2 Edirne Turkey
128 C2 Edmonton Can.
131 D2 Edmundston Can.
111 C3 Edremit Turkey
111 C3 Edremit Körfezi b. Turkey
128 C2 Edson Can.
119 C3 Edward, Lake Dem. Rep.
Congo/Uganda
139 C2 Edwards Plateau U.S.A.
100 C1 Eenrum Neth.
48 F4 Efatē i. Vanuatu
140 B3 Effingham U.S.A.
135 C3 Egan Range mts U.S.A.
103 E2 Eger Hungary
93 E4 Egersund Norway
92 ⬜C2 Egilsstaðir Iceland
80 B2 Eğirdir Turkey
80 B2 Eğirdir Gölü l. Turkey
104 C2 Egletons France
Egmont, Mount vol. N.Z. see
Taranaki, Mount
83 M2 Egvekinot Russia
116 A2 Egypt country Africa
68 C2 Ehen Hudag China
100 C1 Eibergen Neth.
100 C2 Eifel hills Ger.
96 A2 Eigg i. U.K.
73 B4 Eight Degree Channel India/
Maldives
50 B1 Eighty Mile Beach Austr.
80 B3 Eilat Israel
101 F2 Eilenburg Ger.
101 E1 Einbeck Ger.
100 B2 Eindhoven Neth.
150 B2 Eirunepé Brazil
120 B2 Eiseb watercourse Namibia
101 E2 Eisenach Ger.
102 C1 Eisenhüttenstadt Ger.
103 D2 Eisenstadt Austria
101 E2 Eisleben, Lutherstadt Ger.
Eivissa Spain see Ibiza
Eivissa i. Spain see Ibiza
107 C1 Ejea de los Caballeros Spain
121 ⬜D3 Ejeda Madag.

Ejin Qi China see Dalain Hob
93 H4 Ekenäs Fin.
92 J2 Ekostrovskaya Imandra, Ozero l. Russia
93 F4 Eksjö Sweden
122 A4 Eksteenfontein S. Africa
130 B1 Ekwan r. Can.
62 A1 Ela Myanmar
123 C2 Elandsdoorn S. Africa
111 B3 Elassona Greece
80 B2 Elazığ Turkey
108 B2 Elba, Isola d' i. Italy
150 A1 El Banco Col.
138 B2 El Barreal salt l. Mex.
109 D2 Elbasan Albania
150 B1 El Baúl Venez.
114 C1 El Bayadh Alg.
101 D1 Elbe r. Ger.
136 B3 Elbert, Mount U.S.A.
143 D2 Elberton U.S.A.
104 C2 Elbeuf France
80 B2 Elbistan Turkey
103 D1 Elbląg Pol.
87 D4 El'brus mt. Russia
81 C2 Elburz Mountains Iran
150 B1 El Callao Venez.
139 D3 El Campo U.S.A.
135 C4 El Centro U.S.A.
152 B2 El Cerro Bol.
107 C2 Elche-Elx Spain
107 C2 Elda Spain
137 D3 Eldon U.S.A.
144 B2 El Dorado Mex.
142 B2 El Dorado AR U.S.A.
137 D3 El Dorado KS U.S.A.
114 B2 El Eglab plat. Alg.
106 C2 El Ejido Spain
89 E2 Elektrostal' Russia
Elemi Triangle disp. terr. Africa see Ilemi Triangle
150 A2 El Encanto Col.
146 C2 Eleuthera i. Bahamas
117 A3 El Fasher Sudan
144 B2 El Fuerte Mex.
117 A3 El Geneina Sudan
116 B3 El Geteina Sudan
96 C2 Elgin U.K.
140 B2 Elgin U.S.A.
115 C1 El Goléa Alg.
144 A1 El Golfo de Santa Clara Mex.
119 D2 Elgon, Mount Kenya/Uganda
114 A2 El Hammâmi reg. Maur.
114 A2 El Hierro i. Canary Islands
145 C2 El Higo Mex.
114 C2 El Homr Alg.
110 C2 Elhovo Bulg.
87 D4 Elista Russia
141 E2 Elizabeth U.S.A.
143 E1 Elizabeth City U.S.A.
140 B3 Elizabethtown U.S.A.
114 B1 El Jadida Morocco
103 E1 Ełk Pol.
139 C1 Elk City U.S.A.
128 C2 Elkford Can.
140 B2 Elkhart U.S.A.
El Khartum Sudan see Khartoum
140 D3 Elkins U.S.A.
128 C3 Elko Can.
134 C2 Elko U.S.A.
129 C2 Elk Point Can.
126 E1 Ellef Ringnes Island Can.
137 D1 Ellendale U.S.A.
134 B1 Ellensburg U.S.A.
54 B2 Ellesmere, Lake N.Z.
127 F1 Ellesmere Island Can.
98 B2 Ellesmere Port U.K.

126 E2 Ellice r. Can.
Ellice Islands country S. Pacific Ocean see Tuvalu
123 C3 Elliotdale S. Africa
96 C2 Ellon U.K.
141 F2 Ellsworth U.S.A.
55 O2 Ellsworth Mountains Antarctica
111 C3 Elmalı Turkey
115 C1 El Meghaïer Alg.
141 D2 Elmira U.S.A.
107 C2 El Moral Spain
101 D1 Elmshorn Ger.
117 A3 El Muglad Sudan
64 A1 El Nido Phil.
117 B3 El Obeid Sudan
144 B2 El Oro Mex.
115 C1 El Oued Alg.
138 A2 Eloy U.S.A.
138 B2 El Paso U.S.A.
144 B1 El Porvenir Mex.
107 D1 El Prat de Llobregat Spain
139 D1 El Reno U.S.A.
145 B2 El Salado Mex.
144 B2 El Salto Mex.
146 B3 El Salvador country Central America
145 B2 El Salvador Mex.
138 B3 El Sauz Mex.
144 A1 El Socorro Mex.
145 C2 El Temascal Mex.
150 B1 El Tigre Venez.
147 D4 El Tocuyo Venez.
88 C2 Elva Estonia
106 B2 Elvas Port.
93 F3 Elverum Norway
119 E2 El Wak Kenya
99 D2 Ely U.K.
137 E1 Ely MN U.S.A.
135 D3 Ely NV U.S.A.
123 C2 eMalahleni S. Africa
93 G4 Emån r. Sweden
123 D3 eMazimtoti S. Africa
76 B2 Emba r. Kazakh.
123 C2 Embalenhle S. Africa
154 C1 Emborcação, Represa de resr Brazil
119 D3 Embu Kenya
100 C1 Emden Ger.
51 D2 Emerald Austr.
129 E3 Emerson Can.
111 C3 Emet Turkey
123 D2 eMgwenya S. Africa
115 D3 Emi Koussi mt. Chad
110 C2 Emine, Nos pt Bulg.
80 B2 Emirdağ Turkey
123 D2 eMjindini S. Africa
88 B2 Emmaste Estonia
100 B1 Emmeloord Neth.
100 C2 Emmelshausen Ger.
100 C1 Emmen Neth.
123 D2 eMondlo S. Africa
139 C3 Emory Peak U.S.A.
144 A2 Empalme Mex.
123 D2 Empangeni S. Africa
108 B2 Empoli Italy
137 D3 Emporia KS U.S.A.
141 D3 Emporia VA U.S.A.
Empty Quarter des. Saudi Arabia see Rub' al Khali
100 C1 Ems r. Ger.
100 C1 Emsdetten Ger.
123 C2 eMzinoni S. Africa
59 D3 Enarotali Indon.
144 B2 Encarnación Mex.
152 C3 Encarnación Para.
155 D1 Encruzilhada Brazil
58 C3 Ende Indon.

126 A2 Endicott Mountains U.S.A.
91 C2 Enerhodar Ukr.
87 D3 Engel's Russia
60 B2 Enggano i. Indon.
98 C2 England admin. div. U.K.
130 A1 English r. Can.
95 C4 English Channel France/U.K.
139 D1 Enid U.S.A.
100 B1 Enkhuizen Neth.
93 G4 Enköping Sweden
108 B3 Enna Sicily Italy
129 D1 Ennadai Lake Can.
117 A3 En Nahud Sudan
115 E3 Ennedi, Massif mts Chad
53 C1 Enngonia Austr.
97 B2 Ennis Ireland
139 D2 Ennis U.S.A.
97 C2 Enniscorthy Ireland
97 C1 Enniskillen U.K.
97 B2 Ennistymon Ireland
102 C2 Enns r. Austria
92 H2 Enontekiö Fin.
53 C3 Ensay Austr.
100 C1 Enschede Neth.
144 A1 Ensenada Mex.
70 A2 Enshi China
128 C1 Enterprise Can.
142 C2 Enterprise AL U.S.A.
134 C1 Enterprise OR U.S.A.
152 B3 Entre Ríos Bol.
106 B2 Entroncamento Port.
115 C4 Enugu Nigeria
150 A2 Envira Brazil
135 D3 Ephraim U.S.A.
134 C1 Ephrata U.S.A.
105 D2 Épinal France
99 C3 Epsom U.K.
118 A2 Equatorial Guinea country Africa
101 F3 Erbendorf Ger.
100 C3 Erbeskopf h. Ger.
81 C2 Erciş Turkey
65 B1 Erdao Jiang r. China
111 C2 Erdek Turkey
80 B2 Erdemli Turkey
152 C3 Erechim Brazil
69 D1 Ereentsav Mongolia
80 B2 Ereğli Konya Turkey
80 B1 Ereğli Zonguldak Turkey
69 D2 Erenhot China
Erevan Armenia see Yerevan
101 E2 Erfurt Ger.
80 B2 Ergani Turkey
114 B2 'Erg Chech des. Alg./Mali
111 C2 Ergene r. Turkey
140 C2 Erie U.S.A.
140 C2 Erie, Lake Can./U.S.A.
66 D2 Erimo-misaki c. Japan
116 B3 Eritrea country Africa
101 E3 Erlangen Ger.
50 C2 Erldunda Austr.
123 C2 Ermelo S. Africa
80 B2 Ermenek Turkey
111 B3 Ermoupoli Greece
73 B4 Ernakulam India
73 B3 Erode India
100 B2 Erp Neth.
114 B1 Er Rachidia Morocco
117 B3 Er Rahad Sudan
97 B1 Errigal h. Ireland
97 A1 Erris Head Ireland
109 D2 Ersekë Albania
91 E1 Ertil' Russia
101 D1 Erwitte Ger.
101 F2 Erzgebirge mts Czech Rep./Ger.
80 B2 Erzincan Turkey
81 C2 Erzurum Turkey

93	E4	Esbjerg Denmark
135	D3	Escalante U.S.A.
101	C2	Escalón Mex.
140	B1	Escanaba U.S.A.
131	C1	Escárcega Mex.
107	C1	Escatrón Spain
100	A2	Escaut r. Belgium/France
101	E1	Eschede Ger.
101	D4	Esch-sur-Alzette Lux.
101	E2	Eschwege Ger.
100	C2	Eschweiler Ger.
135	C4	Escondido U.S.A.
144	B2	Escuinapa Mex.
111	C3	Eşen Turkey
81	D2	Eşfahān Iran
123	D2	eSikhaleni S. Africa
98	B1	Esk r. U.K.
131	C1	Esker Can.
92	□C2	Eskifjörður Iceland
93	G4	Eskilstuna Sweden
		Eskimo Point Can. see Arviat
80	B2	Eskişehir Turkey
81	C2	Eslāmābād-e Gharb Iran
111	C3	Esler Dağı mt. Turkey
111	C3	Eşme Turkey
150	A1	Esmeraldas Ecuador
79	D2	Espakeh Iran
104	C3	Espalion France
130	B2	Espanola Can.
138	B1	Espanola U.S.A.
50	B3	Esperance Austr.
144	B2	Esperanza Mex.
106	B2	Espichel, Cabo c. Port.
155	D1	Espinhaço, Serra do mts Brazil
151	D3	Espinosa Brazil
48	F4	Espíritu Santo i. Vanuatu
144	A2	Espíritu Santo, Isla i. Mex.
93	H3	Espoo Fin.
153	A5	Esquel Arg.
114	B1	Essaouira Morocco
100	C2	Essen Ger.
150	C1	Essequibo r. Guyana
114	A2	Es-Smara Western Sahara
83	L3	Esso Russia
79	C2	Estahbān Iran
151	E3	Estância Brazil
123	C2	Estcourt S. Africa
107	C1	Estella Spain
106	B2	Estepona Spain
106	C1	Esteras de Medinaceli Spain
152	B3	Esterhazy Can.
152	B3	Esteros Para.
136	B2	Estes Park U.S.A.
129	D3	Estevan Can.
137	E2	Estherville U.S.A.
129	D2	Eston Can.
88	C2	Estonia country Europe
106	B1	Estrela, Serra da mts Port.
106	B2	Estremoz Port.
52	A1	Etadunna Austr.
104	C2	Étampes France
104	C1	Étaples France
75	B2	Etawah India
123	D2	Ethandakukhanya S. Africa
117	B4	Ethiopia country Africa
109	C3	Etna, Mount vol. Sicily Italy
128	A2	Etolin Island U.S.A.
120	A2	Etosha Pan salt pan Namibia
100	C3	Ettelbruck Lux.
100	B2	Etten-Leur Neth.
107	C1	Etxarri-Aranatz Spain
99	D3	Eu France
53	C2	Euabalong Austr.
		Euboea i. Greece see Evvoia
50	B3	Eucla Austr.
143	C2	Eufaula U.S.A.
139	D1	Eufaula Lake resr U.S.A.
134	B2	Eugene U.S.A.
144	A2	Eugenia, Punta pt Mex.
53	C1	Eulo Austr.
53	C2	Eumungerie Austr.
139	C2	Eunice U.S.A.
80	C2	Euphrates r. Asia
134	B2	Eureka CA U.S.A.
134	C1	Eureka MT U.S.A.
135	C3	Eureka NV U.S.A.
52	B2	Euriowie Austr.
53	C3	Euroa Austr.
106	B2	Europa Point Gibraltar
128	B2	Eutsuk Lake Can.
130	C1	Evans, Lac l. Can.
53	D1	Evans Head Austr.
127	F2	Evans Strait Can.
136	A2	Evanston U.S.A.
140	B3	Evansville U.S.A.
123	C2	Evaton S. Africa
79	C2	Evaz Iran
83	L2	Evensk Russia
50	C2	Everard Range hills Austr.
75	C2	Everest, Mount China/Nepal
134	B1	Everett U.S.A.
100	A2	Evergem Belgium
143	D3	Everglades swamp U.S.A.
118	B2	Evinayong Equat. Guinea
93	E4	Evje Norway
106	B2	Évora Port.
104	C2	Évreux France
111	C2	Evros r. Greece/Turkey
111	B3	Evrotas r. Greece
80	B2	Evrychou Cyprus
111	B3	Evvoia i. Greece
119	E2	Ewaso Ngiro r. Kenya
152	B2	Exaltación Bol.
99	B3	Exe r. U.K.
99	B3	Exeter U.K.
99	B3	Exmoor hills U.K.
99	B3	Exmouth U.K.
50	A2	Exmouth Gulf Austr.
146	C2	Exuma Cays is Bahamas
117	C4	Eyl Somalia
51	C2	Eyre Peninsula Austr.
94	B1	Eysturoy i. Faroe Is
123	D2	Ezakheni S. Africa
123	C2	Ezenzeleni S. Africa
70	B2	Ezhou China
86	E2	Ezhva Russia
111	C3	Ezine Turkey

F

138	B2	Fabens U.S.A.
108	B2	Fabriano Italy
115	D3	Fachi Niger
114	C3	Fada-N'Gourma Burkina Faso
108	B2	Faenza Italy
59	C3	Fafanlap Indon.
110	B1	Făgăraş Romania
49	G3	Fagatogo American Samoa
93	E3	Fagernes Norway
93	G4	Fagersta Sweden
153	B6	Fagnano, Lago l. Arg./Chile
114	B3	Faguibine, Lac l. Mali
92	□B3	Fagurhólsmýri Iceland
126	B2	Fairbanks U.S.A.
137	D2	Fairbury U.S.A.
135	B3	Fairfield U.S.A.
96	□	Fair Isle i. U.K.
137	E2	Fairmont MN U.S.A.
140	C3	Fairmont WV U.S.A.
128	C2	Fairview Can.
59	D2	Fais i. Micronesia
74	B1	Faisalabad Pak.
136	C1	Faith U.S.A.
77	D3	Faïzābād Afgh.
75	C2	Faizabad India
59	C3	Fakfak Indon.
65	A1	Faku China
114	A4	Falaba Sierra Leone
139	D3	Falcon Lake Mex./U.S.A.
139	D3	Falfurrias U.S.A.
128	C2	Falher Can.
93	F4	Falkenberg Sweden
101	F2	Falkenberg/Elster Ger.
101	F1	Falkensee Ger.
96	C3	Falkirk U.K.
153	C6	Falkland Islands terr. S. Atlantic Ocean
93	F4	Falköping Sweden
135	C3	Fallon U.S.A.
141	E2	Fall River U.S.A.
137	D2	Falls City U.S.A.
99	A3	Falmouth U.K.
122	A3	False Bay S. Africa
93	F4	Falster i. Denmark
110	C1	Fălticeni Romania
93	G3	Falun Sweden
71	A3	Fangcheng China
71	C3	Fangshan Taiwan
69	E1	Fangzheng China
108	B2	Fano Italy
119	C2	Faradje Dem. Rep. Congo
121	□D3	Farafangana Madag.
116	A2	Farāfirah, Wāḩāt al oasis Egypt
76	C3	Farāh Afgh.
114	A3	Faranah Guinea
78	B3	Farasān, Jazā'ir is Saudi Arabia
127	J3	Farewell, Cape Greenland
54	B2	Farewell, Cape N.Z.
137	D1	Fargo U.S.A.
77	D2	Farg'ona Uzbek.
137	E2	Faribault U.S.A.
141	E2	Farmington ME U.S.A.
138	B1	Farmington NM U.S.A.
141	D3	Farmville U.S.A.
99	C3	Farnborough U.K.
128	C2	Farnham, Mount Can.
128	A1	Faro Can.
106	B2	Faro Port.
88	A2	Fårö i. Sweden
94	B1	Faroe Islands terr. N. Atlantic Ocean
79	C2	Farrāshband Iran
139	C2	Farwell U.S.A.
79	C2	Fāryāb Iran
79	C2	Fasā Iran
109	C2	Fasano Italy
90	B1	Fastiv Ukr.
75	B2	Fatehgarh India
75	C2	Fatehpur India
92	G2	Fauske Norway
92	□A3	Faxaflói b. Iceland
92	G3	Faxälven r. Sweden
115	D3	Faya Chad
142	B1	Fayetteville AR U.S.A.
143	E1	Fayetteville NC U.S.A.
142	C1	Fayetteville TN U.S.A.
74	B1	Fazilka India
114	A2	Fdérik Maur.
143	E2	Fear, Cape U.S.A.
54	C2	Featherston N.Z.
104	C2	Fécamp France
61	C2	Fehmarn i. Ger.
101	F1	Fehrbellin Ger.
155	D2	Feia, Lagoa lag. Brazil
150	A2	Feijó Brazil
54	C2	Feilding N.Z.
151	E3	Feira de Santana Brazil
107	D2	Felanitx Spain
145	D3	Felipe C. Puerto Mex.
155	D1	Felixlândia Brazil

99 D3 **Felixstowe** U.K.
101 D2 **Felsberg** Ger.
108 B1 **Feltre** Italy
93 F3 **Femunden** *l.* Norway
71 B3 **Fengcheng** *Jiangxi* China
65 A1 **Fengcheng** *Liaoning* China
62 A1 **Fengqing** China
70 B2 **Fengxian** China
Fengxiang China *see* **Lincang**
71 C3 **Fengyuan** Taiwan
70 B1 **Fengzhen** China
105 D3 **Feno, Capo di** *c. Corsica* France
121 □D2 **Fenoarivo Atsinanana** Madag.
91 D2 **Feodosiya** Ukr.
108 A3 **Fer, Cap de** *c.* Alg.
137 D1 **Fergus Falls** U.S.A.
109 D2 **Ferizaj** Kosovo
114 B4 **Ferkessédougou** Côte d'Ivoire
108 B2 **Fermo** Italy
131 D1 **Fermont** Can.
106 B1 **Fermoselle** Spain
97 B2 **Fermoy** Ireland
143 D2 **Fernandina Beach** U.S.A.
154 B2 **Fernandópolis** Brazil
128 C3 **Fernie** Can.
108 B2 **Ferrara** Italy
154 B2 **Ferreiros** Brazil
108 A2 **Ferro, Capo** *c. Sardinia* Italy
106 B1 **Ferrol** Spain
100 B1 **Ferwerd** Neth.
114 B1 **Fès** Morocco
118 B3 **Feshi** Dem. Rep. Congo
137 E3 **Festus** U.S.A.
111 C3 **Fethiye** Turkey
96 □ **Fetlar** *i.* U.K.
130 C1 **Feuilles, Rivière aux** *r.* Can.
Fez Morocco *see* **Fès**
121 □D3 **Fianarantsoa** Madag.
117 B4 **Fichē** Eth.
109 C2 **Fier** Albania
96 C2 **Fife Ness** *pt* U.K.
104 C3 **Figeac** France
106 B1 **Figueira da Foz** Port.
107 D1 **Figueres** Spain
114 B1 **Figuig** Morocco
48 F3 **Fiji** *country* S. Pacific Ocean
152 B3 **Filadelfia** Para.
55 Q2 **Filchner Ice Shelf** Antarctica
98 C1 **Filey** U.K.
111 B3 **Filippiada** Greece
93 F4 **Filipstad** Sweden
96 C2 **Findhorn** *r.* U.K.
140 C2 **Findlay** U.S.A.
51 D4 **Fingal** Austr.
141 D2 **Finger Lakes** U.S.A.
106 B1 **Finisterre, Cape** Spain
92 I3 **Finland** *country* Europe
93 H4 **Finland, Gulf of** Europe
128 B2 **Finlay** *r.* Can.
53 C3 **Finley** Austr.
101 E2 **Finne** *ridge* Ger.
92 H2 **Finnmarksvidda** *reg.* Norway
92 G2 **Finnsnes** Norway
93 G4 **Finspång** Sweden
96 A2 **Fionnphort** U.K.
Fırat *r.* Turkey *see* **Euphrates**
Firenze Italy *see* **Florence**
105 C2 **Firminy** France
75 B2 **Firozabad** India
74 B1 **Firozpur** India
79 C2 **Fīrūzābād** Iran
122 A2 **Fish** *watercourse* Namibia
122 B3 **Fish** *r.* S. Africa
99 A3 **Fishguard** U.K.
105 C2 **Fismes** France
Fisterra, Cabo *c.* Spain *see* **Finisterre, Cape**
141 E2 **Fitchburg** U.S.A.
129 C2 **Fitzgerald** Can.
50 B1 **Fitzroy Crossing** Austr.
108 B2 **Fivizzano** Italy
119 C3 **Fizi** Dem. Rep. Congo
92 G3 **Fjällsjöälven** *r.* Sweden
123 C3 **Flagstaff** S. Africa
138 A1 **Flagstaff** U.S.A.
130 C1 **Flaherty Island** Can.
98 C1 **Flamborough Head** U.K.
101 F1 **Fläming** *hills* Ger.
136 B2 **Flaming Gorge Reservoir** U.S.A.
134 D1 **Flathead** *r.* U.S.A.
134 D1 **Flathead Lake** U.S.A.
51 D1 **Flattery, Cape** Austr.
134 B1 **Flattery, Cape** U.S.A.
98 B2 **Fleetwood** U.K.
93 E4 **Flekkefjord** Norway
102 B1 **Flensburg** Ger.
104 B2 **Flers** France
51 D1 **Flinders** *r.* Austr.
50 A3 **Flinders Bay** Austr.
51 D3 **Flinders Island** Austr.
52 A2 **Flinders Ranges** *mts* Austr.
129 D2 **Flin Flon** Can.
140 C2 **Flint** U.S.A.
107 D1 **Florac** France
108 B2 **Florence** Italy
142 C2 **Florence** *AL* U.S.A.
138 A2 **Florence** *AZ* U.S.A.
134 B2 **Florence** *OR* U.S.A.
143 E2 **Florence** *SC* U.S.A.
150 A1 **Florencia** Col.
146 B3 **Flores** Guat.
58 C3 **Flores** *i.* Indon.
58 B3 **Flores, Laut** *sea* Indon.
Flores Sea Indon. *see* **Flores, Laut**
151 E2 **Floresta** Brazil
139 D3 **Floresville** U.S.A.
151 D2 **Floriano** Brazil
152 D3 **Florianópolis** Brazil
153 C4 **Florida** Uru.
143 D2 **Florida** *state* U.S.A.
143 D4 **Florida, Straits of** N. Atlantic Ocean
143 D4 **Florida Keys** *is* U.S.A.
111 B2 **Florina** Greece
93 E3 **Florø** Norway
129 D2 **Foam Lake** Can.
109 C2 **Foča** Bos. & Herz.
110 C1 **Focșani** Romania
71 B3 **Fogang** China
109 C2 **Foggia** Italy
131 E2 **Fogo Island** Can.
104 C3 **Foix** France
89 D3 **Fokino** Russia
130 B2 **Foleyet** Can.
108 B2 **Foligno** Italy
99 D3 **Folkestone** U.K.
143 D2 **Folkston** U.S.A.
108 B2 **Follonica** Italy
129 D2 **Fond-du-Lac** Can.
129 D2 **Fond du Lac** *r.* Can.
140 B2 **Fond du Lac** U.S.A.
106 B1 **Fondevila** Spain
108 B2 **Fondi** Italy
146 B3 **Fonseca, Golfo do** *b.* Central America
150 B2 **Fonte Boa** Brazil
104 B2 **Fontenay-le-Comte** France
92 □C2 **Fontur** *pt* Iceland
53 C2 **Forbes** Austr.
101 E3 **Forchheim** Ger.
93 E3 **Førde** Norway
53 C1 **Fords Bridge** Austr.
142 B2 **Fordyce** U.S.A.
142 C2 **Forest** U.S.A.
53 C3 **Forest Hill** Austr.
131 D2 **Forestville** Can.
96 C2 **Forfar** U.K.
134 B1 **Forks** U.S.A.
108 B2 **Forlì** Italy
107 D2 **Formentera** *i.* Spain
107 D2 **Formentor, Cap de** *c.* Spain
155 C2 **Formiga** Brazil
152 C3 **Formosa** Arg.
Formosa *country* Asia *see* **Taiwan**
154 C1 **Formosa** Brazil
96 C2 **Forres** U.K.
50 B3 **Forrest** Austr.
142 B1 **Forrest City** U.S.A.
51 D1 **Forsayth** Austr.
93 H3 **Forssa** Fin.
53 D2 **Forster** Austr.
136 B1 **Forsyth** U.S.A.
74 B2 **Fort Abbas** Pak.
130 B1 **Fort Albany** Can.
151 E2 **Fortaleza** Brazil
128 C2 **Fort Assiniboine** Can.
96 B2 **Fort Augustus** U.K.
134 D1 **Fort Benton** U.S.A.
135 B3 **Fort Bragg** U.S.A.
Fort Chimo Can. *see* **Kuujjuaq**
129 C2 **Fort Chipewyan** Can.
136 B2 **Fort Collins** U.S.A.
147 D3 **Fort-de-France** Martinique
137 E2 **Fort Dodge** U.S.A.
130 A2 **Fort Frances** Can.
Fort Franklin Can. *see* **Délı̨ne**
Fort George Can. *see* **Chisasibi**
126 C2 **Fort Good Hope** Can.
96 C2 **Forth** *r.* U.K.
96 C2 **Forth, Firth of** *est.* U.K.
152 C3 **Fortín Madrejón** Para.
143 D3 **Fort Lauderdale** U.S.A.
128 B1 **Fort Liard** Can.
129 C2 **Fort Mackay** Can.
129 C2 **Fort McMurray** Can.
126 C2 **Fort McPherson** Can.
136 C2 **Fort Morgan** U.S.A.
143 D3 **Fort Myers** U.S.A.
128 B2 **Fort Nelson** Can.
128 B2 **Fort Nelson** *r.* Can.
142 C2 **Fort Payne** U.S.A.
136 B1 **Fort Peck** U.S.A.
136 B1 **Fort Peck Reservoir** U.S.A.
143 D3 **Fort Pierce** U.S.A.
128 C1 **Fort Providence** Can.
129 D2 **Fort Qu'Appelle** Can.
128 C1 **Fort Resolution** Can.
Fort Rupert Can. *see* **Waskaganish**
128 B2 **Fort St James** Can.
128 B2 **Fort St John** Can.
128 C2 **Fort Saskatchewan** Can.
137 E3 **Fort Scott** U.S.A.
130 B1 **Fort Severn** Can.
76 B2 **Fort-Shevchenko** Kazakh.
128 B1 **Fort Simpson** Can.
129 C1 **Fort Smith** Can.
142 B1 **Fort Smith** U.S.A.
139 C2 **Fort Stockton** U.S.A.
138 C2 **Fort Sumner** U.S.A.
128 C2 **Fort Vermilion** Can.
142 C2 **Fort Walton Beach** U.S.A.
140 B2 **Fort Wayne** U.S.A.
96 B2 **Fort William** U.K.
139 D2 **Fort Worth** U.S.A.
126 B2 **Fort Yukon** U.S.A.
71 B3 **Foshan** China
92 F3 **Fosna** *pen.* Norway
93 E3 **Fosnavåg** Norway
92 □B3 **Foss** Iceland

92	□A2	Fossá Iceland
108	A2	Fossano Italy
53	C3	Foster Austr.
104	B2	Fougères France
96	□	Foula i. U.K.
111	B2	Fournoi i. Greece
114	A3	Fouta Djallon reg. Guinea
54	A3	Foveaux Strait N.Z.
136	C3	Fowler U.S.A.
50	C3	Fowlers Bay Austr.
127	C4	Fox Creek Can.
127	F2	Foxe Basin g. Can.
127	F2	Foxe Channel Can.
127	F2	Foxe Peninsula Can.
54	B2	Fox Glacier N.Z.
128	C2	Fox Lake Can.
128	A1	Fox Mountain Can.
54	C2	Foxton N.Z.
129	D2	Fox Valley Can.
97	C1	Foyle, Lough b. Ireland/U.K.
154	B3	Foz de Areia, Represa de resr Brazil
120	A2	Foz do Cunene Angola
154	B3	Foz do Iguaçu Brazil
107	D1	Fraga Spain
154	C2	Franca Brazil
109	C2	Francavilla Fontana Italy
104	C2	France country Europe
118	B3	Franceville Gabon
137	D2	Francis Case, Lake U.S.A.
155	D1	Francisco Sá Brazil
120	B3	Francistown Botswana
128	B2	Francois Lake Can.
101	D2	Frankenberg (Eder) Ger.
101	D3	Frankenthal (Pfalz) Ger.
101	E2	Frankenwald mts Ger.
140	C3	Frankfort U.S.A.
101	E2	Frankfurt (Oder) Ger.
101	D2	Frankfurt am Main Ger.
102	C2	Fränkische Alb hills Ger.
101	E3	Fränkische Schweiz reg. Ger.
141	D2	Franklin U.S.A.
126	C2	Franklin Bay Can.
134	C1	Franklin D. Roosevelt Lake U.S.A.
128	B1	Franklin Mountains Can.
126	E2	Franklin Strait Can.
82	E1	Frantsa-Iosifa, Zemlya is Russia
54	B2	Franz Josef Glacier N.Z.
		Franz Josef Land is Russia see Frantsa-Iosifa, Zemlya
108	A3	Frasca, Capo della c. Sardinia Italy
128	B3	Fraser r. B.C. Can.
131	D1	Fraser r. Nfld. and Lab. Can.
122	B3	Fraserburg S. Africa
96	C2	Fraserburgh U.K.
130	B2	Fraserdale Can.
51	E2	Fraser Island Austr.
128	B2	Fraser Lake Can.
153	C4	Fray Bentos Uru.
93	E4	Fredericia Denmark
139	D2	Frederick U.S.A.
139	D3	Fredericksburg TX U.S.A.
141	D3	Fredericksburg VA U.S.A.
128	A2	Frederick Sound sea chan. U.S.A.
131	D2	Fredericton Can.
93	F4	Frederikshavn Denmark
		Frederikshamn Fin. see Hamina
93	F4	Fredrikstad Norway
140	B2	Freeport IL U.S.A.
139	D3	Freeport TX U.S.A.
146	C2	Freeport City Bahamas
139	D3	Freer U.S.A.
123	C2	Free State prov. S. Africa
114	A4	Freetown Sierra Leone
106	B2	Fregenal de la Sierra Spain
104	B2	Fréhel, Cap c. France
102	B2	Freiburg im Breisgau Ger.
102	C2	Freising Ger.
102	C2	Freistadt Austria
105	D3	Fréjus France
50	A3	Fremantle Austr.
137	D2	Fremont NE U.S.A.
140	C2	Fremont OH U.S.A.
151	C1	French Guiana terr. S. America
134	E1	Frenchman r. U.S.A.
49	I4	French Polynesia terr. S. Pacific Ocean
144	B3	Fresnillo Mex.
135	C3	Fresno U.S.A.
107	D2	Freu, Cap des c. Spain
105	D2	Freyming-Merlebach France
114	A3	Fria Guinea
152	B3	Frías Arg.
102	B2	Friedrichshafen Ger.
101	F1	Friesack Ger.
100	C1	Friesoythe Ger.
		Frobisher Bay Nunavut Can. see Iqaluit
127	G2	Frobisher Bay Can.
101	F2	Frohburg Ger.
87	D4	Frolovo Russia
52	A2	Frome, Lake imp. l. Austr.
52	A2	Frome Downs Austr.
100	C2	Fröndenberg/Ruhr Ger.
145	C3	Frontera Mex.
144	B1	Fronteras Mex.
108	B2	Frosinone Italy
92	E3	Frøya i. Norway
		Frunze Kyrg. see Bishkek
105	D2	Frutigen Switz.
103	D2	Frýdek-Místek Czech Rep.
71	B3	Fu'an China
106	C1	Fuenlabrada Spain
152	C3	Fuerte Olimpo Para.
114	A2	Fuerteventura i. Canary Islands
64	B1	Fuga i. Phil.
79	C2	Fujairah U.A.E.
67	C3	Fuji Japan
71	B3	Fujian prov. China
67	C3	Fujinomiya Japan
67	C3	Fuji-san vol. Japan
67	C3	Fukui Japan
67	B4	Fukuoka Japan
67	D3	Fukushima Japan
101	D2	Fulda Ger.
101	D2	Fulda r. Ger.
70	A3	Fuling China
137	E3	Fulton U.S.A.
105	C2	Fumay France
49	F3	Funafuti atoll Tuvalu
114	A1	Funchal Arquipélago da Madeira
106	B1	Fundão Port.
131	D2	Fundy, Bay of g. Can.
70	B2	Funing Jiangsu China
71	A3	Funing Yunnan China
115	C3	Funtua Nigeria
79	C2	Fürgun, Küh-e mt. Iran
89	F2	Furmanov Russia
155	C2	Furnas, Represa resr Brazil
51	D4	Furneaux Group is Austr.
100	C1	Fürstenau Ger.
101	E3	Fürth Ger.
127	F2	Fury and Hecla Strait Can.
65	A1	Fushun China
65	B1	Fusong China
79	C2	Fuwayriţ Qatar
70	B2	Fuyang China
69	E1	Fuyu China
68	B1	Fuyun China
71	B3	Fuzhou Fujian China
71	B3	Fuzhou Jiangxi China
93	F4	Fyn i. Denmark
		F.Y.R.O.M. country Europe see Macedonia

G

117	C4	Gaalkacyo Somalia
120	A2	Gabela Angola
115	D1	Gabès Tunisia
115	D1	Gabès, Golfe de g. Tunisia
118	B3	Gabon country Africa
123	C1	Gaborone Botswana
110	C2	Gabrovo Bulg.
114	A3	Gabú Guinea-Bissau
73	B3	Gadag-Betigeri India
75	C2	Gadchiroli India
101	E1	Gadebusch Ger.
142	C2	Gadsden U.S.A.
110	C2	Găeşti Romania
108	B2	Gaeta Italy
143	D1	Gaffney U.S.A.
115	C1	Gafsa Tunisia
89	E2	Gagarin Russia
114	B4	Gagnoa Côte d'Ivoire
131	D1	Gagnon Can.
81	C1	Gagra Georgia
122	A2	Gaiab watercourse Namibia
111	C3	Gaidouronisi i. Greece
104	C3	Gaillac France
143	D3	Gainesville FL U.S.A.
143	D2	Gainesville GA U.S.A.
139	D2	Gainesville TX U.S.A.
98	C2	Gainsborough U.K.
52	A2	Gairdner, Lake imp. l. Austr.
96	B2	Gairloch U.K.
119	E3	Galana r. Kenya
103	D2	Galanta Slovakia
148	B3	Galapagos Islands Ecuador
96	C3	Galashiels U.K.
110	C1	Galaţi Romania
93	E3	Galdhøpiggen mt. Norway
145	B2	Galeana Mex.
128	C2	Galena Bay Can.
140	A2	Galesburg U.S.A.
122	B2	Galeshewe S. Africa
86	D3	Galich Russia
116	B1	Galilee, Sea of l. Israel
142	C1	Gallatin U.S.A.
73	C4	Galle Sri Lanka
150	A1	Gallinas, Punta pt Col.
109	C2	Gallipoli Italy
111	C2	Gallipoli Turkey
92	H2	Gällivare Sweden
138	B1	Gallup U.S.A.
117	C4	Galmudug reg. Somalia
114	A2	Galtat-Zemmour Western Sahara
97	B2	Galtymore h. Ireland
139	E3	Galveston U.S.A.
139	E3	Galveston Bay U.S.A.
97	B2	Galway Ireland
97	B2	Galway Bay Ireland
154	C1	Gama Brazil
123	D3	Gamalakhe S. Africa
114	A3	Gambia, The country Africa
52	A3	Gambier Islands Austr.
131	E2	Gambo Can.
118	B3	Gamboma Congo
128	C1	Gamêtì Can.
138	B1	Ganado U.S.A.
81	C1	Gäncä Azer.

Gand Belgium see Ghent
61 C2 Gandadiwata, Bukit mt. Indon.
118 C3 Gandajika Dem. Rep. Congo
131 E2 Gander Can.
131 E2 Gander r. Can.
101 D1 Ganderkesee Ger.
107 C1 Gandesa Spain
74 B2 Gandhidham India
74 B2 Gandhinagar India
74 B2 Gandhi Sagar resr India
107 C2 Gandia Spain
Ganga r. Bangl./India see Ganges
153 B5 Gangán Arg.
74 B2 Ganganagar India
68 C2 Gangca China
75 C1 Gangdisê Shan mts China
75 D2 Ganges r. Bangl./India
105 C3 Ganges France
Ganges, Mouths of the Bangl./India
65 B2 Gangneung S. Korea
75 C2 Gangtok India
105 C3 Gannat France
136 B2 Gannett Peak U.S.A.
122 A3 Gansbaai S. Africa
70 A1 Gansu prov. China
71 B3 Ganzhou China
114 B3 Gao Mali
97 B1 Gaoth Dobhair Ireland
114 A3 Gaoual Guinea
71 C3 Gaoxiong Taiwan
70 B2 Gaoyou China
70 B2 Gaoyou Hu l. China
105 D3 Gap France
75 C1 Gar China
76 B2 Garabogaz Turkm.
76 B2 Garabogazköl Turkm.
76 B2 Garabogazköl Aylagy b. Turkm.
53 C2 Garah Austr.
151 E2 Garanhuns Brazil
135 B3 Garberville U.S.A.
101 D1 Garbsen Ger.
154 C2 Garça Brazil
101 E1 Gardelegen Ger.
136 C3 Garden City U.S.A.
129 E2 Garden Hill Can.
77 C3 Gardēz Afgh.
88 B2 Gargždai Lith.
123 C3 Gariep Dam dam S. Africa
122 A3 Garies S. Africa
119 D3 Garissa Kenya
102 C2 Garmisch-Partenkirchen Ger.
52 B2 Garnpung Lake imp. l. Austr.
104 B3 Garonne r. France
117 C4 Garoowe Somalia
74 B2 Garoth India
118 B2 Garoua Cameroon
96 B2 Garry r. U.K.
126 E2 Garry Lake Can.
119 E3 Garsen Kenya
61 B2 Garut Indon.
140 B2 Gary U.S.A.
145 B2 Garza García Mex.
68 C2 Garzê China
104 B2 Gascony, Gulf of France
50 A2 Gascoyne r. Austr.
115 D3 Gashua Nigeria
131 D2 Gaspé Can.
131 D2 Gaspésie, Péninsule de la pen. Can.
143 E1 Gaston, Lake U.S.A.
143 D1 Gastonia U.S.A.
107 C2 Gata, Cabo de c. Spain
88 D2 Gatchina Russia
98 C1 Gateshead U.K.
139 D2 Gatesville U.S.A.

130 C2 Gatineau r. Can.
53 D1 Gatton Austr.
129 E2 Gauer Lake Can.
93 E4 Gausta mt. Norway
123 C2 Gauteng prov. S. Africa
111 B3 Gavdos i. Greece
93 G3 Gävle Sweden
89 F2 Gavrilov Posad Russia
89 E2 Gavrilov-Yam Russia
62 A1 Gawai Myanmar
52 A2 Gawler Austr.
52 A2 Gawler Ranges hills Austr.
75 C2 Gaya India
114 C3 Gaya Niger
140 C1 Gaylord U.S.A.
86 E2 Gayny Russia
80 B2 Gaza disp. terr. Asia
80 B2 Gaziantep Turkey
76 C2 Gazojak Turkm.
114 B4 Gbarnga Liberia
103 D1 Gdańsk Pol.
88 A3 Gdańsk, Gulf of Pol./Russia
88 C2 Gdov Russia
103 D1 Gdynia Pol.
116 B3 Gedaref Sudan
101 D2 Gedern Ger.
111 C3 Gediz r. Turkey
102 C1 Gedser Denmark
100 B2 Geel Belgium
52 B3 Geelong Austr.
101 E1 Geesthacht Ger.
75 C1 Gê'gyai China
129 D2 Geikie r. Can.
93 E3 Geilo Norway
62 B1 Gejiu China
108 B3 Gela Sicily Italy
91 D3 Gelendzhik Russia
Gelibolu Turkey see Gallipoli
100 C2 Gelsenkirchen Ger.
118 B2 Gemena Dem. Rep. Congo
111 C2 Gemlik Turkey
117 C4 Genalē Wenz r. Eth.
153 B4 General Acha Arg.
153 B4 General Alvear Arg.
154 C4 General Belgrano Arg.
144 B2 General Cepeda Mex.
153 B4 General Pico Arg.
153 B4 General Roca Arg.
64 B2 General Santos Phil.
141 D2 Genesee r. U.S.A.
141 D2 Geneseo U.S.A.
105 D2 Geneva Switz.
141 D2 Geneva U.S.A.
105 D2 Geneva, Lake France/Switz.
Genève Switz. see Geneva
106 B2 Genil r. Spain
100 B2 Genk Belgium
108 A2 Genoa Italy
Genova Italy see Genoa
101 F1 Genthin Ger.
50 A3 Geographe Bay Austr.
65 B2 Geongju S. Korea
131 D1 George r. Can.
122 B3 George S. Africa
143 D3 George, Lake FL U.S.A.
141 E2 George, Lake NY U.S.A.
151 C1 Georgetown Guyana
60 B1 George Town Malaysia
143 E2 Georgetown SC U.S.A.
139 D2 Georgetown TX U.S.A.
55 I2 George V Land reg. Antarctica
81 C1 Georgia country Asia
143 D2 Georgia state U.S.A.
130 B2 Georgian Bay Can.
51 C2 Georgina watercourse Austr.
77 E2 Georgiyevka Kazakh.
87 D4 Georgiyevsk Russia

101 F2 Gera Ger.
151 D3 Geral de Goiás, Serra hills Brazil
54 B2 Geraldine N.Z.
50 A2 Geraldton Austr.
80 B1 Gerede Turkey
135 C2 Gerlach U.S.A.
103 E2 Gerlachovský štít mt. Slovakia
102 C1 Germany country Europe
100 C2 Gerolstein Ger.
101 E3 Gerolzhofen Ger.
101 D2 Gersfeld (Rhön) Ger.
75 C1 Gêrzê China
117 C4 Gestro Wenz, Wabē r. Eth.
141 D3 Gettysburg PA U.S.A.
136 D1 Gettysburg SD U.S.A.
109 D2 Gevgelija Macedonia
111 C3 Geyikli Turkey
122 B2 Ghaap Plateau S. Africa
115 C1 Ghadāmis Libya
75 C2 Ghaghara r. India
114 B4 Ghana country Africa
120 B3 Ghanzi Botswana
115 C1 Ghardaïa Alg.
115 D1 Gharyān Libya
75 C2 Ghatal India
75 B2 Ghaziabad India
75 C2 Ghazipur India
77 C3 Ghaznī Afgh.
100 A2 Ghent Belgium
110 B1 Gherla Romania
105 D3 Ghisonaccia Corsica France
74 B2 Ghotaru India
74 A2 Ghotki Pak.
91 E3 Giaginskaya Russia
97 C1 Giant's Causeway lava field U.K.
61 C2 Gianyar Indon.
109 C3 Giarre Sicily Italy
108 A1 Giaveno Italy
122 A2 Gibeon Namibia
106 B2 Gibraltar Gibraltar
106 B2 Gibraltar, Strait of Morocco/Spain
50 B2 Gibson Desert Austr.
68 C1 Gichgeniyn Nuruu mts Mongolia
117 B4 Gidolē Eth.
104 C2 Gien France
101 D2 Gießen Ger.
101 E1 Gifhorn Ger.
128 C2 Gift Lake Can.
67 C3 Gifu Japan
96 B3 Gigha i. U.K.
106 B1 Gijón/Xixón Spain
138 A2 Gila r. U.S.A.
138 A2 Gila Bend U.S.A.
51 D1 Gilbert r. Austr.
49 F2 Gilbert Islands is Kiribati
134 D1 Gildford U.S.A.
53 C2 Gilgandra Austr.
74 B1 Gilgit Pak.
74 B1 Gilgit r. Pak.
53 C2 Gilgunnia Austr.
129 E2 Gillam Can.
136 B2 Gillette U.S.A.
99 D3 Gillingham U.K.
65 B2 Gimcheon S. Korea
129 E2 Gimli Can.
117 C4 Gīnīr Eth.
109 C2 Ginosa Italy
53 C3 Gippsland reg. Austr.
74 A2 Girdar Dhor r. Pak.
80 B1 Giresun Turkey
74 A1 Girishk Afgh.
107 D1 Girona Spain
96 B3 Girvan U.K.
54 C1 Gisborne N.Z.

93 F4	Gislaved Sweden
119 C3	Gitarama Rwanda
108 B2	Giulianova Italy
110 C2	Giurgiu Romania
110 C3	Giuvala, Pasul *pass* Romania
105 C2	Givors France
123 D1	Giyani S. Africa
116 B2	Giza Egypt
109 D2	Gjakovë Kosovo
109 D2	Gjilan Kosovo
109 D2	Gjirokastër Albania
126 E2	Gjoa Haven Can.
93 F3	Gjøvik Norway
131 E2	Glace Bay Can.
134 B1	Glacier Peak *vol.* U.S.A.
51 E2	Gladstone Austr.
92 □A2	Gláma *mts* Iceland
109 C2	Glamoč Bos. & Herz.
100 C3	Glan *r.* Ger.
97 B2	Glanaruddery Mountains *hills* Ireland
96 B3	Glasgow U.K.
140 B3	Glasgow KY U.S.A.
136 B1	Glasgow MT U.S.A.
99 B3	Glastonbury U.K.
101 F2	Glauchau Ger.
86 E3	Glazov Russia
89 E3	Glazunovka Russia
138 A2	Glen Coe *val.* U.K.
138 A2	Glendale U.S.A.
53 D2	Glen Davis Austr.
136 C1	Glendive U.S.A.
52 B3	Glenelg *r.* Austr.
53 D1	Glen Innes Austr.
126 B2	Glennallen U.S.A.
96 C2	Glenrothes U.K.
141 E2	Glens Falls U.S.A.
96 C2	Glen Shee *val.* U.K.
97 B1	Glenties Ireland
138 B2	Glenwood U.S.A.
136 B3	Glenwood Springs U.S.A.
103 D1	Gliwice Pol.
103 A2	Globe U.S.A.
103 D1	Głogów Pol.
92 F2	Glomfjord Norway
93 F4	Glomma *r.* Norway
53 D2	Gloucester Austr.
99 B3	Gloucester U.K.
101 F1	Glöwen Ger.
77 E1	Glubokoye Kazakh.
101 D1	Glückstadt Ger.
103 C2	Gmünd Austria
102 C2	Gmunden Austria
101 D1	Gnarrenburg Ger.
103 D1	Gniezno Pol.
75 D2	Goalpara India
96 B3	Goat Fell *h.* U.K.
117 C4	Goba Eth.
120 A3	Gobabis Namibia
153 A5	Gobernador Gregores Arg.
69 D2	Gobi *des.* China/Mongolia
100 C2	Goch Ger.
122 A1	Gochas Namibia
74 B3	Godavari *r.* India
73 C3	Godavari, Mouths of the India
130 B2	Goderich Can.
74 B2	Godhra India
129 E2	Gods *r.* Can.
129 E2	Gods Lake Can.
	Godthåb Greenland *see* Nuuk
	Godwin-Austen, Mount China/Pak. *see* K2
130 C2	Goéland, Lac au *l.* Can.
131 D1	Goélands, Lac aux *l.* Can.
100 A2	Goes Neth.
154 C1	Goiandira Brazil
154 C1	Goiânia Brazil
154 B1	Goiás Brazil
154 B2	Goioerê Brazil
111 C2	Gökçeada *i.* Turkey
111 C3	Gökçedağ Turkey
121 B2	Gokwe Zimbabwe
93 E3	Gol Norway
62 A1	Golaghat India
111 C2	Gölcük Turkey
103 E1	Gołdap Pol.
101 F1	Goldberg Ger.
53 D1	Gold Coast Austr.
114 B4	Gold Coast Ghana
128 C2	Golden Can.
54 B2	Golden Bay N.Z.
128 B3	Golden Hinde *mt.* Can.
97 B2	Golden Vale *lowland* Ireland
135 C3	Goldfield U.S.A.
128 B3	Gold River Can.
143 E1	Goldsboro U.S.A.
135 C4	Goleta U.S.A.
68 C2	Golmud China
81 D2	Golpäyegän Iran
96 C2	Golspie U.K.
119 C3	Goma Dem. Rep. Congo
75 C2	Gomati *r.* India
115 D3	Gombe Nigeria
115 D3	Gombi Nigeria
	Gomel' Belarus *see* Homyel'
144 B2	Gómez Palacio Mex.
147 C3	Gonaïves Haiti
147 C3	Gonâve, Île de la *i.* Haiti
81 D2	Gonbad-e Kāvūs Iran
117 B3	Gonder Eth.
75 C2	Gondiya India
111 C2	Gönen Turkey
65 B2	Gongju S. Korea
115 C4	Gongola *r.* Nigeria
53 C2	Gongolgon Austr.
71 C3	Gongtang China
145 C2	Gonzáles Mex.
139 D3	Gonzales U.S.A.
122 A3	Good Hope, Cape of S. Africa
134 C2	Gooding U.S.A.
136 C3	Goodland U.S.A.
53 C1	Goodooga Austr.
98 C2	Goole U.K.
53 C2	Goolgowi Austr.
52 A3	Goolwa Austr.
53 D1	Goondiwindi Austr.
134 B2	Goose Lake U.S.A.
102 B2	Göppingen Ger.
75 C2	Gorakhpur India
109 C2	Goražde Bos. & Herz.
111 C3	Gördes Turkey
89 D3	Gordeyevka Russia
51 D4	Gordon, Lake Austr.
115 D4	Goré Chad
117 B4	Gorē Eth.
54 A3	Gore N.Z.
97 C2	Gorey Ireland
81 D2	Gorgān Iran
81 C1	Gori Georgia
108 B1	Gorizia Italy
	Gor'kiy Russia *see* Nizhniy Novgorod
103 E2	Gorlice Pol.
103 C1	Görlitz Ger.
109 D2	Gornji Milanovac Serbia
109 C2	Gornji Vakuf Bos. & Herz.
77 E1	Gorno-Altaysk Russia
110 C2	Gornotrakiyska Nizina *lowland* Bulg.
77 E1	Gornyak Russia
59 D3	Goroka P.N.G.
114 B3	Gorom Gorom Burkina Faso
59 C2	Gorontalo Indon.
89 E3	Gorshechnoye Russia
97 B2	Gorumna Island Ireland
91 D1	Goryachiy Klyuch Russia
65 B2	Goryeong S. Korea
103 D1	Gorzów Wielkopolski Pol.
53 D2	Gosford Austr.
66 D2	Goshogawara Japan
101 E2	Goslar Ger.
109 C2	Gospić Croatia
99 C3	Gosport U.K.
109 D2	Gostivar Macedonia
	Göteborg Sweden *see* Gothenburg
101 E2	Gotha Ger.
93 F4	Gothenburg Sweden
136 C2	Gothenburg U.S.A.
93 G4	Gotland *i.* Sweden
111 B2	Gotse Delchev Bulg.
93 G4	Gotska Sandön *i.* Sweden
67 B4	Gōtsu Japan
101 D2	Göttingen Ger.
128 B2	Gott Peak Can.
100 B1	Gouda Neth.
114 A3	Goudiri Senegal
115 D3	Goudoumaria Niger
130 C2	Gouin, Réservoir *resr* Can.
53 C2	Goulburn Austr.
53 B3	Goulburn *r.* N.S.W. Austr.
53 B3	Goulburn *r.* Vic. Austr.
114 B3	Goundam Mali
107 D2	Gouraya Alg.
104 C3	Gourdon France
115 D3	Gouré Niger
122 B3	Gourits *r.* S. Africa
114 B3	Gourma-Rharous Mali
53 C3	Gourock Range *mts* Austr.
155 D1	Governador Valadares Brazil
143 E3	Governor's Harbour Bahamas
68 C2	Govi Altayn Nuruu *mts* Mongolia
75 C2	Govind Ballash Pant Sagar *resr* India
99 A3	Gower *pen.* U.K.
152 C3	Goya Arg.
81 C1	Göyçay Azer.
75 C1	Gozha Co *salt l.* China
122 B3	Graaff-Reinet S. Africa
101 E1	Grabow Ger.
109 C2	Gračac Croatia
87 E3	Grachevka Russia
101 F2	Gräfenhainichen Ger.
53 D1	Grafton Austr.
137 D1	Grafton U.S.A.
139 D2	Graham U.S.A.
128 A2	Graham Island Can.
55 P3	Graham Land *pen.* Antarctica
123 C3	Grahamstown S. Africa
151 E2	Grajaú Brazil
111 B2	Grammos *mts* Greece
96 B2	Grampian Mountains U.K.
146 B3	Granada Nic.
106 C2	Granada Spain
141 E1	Granby Can.
114 A2	Gran Canaria *i.* Canary Islands
152 B3	Gran Chaco *reg.* Arg./Para.
136 C1	Grand *r.* U.S.A.
146 C2	Grand Bahama *i.* Bahamas
131 E2	Grand Bank Can.
158 C2	Grand Banks of Newfoundland N. Atlantic Ocean
	Grand Canal *canal* China *see* Jinghang Yunhe
138 A1	Grand Canyon U.S.A.
138 A1	Grand Canyon *gorge* U.S.A.
146 B3	Grand Cayman *i.* Cayman Is
129 C2	Grand Centre Can.
134 C1	Grand Coulee U.S.A.
152 B2	Grande *r.* Bol.

154 B2 Grande r. Brazil
153 B6 Grande, Bahía b. Arg.
155 D2 Grande, Ilha i. Brazil
128 C2 Grande Cache Can.
128 C2 Grande Prairie Can.
114 B1 Grand Erg Occidental des. Alg.
115 C2 Grand Erg Oriental des. Alg.
131 D2 Grande-Rivière Can.
152 B4 Grandes, Salinas salt flat Arg.
156 D7 Grande Terre i. S. Pacific Ocean
131 D2 Grand Falls Can.
131 E2 Grand Falls-Windsor Can.
128 C3 Grand Forks Can.
137 D3 Grand Forks U.S.A.
128 C1 Grandin, Lac l. Can.
137 D2 Grand Island U.S.A.
142 B3 Grand Isle U.S.A.
136 B3 Grand Junction U.S.A.
114 B4 Grand-Lahou Côte d'Ivoire
131 D2 Grand Lake N.B. Can.
131 E2 Grand Lake Nfld. and Lab. Can.
137 E1 Grand Marais U.S.A.
106 B2 Grândola Port.
129 E2 Grand Rapids Can.
140 B2 Grand Rapids MI U.S.A.
137 E1 Grand Rapids MN U.S.A.
136 A2 Grand Teton mt. U.S.A.
147 C2 Grand Turk Turks and Caicos Is
134 C2 Grangeville U.S.A.
128 B2 Granisle Can.
134 E1 Granite Peak U.S.A.
108 B3 Granitola, Capo c. Sicily Italy
93 F4 Gränna Sweden
101 F1 Gransee Ger.
99 C2 Grantham U.K.
96 C2 Grantown-on-Spey U.K.
138 B1 Grants U.S.A.
134 B2 Grants Pass U.S.A.
104 B2 Granville France
129 D2 Granville Lake Can.
155 D2 Grão Mogol Brazil
123 C1 Graskop S. Africa
105 D3 Grasse France
107 D1 Graus Spain
92 F2 Gravdal Norway
104 B2 Grave, Pointe de pt France
129 D3 Gravelbourg Can.
130 C2 Gravenhurst Can.
53 D1 Gravesend Austr.
99 D3 Gravesend U.K.
105 D2 Gray France
103 D2 Graz Austria
146 C2 Great Abaco i. Bahamas
50 B3 Great Australian Bight g. Austr.
54 C1 Great Barrier Island N.Z.
51 D1 Great Barrier Reef Austr.
135 C3 Great Basin U.S.A.
126 D2 Great Bear Lake Can.
93 F4 Great Belt sea chan. Denmark
137 D3 Great Bend U.S.A.
63 A2 Great Coco Island Cocos Is
53 B3 Great Dividing Range mts Austr.
146 B2 Greater Antilles is Caribbean Sea
60 B2 Greater Sunda Islands is Indon.
134 D1 Great Falls U.S.A.
123 C3 Great Fish r. S. Africa
123 C3 Great Fish Point S. Africa
147 C2 Great Inagua i. Bahamas
122 B3 Great Karoo plat. S. Africa
123 C3 Great Kei r. S. Africa

99 B2 Great Malvern U.K.
122 A2 Great Namaqualand reg. Namibia
63 A3 Great Nicobar i. India
99 D3 Great Ouse r. U.K.
119 D3 Great Rift Valley Africa
119 D3 Great Ruaha r. Tanz.
134 D2 Great Salt Lake U.S.A.
135 D2 Great Salt Lake Desert U.S.A.
116 A2 Great Sand Sea des. Egypt/ Libya
50 B2 Great Sandy Desert Austr.
128 C1 Great Slave Lake Can.
143 D1 Great Smoky Mountains U.S.A.
50 B2 Great Victoria Desert Austr.
70 B1 Great Wall tourist site China
99 D2 Great Yarmouth U.K.
106 B1 Gredos, Sierra de mts Spain
111 B3 Greece country Europe
136 C2 Greeley U.S.A.
82 F1 Greem-Bell, Ostrov i. Russia
140 B3 Green r. KY U.S.A.
136 B3 Green r. WY U.S.A.
140 B2 Green Bay U.S.A.
140 B1 Green Bay b. U.S.A.
140 C3 Greenbrier r. U.S.A.
140 B3 Greencastle U.S.A.
143 D1 Greeneville U.S.A.
141 E2 Greenfield U.S.A.
129 D2 Green Lake Can.
127 I2 Greenland terr. N. America
160 R2 Greenland Sea Greenland/ Svalbard
96 B3 Greenock U.K.
135 D3 Green River UT U.S.A.
136 B2 Green River WY U.S.A.
140 B3 Greensburg IN U.S.A.
141 D2 Greensburg PA U.S.A.
143 E2 Green Swamp U.S.A.
138 A2 Green Valley U.S.A.
114 B4 Greenville Liberia
142 C2 Greenville AL U.S.A.
142 B2 Greenville MS U.S.A.
143 E1 Greenville NC U.S.A.
143 D2 Greenville SC U.S.A.
139 D2 Greenville TX U.S.A.
53 D2 Greenwell Point Austr.
143 D2 Greenwood U.S.A.
50 B2 Gregory, Lake imp. l. Austr.
51 D1 Gregory Range hills Austr.
102 C1 Greifswald Ger.
101 F2 Greiz Ger.
93 F4 Grena Denmark
142 C2 Grenada U.S.A.
147 D3 Grenada country West Indies
104 C3 Grenade France
53 C2 Grenfell Austr.
129 D2 Grenfell Can.
105 D2 Grenoble France
51 D1 Grenville, Cape Austr.
134 B1 Gresham U.S.A.
100 C1 Greven Ger.
111 B2 Grevena Greece
100 C2 Grevenbroich Ger.
101 E1 Grevesmühlen Ger.
136 B2 Greybull U.S.A.
128 A1 Grey Hunter Peak Can.
131 E1 Grey Islands Can.
54 B2 Greymouth N.Z.
52 B1 Grey Range hills Austr.
97 C2 Greystones Ireland
143 D2 Griffin U.S.A.
53 C2 Griffith Austr.
101 F2 Grimma Ger.
102 C1 Grimmen Ger.
98 C2 Grimsby U.K.

128 C2 Grimshaw Can.
92 □B2 Grímsstaðir Iceland
93 E4 Grimstad Norway
137 E2 Grinnell U.S.A.
123 C3 Griqualand East reg. S. Africa
122 B2 Griqualand West reg. S. Africa
127 F1 Grise Fiord Can.
96 C1 Gritley U.K.
123 C2 Groblersdal S. Africa
122 B2 Groblershoop S. Africa
Grodno Belarus see Hrodna
104 B2 Groix, Île de i. France
100 C1 Gronau (Westfalen) Ger.
92 F3 Grong Norway
100 C1 Groningen Neth.
122 B2 Grootdrink S. Africa
51 C1 Groote Eylandt i. Austr.
120 A2 Grootfontein Namibia
122 A2 Groot Karas Berg plat. Namibia
122 B3 Groot Swartberge mts S. Africa
123 C3 Groot Winterberg mt. S. Africa
101 D2 Großenlüder Ger.
102 C2 Großer Rachel mt. Ger.
103 C2 Grosser Speikkogel mt. Austria
108 B2 Grosseto Italy
101 D3 Groß-Gerau Ger.
103 C2 Großglockner mt. Austria
100 C1 Groß Hesepe Ger.
101 E2 Großlöhra Ger.
122 A1 Gross Ums Namibia
131 E1 Groswater Bay Can.
130 B2 Groundhog r. Can.
49 I4 Groupe Actéon is Fr. Polynesia
135 B3 Grover Beach U.S.A.
141 E2 Groveton U.S.A.
87 D4 Groznyy Russia
109 C1 Grubišno Polje Croatia
103 D1 Grudziądz Pol.
122 A2 Grünau Namibia
92 □A3 Grundarfjörður Iceland
89 E3 Gryazi Russia
89 F2 Gryazovets Russia
103 D1 Gryfice Pol.
102 C1 Gryfino Pol.
146 C2 Guacanayabo, Golfo de b. Cuba
144 B2 Guadalajara Mex.
106 C1 Guadalajara prov. Spain
49 E3 Guadalcanal i. Solomon Is
107 C1 Guadalope r. Spain
106 B2 Guadalquivir r. Spain
132 B4 Guadalupe i. Mex.
106 B2 Guadalupe, Sierra de mts Spain
138 C2 Guadalupe Peak U.S.A.
144 B2 Guadalupe Victoria Mex.
144 B2 Guadalupe y Calvo Mex.
106 C1 Guadarrama, Sierra de mts Spain
147 D3 Guadeloupe terr. West Indies
106 B2 Guadiana r. Port./Spain
106 C2 Guadix Spain
154 B2 Guaíra Brazil
147 C3 Guajira, Península de la pen. Col.
150 A2 Gualaceo Ecuador
59 D2 Guam terr. N. Pacific Ocean
144 B2 Guamúchil Mex.
144 B2 Guanacevi Mex.
151 D3 Guanambi Brazil
150 B1 Guanare Venez.
146 B2 Guane Cuba
70 A2 Guang'an China
71 B3 Guangchang China
71 B3 Guangdong prov. China

71 A3 **Guangxi Zhuangzu Zizhiqu** *aut. reg.* China
70 A2 **Guangyuan** China
71 B3 **Guangzhou** China
155 D1 **Guanhães** Brazil
147 D4 **Guanipa** *r.* Venez.
71 A3 **Guanling** China
65 A1 **Guanshui** China
Guansuo China *see* **Guanling**
147 C2 **Guantánamo** Cuba
150 B3 **Guaporé** *r.* Bol./Brazil
154 B3 **Guarapuava** Brazil
154 C3 **Guaraqueçaba** Brazil
155 C2 **Guaratinguetá** Brazil
106 B1 **Guarda** Port.
154 C1 **Guarda-Mor** Brazil
106 C1 **Guardo** Spain
155 C2 **Guarujá** Brazil
144 B2 **Guasave** Mex.
146 A3 **Guatemala** *country* Central America
146 A3 **Guatemala City** Guat.
150 B1 **Guaviare** *r.* Col.
155 C2 **Guaxupé** Brazil
150 A2 **Guayaquil** Ecuador
150 B3 **Guayaramerín** Bol.
144 A2 **Guaymas** Mex.
68 C2 **Guazhou** China
117 B3 **Guba** Eth.
86 E1 **Guba Dolgaya** Russia
89 E3 **Gubkin** Russia
115 C1 **Guelma** Alg.
114 A2 **Guelmim** Morocco
130 B2 **Guelph** Can.
145 C2 **Guémez** Mex.
104 C2 **Guéret** France
95 C4 **Guernsey** *terr.* Channel Is
144 A2 **Guerrero Negro** Mex.
131 D1 **Guers, Lac** *l.* Can.
79 C2 **Gūhrān** Iran
118 B2 **Guider** Cameroon
108 B2 **Guidonia Montecelio** Italy
71 A3 **Guigang** China
100 A3 **Guignicourt** France
123 D1 **Guija** Moz.
99 C3 **Guildford** U.K.
71 B3 **Guilin** China
130 C1 **Guillaume-Delisle, Lac** *l.* Can.
106 B1 **Guimarães** Port.
114 A3 **Guinea** *country* Africa
113 F6 **Guinea, Gulf of** Africa
114 A3 **Guinea-Bissau** *country* Africa
104 B2 **Guingamp** France
104 B2 **Guipavas** France
154 B1 **Guiratinga** Brazil
150 B1 **Güiria** Venez.
100 A3 **Guise** France
64 B1 **Guiuan** Phil.
71 A3 **Guiyang** China
71 A3 **Guizhou** *prov.* China
74 B1 **Gujranwala** Pak.
74 B1 **Gujrat** Pak.
91 D2 **Gukovo** Russia
53 C2 **Gulargambone** Austr.
Gulbarga India *see* Kalaburagi
88 C2 **Gulbene** Latvia
79 C2 **Gulf, the** Asia
111 B3 **Gulf of Corinth** *sea chan.* Greece
142 C2 **Gulfport** U.S.A.
69 E1 **Gulian** China
77 C2 **Guliston** Uzbek.
Gulja China *see* **Yining**
129 D2 **Gull Lake** Can.
111 C3 **Güllük** Turkey
119 D2 **Gulu** Uganda
120 B2 **Gumare** Botswana
76 B3 **Gumdag** Turkm.
65 B2 **Gumi** S. Korea
75 C2 **Gumla** India
100 C2 **Gummersbach** Ger.
74 B2 **Guna** India
53 C3 **Gundagai** Austr.
111 C3 **Güney** Turkey
118 B3 **Gungu** Dem. Rep. Congo
129 E2 **Gunisao** *r.* Can.
53 D2 **Gunnedah** Austr.
136 B3 **Gunnison** *CO* U.S.A.
135 D3 **Gunnison** *UT* U.S.A.
136 B3 **Gunnison** *r.* U.S.A.
65 B2 **Gunsan** S. Korea
73 B3 **Guntakal** India
60 A1 **Gunungsitoli** Indon.
60 A1 **Gunungtua** Indon.
102 C2 **Günzburg** Ger.
102 C2 **Gunzenhausen** Ger.
70 B2 **Guojiaba** China
74 B2 **Gurgaon** India
151 D2 **Gurgueia** *r.* Brazil
150 B1 **Guri, Embalse de** *resr* Venez.
154 C1 **Gurinhatã** Brazil
151 D2 **Gurupi** *r.* Brazil
74 B2 **Guru Sikhar** *mt.* India
Gur'yev Kazakh. *see* **Atyrau**
115 C3 **Gusau** Nigeria
65 A2 **Gushan** China
70 B2 **Gushi** China
83 I3 **Gusinoozersk** Russia
89 F2 **Gus'-Khrustal'nyy** Russia
108 A3 **Guspini** *Sardinia* Italy
128 A2 **Gustavus** U.S.A.
101 F1 **Güstrow** Ger.
101 D2 **Gütersloh** Ger.
121 C2 **Gutu Mupandawana** Zimbabwe
75 D2 **Guwahati** India
150 C1 **Guyana** *country* S. America
Guyi China *see* **Sanjiang**
139 C1 **Guymon** U.S.A.
53 D2 **Guyra** Austr.
70 A2 **Guyuan** China
144 B1 **Guzmán** Mex.
74 A2 **Gwadar** Pak.
128 A2 **Gwaii Haanas** Can.
75 B2 **Gwalior** India
121 B3 **Gwanda** Zimbabwe
65 B2 **Gwangju** S. Korea
117 D3 **Gwardafuy, Gees** *c.* Somalia
97 B1 **Gweebarra Bay** Ireland
121 B2 **Gweru** Zimbabwe
115 D3 **Gwoza** Nigeria
53 D2 **Gwydir** *r.* Austr.
75 C2 **Gyangzê** China
75 C1 **Gyaring Co** *l.* China
68 C2 **Gyaring Hu** *l.* China
86 G1 **Gydan Peninsula** Russia
Gydanskiy Poluostrov *pen.* Russia *see* **Gydan Peninsula**
Gyêgu China *see* **Yushu**
65 B2 **Gyeonggi-man** *b.* S. Korea
51 E2 **Gympie** Austr.
103 D2 **Gyöngyös** Hungary
103 D2 **Gyr** Hungary
129 E2 **Gypsumville** Can.
103 E2 **Gyula** Hungary
81 C1 **Gyumri** Armenia

H

88 B2 **Haapsalu** Estonia
100 B1 **Haarlem** Neth.
101 C2 **Haarstrang** *ridge* Ger.
54 A2 **Haast** N.Z.
78 B3 **Ḩabbān** Yemen
81 C2 **Ḩabbānīyah, Hawr al** *l.* Iraq
67 C4 **Hachijō-jima** *i.* Japan
66 D2 **Hachinohe** Japan
81 C1 **Hacıqabul** Azer.
121 C3 **Hacufera** Moz.
79 C2 **Ḩadd, Ra's al** *pt* Oman
96 C3 **Haddington** U.K.
115 D3 **Hadejia** Nigeria
93 E4 **Haderslev** Denmark
91 C1 **Hadyach** Ukr.
65 B2 **Haeju** N. Korea
65 B2 **Haeju-man** *b.* N. Korea
65 B3 **Haenam** S. Korea
78 B2 **Ḩafar al Bāţin** Saudi Arabia
62 A1 **Haflong** India
92 □A3 **Hafnarfjörður** Iceland
78 A3 **Hagar Nish Plateau** Eritrea/Sudan
48 D2 **Hagåtña** Guam
100 C2 **Hagen** Ger.
101 E1 **Hagenow** Ger.
128 B2 **Hagensborg** Can.
141 D3 **Hagerstown** U.S.A.
93 F3 **Hagfors** Sweden
67 B4 **Hagi** Japan
62 B1 **Ha Giang** Vietnam
97 B2 **Hag's Head** Ireland
104 B2 **Hague, Cap de la** *c.* France
119 D3 **Hai** Tanz.
128 A2 **Haida Gwaii** Can.
62 B1 **Hai Dương** Vietnam
80 B2 **Haifa** Israel
71 B3 **Haifeng** China
71 B3 **Haikou** China
78 B2 **Ḩā'il** Saudi Arabia
92 H2 **Hailuoto** *i.* Fin.
71 A4 **Hainan** *prov.* China
69 D3 **Hainan Dao** *i.* China
128 A2 **Haines** U.S.A.
128 A1 **Haines Junction** Can.
101 E2 **Hainich** *ridge* Ger.
101 E2 **Hainleite** *ridge* Ger.
62 B1 **Hai Phong** Vietnam
147 C3 **Haiti** *country* West Indies
116 B3 **Haiya** Sudan
103 E2 **Hajdúböszörmény** Hungary
78 B3 **Ḩajjah** Yemen
79 C2 **Ḩājjīābād** Iran
62 A1 **Hakha** Myanmar
81 C2 **Hakkâri** Turkey
66 D2 **Hakodate** Japan
Ḩalab Syria *see* **Aleppo**
78 B2 **Halabān** Saudi Arabia
81 C2 **Ḩalabjah/Helebce** Iraq
116 B2 **Ḩalā'ib** Sudan
78 A2 **Halaib Triangle** *terr.* Egypt/Sudan
79 C3 **Ḩalāniyāt, Juzur al** *is* Oman
78 A2 **Ḩālat 'Ammār** Saudi Arabia
101 E2 **Halberstadt** Ger.
64 B1 **Halcon, Mount** Phil.
93 F4 **Halden** Norway
101 E1 **Haldensleben** Ger.
75 B2 **Haldwani** India
79 C2 **Hāleh** Iran
54 A3 **Halfmoon Bay** N.Z.
131 D2 **Halifax** Can.
98 C2 **Halifax** U.K.
141 D3 **Halifax** U.S.A.
65 B3 **Halla-san** *mt.* S. Korea
127 F2 **Hall Beach** Can.
100 B2 **Halle** Belgium
101 E2 **Halle (Saale)** Ger.
102 C2 **Hallein** Austria
137 D1 **Hallock** U.S.A.
127 G2 **Hall Peninsula** Can.
50 B1 **Halls Creek** Austr.

59 C2 **Halmahera** i. Indon.
93 F4 **Halmstad** Sweden
62 B1 **Ha Long** Vietnam
67 B4 **Hamada** Japan
81 C2 **Hamadān** Iran
80 B2 **Hamāh** Syria
67 C4 **Hamamatsu** Japan
93 F3 **Hamar** Norway
116 B2 **Hamātah, Jabal** mt. Egypt
73 C4 **Hambantota** Sri Lanka
101 D1 **Hamburg** Ger.
78 A2 **Hamd, Wādī al** watercourse Saudi Arabia
78 B3 **Hamdah** Saudi Arabia
93 H3 **Hämeenlinna** Fin.
101 D1 **Hameln** Ger.
50 A2 **Hamersley Range** mts Austr.
65 B2 **Hamhŭng** N. Korea
68 C2 **Hami** China
116 B2 **Hamid** Sudan
52 B3 **Hamilton** Austr.
130 C2 **Hamilton** Can.
54 C1 **Hamilton** N.Z.
96 B3 **Hamilton** U.K.
142 C2 **Hamilton** AL U.S.A.
134 D1 **Hamilton** MT U.S.A.
140 C3 **Hamilton** OH U.S.A.
93 I3 **Hamina** Fin.
100 C2 **Hamm** Ger.
81 C2 **Hammār, Hawr al** imp. l. Iraq
101 D2 **Hammelburg** Ger.
92 G3 **Hammerdal** Sweden
92 H1 **Hammerfest** Norway
142 B2 **Hammond** U.S.A.
141 E3 **Hammonton** U.S.A.
115 C2 **Hamrā', Al Hamādah al** plat. Libya
78 A2 **Hanak** Saudi Arabia
66 D3 **Hanamaki** Japan
101 D2 **Hanau** Ger.
69 D2 **Hanbogd** Mongolia
70 B2 **Hancheng** China
140 B1 **Hancock** U.S.A.
70 B2 **Handan** China
135 C3 **Hanford** U.S.A.
68 C1 **Hangayn Nuruu** mts Mongolia
Hanggin Houqi China see Xamba
Hangö Fin. see Hanko
70 C2 **Hangzhou** China
70 C2 **Hangzhou Wan** b. China
Hanjia China see Pengshui
Hanjiang China see Yangzhou
93 H4 **Hanko** Fin.
135 D3 **Hanksville** U.S.A.
54 B2 **Hanmer Springs** N.Z.
129 C2 **Hanna** Can.
137 E3 **Hannibal** U.S.A.
101 D1 **Hannover** Ger.
101 D2 **Hannoversch Münden** Ger.
93 F4 **Hanöbukten** b. Sweden
62 B1 **Ha Nôi** Vietnam
Hanoi Vietnam see Ha Nôi
130 B2 **Hanover** Can.
122 B3 **Hanover** S. Africa
93 E4 **Hanstholm** Denmark
88 C3 **Hantsavichy** Belarus
74 B2 **Hanumana** India
74 A2 **Hanumangarh** India
71 A3 **Hanzhong** China
92 H2 **Haparanda** Sweden
100 B2 **Hapert** Neth.
131 D1 **Happy Valley-Goose Bay** Can.
78 A2 **Haql** Saudi Arabia
79 B2 **Haradh** Saudi Arabia
88 C2 **Haradok** Belarus

78 B3 **Harajā** Saudi Arabia
121 C2 **Harare** Zimbabwe
79 C3 **Harāsīs, Jiddat al** des. Oman
69 E1 **Harbin** China
131 E2 **Harbour Breton** Can.
74 B2 **Harda** India
Hardangerfjorden sea chan. Norway
100 C1 **Hardenberg** Neth.
100 B1 **Harderwijk** Neth.
122 A3 **Hardeveld** mts S. Africa
134 E1 **Hardin** U.S.A.
128 C1 **Hardisty Lake** Can.
100 C1 **Haren (Ems)** Ger.
117 C4 **Härer** Eth.
117 C4 **Hargeysa** Somalia
110 C1 **Harghita-Mădăraş, Vârful** mt. Romania
68 C2 **Har Hu** l. China
74 B1 **Haripur** Pak.
74 A1 **Harī Rōd** r. Afgh./Iran
100 B1 **Harlingen** Neth.
139 D3 **Harlingen** U.S.A.
99 D3 **Harlow** U.K.
134 E1 **Harlowton** U.S.A.
110 C2 **Harmanli** Bulg.
134 C2 **Harney Basin** U.S.A.
134 C2 **Harney Lake** U.S.A.
93 G3 **Härnösand** Sweden
69 E1 **Har Nur** China
68 C1 **Har Nuur** l. Mongolia
114 B4 **Harper** Liberia
101 D1 **Harpstedt** Ger.
130 C1 **Harricana, Rivière d'** r. Can.
53 D2 **Harrington** Austr.
131 E1 **Harrington Harbour** Can.
96 A2 **Harris, Sound of** sea chan. U.K.
140 B3 **Harrisburg** IL U.S.A.
141 D2 **Harrisburg** PA U.S.A.
123 C2 **Harrismith** S. Africa
142 B1 **Harrison** U.S.A.
131 E1 **Harrison, Cape** Can.
126 A2 **Harrison Bay** U.S.A.
141 D3 **Harrisonburg** U.S.A.
137 E3 **Harrisonville** U.S.A.
98 C2 **Harrogate** U.K.
110 C2 **Hârşova** Romania
92 G2 **Harstad** Norway
122 B2 **Hartbees** watercourse S. Africa
103 D2 **Hartberg** Austria
141 E2 **Hartford** U.S.A.
99 A3 **Hartland Point** U.K.
98 C1 **Hartlepool** U.K.
128 B2 **Hartley Bay** Can.
123 B2 **Harts** r. S. Africa
143 D2 **Hartwell Reservoir** U.S.A.
68 C1 **Har Us Nuur** l. Mongolia
136 C1 **Harvey** U.S.A.
99 D3 **Harwich** U.K.
101 E2 **Harz** hills Ger.
110 C2 **Haskovo** Bulg.
73 B3 **Hassan** India
100 B2 **Hasselt** Belgium
115 C1 **Hassi Messaoud** Alg.
93 F4 **Hässleholm** Sweden
100 B2 **Hastière-Lavaux** Belgium
53 C3 **Hastings** Austr.
54 C1 **Hastings** N.Z.
99 D3 **Hastings** U.K.
137 D2 **Hastings** MN U.S.A.
137 D2 **Hastings** NE U.S.A.
Hatay Turkey see Antakya
129 D2 **Hatchet Lake** Can.
52 B2 **Hatfield** Austr.
68 C1 **Hatgal** Mongolia
62 B2 **Ha Tinh** Vietnam
143 E1 **Hatteras, Cape** U.S.A.

142 C2 **Hattiesburg** U.S.A.
100 C2 **Hattingen** Ger.
63 B3 **Hat Yai** Thai.
117 C4 **Haud** reg. Eth.
93 E4 **Haugesund** Norway
93 E4 **Haukeligrend** Norway
92 I2 **Haukipudas** Fin.
54 C1 **Hauraki Gulf** N.Z.
114 B1 **Haut Atlas** mts Morocco
131 D2 **Hauterive** Can.
114 B1 **Hauts Plateaux** Alg.
146 B2 **Havana** Cuba
99 C3 **Havant** U.K.
101 E1 **Havel** r. Ger.
101 F1 **Havelberg** Ger.
54 B2 **Havelock** N.Z.
54 C1 **Havelock North** N.Z.
99 A3 **Haverfordwest** U.K.
103 D2 **Havlíčkův Brod** Czech Rep.
92 H1 **Havøysund** Norway
111 C3 **Havran** Turkey
134 E1 **Havre** U.S.A.
131 D2 **Havre-Aubert** Can.
131 D1 **Havre-St-Pierre** Can.
124 E5 **Hawai'i** i. U.S.A.
124 E5 **Hawai'ian Islands** N. Pacific Ocean
98 B2 **Hawarden** U.K.
54 A2 **Hawea, Lake** N.Z.
54 B1 **Hawera** N.Z.
98 B1 **Hawes** U.K.
96 C3 **Hawick** U.K.
54 C1 **Hawke Bay** N.Z.
52 A2 **Hawker** Austr.
52 B1 **Hawkers Gate** Austr.
135 C3 **Hawthorne** U.S.A.
52 B2 **Hay** Austr.
128 C1 **Hay** r. Can.
134 C1 **Hayden** U.S.A.
129 E2 **Hayes** r. Man. Can.
126 E2 **Hayes** r. Nunavut Can.
79 C3 **Haymā'** Oman
77 C2 **Hayotboshi tog'i** mt. Uzbek.
111 C2 **Hayrabolu** Turkey
128 C1 **Hay River** Can.
137 D3 **Hays** U.S.A.
78 B3 **Hays** Yemen
90 B2 **Haysyn** Ukr.
99 C3 **Haywards Heath** U.K.
81 D2 **Hazar** Turkm.
74 A1 **Hazārah Jāt** reg. Afgh.
140 C3 **Hazard** U.S.A.
75 C2 **Hazaribagh** India
75 C2 **Hazaribagh Range** mts India
128 B2 **Hazelton** Can.
141 D2 **Hazleton** U.S.A.
78 B3 **Hazm al Jawf** Yemen
53 C3 **Healesville** Austr.
130 B2 **Hearst** Can.
70 B2 **Hebei** prov. China
53 C1 **Hebel** Austr.
142 B1 **Heber Springs** U.S.A.
131 D1 **Hebron** Can.
128 A2 **Hecate Strait** Can.
71 A3 **Hechi** China
93 F3 **Hede** Sweden
100 B1 **Heerenveen** Neth.
100 B1 **Heerhugowaard** Neth.
100 B2 **Heerlen** Neth.
Hefa Israel see Haifa
70 B2 **Hefei** China
70 B3 **Hefeng** China
69 E1 **Hegang** China
102 B2 **Heide** Ger.
122 A1 **Heide** Namibia
102 B2 **Heidelberg** Ger.
123 B3 **Heidelberg** S. Africa
102 B2 **Heilbronn** Ger.

69 E1 **Heilong Jiang** *r.* China
93 I3 **Heinola** Fin.
92 F3 **Helagsfjället** *mt.* Sweden
142 B2 **Helena** *AR* U.S.A.
134 D1 **Helena** *MT* U.S.A.
96 B2 **Helensburgh** U.K.
102 B1 **Helgoland** *i.* Ger.
102 B1 **Helgoländer Bucht** *g.* Ger.
Helixi China *see* **Ningguo**
92 □A3 **Hella** Iceland
100 B2 **Hellevoetsluis** Neth.
107 C2 **Hellín** Spain
76 C3 **Helmand** *r.* Afgh.
101 E2 **Helmbrechts** Ger.
122 A2 **Helmeringhausen** Namibia
100 B2 **Helmond** Neth.
96 C1 **Helmsdale** U.K.
96 C1 **Helmsdale** *r.* U.K.
101 E1 **Helmstedt** Ger.
65 B1 **Helong** China
93 F4 **Helsingborg** Sweden
Helsingfors Fin. *see* **Helsinki**
93 F4 **Helsingør** Denmark
93 H3 **Helsinki** Fin.
97 C2 **Helvick Head** Ireland
99 C3 **Hemel Hempstead** U.K.
101 D1 **Hemmoor** Ger.
92 F2 **Hemnesberget** Norway
70 B2 **Henan** *prov.* China
140 B3 **Henderson** *KY* U.S.A.
143 E1 **Henderson** *NC* U.S.A.
135 D3 **Henderson** *NV* U.S.A.
139 E2 **Henderson** *TX* U.S.A.
143 D1 **Hendersonville** U.S.A.
99 C3 **Hendon** U.K.
62 A1 **Hengduan Shan** *mts* China
100 C1 **Hengelo** Neth.
Hengnan China *see* **Hengyang**
70 B2 **Hengshui** China
71 A3 **Hengxian** China
71 B3 **Hengyang** China
Hengzhou China *see* **Hengxian**
91 C2 **Heniches'k** Ukr.
100 C2 **Hennef (Sieg)** Ger.
130 B1 **Henrietta Maria, Cape** Can.
139 D1 **Henryetta** U.S.A.
127 G2 **Henry Kater, Cape** Can.
101 D1 **Henstedt-Ulzburg** Ger.
120 A3 **Hentiesbaai** Namibia
71 A3 **Hepu** China
76 C3 **Herāt** Afgh.
129 D2 **Herbert** Can.
101 D2 **Herbstein** Ger.
99 B2 **Hereford** U.K.
139 C2 **Hereford** U.S.A.
101 D1 **Herford** Ger.
100 C2 **Herkenbosch** Neth.
96 □ **Herma Ness** *hd* U.K.
122 A3 **Hermanus** S. Africa
53 C2 **Hermidale** Austr.
134 C1 **Hermiston** U.S.A.
59 D3 **Hermit Islands** P.N.G.
144 A2 **Hermosillo** Mex.
154 B3 **Hernandarias** Para.
100 C2 **Herne** Ger.
93 E4 **Herning** Denmark
99 C3 **Hertford** U.K.
123 C2 **Hertzogville** S. Africa
51 E2 **Hervey Bay** Austr.
101 F2 **Herzberg (Elster)** Ger.
101 E3 **Herzogenaurach** Ger.
71 A3 **Heshan** China
128 A1 **Hess** *r.* Can.
101 D2 **Hessisch Lichtenau** Ger.
136 C1 **Hettinger** U.S.A.
101 E2 **Hettstedt** Ger.
98 B1 **Hexham** U.K.
71 B3 **Heyuan** China
52 B3 **Heywood** Austr.
70 B2 **Heze** China
71 B3 **Hezhou** China
137 D3 **Hiawatha** U.S.A.
137 E1 **Hibbing** U.S.A.
54 C1 **Hicks Bay** N.Z.
66 D2 **Hidaka-sanmyaku** *mts* Japan
145 C2 **Hidalgo** Mex.
144 B2 **Hidalgo del Parral** Mex.
154 C1 **Hidrolândia** Brazil
High Atlas *mts* Morocco *see* **Haut Atlas**
134 B2 **High Desert** U.S.A.
128 C2 **High Level** Can.
143 E1 **High Point** U.S.A.
128 C2 **High Prairie** Can.
128 C2 **High River** Can.
129 D2 **Highrock Lake** Can.
99 C3 **High Wycombe** U.K.
88 B2 **Hiiumaa** *i.* Estonia
78 A2 **Hijaz** *reg.* Saudi Arabia
54 C1 **Hikurangi** *mt.* N.Z.
101 E2 **Hildburghausen** Ger.
101 E2 **Hilders** Ger.
101 D1 **Hildesheim** Ger.
81 C2 **Hillah** Iraq
100 C2 **Hillesheim** Ger.
140 C3 **Hillsboro** *OH* U.S.A.
139 D2 **Hillsboro** *TX* U.S.A.
53 C2 **Hillston** Austr.
143 D2 **Hilton Head Island** U.S.A.
100 B1 **Hilversum** Neth.
68 B2 **Himalaya** *mts* Asia
67 B4 **Himeji** Japan
123 C2 **Himeville** S. Africa
51 D1 **Hinchinbrook Island** Austr.
74 A1 **Hindu Kush** *mts* Afgh./Pak.
143 D2 **Hinesville** U.S.A.
75 B2 **Hinganghat** India
81 C2 **Hınıs** Turkey
92 G2 **Hinnøya** *i.* Norway
106 B2 **Hinojosa del Duque** Spain
62 A2 **Hinthada** Myanmar
128 C2 **Hinton** Can.
75 C2 **Hirakud Reservoir** India
66 D2 **Hirosaki** Japan
67 B4 **Hiroshima** Japan
101 E3 **Hirschaid** Ger.
101 E2 **Hirschberg** Ger.
105 C2 **Hirson** France
93 E4 **Hirtshals** Denmark
74 B2 **Hisar** India
147 C2 **Hispaniola** *i.* Caribbean Sea
81 C2 **Hīt** Iraq
67 D3 **Hitachi** Japan
67 D3 **Hitachinaka** Japan
92 E3 **Hitra** *i.* Norway
49 I3 **Hiva Oa** *i.* Fr. Polynesia
93 G4 **Hjälmaren** *l.* Sweden
129 D1 **Hjalmar Lake** Can.
93 F4 **Hjørring** Denmark
123 D2 **Hlabisa** S. Africa
92 □B2 **Hlíð** Iceland
91 C2 **Hlobyne** Ukr.
123 C2 **Hlohlowane** S. Africa
123 C2 **Hlotse** Lesotho
91 C1 **Hlukhiv** Ukr.
88 C2 **Hlybokaye** Belarus
114 C4 **Ho** Ghana
62 B2 **Hoa Binh** Vietnam
122 A1 **Hoachanas** Namibia
51 D4 **Hobart** Austr.
139 D1 **Hobart** U.S.A.
139 C2 **Hobbs** U.S.A.
93 E4 **Hobro** Denmark
117 C4 **Hobyo** Somalia
63 B2 **Ho Chi Minh City** Vietnam
114 B3 **Ḥôḍ** *reg.* Maur.
78 B3 **Hodeidah** Yemen
103 E2 **Hódmezővásárhely** Hungary
Hoek van Holland Neth. *see* **Hook of Holland**
65 B2 **Hoeyang** N. Korea
101 E2 **Hof** Ger.
101 E2 **Hofheim in Unterfranken** Ger.
92 □B3 **Höfn** *Austurland* Iceland
92 □A2 **Höfn** *Vestfirðir* Iceland
92 □B3 **Hofsjökull** Iceland
67 B4 **Hōfu** Japan
93 G4 **Högsby** Sweden
101 D2 **Hohe Rhön** *mts* Ger.
100 C2 **Hohe Venn** *moorland* Belgium
70 B1 **Hohhot** China
75 C1 **Hoh Xil Shan** *mts* China
63 B2 **Hôi An** Vietnam
62 A1 **Hojai** India
54 B2 **Hokitika** N.Z.
66 D2 **Hokkaidō** *i.* Japan
128 B2 **Holberg** Can.
138 A2 **Holbrook** U.S.A.
137 D2 **Holdrege** U.S.A.
146 C2 **Holguín** Cuba
92 □B2 **Hóll** Iceland
Holland *country* Europe *see* **Netherlands**
140 B2 **Holland** U.S.A.
100 B1 **Hollum** Neth.
142 C2 **Holly Springs** U.S.A.
134 C4 **Hollywood** *CA* U.S.A.
143 D3 **Hollywood** *FL* U.S.A.
92 F2 **Holm** Norway
92 H3 **Holmsund** Sweden
122 A2 **Holoog** Namibia
93 E4 **Holstebro** Denmark
143 D1 **Holston** *r.* U.S.A.
98 A2 **Holyhead** U.K.
98 C1 **Holy Island** *England* U.K.
98 A2 **Holy Island** *Wales* U.K.
136 C2 **Holyoke** U.S.A.
101 D2 **Holzminden** Ger.
62 A1 **Homalin** Myanmar
101 D2 **Homberg (Efze)** Ger.
114 B3 **Hombori** Mali
127 G2 **Home Bay** Can.
143 D3 **Homestead** U.S.A.
92 F3 **Hommelvik** Norway
80 B2 **Homs** Syria
89 D3 **Homyel'** Belarus
122 A3 **Hondeklipbaai** S. Africa
145 D3 **Hondo** *r.* Belize/Mex.
139 D3 **Hondo** U.S.A.
146 B3 **Honduras** *country* Central America
93 F3 **Hønefoss** Norway
135 B2 **Honey Lake** U.S.A.
104 C2 **Honfleur** France
70 B3 **Honghu** China
71 A3 **Hongjiang** China
71 B3 **Hong Kong** China
71 B3 **Hong Kong** *aut. reg.* China
65 B1 **Hongwŏn** N. Korea
70 B2 **Hongze Hu** *l.* China
49 E3 **Honiara** Solomon Is
92 I1 **Honningsvåg** Norway
67 B3 **Honshū** *i.* Japan
134 B1 **Hood, Mount** *vol.* U.S.A.
50 A3 **Hood Point** Austr.
134 B1 **Hood River** U.S.A.
100 C1 **Hoogeveen** Neth.
100 C1 **Hoogezand-Sappemeer** Neth.
100 C2 **Hoog-Keppel** Neth.
100 B2 **Hook of Holland** Neth.
128 A2 **Hoonah** U.S.A.

100 B1 **Hoorn** Neth.
128 B3 **Hope** Can.
142 B2 **Hope** U.S.A.
83 N2 **Hope, Point** U.S.A.
131 D1 **Hopedale** Can.
131 D1 **Hope Mountains** Can.
52 B3 **Hopetoun** Austr.
122 B2 **Hopetown** S. Africa
141 D3 **Hopewell** U.S.A.
130 C1 **Hopewell Islands** Can.
50 B2 **Hopkins, Lake** *imp. l.* Austr.
140 B3 **Hopkinsville** U.S.A.
134 B1 **Hoquiam** U.S.A.
81 C1 **Horasan** Turkey
93 F4 **Hörby** Sweden
89 D3 **Horki** Belarus
91 D2 **Horlivka** Ukr.
79 D2 **Hormak** Iran
79 C2 **Hormuz, Strait of** Iran/Oman
103 D2 **Horn** Austria
92 □A2 **Horn** *c.* Iceland
153 B6 **Horn, Cape** Chile
141 D2 **Hornell** U.S.A.
130 B2 **Hornepayne** Can.
98 C2 **Hornsea** U.K.
90 B2 **Horodenka** Ukr.
91 C1 **Horodnya** Ukr.
90 B2 **Horodok** *Khmel'nyts'ka Oblast'* Ukr.
90 A2 **Horodok** *L'vivs'ka Oblast'* Ukr.
90 A1 **Horokhiv** Ukr.
Horqin Youyi Qianqi China *see* **Ulanhot**
131 E1 **Horse Islands** Can.
52 B3 **Horsham** Austr.
126 C2 **Horton** *r.* Can.
117 B4 **Hosa'ina** Eth.
74 A2 **Hoshab** Pak.
74 B1 **Hoshiarpur** India
77 E3 **Hotan** China
122 B2 **Hotazel** S. Africa
142 B2 **Hot Springs** *AR* U.S.A.
136 C2 **Hot Springs** *SD* U.S.A.
128 C1 **Hottah Lake** Can.
62 B1 **Houayxay** Laos
100 B2 **Houffalize** Belgium
70 B2 **Houma** China
142 B3 **Houma** U.S.A.
128 B2 **Houston** Can.
139 D3 **Houston** U.S.A.
122 B3 **Houwater** S. Africa
68 C1 **Hovd** Mongolia
99 C3 **Hove** U.K.
68 C1 **Hövsgöl Nuur** *l.* Mongolia
116 A3 **Howar, Wadi** *watercourse* Sudan
53 C3 **Howe, Cape** Austr.
49 G2 **Howland Island** N. Pacific Ocean
53 C3 **Howlong** Austr.
101 D2 **Höxter** Ger.
96 C1 **Hoy** *i.* U.K.
93 E3 **Høyanger** Norway
102 C1 **Hoyerswerda** Ger.
62 A2 **Hpapun** Myanmar
103 D1 **Hradec Králové** Czech Rep.
109 C2 **Hrasnica** Bos. & Herz.
91 C1 **Hrebinka** Ukr.
88 B3 **Hrodna** Belarus
62 A1 **Hsi-hseng** Myanmar
62 A1 **Hsipaw** Myanmar
70 A2 **Huachi** China
150 A3 **Huacho** Peru
69 D2 **Huade** China
65 B1 **Huadian** China
70 B2 **Huai'an** China
70 B2 **Huaibei** China
71 A3 **Huaihua** China
70 B2 **Huainan** China
70 B2 **Huaiyang** China
145 C3 **Huajuápan de León** Mex.
59 C3 **Huaki** Indon.
71 C3 **Hualian** Taiwan
150 A2 **Huallaga** *r.* Peru
120 A2 **Huambo** Angola
150 A3 **Huancayo** Peru
Huangcaoba China *see* **Xingyi**
70 B2 **Huangchuan** China
Huang Hai *sea* N. Pacific Ocean *see* **Yellow Sea**
Huang He *r.* China *see* **Yellow River**
71 A4 **Huangliu** China
70 B3 **Huangshan** China
70 B2 **Huangshi** China
70 A2 **Huangtu Gaoyuan** *plat.* China
71 C3 **Huangyan** China
65 B1 **Huanren** China
150 A2 **Huánuco** Peru
152 B2 **Huanuni** Bol.
150 A2 **Huaraz** Peru
150 A3 **Huarmey** Peru
152 A3 **Huasco** Chile
152 A3 **Huasco** *r.* Chile
144 B2 **Huatabampo** Mex.
145 C3 **Huatusco** Mex.
71 A3 **Huayuan** China
73 B3 **Hubballi** India
70 B2 **Hubei** *prov.* China
Hubli India *see* **Hubballi**
100 C2 **Hückelhoven** Ger.
98 C2 **Hucknall** U.K.
98 C2 **Huddersfield** U.K.
93 G3 **Hudiksvall** Sweden
141 E2 **Hudson** *r.* U.S.A.
129 D2 **Hudson Bay** Can.
127 F3 **Hudson Bay** *sea* Can.
128 B2 **Hudson's Hope** Can.
127 G2 **Hudson Strait** Can.
63 B2 **Huê** Vietnam
146 A3 **Huehuetenango** Guat.
144 B2 **Huehueto, Cerro** *mt.* Mex.
145 C2 **Huejutla** Mex.
106 B2 **Huelva** Spain
107 C2 **Huércal-Overa** Spain
107 C1 **Huesca** Spain
106 C2 **Huéscar** Spain
50 B3 **Hughes (abandoned)** Austr.
139 D2 **Hugo** U.S.A.
122 B2 **Huhudi** S. Africa
122 A2 **Huib-Hoch Plateau** Namibia
71 B3 **Huichang** China
65 B1 **Huich'ŏn** N. Korea
120 A2 **Huíla, Planalto da** Angola
71 B3 **Huilai** China
62 B1 **Huili** China
65 B1 **Huinan** China
93 H3 **Huittinen** Fin.
145 C3 **Huixtla** Mex.
Huiyang China *see* **Huizhou**
62 B1 **Huize** China
71 B3 **Huizhou** China
78 B2 **Hujr** Saudi Arabia
122 B1 **Hukuntsi** Botswana
78 B2 **Ḩulayfah** Saudi Arabia
66 B1 **Hulin** China
130 C2 **Hull** Can.
Hulun China *see* **Hulun Buir**
69 D1 **Hulun Buir** China
69 D1 **Hulun Nur** *l.* China
91 D2 **Hulyaypole** Ukr.
69 E1 **Huma** China
150 B2 **Humaitá** Brazil
122 B3 **Humansdorp** S. Africa
98 C2 **Humber** *est.* U.K.
126 D3 **Humboldt** Can.
142 C1 **Humboldt** U.S.A.
135 C2 **Humboldt** *r.* U.S.A.
103 E2 **Humenné** Slovakia
53 C3 **Hume Reservoir** Austr.
138 A1 **Humphreys Peak** U.S.A.
92 □A2 **Húnaflói** *b.* Iceland
71 B3 **Hunan** *prov.* China
65 C1 **Hunchun** China
110 B1 **Hunedoara** Romania
101 D2 **Hünfeld** Ger.
103 D2 **Hungary** *country* Europe
52 B1 **Hungerford** Austr.
65 B2 **Hŭngnam** N. Korea
65 A1 **Hun He** *r.* China
99 D2 **Hunstanton** U.K.
101 D1 **Hunte** *r.* Ger.
51 D4 **Hunter Islands** Austr.
99 C2 **Huntingdon** U.K.
140 B2 **Huntington** *IN* U.S.A.
140 C3 **Huntington** *WV* U.S.A.
54 C1 **Huntly** N.Z.
96 C2 **Huntly** U.K.
130 C2 **Huntsville** Can.
142 C2 **Huntsville** *AL* U.S.A.
139 D2 **Huntsville** *TX* U.S.A.
59 D3 **Huon Peninsula** P.N.G.
70 B2 **Huozhou** China
116 B1 **Hurghada** Egypt
137 D2 **Huron** U.S.A.
140 C2 **Huron, Lake** Can./U.S.A.
135 D3 **Hurricane** U.S.A.
92 □B2 **Húsavík** Iceland
110 C1 **Huşi** Romania
126 A2 **Huslia** U.S.A.
78 B3 **Ḩusn Āl 'Abr** Yemen
102 B1 **Husum** Ger.
69 C1 **Hutag-Öndör** Mongolia
60 A1 **Hutanopan** Indon.
137 D3 **Hutchinson** U.S.A.
70 C2 **Huzhou** China
92 □C3 **Hvalnes** Iceland
92 □B3 **Hvannadalshnúkur** *vol.* Iceland
109 C2 **Hvar** *i.* Croatia
120 B2 **Hwange** Zimbabwe
136 C2 **Hyannis** U.S.A.
68 C1 **Hyargas Nuur** *salt l.* Mongolia
50 A3 **Hyden** Austr.
73 B3 **Hyderabad** India
74 A2 **Hyderabad** Pak.
105 D3 **Hyères** France
105 D3 **Hyères, Îles d'** *is* France
65 B1 **Hyesan** N. Korea
128 B2 **Hyland Post** Can.
67 B3 **Hyōno-sen** *mt.* Japan
99 D3 **Hythe** U.K.
93 H3 **Hyvinkää** Fin.

I

150 B2 **Iaco** *r.* Brazil
110 C2 **Ialomiţa** *r.* Romania
110 C1 **Ianca** Romania
110 C1 **Iaşi** Romania
64 A1 **Iba** Phil.
115 C4 **Ibadan** Nigeria
150 A1 **Ibagué** Col.
150 A1 **Ibarra** Ecuador
78 B3 **Ibb** Yemen
100 C1 **Ibbenbüren** Ger.
115 C4 **Ibi** Nigeria
155 C1 **Ibiá** Brazil
155 D1 **Ibiraçu** Brazil
107 D2 **Ibiza** Spain

107 D2 **Ibiza** *i.* Spain
151 D3 **Ibotirama** Brazil
79 C2 **Ibrā'** Oman
79 C2 **Ibrī** Oman
150 A3 **Ica** Peru
92 □B2 **Iceland** *country* Europe
66 D3 **Ichinoseki** Japan
91 C1 **Ichnya** Ukr.
65 B2 **Ich'ŏn** N. Korea
139 E2 **Idabel** U.S.A.
134 D2 **Idaho** *state* U.S.A.
134 D2 **Idaho Falls** U.S.A.
100 C3 **Idar-Oberstein** Ger.
116 B2 **Idfū** Egypt
118 B3 **Idiofa** Dem. Rep. Congo
80 B2 **Idlib** Syria
154 B2 **Iepê** Brazil
100 A2 **Ieper** Belgium
119 D3 **Ifakara** Tanz.
121 □D3 **Ifanadiana** Madag.
115 C4 **Ife** Nigeria
114 C3 **Ifôghas, Adrar des** *hills* Mali
61 C1 **Igan** *Sarawak* Malaysia
154 C2 **Igarapava** Brazil
82 G2 **Igarka** Russia
74 B3 **Igatpuri** India
81 C2 **Iğdır** Turkey
108 A3 **Iglesias** *Sardinia* Italy
127 F2 **Igloolik** Can.
Igluligaarjuk Can. *see* **Chesterfield Inlet**
130 A2 **Ignace** Can.
88 C2 **Ignalina** Lith.
110 C2 **İğneada** Turkey
111 B3 **Igoumenitsa** Greece
86 E3 **Igra** Russia
86 F2 **Igrim** Russia
154 B3 **Iguaçu** *r.* Brazil
154 B3 **Iguaçu Falls** Arg./Brazil
145 C3 **Iguala** Mex.
107 D1 **Igualada** Spain
154 C2 **Iguape** Brazil
154 B2 **Iguatemi** Brazil
154 B2 **Iguatemi** *r.* Brazil
151 E2 **Iguatu** Brazil
118 A3 **Iguéla** Gabon
119 D3 **Igunga** Tanz.
121 □D3 **Ihosy** Madag.
110 B2 **Ihtiman** Bulg.
92 I3 **Iisalmi** Fin.
115 C4 **Ijebu-Ode** Nigeria
100 B1 **IJmuiden** Neth.
100 B1 **IJssel** *r.* Neth.
100 B1 **IJsselmeer** *l.* Neth.
123 C2 **Ikageng** S. Africa
111 C3 **Ikaria** *i.* Greece
118 C3 **Ikela** Dem. Rep. Congo
67 A4 **Iki-shima** *i.* Japan
121 □D3 **Ikongo** Madag.
65 B2 **Iksan** S. Korea
64 B1 **Ilagan** Phil.
81 C2 **Īlām** Iran
75 C2 **Ilam** Nepal
103 D1 **Iława** Pol.
79 C2 **Ilazārān, Kūh-e** *mt.* Iran
129 D2 **Île-à-la-Crosse** Can.
129 D2 **Île-à-la-Crosse, Lac** *l.* Can.
118 C3 **Ilebo** Dem. Rep. Congo
124 H6 **Île Clipperton** *terr.* N. Pacific Ocean
119 D2 **Ilemi Triangle** *disp. terr.* Africa
119 D2 **Ileret** Kenya
113 K11 **Îles Crozet** *is* Indian Ocean
49 I3 **Îles du Désappointement** *is* Fr. Polynesia
49 I4 **Îles Gambier** *is* Fr. Polynesia
99 D3 **Ilford** U.K.
99 A3 **Ilfracombe** U.K.
154 B2 **Ilha Grande** *resr* Brazil
155 D2 **Ilha Grande, Baía da** *b.* Brazil
154 B2 **Ilha Solteíra, Represa** *resr* Brazil
106 B1 **Ílhavo** Port.
151 E3 **Ilhéus** Brazil
64 B2 **Iligan** Phil.
152 A4 **Illapel** Chile
90 C2 **Illichivs'k** Ukr.
140 A3 **Illinois** *r.* U.S.A.
140 B3 **Illinois** *state* U.S.A.
90 B2 **Illintsi** Ukr.
115 C2 **Illizi** Alg.
89 D2 **Il'men', Ozero** *l.* Russia
101 E2 **Ilmenau** Ger.
150 A3 **Ilo** Peru
64 B1 **Iloilo** Phil.
92 J3 **Ilomantsi** Fin.
115 C4 **Ilorin** Nigeria
53 D1 **Iluka** Austr.
127 H2 **Ilulissat** Greenland
67 A4 **Imari** Japan
117 C4 **Īmī** Eth.
108 B2 **Imola** Italy
151 D2 **Imperatriz** Brazil
136 C2 **Imperial** U.S.A.
118 B2 **Impfondo** Congo
62 A1 **Imphal** India
111 C2 **İmroz** Turkey
150 B3 **Inambari** *r.* Peru
115 C2 **In Aménas** Alg.
59 C3 **Inanwatan** Indon.
92 I2 **Inari** Fin.
92 I2 **Inarijärvi** *l.* Fin.
67 D3 **Inawashiro-ko** *l.* Japan
80 B1 **İnce Burun** *pt* Turkey
65 B2 **Incheon** S. Korea
123 D2 **Incomati** *r.* Moz.
78 A3 **Inda Silasē** Eth.
144 B2 **Indé** Mex.
135 C3 **Independence** *CA* U.S.A.
137 E2 **Independence** *IA* U.S.A.
137 D3 **Independence** *KS* U.S.A.
137 E3 **Independence** *MO* U.S.A.
134 C2 **Independence Mountains** U.S.A.
76 B2 **Inderbor** Kazakh.
72 B2 **India** *country* Asia
141 D2 **Indiana** U.S.A.
140 B2 **Indiana** *state* U.S.A.
140 B3 **Indianapolis** U.S.A.
129 D2 **Indian Head** Can.
159 **Indian Ocean**
137 E2 **Indianola** *IA* U.S.A.
142 B2 **Indianola** *MS* U.S.A.
135 C3 **Indian Springs** U.S.A.
86 D2 **Indiga** Russia
83 K2 **Indigirka** *r.* Russia
109 D1 **Inđija** Serbia
135 C4 **Indio** U.S.A.
58 B3 **Indonesia** *country* Asia
74 B2 **Indore** India
61 B2 **Indramayu, Tanjung** *pt* Indon.
104 C2 **Indre** *r.* France
74 A2 **Indus** *r.* China/Pak.
74 A2 **Indus, Mouths of the** Pak.
80 B1 **İnebolu** Turkey
111 C2 **İnegöl** Turkey
144 B3 **Infiernillo, Presa** *resr* Mex.
53 D1 **Inglewood** Austr.
102 C2 **Ingolstadt** Ger.
75 C2 **Ingraj Bazar** India
123 D2 **Inhaca** Moz.
121 C3 **Inhambane** Moz.
97 A2 **Inishbofin** *i.* Ireland
97 B2 **Inishmore** *i.* Ireland
97 C1 **Inishowen** *pen.* Ireland
54 B2 **Inland Kaikoura Range** *mts* N.Z.
102 C2 **Inn** *r.* Europe
127 G1 **Innaanganeq** *c.* Greenland
96 B2 **Inner Sound** *sea chan.* U.K.
51 D1 **Innisfail** Austr.
102 C2 **Innsbruck** Austria
154 B1 **Inocência** Brazil
118 B3 **Inongo** Dem. Rep. Congo
103 D1 **Inowrocław** Pol.
114 C2 **In Salah** Alg.
62 A2 **Insein** Myanmar
86 F2 **Inta** Russia
137 E1 **International Falls** U.S.A.
130 C1 **Inukjuak** Can.
126 C2 **Inuvik** Can.
96 B2 **Inveraray** U.K.
54 A3 **Invercargill** N.Z.
53 D1 **Inverell** Austr.
96 B2 **Invergordon** U.K.
128 C2 **Invermere** Can.
131 D2 **Inverness** Can.
96 B2 **Inverness** U.K.
96 C2 **Inverurie** U.K.
52 A3 **Investigator Strait** Austr.
77 E1 **Inya** Russia
119 D3 **Inyonga** Tanz.
87 D3 **Inza** Russia
111 B3 **Ioannina** Greece
137 D3 **Iola** U.S.A.
96 A2 **Iona** *i.* U.K.
111 B3 **Ionian Islands** Greece
109 C3 **Ionian Sea** Greece/Italy
Ionioi Nisoi *is* Greece *see* **Ionian Islands**
111 C3 **Ios** *i.* Greece
137 E2 **Iowa** *state* U.S.A.
137 E2 **Iowa City** U.S.A.
154 C1 **Ipameri** Brazil
155 D1 **Ipatinga** Brazil
81 C1 **Ipatovo** Russia
123 C2 **Ipelegeng** S. Africa
150 A1 **Ipiales** Col.
154 B3 **Ipiranga** Brazil
60 B1 **Ipoh** Malaysia
154 B1 **Iporá** Brazil
118 C2 **Ippy** C.A.R.
111 C2 **İpsala** Turkey
53 D1 **Ipswich** Austr.
99 D2 **Ipswich** U.K.
127 G2 **Iqaluit** Can.
152 A3 **Iquique** Chile
150 A2 **Iquitos** Peru
Irakleio Greece *see* **Iraklion**
111 C3 **Iraklion** Greece
76 B3 **Iran** *country* Asia
61 C1 **Iran, Pegunungan** *mts* Indon.
79 D2 **Īrānshahr** Iran
144 B2 **Irapuato** Mex.
81 C2 **Iraq** *country* Asia
154 B3 **Irati** Brazil
80 B2 **Irbid** Jordan
86 F3 **Irbit** Russia
151 D3 **Irecê** Brazil
97 C2 **Ireland** *country* Europe
118 C3 **Irema** Dem. Rep. Congo
114 B3 **Irîgui** *reg.* Mali/Maur.
119 D3 **Iringa** Tanz.
151 C2 **Iriri** *r.* Brazil
95 B3 **Irish Sea** Ireland/U.K.
69 C1 **Irkutsk** Russia
52 A2 **Iron Knob** Austr.
140 B1 **Iron Mountain** U.S.A.
140 A1 **Ironwood** U.S.A.
64 B1 **Irosin** Phil.
67 C4 **Irō-zaki** *pt* Japan
90 C1 **Irpin'** Ukr.

62 A2 **Irrawaddy** *r.* Myanmar
63 A2 **Irrawaddy, Mouths of the** Myanmar
86 F2 **Irtysh** *r.* Kazakh./Russia
107 C1 **Irun** Spain
96 B3 **Irvine** U.K.
64 B2 **Isabela** Phil.
146 B3 **Isabelia, Cordillera** *mts* Nic.
92 □A2 **Ísafjörður** Iceland
67 B4 **Isahaya** Japan
105 E2 **Isar** *r.* Ger.
96 □ **Isbister** U.K.
108 B2 **Ischia, Isola d'** *i.* Italy
67 C4 **Ise** Japan
118 C2 **Isengi** Dem. Rep. Congo
100 C2 **Iserlohn** Ger.
101 D1 **Isernhagen** Ger.
67 C4 **Ise-wan** *b.* Japan
114 C4 **Iseyin** Nigeria
Isfahan Iran *see* **Eşfahān**
66 D2 **Ishikari-wan** *b.* Japan
66 D3 **Ishinomaki** Japan
67 D3 **Ishioka** Japan
140 B1 **Ishpeming** U.S.A.
111 C2 **Işıklar Dağı** *mts* Turkey
111 C3 **Işıklı** Turkey
123 D2 **Isipingo** S. Africa
119 C2 **Isiro** Dem. Rep. Congo
110 B2 **Iskar** *r.* Bulg.
80 B2 **İskenderun** Turkey
82 G3 **Iskitim** Russia
128 A2 **Iskut** *r.* Can.
74 B1 **Islamabad** Pak.
52 A2 **Island Lagoon** *imp. l.* Austr.
129 E2 **Island Lake** Can.
54 B1 **Islands, Bay of** N.Z.
Islas Malvinas *terr.* S. Atlantic Ocean *see* **Falkland Islands**
96 A3 **Islay** *i.* U.K.
98 A1 **Isle of Man** *i.* Irish Sea.
77 D3 **Ismoili Somoní, Qullai** *mt.* Tajik.
109 C3 **Isola di Capo Rizzuto** Italy
110 C2 **Isperih** Bulg.
80 B2 **Israel** *country* Asia
105 C2 **Issoire** France
111 C2 **İstanbul** Turkey
İstanbul Boğazı *str.* Turkey *see* **Bosporus**
111 B3 **Istiaia** Greece
105 C3 **Istres** France
108 B1 **Istria** *pen.* Croatia
151 D3 **Itaberaba** Brazil
155 D1 **Itabira** Brazil
155 D2 **Itabirito** Brazil
151 E3 **Itabuna** Brazil
150 C2 **Itacoatiara** Brazil
154 B2 **Itaguajé** Brazil
154 C2 **Itaí** Brazil
154 B3 **Itaipu, Represa de** *resr* Brazil
151 C2 **Itaituba** Brazil
152 D3 **Itajaí** Brazil
155 C2 **Itajubá** Brazil
108 B2 **Italy** *country* Europe
155 D1 **Itamarandiba** Brazil
155 D1 **Itambacuri** Brazil
155 D1 **Itambé, Pico de** *mt.* Brazil
62 A1 **Itanagar** India
154 C2 **Itanhaém** Brazil
155 D1 **Itanhém** Brazil
155 D1 **Itaobim** Brazil
154 C1 **Itapajipe** Brazil
155 E1 **Itapebi** Brazil
151 D2 **Itapecuru Mirim** Brazil
155 D2 **Itapemirim** Brazil
155 D2 **Itaperuna** Brazil
155 D1 **Itapetinga** Brazil
154 C2 **Itapetininga** Brazil
154 C2 **Itapeva** Brazil
151 E3 **Itapicuru** *r.* Brazil
154 C1 **Itapuranga** Brazil
154 C2 **Itararé** Brazil
75 B2 **Itarsi** India
154 B1 **Itarumã** Brazil
155 D2 **Itaúna** Brazil
71 C3 **Itbayat** *i.* Phil.
141 D2 **Ithaca** U.S.A.
101 D1 **Ith Hils** *ridge* Ger.
118 C2 **Itimbiri** *r.* Dem. Rep. Congo
155 D1 **Itinga** Brazil
154 B1 **Itiquira** Brazil
154 A1 **Itiquira** *r.* Brazil
67 C4 **Itō** Japan
154 C2 **Itu** Brazil
150 A2 **Ituí** *r.* Brazil
154 C1 **Ituiutaba** Brazil
119 C3 **Itula** Dem. Rep. Congo
154 C1 **Itumbiara** Brazil
154 B1 **Iturama** Brazil
101 D1 **Itzehoe** Ger.
83 N2 **Iul'tin** Russia
154 B2 **Ivaí** *r.* Brazil
92 I2 **Ivalo** Fin.
88 C3 **Ivanava** Belarus
52 B2 **Ivanhoe** Austr.
90 B1 **Ivankiv** Ukr.
90 A2 **Ivano-Frankivs'k** Ukr.
89 F2 **Ivanovo** Russia
88 C3 **Ivatsevichy** Belarus
111 C2 **Ivaylovgrad** Bulg.
86 F2 **Ivdel'** Russia
154 B2 **Ivinheima** *r.* Brazil
154 B2 **Ivinhema** Brazil
127 H2 **Ivittuut** Greenland
Ivory Coast *country* Africa *see* **Côte d'Ivoire**
108 A1 **Ivrea** Italy
111 C3 **İvrindi** Turkey
127 F2 **Ivujivik** Can.
67 D3 **Iwaki** Japan
67 B4 **Iwakuni** Japan
66 D2 **Iwamizawa** Japan
66 D2 **Iwanai** Japan
88 C3 **Iwye** Belarus
123 D3 **Ixopo** S. Africa
144 B2 **Ixtlán** Mex.
145 D2 **Izamal** Mex.
81 C1 **Izberbash** Russia
86 E3 **Izhevsk** Russia
86 E2 **Izhma** Russia
89 E3 **Izmalkovo** Russia
90 B2 **Izmayil** Ukr.
111 C3 **İzmir** Turkey
111 C2 **İznik Gölü** *l.* Turkey
152 B2 **Izozog, Bañados del** *swamp* Bol.
67 B3 **Izumo** Japan
90 B1 **Izyaslav** Ukr.
91 D2 **Izyum** Ukr.

J

Jabal, Bahr el *r.* South Sudan/Uganda *see* **White Nile**
106 C2 **Jabalón** *r.* Spain
75 B2 **Jabalpur** India
50 C1 **Jabiru** Austr.
109 C2 **Jablanica** Bos. & Herz.
151 E2 **Jaboatão dos Guararapes** Brazil
154 C2 **Jaboticabal** Brazil
107 C1 **Jaca** Spain
145 C2 **Jacala** Mex.
151 C2 **Jacareacanga** Brazil
155 C2 **Jacareí** Brazil
155 D1 **Jacinto** Brazil
141 E1 **Jackman** U.S.A.
142 C2 **Jackson** *AL* U.S.A.
140 C2 **Jackson** *MI* U.S.A.
142 B2 **Jackson** *MS* U.S.A.
142 C1 **Jackson** *TN* U.S.A.
136 A2 **Jackson** *WY* U.S.A.
54 A2 **Jackson Head** N.Z.
142 B2 **Jacksonville** *AR* U.S.A.
143 D2 **Jacksonville** *FL* U.S.A.
140 A3 **Jacksonville** *IL* U.S.A.
143 E2 **Jacksonville** *NC* U.S.A.
139 D2 **Jacksonville** *TX* U.S.A.
147 C3 **Jacmel** Haiti
74 A2 **Jacobabad** Pak.
151 D3 **Jacobina** Brazil
131 D2 **Jacques-Cartier, Mont** *mt.* Can.
151 D2 **Jacundá** Brazil
154 C2 **Jacupiranga** Brazil
109 C2 **Jadovnik** *mt.* Bos. & Herz.
150 A2 **Jaén** Peru
106 C2 **Jaén** Spain
52 A3 **Jaffa, Cape** Austr.
73 B4 **Jaffna** Sri Lanka
73 C3 **Jagdalpur** India
123 C2 **Jagersfontein** S. Africa
154 C2 **Jaguariaíva** Brazil
79 C2 **Jahrom** Iran
74 B2 **Jaipur** India
74 B2 **Jaisalmer** India
75 C2 **Jajarkot** Nepal
109 C2 **Jajce** Bos. & Herz.
60 B2 **Jakarta** Indon.
128 A1 **Jakes Corner** Can.
92 G2 **Jäkkvik** Sweden
92 H3 **Jakobstad** Fin.
77 D3 **Jalālābād** Afgh.
77 D2 **Jalal-Abad** Kyrg.
74 B1 **Jalandhar** India
Jalapa Mex. *see* **Xalapa**
154 B2 **Jales** Brazil
74 B2 **Jalgaon** India
115 D4 **Jalingo** Nigeria
74 B3 **Jalna** India
144 B2 **Jalpa** Mex.
75 C2 **Jalpaiguri** India
145 C2 **Jalpan** Mex.
115 E2 **Jālū** Libya
146 C3 **Jamaica** *country* West Indies
146 C3 **Jamaica Channel** Haiti/Jamaica
75 C2 **Jamalpur** Bangl.
60 B2 **Jambi** Indon.
137 D2 **James** *r. ND/SD* U.S.A.
141 D3 **James** *r. VA* U.S.A.
130 B1 **James Bay** Can.
52 A2 **Jamestown** Austr.
137 D1 **Jamestown** *ND* U.S.A.
141 D2 **Jamestown** *NY* U.S.A.
74 B1 **Jammu** India
74 B1 **Jammu and Kashmir** *state* India
74 B2 **Jamnagar** India
93 I3 **Jämsä** Fin.
75 C2 **Jamshedpur** India
75 C2 **Jamuna** *r.* Bangl.
75 C2 **Janakpur** Nepal
155 D1 **Janaúba** Brazil
81 D2 **Jandaq** Iran
140 B2 **Janesville** U.S.A.
65 B3 **Jangheung** S. Korea
65 B2 **Jangseong** S. Korea
80 B2 **Janīn** West Bank
114 A3 **Janjanbureh** Gambia
84 F1 **Jan Mayen** *terr.* Arctic Ocean
81 D1 **Jaňňa** Turkm.

122	B3	Jansenville S. Africa
155	D1	Januária Brazil
74	B2	Jaora India
67	C3	Japan country Asia
156	C3	Japan, Sea of N. Pacific Ocean
150	B2	Japurá r. Brazil
154	C1	Jaraguá Brazil
154	B2	Jaraguari Brazil
70	A2	Jarantai China
152	B2	Jardim Brazil
103	D1	Jarocin Pol.
103	E1	Jarosław Pol.
92	F3	Järpen Sweden
150	B3	Jaru Brazil
		Jarud China see Lubei
49	H3	Jarvis Island S. Pacific Ocean
103	E2	Jasło Pol.
128	C2	Jasper Can.
140	B3	Jasper IN U.S.A.
139	E2	Jasper TX U.S.A.
103	D2	Jastrzębie-Zdrój Pol.
103	D2	Jászberény Hungary
154	B1	Jataí Brazil
74	A2	Jati Pak.
154	C2	Jaú Brazil
150	B2	Jaú r. Brazil
145	C2	Jaumave Mex.
75	C2	Jaunpur India
154	B1	Jauru Brazil
61	B2	Java i. Indon.
		Jawa i. Indon. see Java
		Jawa, Laut Indon. see Java
		Jawa, Laut sea Indon. see Java
159	D4	Jawa, Laut sea Indon.
117	C4	Jawhar Somalia
103	D1	Jawor Pol.
103	D1	Jaworzno Pol.
59	D3	Jaya, Puncak mt. Indon.
59	D3	Jayapura Indon.
78	B3	Jāzān Saudi Arabia
79	C2	Jaz Mūrīān, Hāmūn-e imp. l. Iran
128	B1	Jean Marie River Can.
131	D1	Jeannin, Lac l. Can.
116	A3	Jebel Abyad Plateau Sudan
65	B2	Jecheon S. Korea
96	C3	Jedburgh U.K.
78	A2	Jeddah Saudi Arabia
101	E1	Jeetze r. Ger.
135	C3	Jefferson, Mount U.S.A.
137	E3	Jefferson City U.S.A.
65	B3	Jeju S. Korea
65	B3	Jeju-do i. S. Korea
65	B3	Jeju-haehyeop sea chan. S. Korea
88	C2	Jēkabpils Latvia
103	D1	Jelenia Góra Pol.
88	B2	Jelgava Latvia
62	B2	Jember Indon.
101	E2	Jena Ger.
		Jengish Chokusu mt. China/ Kyrg. see Pobeda Peak
142	B2	Jennings U.S.A.
65	B2	Jeonju S. Korea
155	D1	Jequié Brazil
155	D1	Jequitaí Brazil
155	D1	Jequitinhonha Brazil
155	E1	Jequitinhonha r. Brazil
147	C3	Jérémie Haiti
144	B2	Jerez Mex.
106	B2	Jerez de la Frontera Spain
109	D3	Jergucat Albania
115	C1	Jerid, Chott el salt l. Tunisia
95	C4	Jerome U.S.A.
151	D2	Jersey terr. Channel Is
80	B2	Jerusalem Israel/West Bank

53	D3	Jervis Bay Territory admin. div. Austr.
108	B1	Jesenice Slovenia
108	B2	Jesi Italy
101	F2	Jessen (Elster) Ger.
75	C2	Jessore Bangl.
143	D2	Jesup U.S.A.
145	C3	Jesús Carranza Mex.
109	C2	Jezercë, Maja mt. Albania
74	B2	Jhalawar India
74	B1	Jhang Pak.
75	B2	Jhansi India
75	C2	Jharsuguda India
74	B1	Jhelum Pak.
69	E1	Jiading China
71	B3	Jiamusi China
65	B1	Ji'an Jiangxi China
62	A1	Ji'an Jilin China
		Jianchuan China
		Jiandaoyu China see Guojiaba
70	B2	Jiangsu prov. China
71	B3	Jiangxi prov. China
70	A2	Jiangyou China
70	B3	Jianli China
70	B3	Jianqiao China
71	B3	Jianyang Fujian China
70	A2	Jianyang Sichuan China
70	C2	Jiaozhou China
70	B2	Jiaozuo China
70	C2	Jiaxing China
71	C3	Jiayi Taiwan
68	C2	Jiayuguan China
		Jiddah Saudi Arabia see Jeddah
92	G2	Jiehkkevárri mt. Norway
70	B2	Jiexiu China
68	C2	Jigzhi China
103	D2	Jihlava Czech Rep.
117	C4	Jijiga Eth.
116	A2	Jilf al Kabīr, Hadabat al plat. Egypt
117	C4	Jilib Somalia
69	E1	Jilin China
65	B1	Jilin China
65	A1	Jilin Hada Ling mts China
117	B4	Jīma Eth.
144	B2	Jiménez Chihuahua Mex.
145	C2	Jiménez Tamaulipas Mex.
70	B2	Jinan China
70	B2	Jincheng China
53	C3	Jindabyne Austr.
65	B3	Jin-do i. S. Korea
103	D2	Jindřichův Hradec Czech Rep.
71	B3	Jingdezhen China
70	B2	Jinghang Yunhe canal China
62	B1	Jinghong China
70	B2	Jingmen China
70	A2	Jingning China
70	A2	Jingtai China
71	A3	Jingxi China
65	B1	Jingyu China
70	B2	Jingyuan China
70	B2	Jingzhou China
65	B2	Jinhae S. Korea
71	B3	Jinhua China
70	B2	Jining China
119	D2	Jinja Uganda
117	B4	Jinka Eth.
71	B3	Jinmen Taiwan
146	B3	Jinotepe Nic.
71	A3	Jinping China
		Jinsha Jiang r. China see Yangtze
70	B3	Jinshi China
70	B2	Jinzhong China
70	C1	Jinzhou China
150	B2	Ji-Paraná r. Brazil

75	C1	Jirang China
65	B2	Jiri-san mt. S. Korea
79	C2	Jīroft Iran
71	A3	Jishou China
110	B2	Jiu r. Romania
70	A2	Jiuding Shan mt. China
70	B3	Jiujiang China
66	B1	Jixi China
77	C2	Jizzax Uzbek.
151	E2	João Pessoa Brazil
155	C1	João Pinheiro Brazil
74	B2	Jodhpur India
92	I3	Joensuu Fin.
67	C3	Jōetsu Japan
121	C3	Jofane Moz.
88	C2	Jõgeva Estonia
123	C2	Johannesburg S. Africa
134	C2	John Day U.S.A.
134	B1	John Day r. U.S.A.
128	C2	John D'Or Prairie Can.
143	E1	John H. Kerr Reservoir U.S.A.
96	C1	John o' Groats U.K.
143	D1	Johnson City U.S.A.
128	A1	Johnson's Crossing Can.
49	H1	Johnston Atoll N. Pacific Ocean
96	B3	Johnstone U.K.
141	D2	Johnstown U.S.A.
60	B1	Johor Bahru Malaysia
88	C2	Jõhvi Estonia
154	C3	Joinville Brazil
105	D2	Joinville France
92	G2	Jokkmokk Sweden
92	☐B2	Jökulsá á Fjöllum r. Iceland
140	B2	Joliet U.S.A.
130	C2	Joliette Can.
64	B2	Jolo i. Phil.
64	B2	Jolo i. Phil.
61	C2	Jombang Indon.
75	C2	Jomsom Nepal
88	B2	Jonava Lith.
142	B1	Jonesboro AR U.S.A.
142	B2	Jonesboro LA U.S.A.
127	F1	Jones Sound sea chan. Can.
93	F4	Jönköping Sweden
131	C2	Jonquière Can.
145	C3	Jonuta Mex.
137	E3	Joplin U.S.A.
80	B2	Jordan country Asia
80	B2	Jordan r. Asia
136	B1	Jordan U.S.A.
134	C2	Jordan Valley U.S.A.
75	C2	Jorhat India
93	E4	Jørpeland Norway
115	C4	Jos Nigeria
145	C3	José Cardel Mex.
131	D1	Joseph, Lac l. Can.
50	B1	Joseph Bonaparte Gulf Austr.
115	C4	Jos Plateau Nigeria
93	E3	Jotunheimen mts Norway
122	B3	Joubertina S. Africa
123	C2	Jouberton S. Africa
93	I3	Joutseno Fin.
134	B2	Juan de Fuca Strait Can./ U.S.A.
		Juanshui China see Tongcheng
145	B2	Juárez Mex.
151	D2	Juazeiro Brazil
151	E2	Juazeiro do Norte Brazil
117	B4	Juba South Sudan
117	C4	Jubaland reg. Somalia
117	C5	Jubba r. Somalia
78	B2	Jubbah Saudi Arabia
145	C3	Juchitán Mex.
102	C2	Judenburg Austria
101	D2	Jühnde Ger.

146 B3 **Juigalpa** Nic.
100 C1 **Juist** *i.* Ger.
155 D2 **Juiz de Fora** Brazil
150 A3 **Juliaca** Peru
75 C2 **Jumla** Nepal
74 B2 **Junagadh** India
139 D2 **Junction** U.S.A.
137 D3 **Junction City** U.S.A.
154 C2 **Jundiaí** Brazil
128 A2 **Juneau** U.S.A.
53 C2 **Junee** Austr.
105 D2 **Jungfrau** *mt.* Switz.
141 D2 **Juniata** *r.* U.S.A.
92 G3 **Junsele** Sweden
134 C2 **Juntura** U.S.A.
154 C2 **Juquiá** Brazil
117 A4 **Jur** *r.* South Sudan
105 D2 **Jura** *mts* France/Switz.
96 B2 **Jura** *i.* U.K.
96 B3 **Jura, Sound of** *sea chan.* U.K.
88 B2 **Jurbarkas** Lith.
88 B2 **Jūrmala** Latvia
150 B2 **Juruá** *r.* Brazil
150 C2 **Juruena** *r.* Brazil
150 B2 **Jutaí** *r.* Brazil
101 F2 **Jüterbog** Ger.
146 B2 **Juventud, Isla de la** *i.* Cuba
70 B2 **Juxian** China
79 C2 **Jūyom** Iran
122 B1 **Jwaneng** Botswana
93 I3 **Jyväskylä** Fin.

K

74 B1 **K2** *mt.* China/Pak.
119 C3 **Kabalo** Dem. Rep. Congo
119 C3 **Kabare** Dem. Rep. Congo
130 B2 **Kabinakagami Lake** Can.
118 C3 **Kabinda** Dem. Rep. Congo
118 B2 **Kabo** C.A.R.
120 B2 **Kabompo** Zambia
119 C3 **Kabongo** Dem. Rep. Congo
77 C3 **Kābul** Afgh.
64 B2 **Kaburuang** *i.* Indon.
121 B2 **Kabwe** Zambia
74 A2 **Kachchh, Gulf of** India
74 B2 **Kachchh, Rann of** *marsh* India
83 I3 **Kachug** Russia
81 C1 **Kaçkar Dağı** *mt.* Turkey
73 B3 **Kadapa** India
111 C2 **Kadıköy** Turkey
114 B3 **Kadiolo** Mali
73 B3 **Kadmat** *atoll* India
121 B2 **Kadoma** Zimbabwe
63 A2 **Kadonkani** Myanmar
117 A3 **Kadugli** Sudan
115 C3 **Kaduna** Nigeria
89 E2 **Kaduy** Russia
86 E2 **Kadzherom** Russia
114 A3 **Kaédi** Maur.
118 B1 **Kaélé** Cameroon
65 B2 **Kaesŏng** N. Korea
114 A3 **Kaffrine** Senegal
121 B2 **Kafue** Zambia
120 B2 **Kafue** *r.* Zambia
118 B2 **Kaga Bandoro** C.A.R.
91 E2 **Kagal'nitskaya** Russia
67 B4 **Kagoshima** Japan
90 C2 **Kaharlyk** Ukr.
61 C2 **Kahayan** *r.* Indon.
118 B3 **Kahemba** Dem. Rep. Congo
101 E2 **Kahla** Ger.
79 C2 **Kahnūj** Iran
92 H2 **Kahperusvaarat** *mts* Fin.
80 B2 **Kahramanmaraş** Turkey
79 C2 **Kahūrak** Iran
59 C3 **Kai, Kepulauan** *is* Indon.
115 C4 **Kaiama** Nigeria
54 B2 **Kaiapoi** N.Z.
59 C3 **Kai Besar** *i.* Indon.
70 B2 **Kaifeng** China
Kaihua China *see* **Wenshan**
122 B2 **Kaiingveld** *reg.* S. Africa
59 C3 **Kai Kecil** *i.* Indon.
54 B2 **Kaikoura** N.Z.
114 A4 **Kailahun** Sierra Leone
71 A3 **Kaili** China
59 C3 **Kaimana** Indon.
54 C1 **Kaimanawa Mountains** N.Z.
67 C4 **Kainan** Japan
115 C3 **Kainji Reservoir** Nigeria
54 B1 **Kaipara Harbour** N.Z.
74 B2 **Kairana** India
115 D1 **Kairouan** Tunisia
54 B1 **Kaitaia** N.Z.
54 C1 **Kaitawa** N.Z.
Kaitong China *see* **Tongyu**
59 C3 **Kaiwatu** Indon.
65 A1 **Kaiyuan** *Liaoning* China
62 B1 **Kaiyuan** *Yunnan* China
92 I3 **Kajaani** Fin.
51 D2 **Kajabbi** Austr.
122 B2 **Kakamas** S. Africa
119 D2 **Kakamega** Kenya
91 C2 **Kakhovka** Ukr.
91 C2 **Kakhovs'ke Vodoskhovyshche** *resr* Ukr.
73 C3 **Kakinada** India
128 C1 **Kakisa** Can.
119 C3 **Kakoswa** Dem. Rep. Congo
126 B2 **Kaktovik** U.S.A.
Kalaallit Nunaat *terr.* N. America *see* **Greenland**
59 C3 **Kalabahi** Indon.
120 B2 **Kalabo** Zambia
73 B3 **Kalaburagi** India
91 E1 **Kalach** Russia
119 D2 **Kalacha Dida** Kenya
62 A1 **Kaladan** *r.* India/Myanmar
120 B3 **Kalahari Desert** Africa
92 H3 **Kalajoki** Fin.
111 B2 **Kalamaria** Greece
111 B3 **Kalamata** Greece
140 B2 **Kalamazoo** U.S.A.
111 B3 **Kalampaka** Greece
91 C2 **Kalanchak** Ukr.
115 E1 **Kalanshiyū ar Ramlī al Kabīr, Sarīr** *des.* Libya
63 B2 **Kalasin** Thai.
79 C2 **Kalāt** Iran
74 A2 **Kalat** Pak.
50 A2 **Kalbarri** Austr.
Kalburgi India *see* **Gulbarga**
111 C3 **Kale** Turkey
80 B1 **Kalecik** Turkey
118 C3 **Kalema** Dem. Rep. Congo
119 C3 **Kalemie** Dem. Rep. Congo
62 A1 **Kalemyo** Myanmar
Kalgan China *see* **Zhangjiakou**
50 B3 **Kalgoorlie** Austr.
109 C2 **Kali** Croatia
110 C2 **Kaliakra, Nos** *pt* Bulg.
119 C3 **Kalima** Dem. Rep. Congo
61 C2 **Kalimantan** *reg.* Indon.
88 B3 **Kaliningrad** Russia
91 D2 **Kalininskaya** Russia
88 C3 **Kalinkavichy** Belarus
134 D1 **Kalispell** U.S.A.
103 D1 **Kalisz** Pol.
91 E2 **Kalitva** *r.* Russia
92 H2 **Kalix** Sweden
92 H2 **Kalixälven** *r.* Sweden
111 C3 **Kalkan** Turkey
120 A3 **Kalkfeld** Namibia
92 I3 **Kallavesi** *l.* Fin.
92 F3 **Kallsjön** *l.* Sweden
93 G4 **Kalmar** Sweden
93 G4 **Kalmarsund** *sea chan.* Sweden
120 B2 **Kalomo** Zambia
128 B2 **Kalone Peak** Can.
75 B1 **Kalpa** India
73 B3 **Kalpeni** *atoll* India
75 B2 **Kalpi** India
101 D1 **Kaltenkirchen** Ger.
89 E3 **Kaluga** Russia
93 F4 **Kalundborg** Denmark
90 A2 **Kalush** Ukr.
89 E2 **Kalyazin** Russia
111 C3 **Kalymnos** *i.* Greece
119 C3 **Kama** Dem. Rep. Congo
62 A2 **Kama** Myanmar
86 E3 **Kama** *r.* Russia
66 D3 **Kamaishi** Japan
80 B2 **Kaman** Turkey
120 A2 **Kamanjab** Namibia
78 B3 **Kamarān** *i.* Yemen
74 A2 **Kamarod** Pak.
50 B3 **Kambalda** Austr.
119 C4 **Kambove** Dem. Rep. Congo
183 L3 **Kamchatka Peninsula** Russia
110 C2 **Kamchia** *r.* Bulg.
110 B2 **Kamenitsa** *mt.* Bulg.
108 B2 **Kamenjak, Rt** *pt* Croatia
91 D1 **Kamenka** Russia
91 E2 **Kamenolomni** Russia
83 M2 **Kamenskoye** Russia
91 E2 **Kamensk-Shakhtinskiy** Russia
86 F3 **Kamensk-Ural'skiy** Russia
122 A3 **Kamiesberge** *mts* S. Africa
122 A3 **Kamieskroon** S. Africa
129 D1 **Kamilukuak Lake** Can.
119 C3 **Kamina** Dem. Rep. Congo
129 E1 **Kaminak Lake** Can.
90 A1 **Kamin'-Kashyrs'kyy** Ukr.
128 B2 **Kamloops** Can.
118 C3 **Kamonia** Dem. Rep. Congo
119 D2 **Kampala** Uganda
60 B1 **Kampar** *r.* Indon.
60 B1 **Kampar** Malaysia
100 B1 **Kampen** Neth.
119 C3 **Kampene** Dem. Rep. Congo
63 A2 **Kamphaeng Phet** Thai.
63 B2 **Kâmpóng Cham** Cambodia
63 B2 **Kâmpóng Chhnăng** Cambodia
Kâmpóng Saôm Cambodia *see* **Sihanoukville**
63 B2 **Kâmpóng Spœ** Cambodia
63 B2 **Kâmpôt** Cambodia
129 D2 **Kamsack** Can.
86 E3 **Kamskoye Vodokhranilishche** *resr* Russia
117 C4 **Kamsuuma** Somalia
90 B2 **Kam"yanets'-Podil's'kyy** Ukr.
90 A1 **Kam"yanka-Buz'ka** Ukr.
88 B3 **Kamyanyets** Belarus
91 D2 **Kamyshevatskaya** Russia
87 D3 **Kamyshin** Russia
135 D3 **Kanab** U.S.A.
118 C3 **Kananga** Dem. Rep. Congo
87 D3 **Kanash** Russia
140 C3 **Kanawha** *r.* U.S.A.
67 C3 **Kanazawa** Japan
62 A1 **Kanbalu** Myanmar
63 A2 **Kanchanaburi** Thai.
73 B3 **Kanchipuram** India
77 C3 **Kandahār** Afgh.
86 C2 **Kandalaksha** Russia
61 C2 **Kandangan** Indon.
74 A2 **Kandh Kot** Pak.
114 C3 **Kandi** Benin
74 B2 **Kandla** India

53 C2 Kandos Austr.
121 □D2 Kandreho Madag.
73 C4 Kandy Sri Lanka
76 B2 Kandyagash Kazakh.
127 G1 Kane Bassin *b.* Greenland
91 D2 Kanevskaya Russia
122 B1 Kang Bhutan
127 H2 Kangaatsiaq Greenland
114 B3 Kangaba Mali
80 B2 Kangal Turkey
60 B1 Kangar Malaysia
52 A3 Kangaroo Island Austr.
75 C2 Kangchenjunga *mt.* India/
       Nepal
65 B2 Kangdong N. Korea
61 C2 Kangean, Kepulauan *is* Indon.
127 I2 Kangeq *c.* Greenland
127 H2 Kangerlussuaq *inlet* Greenland
127 I2 Kangerlussuaq *inlet* Greenland
127 H2 Kangersuatsiaq Greenland
65 B1 Kanggye N. Korea
131 D1 Kangiqsualujjuaq Can.
127 G2 Kangiqsujuaq Can.
127 G2 Kangirsuk Can.
75 C2 Kangmar China
65 A1 Kangping China
72 D2 Kangto *mt.* China/India
62 A1 Kani Myanmar
86 D2 Kanin, Poluostrov *pen.* Russia
86 D2 Kanin Nos Russia
91 C2 Kaniv Ukr.
93 H3 Kankaanpää Fin.
140 B2 Kankakee U.S.A.
114 B3 Kankan Guinea
75 C2 Kanker India
73 B3 Kannur India
115 C3 Kano Nigeria
122 B3 Kanonpunt *pt* S. Africa
67 B4 Kanoya Japan
75 C2 Kanpur India
136 C2 Kansas *r.* U.S.A.
137 D3 Kansas *state* U.S.A.
137 E3 Kansas City U.S.A.
83 H3 Kansk Russia
114 C3 Kantchari Burkina Faso
91 D2 Kantemirovka Russia
97 B2 Kanturk Ireland
123 D2 Kanyamazane S. Africa
123 C1 Kanye Botswana
120 A2 Kaokoveld *plat.* Namibia
114 A3 Kaolack Senegal
119 C2 Kaoma Zambia
118 C3 Kapanga Dem. Rep. Congo
100 B2 Kapellen Belgium
121 B2 Kapiri Mposhi Zambia
127 H2 Kapisillit Greenland
130 B1 Kapiskau *r.* Can.
61 C1 Kapit *Sarawak* Malaysia
63 A3 Kapoe Thai.
81 B4 Kapoeta South Sudan
103 D2 Kaposvár Hungary
77 D2 Kapshagay Kazakh.
77 D2 Kapshagay,
       Vodokhranilishche *resr*
       Kazakh.
61 B2 Kapuas *r.* Indon.
52 A2 Kapunda Austr.
130 B2 Kapuskasing Can.
103 D2 Kapuvár Hungary
88 C3 Kapyl' Belarus
114 C4 Kara Togo
111 C3 Kara Ada *i.* Turkey
76 C1 Karabalyk Kazakh.
80 B1 Karabük Turkey
76 C2 Karabutak Kazakh.
110 C2 Karacaköy Turkey
89 D3 Karachev Russia
74 A2 Karachi Pak.

       Kara Deniz *sea* Asia/Europe
       see Black Sea
77 D2 Karagandy Kazakh.
77 D2 Karagayly Kazakh.
83 L3 Karaginskiy Zaliv *b.* Russia
81 D2 Karaj Iran
64 B2 Karakelong *i.* Indon.
77 D2 Kara-Köl Kyrg.
77 D2 Karakol Kyrg.
74 B1 Karakoram Range *mts* Asia
       Karakum, Peski *des.* Kazakh.
       see Karakum Desert
76 B2 Karakum Desert Kazakh.
76 C3 Karakum Desert Turkm.
       Karakumy, Peski *des.* Turkm.
       see Karakum Desert
80 B2 Karaman Turkey
77 E2 Karamay China
54 B2 Karamea N.Z.
54 B2 Karamea Bight *b.* N.Z.
80 B2 Karapınar Turkey
122 A2 Karasburg Namibia
86 F1 Kara Sea Russia
92 I2 Karasjok Norway
82 G3 Karasuk Russia
77 D2 Karatau Kazakh.
77 C2 Karatau, Khrebet *mts* Kazakh.
86 F2 Karatayka Russia
67 A4 Karatsu Japan
60 B2 Karawang Indon.
81 C2 Karbalā' Iraq
103 E2 Karcag Hungary
111 B3 Karditsa Greece
88 B2 Kärdla Estonia
110 C2 Kardzhali Bulg.
122 B3 Kareeberge *mts* S. Africa
75 B2 Kareli India
74 B1 Kargil India
121 B2 Kariba Zimbabwe
121 B2 Kariba, Lake *resr* Zambia/
       Zimbabwe
61 B2 Karimata, Pulau-pulau *is*
       Indon.
61 B2 Karimata, Selat *str.* Indon.
73 B3 Karimnagar India
61 C2 Karimunjawa, Pulau-pulau *is*
       Indon.
91 C2 Karkinits'ka Zatoka *g.* Ukr.
91 D2 Karlivka Ukr.
109 C1 Karlovac Croatia
102 C1 Karlovy Vary Czech Rep.
93 F4 Karlshamn Sweden
93 G4 Karlskrona Sweden
102 B2 Karlsruhe Ger.
93 F4 Karlstad Sweden
101 D3 Karlstadt Ger.
89 D3 Karma Belarus
93 E4 Karmøy *i.* Norway
74 B2 Karnal India
75 D2 Karnaphuli Reservoir Bangl.
110 C2 Karnobat Bulg.
74 A2 Karodi Pak.
121 B2 Karoi Zimbabwe
121 C1 Karonga Malawi
116 B3 Karora Eritrea
111 C3 Karpathos *i.* Greece
111 B3 Karpenisi Greece
86 D2 Karpogory Russia
50 A2 Karratha Austr.
81 C1 Kars Turkey
88 C2 Kärsava Latvia
86 E2 Karskiye Vorota, Proliv *str.*
       Russia
       Karskoye More *sea* Russia *see*
       Kara Sea
101 E1 Karstädt Ger.
111 C2 Kartal Turkey
87 F3 Kartaly Russia

73 B3 Karwar India
83 I3 Karymskoye Russia
111 B3 Karystos Greece
80 A2 Kaş Turkey
130 B1 Kasabonika Lake Can.
118 C4 Kasaji Dem. Rep. Congo
121 C2 Kasama Zambia
120 B2 Kasane Botswana
118 B3 Kasangulu Dem. Rep. Congo
73 B3 Kasaragod India
129 C1 Kasba Lake Can.
120 B2 Kasempa Zambia
119 C4 Kasenga Dem. Rep. Congo
119 C3 Kasese Dem. Rep. Congo
119 D2 Kasese Uganda
81 D2 Kāshān Iran
       Kashgar China *see* Kashi
77 D3 Kashi China
67 D3 Kashima-nada *b.* Japan
89 E2 Kashin Russia
89 E3 Kashira Russia
89 E3 Kashirskoye Russia
67 C3 Kashiwazaki Japan
76 B3 Kāshmar Iran
74 B1 Kashmir *reg.* Asia
74 A2 Kashmore Pak.
119 C3 Kashyukulu Dem. Rep. Congo
87 D3 Kasimov Russia
93 H3 Kaskinen Fin.
119 C3 Kasongo Dem. Rep. Congo
118 B3 Kasongo-Lunda
       Dem. Rep. Congo
111 C3 Kasos *i.* Greece
       Kaspiyskoye More *l.* Asia/
       Europe *see* Caspian Sea
116 B3 Kassala Sudan
101 D2 Kassel Ger.
80 B1 Kastamonu Turkey
111 B2 Kastoria Greece
89 D3 Kastsyukovichy Belarus
119 D3 Kasulu Tanz.
121 C2 Kasungu Malawi
141 F1 Katahdin, Mount U.S.A.
118 C3 Katako-Kombe
       Dem. Rep. Congo
50 A3 Katanning Austr.
63 A3 Katchall *i.* India
111 B2 Katerini Greece
128 A2 Kate's Needle *mt.* Can./U.S.A.
121 C2 Katete Zambia
62 A1 Katha Myanmar
50 C1 Katherine *r.* Austr.
74 B2 Kathiawar *pen.* India
75 C2 Kathmandu Nepal
122 B2 Kathu S. Africa
74 B1 Kathua India
114 B3 Kati Mali
75 C2 Katihar India
54 C1 Katikati N.Z.
123 C3 Katikati S. Africa
120 B2 Katima Mulilo Namibia
52 A1 Kati Thanda-Lake Eyre
       (North) *imp. l.* Austr.
52 A1 Kati Thanda-Lake Eyre
       (South) *imp. l.* Austr.
123 C2 Katlehong S. Africa
75 C2 Katni India
111 B3 Kato Achaïa Greece
53 D2 Katoomba Austr.
103 D1 Katowice Pol.
75 C2 Kātrīnā, Jabal *mt.* Egypt
93 G4 Katrineholm Sweden
115 C3 Katsina Nigeria
115 C4 Katsina-Ala Nigeria
77 C3 Kattaqo'rg'on Uzbek.
93 F4 Kattegat *str.* Denmark/
       Sweden
100 B1 Katwijk aan Zee Neth.

124 E5 **Kaua'i** *i.* U.S.A.
93 H3 **Kauhajoki** Fin.
88 B3 **Kaunas** Lith.
115 C3 **Kaura-Namoda** Nigeria
92 H2 **Kautokeino** Norway
111 B2 **Kavala** Greece
66 C2 **Kavalerovo** Russia
73 C3 **Kavali** India
73 B3 **Kavaratti** *atoll* India
110 C2 **Kavarna** Bulg.
81 D2 **Kavīr, Dasht-e** *des.* Iran
67 C3 **Kawagoe** Japan
54 B1 **Kawakawa** N.Z.
67 C3 **Kawanishi** Japan
130 C2 **Kawartha Lakes** Can.
67 C3 **Kawasaki** Japan
54 C1 **Kawerau** N.Z.
63 A2 **Kawkareik** Myanmar
62 A1 **Kawlin** Myanmar
63 A2 **Kawmapyin** Myanmar
63 A2 **Kawthaung** Myanmar
77 D3 **Kaxgar He** *r.* China
114 B3 **Kaya** Burkina Faso
61 C1 **Kayan** *r.* Indon.
136 B2 **Kaycee** U.S.A.
138 A1 **Kayenta** U.S.A.
114 A3 **Kayes** Mali
77 D2 **Kaynar** Kazakh.
80 B2 **Kayseri** Turkey
81 D1 **Kazakhskiy Zaliv** *b.* Kazakh.
76 C2 **Kazakhstan** *country* Asia
87 D3 **Kazan'** Russia
80 A1 **Kazanlak** Bulg.
76 A2 **Kazbek** *mt.* Georgia/Russia
81 D3 **Kāzerūn** Iran
103 E2 **Kazincbarcika** Hungary
86 F2 **Kazym-Mys** Russia
111 B3 **Kea** *i.* Greece
97 C1 **Keady** U.K.
137 D2 **Kearney** U.S.A.
138 A2 **Kearny** U.S.A.
116 A3 **Kebkabiya** Sudan
92 G2 **Kebnekaise** *mt.* Sweden
117 C4 **K'ebrī Dehar** Eth.
61 B2 **Kebumen** Indon.
128 B2 **Kechika** *r.* Can.
103 D2 **Kecskemét** Hungary
88 B2 **Kdainiai** Lith.
114 A3 **Kédougou** Senegal
103 D1 **Kędzierzyn-Koźle** Pol.
128 B1 **Keele** *r.* Can.
128 A1 **Keele Peak** Can.
Keelung Taiwan *see* **Chilung**
141 E2 **Keene** U.S.A.
122 A2 **Keetmanshoop** Namibia
129 E3 **Keewatin** Can.
Kefallonia *i.* Greece *see* **Cephalonia**
59 C3 **Kefamenanu** Indon.
92 □A3 **Keflavík** Iceland
77 D2 **Kegen** Kazakh.
128 C2 **Keg River** Can.
62 A1 **Kehsi Mansam** Myanmar
98 C2 **Keighley** U.K.
88 B2 **Keila** Estonia
122 B2 **Keimoes** S. Africa
92 I3 **Keitele** *l.* Fin.
52 B3 **Keith** Austr.
103 E2 **Kékes** *mt.* Hungary
92 J2 **Kelesuayv, Gora** *h.* Russia
102 C2 **Kelheim** Ger.
80 B1 **Kelkit** *r.* Turkey
128 B1 **Keller Lake** Can.
134 C1 **Kellogg** U.S.A.
92 I2 **Kelloselkä** Fin.
97 C2 **Kells** Ireland
88 B2 **Kelm** Lith.
115 D4 **Kélo** Chad
128 C3 **Kelowna** Can.
96 C3 **Kelso** U.K.
134 B1 **Kelso** U.S.A.
60 B1 **Keluang** Malaysia
129 D2 **Kelvington** Can.
86 C2 **Kem'** Russia
128 B2 **Kemano (abandoned)** Can.
111 C3 **Kemer** Turkey
82 G3 **Kemerovo** Russia
92 H2 **Kemi** Fin.
92 I2 **Kemijärvi** Fin.
92 I2 **Kemijoki** *r.* Fin.
136 A2 **Kemmerer** U.S.A.
92 I3 **Kempele** Fin.
53 D2 **Kempsey** Austr.
130 C2 **Kempt, Lac** *l.* Can.
102 C2 **Kempten (Allgäu)** Ger.
123 C2 **Kempton Park** S. Africa
61 C2 **Kemujan** *i.* Indon.
129 D2 **Kenaston** Can.
98 B1 **Kendal** U.K.
59 C3 **Kendari** Indon.
61 C2 **Kendawangan** Indon.
115 D3 **Kendégué** Chad
75 C2 **Kendujhargarh** India
114 A4 **Kenema** Sierra Leone
118 B3 **Kenge** Dem. Rep. Congo
62 A1 **Kengtung** Myanmar
122 B2 **Kenhardt** S. Africa
114 B1 **Kenitra** Morocco
97 B3 **Kenmare** Ireland
136 C1 **Kenmare** U.S.A.
97 A3 **Kenmare River** *inlet* Ireland
100 C3 **Kenn** Ger.
137 E3 **Kennett** U.S.A.
134 C1 **Kennewick** U.S.A.
129 E3 **Kenora** Can.
140 B2 **Kenosha** U.S.A.
138 C2 **Kent** U.S.A.
77 C2 **Kentau** Kazakh.
140 B3 **Kentucky** *r.* U.S.A.
140 C3 **Kentucky** *state* U.S.A.
140 B3 **Kentucky Lake** U.S.A.
119 D2 **Kenya** *country* Africa
119 D3 **Kenya, Mount** Kenya
137 E2 **Keokuk** U.S.A.
111 C3 **Kepsut** Turkey
52 B3 **Kerang** Austr.
91 D2 **Kerch** Ukr.
59 D3 **Kerema** P.N.G.
128 C3 **Keremeos** Can.
116 B3 **Keren** Eritrea
81 C2 **Kerend-e Gharb** Iran
Kerepakupai Merú Venez. *see* **Angel Falls**
159 C7 **Kerguelen Plateau** Indian Ocean
119 D3 **Kericho** Kenya
54 B1 **Kerikeri** N.Z.
60 B2 **Kerinci, Gunung** *vol.* Indon.
111 A3 **Kerkyra** Greece
Kerkyra *i.* Greece *see* **Corfu**
116 B3 **Kerma** Sudan
49 G4 **Kermadec Islands** S. Pacific Ocean
79 C1 **Kermān** Iran
81 C2 **Kermānshāh** Iran
139 C2 **Kermit** U.S.A.
135 C3 **Kern** *r.* U.S.A.
100 C2 **Kerpen** Ger.
129 D2 **Kerrobert** Can.
139 D2 **Kerrville** U.S.A.
Keryneia Cyprus *see* **Kyrenia**
130 B1 **Kesagami Lake** Can.
111 C2 **Keşan** Turkey
66 D3 **Kesennuma** Japan
74 B2 **Keshod** India
100 C2 **Kessel** Neth.
98 B1 **Keswick** U.K.
103 D2 **Keszthely** Hungary
82 G3 **Ket'** *r.* Russia
61 C2 **Ketapang** Indon.
128 A2 **Ketchikan** U.S.A.
134 D2 **Ketchum** U.S.A.
99 C2 **Kettering** U.K.
93 H3 **Keuruu** Fin.
100 C2 **Kevelaer** Ger.
140 B2 **Kewanee** U.S.A.
140 B1 **Keweenaw Bay** U.S.A.
140 B1 **Keweenaw Peninsula** U.S.A.
143 D3 **Key Largo** U.S.A.
141 D3 **Keyser** U.S.A.
143 D4 **Key West** U.S.A.
123 C2 **Kgotsong** S. Africa
69 E1 **Khabarovsk** Russia
80 B1 **Khadyzhensk** Russia
62 A1 **Khagrachari** Bangl.
74 A2 **Khairpur** Pak.
122 B1 **Khakhea** Botswana
68 C1 **Khamar-Daban, Khrebet** *mts* Russia
74 B2 **Khambhat** India
73 B3 **Khambhat, Gulf of** India
74 B2 **Khamgaon** India
78 B3 **Khamis Mushayţ** Saudi Arabia
78 B3 **Khamr** Yemen
77 C3 **Khānābād** Afgh.
74 B2 **Khandwa** India
83 K2 **Khandyga** Russia
74 B1 **Khanewal** Pak.
66 B2 **Khanka, Lake** China/Russia
74 B2 **Khanpur** Pak.
77 D2 **Khantau** Kazakh.
83 H2 **Khantayskoye, Ozero** *l.* Russia
86 F2 **Khanty-Mansiysk** Russia
63 A3 **Khao Chum Thong** Thai.
63 A2 **Khao Laem, Ang Kep Nam** Thai.
87 D4 **Kharabali** Russia
75 C2 **Kharagpur** India
81 D3 **Kharāmeh** Iran
79 C2 **Khārān** *r.* Iran
116 B2 **Khārijah, Wāḩāt al** *oasis* Egypt
91 D2 **Kharkiv** Ukr.
Khar'kov Ukr. *see* **Kharkiv**
86 D3 **Kharovsk** Russia
116 B3 **Khartoum** Sudan
87 D4 **Khasavyurt** Russia
79 D2 **Khāsh** Iran
78 A3 **Khashm el Girba** Sudan
78 A3 **Khashm el Girba Dam** Sudan
74 D2 **Khasi Hills** India
123 C3 **Khayamnandi** S. Africa
78 A2 **Khaybar** Saudi Arabia
122 A3 **Khayelitsha** S. Africa
107 D2 **Khemis Miliana** Alg.
63 B2 **Khemmarat** Thai.
115 C1 **Khenchela** Alg.
91 C2 **Kherson** Ukr.
83 H2 **Kheta** *r.* Russia
69 D1 **Khilok** Russia
89 E2 **Khimki** Russia
74 A2 **Khipro** Pak.
89 E3 **Khlevnoye** Russia
90 B2 **Khmel'nyts'kyy** Ukr.
89 E3 **Khokhol'skiy** Russia
74 B2 **Khokhropar** Pak.
89 D2 **Kholm** Russia
89 D2 **Kholm-Zhirkovskiy** Russia
79 C2 **Khonj** Iran
63 B2 **Khon Kaen** Thai.
62 A1 **Khonsa** India
83 K2 **Khonuu** Russia

86 E2 **Khorey-Ver** Russia
69 D1 **Khorinsk** Russia
91 C2 **Khorol** Ukr.
81 C2 **Khorramābād** Iran
81 C2 **Khorramshahr** Iran
77 D3 **Khorugh** Tajik.
74 A1 **Khōst** Afgh.
62 A1 **Khreum** Myanmar
76 B1 **Khromtau** Kazakh.
90 B2 **Khrystynivka** Ukr.
77 C2 **Khŭjand** Tajik.
63 B2 **Khu Khan** Thai.
78 A2 **Khulays** Saudi Arabia
75 C2 **Khulna** Bangl.
79 B2 **Khurayş** Saudi Arabia
74 B1 **Khushab** Pak.
90 A2 **Khust** Ukr.
123 C2 **Khutsong** S. Africa
74 A2 **Khuzdar** Pak.
81 D2 **Khvānsār** Iran
79 C2 **Khvormūj** Iran
81 C2 **Khvoy** Iran
89 D2 **Khvoynaya** Russia
77 D3 **Khyber Pass** Afgh./Pak.
53 D2 **Kiama** Austr.
119 C3 **Kiambi** Dem. Rep. Congo
119 D3 **Kibiti** Tanz.
109 D2 **Kičevo** Macedonia
114 C3 **Kidal** Mali
99 B2 **Kidderminster** U.K.
114 A3 **Kidira** Senegal
54 C1 **Kidnappers, Cape** N.Z.
102 C1 **Kiel** Ger.
103 E1 **Kielce** Pol.
98 B1 **Kielder Water** U.K.
90 C1 **Kiev** Ukr.
114 A3 **Kiffa** Maur.
119 D3 **Kigali** Rwanda
119 C3 **Kigoma** Tanz.
88 B2 **Kihnu** *i.* Estonia
92 I2 **Kiiminki** Fin.
67 B4 **Kii-suidō** *sea chan.* Japan
109 D1 **Kikinda** Serbia
119 C3 **Kikondja** Dem. Rep. Congo
59 D3 **Kikori** P.N.G.
59 D3 **Kikori** *r.* P.N.G.
118 B3 **Kikwit** Dem. Rep. Congo
65 B1 **Kilchu** N. Korea
118 B3 **Kilembe** Dem. Rep. Congo
139 E2 **Kilgore** U.S.A.
119 D3 **Kilimanjaro** *vol.* Tanz.
80 B2 **Kilis** Turkey
90 B2 **Kiliya** Ukr.
97 B2 **Kilkee** Ireland
97 D1 **Kilkeel** U.K.
97 C2 **Kilkenny** Ireland
111 B2 **Kilkis** Greece
97 B1 **Killala Bay** Ireland
97 B2 **Killaloe** Ireland
129 C2 **Killam** Can.
97 B2 **Killarney** Ireland
139 D2 **Killeen** U.S.A.
96 B2 **Killin** U.K.
97 B1 **Killybegs** Ireland
96 B3 **Kilmarnock** U.K.
53 B3 **Kilmore** Austr.
119 D3 **Kilosa** Tanz.
97 B2 **Kilrush** Ireland
119 C3 **Kilwa** Dem. Rep. Congo
119 D3 **Kilwa Masoko** Tanz.
119 D3 **Kimambi** Tanz.
52 A2 **Kimba** Austr.
136 C2 **Kimball** U.S.A.
128 C3 **Kimberley** Can.
122 B2 **Kimberley** S. Africa
50 B1 **Kimberley Plateau** Austr.
65 B1 **Kimch'aek** N. Korea
127 G2 **Kimmirut** Can.
89 E3 **Kimovsk** Russia
118 B3 **Kimpese** Dem. Rep. Congo
89 E2 **Kimry** Russia
61 C1 **Kinabalu, Gunung** *mt. Sabah* Malaysia
128 C2 **Kinbasket Lake** Can.
130 B2 **Kincardine** Can.
62 A1 **Kinchang** Myanmar
119 C3 **Kinda** Dem. Rep. Congo
98 C2 **Kinder Scout** *h.* U.K.
129 D2 **Kindersley** Can.
114 A3 **Kindia** Guinea
119 C3 **Kindu** Dem. Rep. Congo
86 D3 **Kineshma** Russia
78 A2 **King Abdullah Economic City** Saudi Arabia
51 E2 **Kingaroy** Austr.
135 B3 **King City** U.S.A.
130 C1 **King George Islands** Can.
88 C2 **Kingisepp** Russia
51 D3 **King Island** Austr.
50 B1 **King Leopold Ranges** *hills* Austr.
138 A1 **Kingman** U.S.A.
135 B3 **Kings** *r.* U.S.A.
52 A3 **Kingscote** Austr.
99 D2 **King's Lynn** U.K.
48 F3 **Kingsmill Group** *is* Kiribati
50 B1 **King Sound** *b.* Austr.
134 D2 **Kings Peak** U.S.A.
143 D1 **Kingsport** U.S.A.
130 C2 **Kingston** Can.
146 C3 **Kingston** Jamaica
141 E2 **Kingston** U.S.A.
52 A3 **Kingston South East** Austr.
98 C2 **Kingston upon Hull** U.K.
147 D3 **Kingstown** St Vincent
139 D3 **Kingsville** U.S.A.
99 B3 **Kingswood** U.K.
96 B2 **Kingussie** U.K.
126 E2 **King William Island** Can.
123 C3 **King William's Town** S. Africa
67 D3 **Kinka-san** *i.* Japan
93 F4 **Kinna** Sweden
97 B3 **Kinsale** Ireland
118 B3 **Kinshasa** Dem. Rep. Congo
143 E1 **Kinston** U.S.A.
88 B2 **Kintai** Lith.
114 B4 **Kintampo** Ghana
96 B3 **Kintyre** *pen.* U.K.
62 A1 **Kin-U** Myanmar
130 C2 **Kipawa, Lac** *l.* Can.
83 I3 **Kirensk** Russia
89 E3 **Kireyevsk** Russia
Kirghizia *country* Asia *see* **Kyrgyzstan**
49 G3 **Kiribati** *country* Pacific Ocean
80 B2 **Kırıkkale** Turkey
89 E2 **Kirillov** Russia
Kirin China *see* **Jilin**
Kirinyaga *mt.* Kenya *see* **Kenya, Mount**
89 D2 **Kirishi** Russia
49 H2 **Kiritimati** *atoll* Kiribati
111 C3 **Kırkağaç** Turkey
96 C2 **Kirkcaldy** U.K.
96 B3 **Kirkcudbright** U.K.
92 J2 **Kirkenes** Norway
130 B2 **Kirkland Lake** Can.
137 E2 **Kirksville** U.S.A.
81 C2 **Kirkūk** Iraq
96 C1 **Kirkwall** U.K.
89 D3 **Kirov** *Kaluzhskaya Oblast'* Russia
86 D3 **Kirov** *Kirovskaya Oblast'* Russia
Kirovabad Azer. *see* **Gäncä**
86 E3 **Kirovo-Chepetsk** Russia
91 C2 **Kirovohrad** Ukr.
86 C2 **Kirovsk** Russia
91 D2 **Kirovs'ke** Ukr.
96 C2 **Kirriemuir** U.K.
86 E3 **Kirs** Russia
74 A2 **Kirthar Range** *mts* Pak.
92 H2 **Kiruna** Sweden
67 C3 **Kiryū** Japan
89 E2 **Kirzhach** Russia
119 C2 **Kisangani** Dem. Rep. Congo
118 B3 **Kisantu** Dem. Rep. Congo
60 A1 **Kisaran** Indon.
82 G3 **Kiselevsk** Russia
75 C2 **Kishanganj** India
115 C4 **Kishi** Nigeria
Kishinev Moldova *see* **Chişinău**
77 D1 **Kishkenekol'** Kazakh.
74 B1 **Kishtwar** India
119 D3 **Kisii** Kenya
103 D2 **Kiskunfélegyháza** Hungary
103 D2 **Kiskunhalas** Hungary
87 D4 **Kislovodsk** Russia
117 C5 **Kismaayo** Somalia
119 C3 **Kisoro** Uganda
111 B3 **Kissamos** Greece
114 A4 **Kissidougou** Guinea
143 D3 **Kissimmee** U.S.A.
143 D3 **Kissimmee, Lake** U.S.A.
129 D2 **Kississing Lake** Can.
119 D3 **Kisumu** Kenya
114 B3 **Kita** Mali
66 D3 **Kitakami** Japan
66 D3 **Kitakami-gawa** *r.* Japan
67 B4 **Kita-Kyūshū** Japan
119 D2 **Kitale** Kenya
66 D2 **Kitami** Japan
130 B2 **Kitchener** Can.
93 J3 **Kitee** Fin.
119 D2 **Kitgum** Uganda
128 B2 **Kitimat** Can.
118 B3 **Kitona** Dem. Rep. Congo
119 D3 **Kitunda** Tanz.
128 B2 **Kitwanga** Can.
121 B2 **Kitwe** Zambia
101 E3 **Kitzingen** Ger.
92 I3 **Kiuruvesi** Fin.
119 C3 **Kivu, Lac** Dem. Rep. Congo/ Rwanda
110 C2 **Kıyıköy** Turkey
86 E3 **Kizel** Russia
111 C3 **Kızılca Dağ** *mt.* Turkey
87 D4 **Kizlyar** Russia
92 I1 **Kjøllefjord** Norway
92 G2 **Kjøpsvik** Norway
102 C1 **Kladno** Czech Rep.
102 C2 **Klagenfurt am Wörthersee** Austria
88 B2 **Klaipda** Lith.
94 B1 **Klaksvík** Faroe Is
134 B2 **Klamath** *r.* U.S.A.
134 B2 **Klamath Falls** U.S.A.
134 B2 **Klamath Mountains** U.S.A.
60 B1 **Klang** Malaysia
102 C2 **Klatovy** Czech Rep.
122 A3 **Klawer** S. Africa
128 A2 **Klawock** U.S.A.
128 B2 **Kleena Kleene** Can.
122 B2 **Kleinbegin** S. Africa
122 A2 **Kleinsee** S. Africa
123 C2 **Klerksdorp** S. Africa
89 D3 **Kletnya** Russia
100 C2 **Kleve** Ger.
89 D3 **Klimavichy** Belarus
89 D3 **Klimovo** Russia
89 E2 **Klimovsk** Russia
89 E2 **Klin** Russia
101 F2 **Klínovec** *mt.* Czech Rep.

93 G4 **Klintehamn** Sweden
89 D3 **Klintsy** Russia
109 C2 **Ključ** Bos. & Herz.
103 D1 **Kłodzko** Pol.
100 C1 **Kloosterhaar** Neth.
103 D2 **Klosterneuburg** Austria
101 E1 **Klötze (Altmark)** Ger.
103 D1 **Kluczbork** Pol.
128 A2 **Klukwan** U.S.A.
88 C3 **Klyetsk** Belarus
98 C1 **Knaresborough** U.K.
93 F3 **Knästen** *h.* Sweden
129 E2 **Knee Lake** Can.
101 E1 **Knesebeck** Ger.
101 E3 **Knetzgau** Ger.
109 C2 **Knin** Croatia
103 C2 **Knittelfeld** Austria
109 D2 **Knjaževac** Serbia
100 A2 **Knokke-Heist** Belgium
143 D1 **Knoxville** U.S.A.
122 B3 **Knysna** S. Africa
76 B1 **Kobda** Kazakh.
67 C4 **Kōbe** Japan
København Denmark *see* **Copenhagen**
100 C2 **Koblenz** Ger.
59 C3 **Kobroör** *i.* Indon.
88 B3 **Kobryn** Belarus
109 D2 **Kočani** Macedonia
111 C2 **Kocasu** *r.* Turkey
75 C2 **Koch Bihar** India
73 B4 **Kochi** India
67 B4 **Kōchi** Japan
87 D4 **Kochubey** Russia
75 C2 **Kodarma** India
117 B4 **Kodok** South Sudan
90 B2 **Kodyma** Ukr.
111 C2 **Kodzhaele** *mt.* Bulg./Greece
122 A2 **Koës** Namibia
122 C2 **Koffiefontein** S. Africa
114 B4 **Koforidua** Ghana
67 C3 **Kōfu** Japan
131 D1 **Kogaluk** *r.* Can.
117 B5 **Kogelo** Kenya
74 B1 **Kohat** Pak.
62 A1 **Kohima** India
88 C2 **Kohtla-Järve** Estonia
114 A4 **Koidu-Sefadu** Sierra Leone
92 H3 **Kokkola** Fin.
140 B2 **Kokomo** U.S.A.
77 E2 **Kokpekty** Kazakh.
77 C1 **Kokshetau** Kazakh.
131 D1 **Koksoak** *r.* Can.
123 C3 **Kokstad** S. Africa
Koktokay China *see* **Fuyun**
58 C3 **Kolaka** Indon.
86 C2 **Kola Peninsula** Russia
92 H2 **Kolari** Fin.
114 A3 **Kolda** Senegal
93 E4 **Kolding** Denmark
107 D2 **Koléa** Alg.
86 D2 **Kolguyev, Ostrov** *i.* Russia
73 B3 **Kolhapur** India
88 B2 **Kolkasrags** *pt* Latvia
75 C2 **Kolkata** India
73 B4 **Kollam** India
Köln Ger. *see* **Cologne**
103 D1 **Kołobrzeg** Pol.
114 B3 **Kolokani** Mali
89 E2 **Kolomna** Russia
90 B2 **Kolomyya** Ukr.
122 B2 **Kolonkwaneng** Botswana
82 G3 **Kolpashevo** Russia
89 E3 **Kolpny** Russia
Kol'skiy Poluostrov *pen.* Russia *see* **Kola Peninsula**
78 B3 **Koluli** Eritrea
119 C4 **Kolwezi** Dem. Rep. Congo
83 L2 **Kolyma** *r.* Russia
83 L2 **Kolymskaya Nizmennost'** *lowland* Russia
83 M2 **Kolymskiy, Khrebet** *mts* Russia
67 C3 **Komaki** Japan
83 M3 **Komandorskiye Ostrova** *is* Russia
103 D2 **Komárno** Slovakia
123 D2 **Komati** *r.* S. Africa/Swaziland
67 C3 **Komatsu** Japan
119 C3 **Kombe** Dem. Rep. Congo
90 C2 **Kominternivs'ke** Ukr.
109 C2 **Komiža** Croatia
103 D2 **Komló** Hungary
118 B3 **Komono** Congo
111 C2 **Komotini** Greece
122 B3 **Komsberg** *mts* S. Africa
83 H1 **Komsomolets, Ostrov** *i.* Russia
89 F2 **Komsomol'sk** Russia
91 C2 **Komsomol's'k** Ukr.
87 D4 **Komsomol'skiy** Russia
69 E1 **Komsomol'sk-na-Amure** Russia
89 E2 **Konakovo** Russia
75 C3 **Kondagaon** India
86 F2 **Kondinskoye** Russia
119 D3 **Kondoa** Tanz.
86 C2 **Kondopoga** Russia
89 E3 **Kondrovo** Russia
127 I2 **Kong Christian IX Land** *reg.* Greenland
127 I2 **Kong Frederik VI Kyst** *coastal area* Greenland
119 C3 **Kongolo** Dem. Rep. Congo
93 E4 **Kongsberg** Norway
93 F3 **Kongsvinger** Norway
77 D3 **Kongur Shan** *mt.* China
100 C2 **Königswinter** Ger.
103 D1 **Konin** Pol.
109 C2 **Konjic** Bos. & Herz.
122 A2 **Konkiep** *watercourse* Namibia
91 C1 **Konotop** Ukr.
102 B2 **Konstanz** Ger.
115 C3 **Kontagora** Nigeria
63 B2 **Kon Tum** Vietnam
63 B2 **Kon Tum, Cao Nguyên** *plat.* Vietnam
80 B2 **Konya** Turkey
100 C3 **Konz** Ger.
86 E3 **Konzhakovskiy Kamen', Gora** *mt.* Russia
134 C1 **Kooskia** U.S.A.
128 C3 **Kootenay Lake** Can.
122 B3 **Kootjieskolk** S. Africa
92 □B2 **Kópasker** Iceland
108 B1 **Koper** Slovenia
93 G4 **Kopparberg** Sweden
109 C1 **Koprivnica** Croatia
89 F3 **Korablino** Russia
73 C3 **Koraput** India
101 D2 **Korbach** Ger.
109 D2 **Korçë** Albania
109 C2 **Korčula** Croatia
109 C2 **Korčula** *i.* Croatia
70 C2 **Korea Bay** *g.* China/N. Korea
65 B2 **Korea, North** *country* Asia
65 B2 **Korea, South** *country* Asia
65 B3 **Korea Strait** Japan/S. Korea
91 D2 **Korenovsk** Russia
90 B1 **Korets'** Ukr.
111 C2 **Körfez** Turkey
114 B4 **Korhogo** Côte d'Ivoire
Korinthos Greece *see* **Corinth**
103 D2 **Kris-hegy** *h.* Hungary
109 D2 **Koritnik** *mt.* Albania/Kosovo
67 D3 **Kōriyama** Japan
80 B2 **Korkuteli** Turkey
68 B2 **Korla** China
103 D2 **Körmend** Hungary
114 B3 **Koro** Mali
131 D1 **Koroc** *r.* Can.
91 D1 **Korocha** Russia
119 D3 **Korogwe** Tanz.
59 C2 **Koror** Palau
90 B1 **Korosten'** Ukr.
90 B1 **Korostyshiv** Ukr.
115 D3 **Koro Toro** Chad
91 C2 **Korsun'-Shevchenkivs'kyy** Ukr.
103 E1 **Korsze** Pol.
100 A2 **Kortrijk** Belgium
83 L3 **Koryakskaya Sopka, Vulkan** *vol.* Russia
83 M2 **Koryakskoye Nagor'ye** *mts* Russia
86 D2 **Koryazhma** Russia
91 C1 **Koryukivka** Ukr.
111 C3 **Kos** Greece
111 C3 **Kos** *i.* Greece
65 B2 **Kosan** N. Korea
103 D1 **Kościan** Pol.
53 C3 **Kosciuszko, Mount** Austr.
68 B1 **Kosh-Agach** Russia
103 E2 **Košice** Slovakia
92 H2 **Koskullskulle** Sweden
65 B2 **Kosŏng** N. Korea
109 D2 **Kosovo** *country* Europe
Kosovska Mitrovica Kosovo *see* **Mitrovicë**
48 E2 **Kosrae** *atoll* Micronesia
114 B4 **Kossou, Lac de** *l.* Côte d'Ivoire
76 C1 **Kostanay** Kazakh.
110 B2 **Kostenets** Bulg.
123 C2 **Koster** S. Africa
117 B3 **Kosti** Sudan
86 C2 **Kostomuksha** Russia
90 B1 **Kostopil'** Ukr.
89 F2 **Kostroma** Russia
89 F2 **Kostroma** *r.* Russia
102 C1 **Kostrzyn nad Odrą** Pol.
103 D1 **Koszalin** Pol.
103 D2 **Kszeg** Hungary
74 B2 **Kota** India
60 B2 **Kotaagung** Indon.
61 C2 **Kotabaru** Indon.
61 C1 **Kota Belud** *Sabah* Malaysia
60 B1 **Kota Bharu** Malaysia
60 B2 **Kotabumi** Indon.
61 C1 **Kota Kinabalu** *Sabah* Malaysia
61 C1 **Kota Samarahan** *Sarawak* Malaysia
86 D3 **Kotel'nich** Russia
87 D4 **Kotel'nikovo** Russia
83 K1 **Kotel'nyy, Ostrov** *i.* Russia
101 E2 **Köthen (Anhalt)** Ger.
93 I3 **Kotka** Fin.
86 D2 **Kotlas** Russia
109 C2 **Kotor Varoš** Bos. & Herz.
87 D3 **Kotovo** Russia
91 E1 **Kotovsk** Russia
90 B2 **Kotovs'k** Ukr.
83 H2 **Kotuy** *r.* Russia
100 A2 **Koudekerke** Neth.
114 B3 **Koudougou** Burkina Faso
122 B3 **Kougaberge** *mts* S. Africa
118 B3 **Koulamoutou** Gabon
114 B3 **Koulikoro** Mali
114 A3 **Koundâra** Guinea
151 C1 **Kourou** Fr. Guiana
114 B3 **Kouroussa** Guinea
115 D3 **Kousséri** Cameroon
114 B3 **Koutiala** Mali
93 I3 **Kouvola** Fin.
92 J2 **Kovdor** Russia

90 A1 **Kovel'** Ukr.
89 F2 **Kovrov** Russia
54 B2 **Kowhitirangi** N.Z.
Koyamutthoor India *see* **Coimbatore**
111 C3 **Köyceğiz** Turkey
86 D2 **Koyda** Russia
126 A2 **Koyukuk** *r.* U.S.A.
111 B2 **Kozani** Greece
90 C1 **Kozelets'** Ukr.
89 E3 **Kozel'sk** Russia
73 B3 **Kozhikode** India
90 B2 **Kozyatyn** Ukr.
63 A3 **Krabi** Thai.
63 A2 **Kra Buri** Thai.
63 B2 **Krâchéh** Cambodia
93 E4 **Kragerø** Norway
100 B1 **Kraggenburg** Neth.
109 D2 **Kragujevac** Serbia
60 B2 **Krakatau** *i.* Indon.
103 D1 **Kraków** Pol.
91 D2 **Kramators'k** Ukr.
93 G3 **Kramfors** Sweden
111 B3 **Kranidi** Greece
86 E1 **Krasino** Russia
88 C2 **Krāslava** Latvia
101 F2 **Kraslice** Czech Rep.
89 D3 **Krasnapollye** Belarus
89 D3 **Krasnaya Gora** Russia
87 D3 **Krasnoarmeysk** Russia
91 D2 **Krasnoarmiys'k** Ukr.
86 D2 **Krasnoborsk** Russia
91 D2 **Krasnodar** Russia
91 D2 **Krasnodarskoye Vodokhranilishche** *resr* Russia
91 D2 **Krasnodon** Ukr.
88 C2 **Krasnogorodsk** Russia
91 D2 **Krasnohrad** Ukr.
91 C2 **Krasnohvardiys'ke** Ukr.
86 E3 **Krasnokamsk** Russia
89 D2 **Krasnomayskiy** Russia
91 C2 **Krasnoperekops'k** Ukr.
87 D3 **Krasnoslobodsk** Russia
86 E3 **Krasnoufimsk** Russia
83 H3 **Krasnoyarsk** Russia
89 F2 **Krasnoye-na-Volge** Russia
89 D3 **Krasnyy** Russia
89 E2 **Krasnyy Kholm** Russia
87 C4 **Krasnyy Luch** Ukr.
91 E2 **Krasnyy Sulin** Russia
90 B2 **Krasyliv** Ukr.
100 C2 **Krefeld** Ger.
91 C2 **Kremenchuk** Ukr.
91 C2 **Kremenchuts'ke Vodoskhovyshche** *resr* Ukr.
103 D2 **Křemešník** *h.* Czech Rep.
91 D2 **Kreminna** Ukr.
103 D2 **Krems an der Donau** Austria
89 D2 **Kresttsy** Russia
88 B2 **Kretinga** Lith.
100 C2 **Kreuzau** Ger.
100 C2 **Kreuztal** Ger.
118 A2 **Kribi** Cameroon
111 B3 **Krikellos** Greece
66 D1 **Kril'on, Mys** *c.* Russia
73 C3 **Krishna** *r.* India
73 C3 **Krishna, Mouths of the** India
75 C2 **Krishnanagar** India
93 E4 **Kristiansand** Norway
93 F4 **Kristianstad** Sweden
92 E3 **Kristiansund** Norway
93 F4 **Kristinehamn** Sweden
Kriti *i.* Greece *see* **Crete**
111 C3 **Kritiko Pelagos** *sea* Greece
Krivoy Rog Ukr. *see* **Kryvyy Rih**
109 C1 **Križevci** Croatia
108 B1 **Krk** *i.* Croatia
92 F3 **Krokom** Sweden
91 C1 **Krolevets'** Ukr.
101 E2 **Kronach** Ger.
63 B2 **Krŏng Kaôh Kŏng** Cambodia
127 I2 **Kronprins Frederik Bjerge** *nunataks* Greenland
123 C2 **Kroonstad** S. Africa
87 D4 **Kropotkin** Russia
103 E2 **Krosno** Pol.
103 D1 **Krotoszyn** Pol.
60 B2 **Krui** Indon.
109 C2 **Krujë** Albania
111 C2 **Krumovgrad** Bulg.
Krung Thep Thai. *see* **Bangkok**
88 C3 **Krupki** Belarus
109 D2 **Kruševac** Serbia
101 F2 **Krušné hory** *mts* Czech Rep.
128 A2 **Kruzof Island** U.S.A.
89 D3 **Krychaw** Belarus
91 D3 **Krymsk** Russia
91 C2 **Kryvyy Rih** Ukr.
114 B2 **Ksabi** Alg.
107 D2 **Ksar el Boukhari** Alg.
114 B1 **Ksar el Kebir** Morocco
89 E3 **Kshenskiy** Russia
78 B2 **Kū', Jabal al** *h.* Saudi Arabia
61 C1 **Kuala Belait** Brunei
60 B1 **Kuala Kerai** Malaysia
60 B1 **Kuala Lipis** Malaysia
60 B1 **Kuala Lumpur** Malaysia
61 C2 **Kualapembuang** Indon.
60 B1 **Kuala Terengganu** Malaysia
60 B2 **Kualatungal** Indon.
61 C1 **Kuamut** *Sabah* Malaysia
65 A1 **Kuandian** China
60 B1 **Kuantan** Malaysia
91 D2 **Kuban'** *r.* Russia
89 E2 **Kubenskoye, Ozero** *l.* Russia
110 C2 **Kubrat** Bulg.
61 C1 **Kubuang** Indon.
61 C1 **Kuching** *Sarawak* Malaysia
109 C2 **Kuçovë** Albania
61 C1 **Kudat** *Sabah* Malaysia
61 C2 **Kudus** Indon.
102 C2 **Kufstein** Austria
127 F2 **Kugaaruk** Can.
126 D2 **Kugluktuk** Can.
92 I3 **Kuhmo** Fin.
79 C2 **Kūhrān, Kūh-e** *mt.* Iran
Kuitin China *see* **Kuytun**
120 A2 **Kuito** Angola
92 I2 **Kuivaniemi** Fin.
65 B2 **Kujang** N. Korea
109 D2 **Kukës** Albania
111 C3 **Kula** Turkey
75 D2 **Kula Kangri** *mt.* Bhutan/China
76 B2 **Kulandy** Kazakh.
88 B2 **Kuldīga** Latvia
122 B1 **Kule** Botswana
101 E2 **Kulmbach** Ger.
77 C3 **Kŭlob** Tajik.
76 B2 **Kul'sary** Kazakh.
77 D1 **Kulunda** Russia
127 I2 **Kulusuk** Greenland
67 C3 **Kumagaya** Japan
67 B4 **Kumamoto** Japan
109 D2 **Kumanovo** Macedonia
114 B4 **Kumasi** Ghana
118 A2 **Kumba** Cameroon
78 B2 **Kumdah** Saudi Arabia
87 E3 **Kumertau** Russia
93 G4 **Kumla** Sweden
115 D3 **Kumo** Nigeria
62 A1 **Kumon Range** *mts* Myanmar
62 B2 **Kumphawapi** Thai.
Kumul China *see* **Hami**
66 D2 **Kunashir, Ostrov** *i.* Russia
120 A2 **Kunene** *r.* Angola/Namibia
77 D2 **Kungei Alatau** *mts* Kazakh./Kyrg.
93 F4 **Kungsbacka** Sweden
118 B2 **Kungu** Dem. Rep. Congo
86 E3 **Kungur** Russia
62 A1 **Kunhing** Myanmar
77 D3 **Kunlun Shan** *mts* China
62 B1 **Kunming** China
50 B1 **Kununurra** Austr.
92 I3 **Kuopio** Fin.
109 C1 **Kupa** *r.* Croatia/Slovenia
59 C3 **Kupang** Indon.
88 B2 **Kupiškis** Lith.
111 C2 **Küplü** Turkey
128 A2 **Kupreanof Island** U.S.A.
91 D2 **Kup"yans'k** Ukr.
77 E2 **Kuqa** China
67 B4 **Kurashiki** Japan
67 B3 **Kurayoshi** Japan
89 E3 **Kurchatov** Russia
81 C2 **Kurdistan** *reg.* Asia
67 B4 **Kure** Japan
88 B2 **Kuressaare** Estonia
86 F3 **Kurgan** Russia
93 H3 **Kurikka** Fin.
156 C3 **Kuril Trench** N. Pacific Ocean
89 E3 **Kurkino** Russia
117 B3 **Kurmuk** Sudan
73 B3 **Kurnool** India
53 D2 **Kurri Kurri** Austr.
89 E3 **Kursk** Russia
122 B2 **Kuruman** S. Africa
122 B2 **Kuruman** *watercourse* S. Africa
67 B4 **Kurume** Japan
83 I3 **Kurumkan** Russia
73 C4 **Kurunegala** Sri Lanka
111 C3 **Kuşadası** Turkey
111 C3 **Kuşadası, Gulf of** *b.* Turkey
111 C2 **Kuş Gölü** *l.* Turkey
91 D2 **Kushchevskaya** Russia
66 D2 **Kushiro** Japan
75 C2 **Kushtia** Bangl.
76 C1 **Kusmuryn** Kazakh.
66 D2 **Kussharo-ko** *l.* Japan
111 C3 **Kütahya** Turkey
81 C1 **Kutaisi** Georgia
109 C1 **Kutjevo** Croatia
103 D1 **Kutno** Pol.
118 B3 **Kutu** Dem. Rep. Congo
126 D2 **Kuujjua** *r.* Can.
131 D1 **Kuujjuaq** Can.
130 C1 **Kuujjuarapik** Can.
92 I2 **Kuusamo** Fin.
120 A2 **Kuvango** Angola
89 D2 **Kuvshinovo** Russia
78 B2 **Kuwait** *country* Asia
79 B2 **Kuwait** Kuwait
Kuybyshev Russia *see* **Samara**
91 D2 **Kuybysheve** Ukr.
87 D3 **Kuybyshevskoye Vodokhranilishche** *resr* Russia
77 E2 **Kuytun** China
111 C3 **Kuyucak** Turkey
87 D3 **Kuznetsk** Russia
90 B1 **Kuznetsovs'k** Ukr.
92 H1 **Kvalsund** Norway
123 D2 **KwaDukuza** S. Africa
123 D2 **KwaMashu** S. Africa
Kwangju S. Korea *see* **Gwangju**
65 B1 **Kwanmo-bong** *mt.* N. Korea
123 C3 **KwaNobuhle** S. Africa
122 B3 **KwaNonzame** S. Africa
123 C3 **Kwatinidubu** S. Africa
123 C2 **KwaZamokuhle** S. Africa
123 D2 **KwaZulu-Natal** *prov.* S. Africa
121 B2 **Kwekwe** Zimbabwe
118 B3 **Kwenge** *r.* Dem. Rep. Congo

103 D1 **Kwidzyn** Pol.
118 B3 **Kwilu** *r.* Angola/Dem. Rep. Congo
59 C3 **Kwoka** *mt.* Indon.
53 C3 **Kyabram** Austr.
62 A2 **Kyaikto** Myanmar
69 D1 **Kyakhta** Russia
52 A2 **Kyancutta** Austr.
62 A1 **Kyaukpadaung** Myanmar
62 A2 **Kyaukpyu** Myanmar
88 B3 **Kybartai** Lith.
62 A2 **Kyebogyi** Myanmar
62 A2 **Kyeintali** Myanmar
74 B1 **Kyelang** India
Kyiv Ukr. *see* **Kiev**
90 C1 **Kyivs'ke Vodoskhovyshche** *resr* Ukr.
Kyklades *is* Greece *see* **Cyclades**
129 D2 **Kyle** Can.
96 B2 **Kyle of Lochalsh** U.K.
100 C3 **Kyll** *r.* Ger.
111 B3 **Kyllini** *mt.* Greece
111 B3 **Kymi** Greece
52 B3 **Kyneton** Austr.
119 D2 **Kyoga, Lake** Uganda
53 D1 **Kyogle** Austr.
67 C4 **Kyōto** Japan
111 B3 **Kyparissia** Greece
111 B3 **Kyra Panagia** *i.* Greece
80 B2 **Kyrenia** Cyprus
77 D2 **Kyrgyzstan** *country* Asia
101 F1 **Kyritz** Ger.
92 H3 **Kyrönjoki** *r.* Fin.
86 E2 **Kyrta** Russia
86 D2 **Kyssa** Russia
111 B3 **Kythira** *i.* Greece
111 B3 **Kythnos** *i.* Greece
128 B2 **Kyuquot** Can.
67 B4 **Kyūshū** *i.* Japan
110 B2 **Kyustendil** Bulg.
92 H3 **Kyyjärvi** Fin.
68 C1 **Kyzyl** Russia
76 C2 **Kyzylkum Desert** Kazakh./Uzbek.
77 C2 **Kyzylorda** Kazakh.

L

145 C3 **La Angostura, Presa de** *resr* Mex.
117 C4 **Laascaanood** Somalia
150 B1 **La Asunción** Venez.
114 A2 **Laâyoune** Western Sahara
87 D4 **Laba** *r.* Russia
144 B2 **La Babia** Mex.
146 B3 **La Bahía, Islas de** *is* Hond.
152 B3 **La Banda** Arg.
104 B2 **La Baule-Escoublac** France
114 A3 **Labé** Guinea
128 A1 **Laberge, Lake** Can.
129 C2 **La Biche, Lac** *l.* Can.
87 D4 **Labinsk** Russia
64 B1 **Labo** Phil.
104 B3 **Labouheyre** France
131 D1 **Labrador** *reg.* Can.
131 D1 **Labrador City** Can.
158 C1 **Labrador Sea** Can./Greenland
150 B2 **Lábrea** Brazil
61 C1 **Labuan** Malaysia
60 B1 **Labuhanbilik** Indon.
59 C3 **Labuna** Indon.
63 A2 **Labutta** Myanmar
86 F2 **Labytnangi** Russia
109 C2 **Laç** Albania
105 C2 **La Capelle** France
Laccadive Islands India *see* **Lakshadweep**
129 E2 **Lac du Bonnet** Can.
146 B3 **La Ceiba** Hond.
53 B2 **Lachlan** *r.* Austr.
141 E1 **Lachute** Can.
105 D3 **La Ciotat** France
129 C2 **Lac La Biche** Can.
Lac la Martre Can. *see* **Whatî**
141 E1 **Lac-Mégantic** Can.
128 C2 **Lacombe** Can.
146 B4 **La Concepción** Panama
145 C3 **La Concordia** Mex.
108 A3 **Laconi** *Sardinia* Italy
141 E2 **Laconia** U.S.A.
128 C2 **La Crete** Can.
140 A2 **La Crosse** U.S.A.
144 B2 **La Cruz** Mex.
144 B2 **La Cuesta** Mex.
75 B1 **Ladakh Range** *mts* India/Pak.
122 B3 **Ladismith** S. Africa
79 D2 **Lādīz** Iran
86 C2 **Ladoga, Lake** Russia
Ladozhskoye Ozero *l.* Russia *see* **Ladoga, Lake**
128 B3 **Ladysmith** Can.
123 C2 **Ladysmith** S. Africa
59 D3 **Lae** P.N.G.
152 B3 **La Esmeralda** Bol.
93 F4 **Læsø** *i.* Denmark
143 C2 **La Fayette** U.S.A.
140 B2 **Lafayette** *IN* U.S.A.
142 B2 **Lafayette** *LA* U.S.A.
115 C4 **Lafia** Nigeria
104 B2 **La Flèche** France
143 D1 **La Follette** U.S.A.
131 C1 **Laforge** Can.
79 C2 **Lāft** Iran
108 A3 **La Galite** *i.* Tunisia
81 C1 **Lagan'** Russia
118 B2 **Lagdo, Lac de** *l.* Cameroon
152 C3 **Lages** Brazil
115 C1 **Laghouat** Alg.
155 D1 **Lagoa Santa** Brazil
114 A2 **La Gomera** *i.* Canary Islands
114 C4 **Lagos** Nigeria
106 B2 **Lagos** Port.
134 C1 **La Grande** U.S.A.
130 C1 **La Grande 3, Réservoir** *resr* Can.
131 C1 **La Grande 4, Réservoir** *resr* Can.
50 B1 **La Grange** Austr.
143 C2 **La Grange** U.S.A.
150 B1 **La Gran Sabana** *plat.* Venez.
152 D3 **Laguna** Brazil
144 A2 **Laguna, Picacho de la** *mt.* Mex.
150 A2 **Lagunas** Peru
La Habana Cuba *see* **Havana**
61 C1 **Lahad Datu** *Sabah* Malaysia
60 B2 **Lahat** Indon.
78 B3 **Laḥij** Yemen
81 C2 **Lāhījān** Iran
100 C2 **Lahnstein** Ger.
74 B1 **Lahore** Pak.
74 A2 **Lahri** Pak.
93 I3 **Lahti** Fin.
115 D4 **Laï** Chad
53 D1 **Laidley** Austr.
104 C2 **L'Aigle** France
111 B3 **Laimos, Akrotirio** *pt* Greece
122 B3 **Laingsburg** S. Africa
92 H2 **Lainioälven** *r.* Sweden
96 B1 **Lairg** U.K.
108 B1 **Laives** Italy
70 B2 **Laiwu** China
70 C2 **Laiyang** China
70 B2 **Laizhou** China
70 B2 **Laizhou Wan** *b.* China
50 C1 **Lajamanu** Austr.
144 B2 **La Junta** Mex.
136 C3 **La Junta** U.S.A.
121 C2 **Lake Cabora Bassa** *resr* Moz.
53 C2 **Lake Cargelligo** Austr.
53 D2 **Lake Cathie** Austr.
142 B2 **Lake Charles** U.S.A.
143 D2 **Lake City** *FL* U.S.A.
143 E2 **Lake City** *SC* U.S.A.
108 A1 **Lake Como** Italy
128 B3 **Lake Cowichan** Can.
108 B1 **Lake Garda** Italy
138 A2 **Lake Havasu City** U.S.A.
143 D3 **Lakeland** U.S.A.
128 C2 **Lake Louise** Can.
134 B1 **Lake Oswego** U.S.A.
54 A2 **Lake Paringa** N.Z.
142 B2 **Lake Providence** U.S.A.
54 B2 **Lake Pukaki** N.Z.
53 C3 **Lakes Entrance** Austr.
134 B2 **Lakeview** U.S.A.
136 B3 **Lakewood** *CO* U.S.A.
141 E2 **Lakewood** *NJ* U.S.A.
143 D3 **Lake Worth** U.S.A.
74 A2 **Lakhpat** India
74 B1 **Lakki Marwat** Pak.
111 B3 **Lakonikos Kolpos** *b.* Greece
92 H1 **Lakselv** Norway
73 B3 **Lakshadweep** India
107 C1 **L'Alcora** Spain
145 C3 **La Libertad** Guat.
106 B1 **Lalín** Spain
75 B2 **Lalitpur** India
129 D2 **La Loche** Can.
100 B2 **La Louvière** Belgium
108 A2 **La Maddalena** *Sardinia* Italy
61 C1 **Lamag** *Sabah* Malaysia
La Manche *str.* France/U.K. *see* **English Channel**
136 C3 **Lamar** U.S.A.
108 A3 **La Marmora, Punta** *mt.* *Sardinia* Italy
128 C1 **La Martre, Lac** *l.* Can.
118 B3 **Lambaréné** Gabon
122 A3 **Lambert's Bay** S. Africa
92 □A2 **Lambeyri** Iceland
106 B1 **Lamego** Port.
150 A3 **La Merced** Peru
79 C2 **Lāmerd** Iran
52 B3 **Lameroo** Austr.
139 C2 **Lamesa** U.S.A.
111 B3 **Lamia** Greece
137 E2 **Lamoni** U.S.A.
62 A2 **Lampang** Thai.
139 D2 **Lampasas** U.S.A.
145 B2 **Lampazos** Mex.
99 A2 **Lampeter** U.K.
62 A2 **Lamphun** Thai.
119 E3 **Lamu** Kenya
62 A2 **Lan, Loi** *mt.* Myanmar/Thai.
96 C3 **Lanark** U.K.
63 A2 **Lanbi Kyun** *i.* Myanmar
Lancang Jiang *r.* China *see* **Mekong**
98 B1 **Lancaster** U.K.
135 C4 **Lancaster** *CA* U.S.A.
140 C3 **Lancaster** *OH* U.S.A.
141 D2 **Lancaster** *PA* U.S.A.
143 D2 **Lancaster** *SC* U.S.A.
127 F2 **Lancaster Sound** *str.* Can.
Lanchow China *see* **Lanzhou**
102 C2 **Landeck** Austria
136 B2 **Lander** U.S.A.
99 A3 **Land's End** *pt* U.K.
102 C2 **Landshut** Ger.

93 F4 **Landskrona** Sweden
143 C2 **Lanett** U.S.A.
122 B2 **Langberg** *mts* S. Africa
137 D1 **Langdon** U.S.A.
93 F4 **Langeland** *i.* Denmark
101 D1 **Langen** Ger.
100 C1 **Langeoog** Ger.
100 C1 **Langeoog** *i.* Ger.
101 D2 **Langgöns** Ger.
92 □A3 **Langjökull** Iceland
105 C3 **Langogne** France
104 B3 **Langon** France
105 D2 **Langres** France
60 A1 **Langsa** Indon.
62 B1 **Lang Sơn** Vietnam
101 D1 **Langwedel** Ger.
129 D2 **Lanigan** Can.
153 A4 **Lanín, Volcán** *vol.* Arg./Chile
81 C2 **Länkäran** Azer.
104 B2 **Lannion** France
140 C2 **Lansing** U.S.A.
71 B3 **Lanxi** China
117 B4 **Lanya** South Sudan
114 A2 **Lanzarote** *i.* Canary Islands
70 A2 **Lanzhou** China
64 B1 **Laoag City** Phil.
62 B1 **Lao Cai** Vietnam
70 B2 **Laohekou** China
68 C2 **Laojunmiao** China
65 B1 **Laoling** China
65 B1 **Lao Ling** *mts* China
68 C2 **Lao Mangnai** China
105 C2 **Laon** France
62 B2 **Laos** *country* Asia
65 B1 **Laotougou** China
154 C3 **Lapa** Brazil
114 A2 **La Palma** *i.* Canary Islands
146 C4 **La Palma** Panama
150 B1 **La Paragua** Venez.
152 B2 **La Paz** Bol.
144 A2 **La Paz** Mex.
150 B2 **La Pedrera** Col.
66 D1 **La Pérouse Strait** Japan/Russia
145 C2 **La Pesca** Mex.
144 B2 **La Piedad** Mex.
153 C4 **La Plata** Arg.
153 C4 **La Plata, Río de** *sea chan.* Arg./Uru.
106 B1 **La Pola** Spain
93 I3 **Lappeenranta** Fin.
92 G2 **Lappland** *reg.* Europe
111 C2 **Lâpseki** Turkey
83 J1 **Laptev Sea** Russia
Laptevykh, More *sea* Russia *see* **Laptev Sea**
92 H3 **Lapua** Fin.
152 B3 **La Quiaca** Arg.
79 C2 **Lār** Iran
114 B1 **Larache** Morocco
136 B2 **Laramie** U.S.A.
136 B2 **Laramie Mountains** U.S.A.
154 B3 **Laranjeiras do Sul** Brazil
59 C3 **Larantuka** Indon.
59 C3 **Larat** *i.* Indon.
107 D2 **Larba** Alg.
106 C1 **Laredo** Spain
139 D3 **Laredo** U.S.A.
96 B3 **Largs** U.K.
152 B3 **La Rioja** Arg.
111 B3 **Larisa** Greece
74 A2 **Larkana** Pak.
80 B2 **Larnaca** Cyprus
97 D1 **Larne** U.K.
100 B2 **La Roche-en-Ardenne** Belgium
104 B2 **La Rochelle** France
104 B2 **La Roche-sur-Yon** France
107 C2 **La Roda** Spain
147 D3 **La Romana** Dom. Rep.
129 D2 **La Ronge** Can.
129 D2 **La Ronge, Lac** *l.* Can.
50 C1 **Larrimah** Austr.
55 P3 **Larsen Ice Shelf** Antarctica
93 F4 **Larvik** Norway
130 C2 **La Sarre** Can.
138 B2 **Las Cruces** U.S.A.
153 C4 **Las Flores** Arg.
153 B4 **Las Heras** Arg.
62 A1 **Lashio** Myanmar
74 A1 **Lashkar Gāh** Afgh.
109 C3 **La Sila** *reg.* Italy
152 B3 **Las Lomitas** Arg.
144 B2 **Las Nieves** Mex.
114 A2 **Las Palmas de Gran Canaria** Canary Islands
108 A2 **La Spezia** Italy
153 B5 **Las Plumas** Arg.
129 D2 **Last Mountain Lake** Can.
152 B3 **Las Tórtolas, Cerro** *mt.* Arg./Chile
118 B3 **Lastoursville** Gabon
109 C2 **Lastovo** *i.* Croatia
144 A2 **Las Tres Vírgenes, Volcán** *vol.* Mex.
146 C2 **Las Tunas** Cuba
144 B2 **Las Varas** *Chihuahua* Mex.
144 B2 **Las Varas** *Nayarit* Mex.
138 B1 **Las Vegas** *NM* U.S.A.
135 C3 **Las Vegas** *NV* U.S.A.
131 E1 **La Tabatière** Can.
80 B2 **Latakia** Syria
104 B3 **La Teste-de-Buch** France
108 B2 **Latina** Italy
147 D3 **La Tortuga, Isla** *i.* Venez.
64 B1 **La Trinidad** Phil.
89 E2 **Latskoye** Russia
131 C2 **La Tuque** Can.
88 B2 **Latvia** *country* Europe
102 C1 **Lauchhammer** Ger.
106 C1 **Laudio** Spain
101 E3 **Lauf an der Pegnitz** Ger.
105 D2 **Laufen** Switz.
51 D4 **Launceston** Austr.
99 A3 **Launceston** U.K.
62 A1 **Launggyaung** Myanmar
51 D1 **Laura** Austr.
142 C2 **Laurel** *MS* U.S.A.
134 E1 **Laurel** *MT* U.S.A.
109 C2 **Lauria** Italy
143 E2 **Laurinburg** U.S.A.
105 D2 **Lausanne** Switz.
61 C2 **Laut** *i.* Indon.
61 C2 **Laut Bali** *sea* Indon.
101 D2 **Lauterbach (Hessen)** Ger.
61 C2 **Laut Kecil, Kepulauan** *is* Indon.
100 C1 **Lauwersmeer** *l.* Neth.
104 B2 **Laval** France
107 C2 **La Vall d'Uixó** Spain
50 B2 **Laverton** Austr.
155 D2 **Lavras** Brazil
123 D2 **Lavumisa** Swaziland
61 C1 **Lawas** *Sarawak* Malaysia
62 A1 **Lawksawk** Myanmar
114 B3 **Lawra** Ghana
137 D3 **Lawrence** U.S.A.
142 C1 **Lawrenceburg** U.S.A.
139 D2 **Lawton** U.S.A.
78 A2 **Lawz, Jabal al** *mt.* Saudi Arabia
93 F4 **Laxå** Sweden
74 B1 **Layyah** Pak.
109 D2 **Lazarevac** Serbia
144 A1 **Lázaro Cárdenas** *Baja California* Mex.
144 B3 **Lázaro Cárdenas** *Michoacán* Mex.
88 B3 **Lazdijai** Lith.
136 C2 **Lead** U.S.A.
129 D2 **Leader** Can.
136 B3 **Leadville** U.S.A.
Leaf Bay Can. *see* **Tasiujaq**
129 D2 **Leaf Rapids** Can.
97 B2 **Leane, Lough** *l.* Ireland
137 E3 **Leavenworth** U.S.A.
80 B2 **Lebanon** *country* Asia
137 E3 **Lebanon** *MO* U.S.A.
141 E2 **Lebanon** *NH* U.S.A.
134 B2 **Lebanon** *OR* U.S.A.
141 D2 **Lebanon** *PA* U.S.A.
89 E3 **Lebedyan'** Russia
91 C1 **Lebedyn** Ukr.
104 C2 **Le Blanc** France
103 D1 **Lębork** Pol.
123 C1 **Lebowakgomo** S. Africa
106 B2 **Lebrija** Spain
153 A4 **Lebu** Chile
104 C3 **Le Bugue** France
109 C2 **Lecce** Italy
108 A1 **Lecco** Italy
102 C2 **Lech** *r.* Austria/Ger.
111 B3 **Lechaina** Greece
71 B3 **Lechang** China
102 B1 **Leck** Ger.
104 C3 **Lectoure** France
106 B1 **Ledesma** Spain
104 C2 **Le Dorat** France
128 C2 **Leduc** Can.
137 E1 **Leech Lake** U.S.A.
98 C2 **Leeds** U.K.
98 B2 **Leek** U.K.
100 C1 **Leer (Ostfriesland)** Ger.
143 D3 **Leesburg** U.S.A.
142 B2 **Leesville** U.S.A.
53 C2 **Leeton** Austr.
122 B3 **Leeu-Gamka** S. Africa
100 B1 **Leeuwarden** Neth.
50 A3 **Leeuwin, Cape** Austr.
147 D3 **Leeward Islands** Caribbean Sea
111 B3 **Lefkada** Greece
111 B3 **Lefkada** *i.* Greece
Lefkosia Cyprus *see* **Nicosia**
64 B1 **Legazpi** Phil.
108 B1 **Legnago** Italy
103 D1 **Legnica** Pol.
75 B1 **Leh** India
104 C2 **Le Havre** France
122 B1 **Lehututu** Botswana
103 D2 **Leibnitz** Austria
99 C2 **Leicester** U.K.
51 C1 **Leichhardt** *r.* Austr.
100 B1 **Leiden** Neth.
100 A2 **Leie** *r.* Belgium
52 A2 **Leigh Creek** Austr.
93 E3 **Leikanger** Norway
101 D1 **Leine** *r.* Ger.
97 C2 **Leinster** *reg.* Ireland
101 F2 **Leipzig** Ger.
92 F2 **Leiranger** Norway
106 B2 **Leiria** Port.
93 E4 **Leirvik** Norway
97 C2 **Leixlip** Ireland
71 B3 **Leiyang** China
71 B3 **Leizhou** China
71 A3 **Leizhou Bandao** *pen.* China
118 B3 **Lékana** Congo
100 B1 **Lelystad** Neth.
153 B6 **Le Maire, Estrecho de** *sea chan.* Arg.
Léman, Lac *l.* France/Switz. *see* **Geneva, Lake**
104 C2 **Le Mans** France

Le Mars

137 D2 Le Mars U.S.A.
154 C2 Leme Brazil
Lemesos Cyprus see Limassol
101 D1 Lemförde Ger.
127 G2 Lemieux Islands Can.
136 C1 Lemmon U.S.A.
135 C3 Lemoore U.S.A.
131 D1 Le Moyne, Lac l. Can.
62 A1 Lemro r. Myanmar
109 C2 Le Murge hills Italy
83 J2 Lena r. Russia
100 C1 Lengerich Ger.
71 B3 Lengshuijiang China
91 D2 Lenine Ukr.
Leningrad Russia see St Petersburg
91 D2 Leningradskaya Russia
77 D3 Lenin Peak Kyrg./Tajik.
Leninsk Kazakh. see Baykonyr
89 E3 Leninskiy Russia
100 A2 Lens Belgium
104 C1 Lens France
83 I2 Lensk Russia
103 D2 Lenti Hungary
109 C3 Lentini Sicily Italy
114 B3 Léo Burkina Faso
103 D2 Leoben Austria
99 B2 Leominster U.K.
144 B2 León Mex.
146 B3 León Nic.
106 B1 León Spain
122 A1 León Namibia
50 B2 Leonora Austr.
155 D2 Leopoldina Brazil
123 C1 Lephalale S. Africa
123 C1 Lephepe Botswana
123 C3 Lephoi S. Africa
71 B3 Leping China
105 C2 Le Puy-en-Velay France
150 A2 Lérida Col.
106 C1 Lerma Spain
111 C3 Leros i. Greece
131 C1 Le Roy, Lac l. Can.
93 F4 Lerum Sweden
96 □ Lerwick U.K.
111 C3 Lesbos i. Greece
147 C3 Les Cayes Haiti
69 C3 Leshan China
86 D2 Leshukonskoye Russia
109 D2 Leskovac Serbia
104 B2 Lesneven France
89 E2 Lesnoye Russia
83 H3 Lesosibirsk Russia
123 C2 Lesotho country Africa
66 B1 Lesozavodsk Russia
104 B2 Les Sables-d'Olonne France
147 D3 Lesser Antilles is Caribbean Sea
81 C1 Lesser Caucasus mts Asia
128 C2 Lesser Slave Lake l. Can.
58 B3 Lesser Sunda Islands is Indon.
105 C3 Les Vans France
Lesvos i. Greece see Lesbos
103 D1 Leszno Pol.
99 C3 Letchworth Garden City U.K.
128 C3 Lethbridge Can.
150 C1 Lethem Guyana
59 C3 Leti, Kepulauan is Indon.
150 B2 Leticia Col.
123 C1 Letlhakeng Botswana
104 C1 Le Touquet-Paris-Plage France
99 D3 Le Tréport France
63 A2 Letsok-aw Kyun i. Myanmar
123 C2 Letsopa S. Africa
97 C1 Letterkenny Ireland
105 C3 Leucate, Étang de l. France

60 A1 Leuser, Gunung mt. Indon.
100 B2 Leuven Belgium
92 F3 Levanger Norway
139 C2 Levelland U.S.A.
50 B1 Lévêque, Cape Austr.
100 C2 Leverkusen Ger.
103 D2 Levice Slovakia
54 C2 Levin N.Z.
131 C2 Lévis Can.
89 E3 Lev Tolstoy Russia
99 D3 Lewes U.K.
96 A1 Lewis, Isle of i. U.K.
140 C3 Lewisburg U.S.A.
134 D1 Lewis Range mts U.S.A.
134 C1 Lewiston ID U.S.A.
141 E2 Lewiston ME U.S.A.
134 E1 Lewistown U.S.A.
140 C3 Lexington KY U.S.A.
141 D3 Lexington VA U.S.A.
64 B1 Leyte i. Phil.
109 C2 Lezhë Albania
89 F2 Lezhnevo Russia
68 C2 L'gov Russia
75 D2 Lharigarbo China
75 C2 Lhasa China
75 C2 Lhazê China
60 A1 Lhokseumawe Indon.
111 B3 Liakoura mt. Greece
67 B3 Liancourt Rocks i. N. Pacific Ocean
71 B3 Lianhua China
71 B3 Lianjiang China
70 C1 Lianshan China
70 B2 Lianyungang China
71 B3 Lianzhou China
70 B2 Liaocheng China
70 C1 Liaodong Bandao pen. China
70 C1 Liaodong Wan b. China
65 A1 Liao He r. China
70 C1 Liaoning prov. China
65 A1 Liaoyang China
65 B1 Liaoyuan China
128 B1 Liard r. Can.
134 C1 Libby U.S.A.
118 B2 Libenge Dem. Rep. Congo
136 C3 Liberal U.S.A.
103 D1 Liberec Czech Rep.
114 B4 Liberia country Africa
146 B3 Liberia Costa Rica
150 B1 Libertad Venez.
100 B3 Libin Belgium
104 B2 Libourne France
118 A2 Libreville Gabon
115 D2 Libya country Africa
115 E2 Libyan Desert Egypt/Libya
116 A1 Libyan Plateau Egypt/Libya
108 B3 Licata Sicily Italy
Licheng China see Lipu
121 C2 Lichinga Moz.
101 E2 Lichte Ger.
123 C2 Lichtenburg S. Africa
101 E2 Lichtenfels Ger.
88 C3 Lida Belarus
93 F4 Lidköping Sweden
50 C2 Liebig, Mount Austr.
105 D2 Liechtenstein country Europe
100 B2 Liège Belgium
92 J3 Lieksa Fin.
119 C2 Lienart Dem. Rep. Congo
63 B2 Liên Nghia Vietnam
102 C2 Lienz Austria
88 B2 Liepāja Latvia
102 C2 Liezen Austria
97 C2 Liffey r. Ireland
97 C1 Lifford Ireland
53 C1 Lightning Ridge Austr.
121 C2 Ligonha r. Moz.

105 D3 Ligurian Sea France/Italy
119 C4 Likasi Dem. Rep. Congo
128 B2 Likely Can.
89 E2 Likhoslavl' Russia
61 B1 Liku Indon.
105 D3 L'Île-Rousse Corsica France
71 B3 Liling China
93 F4 Lilla Edet Sweden
100 B2 Lille Belgium
105 C1 Lille France
Lille Bælt sea chan. Denmark see Little Belt
93 F3 Lillehammer Norway
93 F4 Lillestrøm Norway
128 B2 Lillooet Can.
121 C2 Lilongwe Malawi
64 B2 Liloy Phil.
150 A3 Lima Peru
140 C2 Lima U.S.A.
80 B2 Limassol Cyprus
97 C1 Limavady U.K.
153 B4 Limay r. Arg.
101 F2 Limbach-Oberfrohna Ger.
88 B2 Limbaži Latvia
118 A2 Limbe Cameroon
100 D2 Limburg an der Lahn Ger.
122 B2 Lime Acres S. Africa
154 C2 Limeira Brazil
97 B2 Limerick Ireland
93 E4 Limfjorden sea chan. Denmark
111 C3 Limnos i. Greece
104 C2 Limoges France
136 C3 Limon U.S.A.
104 C3 Limoux France
123 C1 Limpopo prov. S. Africa
121 C3 Limpopo r. S. Africa/Zimbabwe
64 A1 Linapacan i. Phil.
153 A4 Linares Chile
145 C2 Linares Mex.
106 C2 Linares Spain
62 B1 Lincang China
Linchuan China see Fuzhou
98 C2 Lincoln U.K.
140 B2 Lincoln IL U.S.A.
141 F1 Lincoln ME U.S.A.
137 D2 Lincoln NE U.S.A.
151 C1 Linden Guyana
142 C1 Linden U.S.A.
119 C2 Lindi r. Dem. Rep. Congo
119 D3 Lindi Tanz.
Lindisfarne i. U.K. see Holy Island
80 A2 Lindos Greece
141 D2 Lindsay Can.
49 H2 Line Islands Kiribati
70 B2 Linfen China
64 B1 Lingayen Phil.
70 B2 Lingbao China
123 C3 Lingelethu S. Africa
123 C3 Lingelihle S. Africa
100 C1 Lingen (Ems) Ger.
60 B2 Lingga, Kepulauan is Indon.
71 B3 Lingling China
71 A3 Lingshan China
71 A4 Lingshui China
114 A3 Linguère Senegal
63 B2 Linh, Ngok mt. Vietnam
155 D1 Linhares Brazil
93 G4 Linköping Sweden
66 B1 Linkou China
70 B2 Linqing China
154 C2 Lins Brazil
136 C1 Linton U.S.A.
69 D2 Linxi China
69 C2 Linxia China
70 B2 Linyi Shandong China

70 B2 Linyi *Shandong* China
70 B2 Linying China
102 C2 Linz Austria
105 C3 Lion, Golfe du *g.* France
109 B3 Lipari Italy
108 B3 Lipari, Isole *is* Italy
89 E3 Lipetsk Russia
110 B1 Lipova Romania
101 D2 Lippstadt Ger.
71 B3 Lipu China
76 C1 Lisakovsk Kazakh.
118 C2 Lisala Dem. Rep. Congo
Lisboa Port. *see* Lisbon
106 B2 Lisbon Port.
97 C1 Lisburn U.K.
97 B2 Liscannor Bay Ireland
104 C2 Lisieux France
99 A3 Liskeard U.K.
89 E3 Liski Russia
53 D1 Lismore Austr.
99 A3 Lisnaskea U.K.
97 B2 Listowel Ireland
71 A3 Litang *Guangxi* China
68 C2 Litang *Sichuan* China
140 B3 Litchfield *IL* U.S.A.
137 E1 Litchfield *MN* U.S.A.
53 D2 Lithgow Austr.
111 B3 Lithino, Akrotirio *pt* Greece
88 C2 Lithuania *country* Europe
102 C1 Litoměřice Czech Rep.
146 C2 Little Abaco *i.* Bahamas
73 D3 Little Andaman *i.* India
93 E4 Little Belt *sea chan.* Denmark
137 E1 Little Falls U.S.A.
63 C2 Littlefield U.S.A.
99 C3 Littlehampton U.K.
122 A2 Little Karas Berg *plat.*
   Namibia
122 B3 Little Karoo *plat.* S. Africa
96 A2 Little Minch *sea chan.* U.K.
136 C1 Little Missouri *r.* U.S.A.
63 A3 Little Nicobar *i.* India
142 B2 Little Rock U.S.A.
141 E2 Littleton U.S.A.
70 B2 Liujiachang China
71 A3 Liupanshui China
71 A3 Liuzhou China
111 B3 Livadeia Greece
88 C2 Līvāni Latvia
50 B1 Liveringa Austr.
138 C2 Livermore, Mount U.S.A.
131 D2 Liverpool Can.
98 B2 Liverpool U.K.
127 F2 Liverpool, Cape Can.
53 C2 Liverpool Range *mts* Austr.
134 D1 Livingston *MT* U.S.A.
139 E2 Livingston *TX* U.S.A.
139 D2 Livingston, Lake U.S.A.
120 B2 Livingstone Zambia
109 C2 Livno Bos. & Herz.
89 E3 Livny Russia
140 C2 Livonia U.S.A.
108 B2 Livorno Italy
119 D3 Liwale Tanz.
70 B2 Liyang China
99 A4 Lizard Point U.K.
108 B1 Ljubljana Slovenia
93 G3 Ljungan *r.* Sweden
93 F4 Ljungby Sweden
93 G3 Ljusdal Sweden
93 G3 Ljusnan *r.* Sweden
99 B3 Llandeilo U.K.
99 B3 Llandovery U.K.
99 B2 Llandrindod Wells U.K.
98 B2 Llandudno U.K.
99 A3 Llanelli U.K.
98 A2 Llangefni U.K.
139 C2 Llano Estacado *plain* U.S.A.

150 B1 Llanos *plain* Col./Venez.
107 D1 Lleida Spain
99 A2 Lleyn Peninsula U.K.
107 C2 Llíria Spain
128 B2 Lloyd George, Mount Can.
129 D2 Lloyd Lake Can.
129 C2 Lloydminster Can.
152 B3 Llullaillaco, Volcán *vol.*
   Chile
123 C2 Loange Botswana
120 A2 Lobito Angola
101 F1 Loburg Ger.
96 B2 Lochaber *reg.* U.K.
104 C2 Loches France
96 B2 Lochgilphead U.K.
96 B1 Lochinver U.K.
96 A2 Lochmaddy U.K.
96 C2 Lochnagar *mt.* U.K.
52 A2 Lock Austr.
96 C3 Lockerbie U.K.
139 D3 Lockhart U.S.A.
141 D2 Lock Haven U.S.A.
141 D2 Lockport U.S.A.
63 B2 Lộc Ninh Vietnam
105 C3 Lodève France
86 C2 Lodeynoye Pole Russia
92 F2 Løding Norway
119 D2 Lodwar Kenya
103 D1 Łódź Pol.
62 B2 Loei Thai.
122 A3 Loeriesfontein S. Africa
92 F2 Lofoten *is* Norway
134 D2 Logan U.S.A.
126 B2 Logan, Mount Can.
140 B2 Logansport U.S.A.
108 B1 Logatec Slovenia
106 C1 Logroño Spain
101 D1 Lohne Ger.
101 D1 Lohne (Oldenburg) Ger.
62 A2 Loikaw Myanmar
104 A2 Loire *r.* France
150 A2 Loja Ecuador
106 C2 Loja Spain
92 I2 Lokan tekojärvi *resr* Fin.
74 B1 Lōkar Afgh.
100 B2 Lokeren Belgium
122 B1 Lokgwabe Botswana
119 D2 Lokichar Kenya
119 D2 Lokichokio Kenya
89 D2 Loknya Russia
115 C4 Lokoja Nigeria
89 D3 Lokot' Russia
123 E4 Loksa Estonia
127 G2 Loks Land *i.* Can.
119 D3 Loliondo Tanz.
93 F4 Lolland *i.* Denmark
122 B2 Lolwane S. Africa
110 B2 Lom Bulg.
93 E3 Lom Norway
119 C2 Lomami *r.* Dem. Rep. Congo
153 C4 Lomas de Zamora Arg.
61 C2 Lombok *i.* Indon.
61 C2 Lombok, Selat *sea chan.*
   Indon.
114 C4 Lomé Togo
118 C3 Lomela Dem. Rep. Congo
100 B2 Lommel Belgium
96 B2 Lomond, Loch *l.* U.K.
88 C2 Lomonosov Russia
135 B4 Lompoc U.S.A.
63 B2 Lom Sak Thai.
103 E1 Łomża Pol.
130 B2 London Can.
99 C3 London U.K.
140 C3 London U.S.A.
97 C1 Londonderry U.K.
50 B1 Londonderry, Cape Austr.
154 B2 Londrina Brazil

83 M2 Longa, Proliv *sea chan.* Russia
61 C1 Long Akah *Sarawak* Malaysia
143 E2 Long Bay U.S.A.
135 C4 Long Beach U.S.A.
97 C2 Longford Ireland
61 C2 Longiram Indon.
147 C2 Long Island Bahamas
130 C1 Long Island Can.
59 D3 Long Island P.N.G.
141 E2 Long Island U.S.A.
130 B2 Longlac Can.
130 B2 Long Lake Can.
136 B2 Longmont U.S.A.
70 A2 Longnan China
140 C2 Long Point Can.
71 B3 Longquan China
131 E2 Long Range Mountains Can.
51 D2 Longreach Austr.
99 D2 Long Stratton U.K.
98 B1 Longtown U.K.
105 D2 Longuyon France
139 E2 Longview *TX* U.S.A.
134 B1 Longview *WA* U.S.A.
70 A2 Longxi China
71 B3 Longxi Shan *mt.* China
63 B2 Long Xuyên Vietnam
71 B3 Longyan China
82 C1 Longyearbyen Svalbard
102 B2 Lons-le-Saunier France
143 E2 Lookout, Cape U.S.A.
97 B2 Loop Head Ireland
64 B1 Lop Buri Thai.
71 B3 Lopez Phil.
68 C2 Lop Nur *salt flat* China
118 B3 Lopori *r.* Dem. Rep. Congo
74 A2 Lora, Hāmūn-i- *dry lake* Afgh./
   Pak.
106 B2 Lora del Río Spain
140 C2 Lorain U.S.A.
74 A1 Loralai Pak.
107 C2 Lorca Spain
48 E5 Lord Howe Island Austr.
138 B2 Lordsburg U.S.A.
155 B2 Lorena Brazil
152 B2 Loreto Bol.
144 A2 Loreto Mex.
104 B2 Lorient France
96 B2 Lorn, Firth of *est.* U.K.
52 B3 Lorne Austr.
138 B1 Los Alamos U.S.A.
139 D3 Los Aldamas Mex.
153 A4 Los Ángeles Chile
135 C4 Los Angeles U.S.A.
135 B3 Los Banos U.S.A.
152 B3 Los Blancos Arg.
91 E1 Losevo Russia
108 B2 Lošinj *i.* Croatia
144 B2 Los Mochis Mex.
118 B2 Losombo Dem. Rep. Congo
147 D3 Los Roques, Islas *is* Venez.
96 C2 Lossiemouth U.K.
150 B1 Los Teques Venez.
152 A4 Los Vilos Chile
96 C1 Loth U.K.
119 D2 Lotikipi Plain Kenya/Sudan
118 C3 Loto Dem. Rep. Congo
89 E2 Lotoshino Russia
62 B1 Louangnamtha Laos
62 B2 Louangphabang Laos
118 B3 Loubomo Congo
104 B2 Loudéac France
71 B3 Loudi China
118 B3 Loudima Congo
114 A3 Louga Senegal
99 C2 Loughborough U.K.
97 B2 Loughrea Ireland
105 D2 Louhans France

97 B2 Louisburgh Ireland
142 B2 Louisiana *state* U.S.A.
140 B3 Louisville *KY* U.S.A.
142 C2 Louisville *MS* U.S.A.
86 C2 Loukhi Russia
130 C1 Loups Marins, Lacs des *lakes* Can.
104 B3 Lourdes France
151 C3 Lourenço Brazil
106 B1 Lousã Port.
53 C2 Louth Austr.
98 C2 Louth U.K.
Louvain Belgium *see* Leuven
89 D2 Lovat' *r.* Russia
110 B2 Lovech Bulg.
136 B2 Loveland U.S.A.
135 C2 Lovelock U.S.A.
139 C2 Lovington U.S.A.
119 C3 Lowa Dem. Rep. Congo
141 E2 Lowell U.S.A.
128 C3 Lower Arrow Lake Can.
54 B2 Lower Hutt N.Z.
97 C1 Lower Lough Erne *l.* U.K.
128 C2 Lower Post Can.
99 D2 Lowestoft U.K.
103 D1 Łowicz Pol.
141 D2 Lowville U.S.A.
52 B2 Loxton Austr.
Loyang China *see* Luoyang
48 F4 Loyauté, Îles *is* New Caledonia
89 D3 Loyew Belarus
92 F2 Løypskardtinden *mt.* Norway
109 C2 Loznica Serbia
91 D2 Lozova Ukr.
120 B2 Luacano Angola
120 A1 Luanda Angola
121 C2 Luangwa *r.* Zambia
121 B2 Luanshya Zambia
106 B1 Luarca Spain
120 B2 Luau Angola
103 E1 Lubaczów Pol.
120 A2 Lubango Angola
119 C3 Lubao Dem. Rep. Congo
103 E1 Lubartów Pol.
101 D1 Lübbecke Ger.
139 C2 Lubbock U.S.A.
101 E1 Lübeck Ger.
69 E2 Lubei China
87 E3 Lubenka Kazakh.
103 D1 Lublin Pol.
103 E1 Lublin Pol.
91 C1 Lubny Ukr.
61 C1 Lubok Antu *Sarawak* Malaysia
101 E1 Lübow Ger.
101 E1 Lübtheen Ger.
119 C3 Lubudi Dem. Rep. Congo
60 B2 Lubuklinggau Indon.
119 C4 Lubumbashi Dem. Rep. Congo
120 B2 Lubungu Zambia
119 C3 Lubutu Dem. Rep. Congo
120 A1 Lucala Angola
97 C2 Lucan Ireland
120 B1 Lucapa Angola
96 B3 Luce Bay U.K.
154 B2 Lucélia Brazil
64 B1 Lucena Phil.
106 C2 Lucena Spain
103 D2 Lučenec Slovakia
109 C2 Lucera Italy
105 D2 Lucerne Switz.
101 E1 Lüchow (Wendland) Ger.
120 A2 Lucira Angola
101 F1 Luckenwalde Ger.
122 B2 Luckhoff S. Africa
75 C2 Lucknow India

120 B2 Lucusse Angola
Lüda China *see* Dalian
100 C2 Lüdenscheid Ger.
101 E1 Lüder Ger.
120 A3 Lüderitz Namibia
74 B1 Ludhiana India
140 B2 Ludington U.S.A.
99 B2 Ludlow U.K.
135 C4 Ludlow U.S.A.
110 C2 Ludogorie *reg.* Bulg.
93 G3 Ludvika Sweden
102 B2 Ludwigsburg Ger.
101 F1 Ludwigsfelde Ger.
101 D3 Ludwigshafen am Rhein Ger.
101 E1 Ludwigslust Ger.
88 C2 Ludza Latvia
118 C3 Luebo Dem. Rep. Congo
120 A2 Luena Angola
70 A2 Lüeyang China
71 B3 Lufeng China
139 E2 Lufkin U.S.A.
88 C2 Luga Russia
88 C2 Luga *r.* Russia
105 D2 Lugano Switz.
121 C2 Lugenda *r.* Moz.
97 C2 Lugnaquilla *h.* Ireland
106 B1 Lugo Spain
110 B1 Lugoj Romania
91 D2 Luhans'k Ukr.
119 D3 Luhombero Tanz.
90 B1 Luhyny Ukr.
120 B2 Luiana Angola
118 C3 Luilaka *r.* Dem. Rep. Congo
105 D2 Luino Italy
92 I2 Luiro *r.* Fin.
118 C3 Luiza Dem. Rep. Congo
70 B2 Lujiang China
109 C2 Lukavac Bos. & Herz.
118 B3 Lukenie *r.* Dem. Rep. Congo
138 A2 Lukeville U.S.A.
89 E3 Lukhovitsy Russia
110 B2 Lukovit Bulg.
103 E1 Łuków Pol.
120 B2 Lukulu Zambia
92 H2 Luleå Sweden
92 H2 Luleälven *r.* Sweden
110 C2 Lüleburgaz Turkey
70 B2 Lüliang Shan *mts* China
61 C2 Lumajang Indon.
75 C1 Lumajangdong Co *salt l.* China
120 B2 Lumbala Kaquengue Angola
120 B2 Lumbala N'guimbo Angola
143 E2 Lumberton U.S.A.
61 C1 Lumbis Indon.
106 B1 Lumbrales Spain
63 B2 Lumphăt Cambodia
129 D2 Lumsden Can.
54 A3 Lumsden N.Z.
93 F4 Lund Sweden
121 C2 Lundazi Zambia
99 A3 Lundy *i.* U.K.
101 E1 Lüneburg Ger.
101 E1 Lüneburger Heide *reg.* Ger.
100 C2 Lünen Ger.
105 D2 Lunéville France
120 B2 Lunga *r.* Zambia
114 A4 Lungi Sierra Leone
62 A1 Lunglei India
120 B2 Lungwebungu *r.* Zambia
74 B2 Luni *r.* India
88 C3 Luninyets Belarus
114 A4 Lunsar Sierra Leone
77 E2 Luntai China
71 A3 Luodian China
71 B3 Luoding China
70 B2 Luohe China
70 B2 Luoyang China
121 B2 Lupane Zimbabwe

110 B1 Lupeni Romania
121 C2 Lupilichi Moz.
101 F2 Luppa Ger.
121 D2 Lúrio Moz.
121 D2 Lurio *r.* Moz.
121 B2 Lusaka Zambia
118 C3 Lusambo Dem. Rep. Congo
109 C2 Lushnjë Albania
136 C2 Lusk U.S.A.
76 B3 Lüt, Kavīr-e *des.* Iran
99 C3 Luton U.K.
61 C1 Lutong *Sarawak* Malaysia
129 C1 Łutselk'e Can.
90 B1 Luts'k Ukr.
122 B2 Lutzputs S. Africa
122 A3 Lutzville S. Africa
117 C4 Luuq Somalia
137 D2 Luverne U.S.A.
119 C3 Luvua *r.* Dem. Rep. Congo
123 D1 Luvuvhu *r.* S. Africa
119 D2 Luwero Uganda
59 C3 Luwuk Indon.
100 C3 Luxembourg *country* Europe
100 C3 Luxembourg Lux.
105 D2 Luxeuil-les-Bains France
123 C3 Luxolweni S. Africa
116 B2 Luxor Egypt
100 B2 Luyksgestel Neth.
86 D2 Luza Russia
Luzern Switz. *see* Lucerne
71 A3 Luzhou China
154 C1 Luziânia Brazil
151 D2 Luzilândia Brazil
64 B1 Luzon *i.* Phil.
71 C3 Luzon Strait Phil./Taiwan
90 A2 L'viv Ukr.
L'vov Ukr. *see* L'viv
88 C3 Lyakhavichy Belarus
92 G3 Lycksele Sweden
88 C3 Lyel'chytsy Belarus
88 C3 Lyepyel' Belarus
99 B3 Lyme Bay U.K.
141 D3 Lynchburg U.S.A.
129 D2 Lynn Lake Can.
129 D1 Lynx Lake Can.
105 C2 Lyon France
89 D2 Lyozna Belarus
86 E3 Lys'va Russia
91 D2 Lysychans'k Ukr.
98 B2 Lytham St Anne's U.K.
88 C3 Lyuban' Belarus
89 E2 Lyubertsy Russia
90 B1 Lyubeshiv Ukr.
89 F2 Lyubim Russia
91 D2 Lyubotyn Ukr.
89 D2 Lyubytino Russia
89 D3 Lyudinovo Russia

M

80 B2 Ma'ān Jordan
88 C2 Maardu Estonia
100 B1 Maas *r.* Neth.
100 B2 Maaseik Belgium
64 B1 Maasin Phil.
100 B2 Maastricht Neth.
121 C3 Mabalane Moz.
150 C1 Mabaruma Guyana
121 C3 Mabote Moz.
122 B2 Mabule Botswana
122 B1 Mabutsane Botswana
155 D2 Macaé Brazil
121 C2 Macaloge Moz.
126 E2 MacAlpine Lake Can.
71 B3 Macao *aut. reg.* China
151 C1 Macapá Brazil

150 A2 **Macará** Ecuador
155 D1 **Macarani** Brazil
123 D1 **Macarretane** Moz.
151 E2 **Macau** Brazil
98 B2 **Macclesfield** U.K.
50 B2 **Macdonald, Lake** *imp. l.* Austr.
50 C2 **Macdonnell Ranges** *mts* Austr.
130 A1 **MacDowell Lake** Can.
106 B1 **Macedo de Cavaleiros** Port.
52 B3 **Macedon** *mt.* Austr.
109 D2 **Macedonia** *country* Europe
151 E2 **Maceió** Brazil
108 B2 **Macerata** Italy
52 A2 **Macfarlane, Lake** *imp. l.* Austr.
97 B3 **Macgillycuddy's Reeks** *mts* Ireland
74 A2 **Mach** Pak.
155 C2 **Machado** Brazil
121 C3 **Machaila** Moz.
119 D3 **Machakos** Kenya
150 A2 **Machala** Ecuador
68 C2 **Machali** China
121 C3 **Machanga** Moz.
70 B2 **Macheng** China
141 F2 **Machias** U.S.A.
150 A1 **Machiques** Venez.
150 A3 **Machu Picchu** *tourist site* Peru
123 D2 **Macia** Moz.
110 C1 **Măcin** Romania
114 B3 **Macina** Mali
53 D1 **Macintyre** *r.* Austr.
51 D2 **Mackay** Austr.
50 B2 **Mackay, Lake** *imp. l.* Austr.
129 C1 **MacKay Lake** Can.
128 B2 **Mackenzie** Can.
128 A1 **Mackenzie** *r.* Can.
126 B2 **Mackenzie Bay** Can.
126 D1 **Mackenzie King Island** Can.
128 A1 **Mackenzie Mountains** Can.
129 D2 **Macklin** Can.
53 D2 **Macksville** Austr.
53 D1 **Maclean** Austr.
50 A2 **MacLeod, Lake** *dry lake* Austr.
140 A2 **Macomb** U.S.A.
108 A2 **Macomer** *Sardinia* Italy
105 C2 **Mâcon** France
143 D2 **Macon** *GA* U.S.A.
137 E3 **Macon** *MO* U.S.A.
53 C2 **Macquarie** *r.* Austr.
48 E6 **Macquarie Island** S. Pacific Ocean
53 C2 **Macquarie Marshes** Austr.
97 B3 **Macroom** Ireland
52 A1 **Macumba** *watercourse* Austr.
145 C3 **Macuspana** Mex.
144 B2 **Macuzari, Presa** *resr* Mex.
123 D2 **Madadeni** S. Africa
121 □D3 **Madagascar** *country* Africa
115 D2 **Madama** Niger
111 B2 **Madan** Bulg.
59 D3 **Madang** P.N.G.
150 C2 **Madeira** *r.* Brazil
114 A1 **Madeira** *terr.* N. Atlantic Ocean
131 D2 **Madeleine, Îles de la** *is* Can.
144 B2 **Madera** Mex.
135 B3 **Madera** U.S.A.
118 B3 **Madingou** Congo
140 B3 **Madison** *IN* U.S.A.
137 D2 **Madison** *SD* U.S.A.
140 B2 **Madison** *WI* U.S.A.
140 C3 **Madison** *WV* U.S.A.
134 D1 **Madison** *r.* U.S.A.
140 B3 **Madisonville** U.S.A.
61 C2 **Madiun** Indon.
119 D2 **Mado Gashi** Kenya
88 C2 **Madona** Latvia
78 A2 **Madrakah** Saudi Arabia
Madras India *see* **Chennai**
134 B2 **Madras** U.S.A.
145 C2 **Madre, Laguna** *lag.* Mex.
145 B3 **Madre del Sur, Sierra** *mts* Mex.
144 B2 **Madre Occidental, Sierra** *mts* Mex.
145 B2 **Madre Oriental, Sierra** *mts* Mex.
106 C1 **Madrid** Spain
106 C2 **Madridejos** Spain
61 C2 **Madura** *i.* Indon.
61 C2 **Madura, Selat** *sea chan.* Indon.
73 B4 **Madurai** India
67 C3 **Maebashi** Japan
62 A2 **Mae Hong Son** Thai.
62 A1 **Mae Sai** Thai.
62 A2 **Mae Sariang** Thai.
62 A2 **Mae Suai** Thai.
123 C2 **Mafeteng** Lesotho
119 D3 **Mafia Island** Tanz.
119 D3 **Mafinga** Tanz.
154 C3 **Mafra** Brazil
83 L3 **Magadan** Russia
147 C4 **Magangué** Col.
144 A1 **Magdalena** Mex.
138 B2 **Magdalena** U.S.A.
144 A2 **Magdalena, Bahía** *b.* Mex.
101 E1 **Magdeburg** Ger.
153 A6 **Magellan, Strait of** Chile
97 C1 **Magherafelt** U.K.
87 E3 **Magnitogorsk** Russia
142 B2 **Magnolia** U.S.A.
131 D1 **Magpie, Lac** *l.* Can.
114 A3 **Magṭa' Laḥjar** Maur.
151 D2 **Maguarinho, Cabo** *c.* Brazil
123 D2 **Magude** Moz.
62 A1 **Magway** Myanmar
Magwe Myanmar *see* **Magway**
81 C2 **Mahābād** Iran
74 B2 **Mahajan** India
121 □D2 **Mahajanga** Madag.
61 C2 **Mahakam** *r.* Indon.
123 C1 **Mahalapye** Botswana
121 □D2 **Mahalevona** Madag.
75 C2 **Mahanadi** *r.* India
121 □D2 **Mahanoro** Madag.
63 B2 **Maha Sarakham** Thai.
121 □D2 **Mahavavy** *r.* Madag.
78 B2 **Mahd adh Dhahab** Saudi Arabia
150 C1 **Mahdia** Guyana
113 K7 **Mahé** *i.* Seychelles
74 B2 **Mahesana** India
74 B2 **Mahi** *r.* India
54 C1 **Mahia Peninsula** N.Z.
123 C2 **Mahikeng** S. Africa
89 D3 **Mahilyow** Belarus
74 B2 **Mahuva** India
110 C2 **Mahya Dağı** *mt.* Turkey
74 A1 **Maīdān Shahr** Afgh.
129 D2 **Maidstone** Can.
99 D3 **Maidstone** U.K.
115 D3 **Maiduguri** Nigeria
75 C2 **Mailani** India
76 C3 **Maīmanah** Afgh.
101 D2 **Main** *r.* Ger.
118 B3 **Mai-Ndombe, Lac** *l.* Dem. Rep. Congo
101 E3 **Main-Donau-Kanal** *canal* Ger.
141 F1 **Maine** *state* U.S.A.
62 A1 **Maingkwan** Myanmar
96 C1 **Mainland** *i. Scotland* U.K.
96 □ **Mainland** *i. Scotland* U.K.
121 □D2 **Maintirano** Madag.
101 D2 **Mainz** Ger.
150 B1 **Maiquetía** Venez.
53 D2 **Maitland** *N.S.W.* Austr.
52 A2 **Maitland** *S.A.* Austr.
146 B3 **Maíz, Islas del** *is* Nic.
67 C3 **Maizuru** Japan
61 C2 **Majene** Indon.
107 D2 **Majorca** *i.* Spain
48 F2 **Majuro** *atoll* Marshall Is
123 C2 **Majwemasweu** S. Africa
118 B3 **Makabana** Congo
58 B3 **Makale** Indon.
77 E2 **Makanshy** Kazakh.
109 C2 **Makarska** Croatia
58 B3 **Makassar** Indon.
61 C2 **Makassar, Selat** *str.* Indon.
Makassar Strait Indon. *see* **Makassar, Selat**
76 B2 **Makat** Kazakh.
123 D2 **Makatini Flats** *lowland* S. Africa
114 A4 **Makeni** Sierra Leone
120 B3 **Makgadikgadi** *depr.* Botswana
87 D4 **Makhachkala** Russia
123 C1 **Makhado** S. Africa
76 B2 **Makhambet** Kazakh.
119 D3 **Makindu** Kenya
77 D1 **Makinsk** Kazakh.
91 D2 **Makiyivka** Ukr.
Makkah Saudi Arabia *see* **Mecca**
131 E1 **Makkovik** Can.
103 E2 **Makó** Hungary
118 B2 **Makokou** Gabon
119 D3 **Makongolosi** Tanz.
122 B2 **Makopong** Botswana
79 D2 **Makran** *reg.* Iran/Pak.
74 A2 **Makran Coast Range** *mts* Pak.
89 E2 **Maksatikha** Russia
81 C2 **Mākū** Iran
62 A1 **Makum** India
67 B4 **Makurazaki** Japan
115 C4 **Makurdi** Nigeria
92 G2 **Malå** Sweden
146 B4 **Mala, Punta** *pt* Panama
118 A2 **Malabo** Equat. Guinea
60 A1 **Malacca, Strait of** Indon./Malaysia
134 D2 **Malad City** U.S.A.
88 C3 **Maladzyechna** Belarus
106 C2 **Málaga** Spain
97 B1 **Málainn Mhóir** Ireland
48 E3 **Malaita** *i.* Solomon Is
117 B4 **Malakal** South Sudan
48 F4 **Malakula** *i.* Vanuatu
61 C2 **Malang** Indon.
120 A1 **Malanje** Angola
93 G4 **Mälaren** *l.* Sweden
153 B4 **Malargüe** Arg.
80 B2 **Malatya** Turkey
121 C2 **Malawi** *country* Africa
Malawi, Lake Africa *see* **Nyasa, Lake**
89 D2 **Malaya Vishera** Russia
64 B2 **Malaybalay** Phil.
81 C2 **Malāyer** Iran
60 B1 **Malaysia** *country* Asia
81 C2 **Malazgirt** Turkey
103 D1 **Malbork** Pol.
101 F1 **Malchin** Ger.
100 A2 **Maldegem** Belgium
49 H3 **Malden Island** Kiribati
56 G6 **Maldives** *country* Indian Ocean
153 C4 **Maldonado** Uru.
56 G6 **Male** Maldives
111 B3 **Maleas, Akrotirio** *pt* Greece

103 D2 **Malé Karpaty** *hills* Slovakia
134 C2 **Malheur Lake** U.S.A.
114 B3 **Mali** *country* Africa
114 A3 **Mali** Guinea
58 C3 **Malili** Indon.
119 E3 **Malindi** Kenya
97 C1 **Malin Head** Ireland
111 C2 **Malkara** Turkey
88 C3 **Mal'kavichy** Belarus
110 C2 **Malko Tarnovo** Bulg.
53 C3 **Mallacoota** Austr.
53 C3 **Mallacoota Inlet** *b.* Austr.
96 B2 **Mallaig** U.K.
129 E1 **Mallery Lake** Can.
Mallorca *i.* Spain *see* **Majorca**
97 B2 **Mallow** Ireland
92 H2 **Malmberget** Sweden
100 C2 **Malmedy** Belgium
122 A3 **Malmesbury** S. Africa
93 F4 **Malmö** Sweden
62 B1 **Malong** China
118 C4 **Malonga** Dem. Rep. Congo
93 E3 **Måløy** Norway
89 E2 **Maloyaroslavets** Russia
89 E2 **Maloye Borisovo** Russia
85 H5 **Malta** *country* Europe
88 C2 **Malta** Latvia
134 E1 **Malta** U.S.A.
122 A1 **Maltahöhe** Namibia
98 C1 **Malton** U.K.
Maluku *is* Indon. *see* **Moluccas**
59 C3 **Maluku, Laut** *sea* Indon.
93 F3 **Malung** Sweden
123 C2 **Maluti Mountains** Lesotho
73 B3 **Malvan** India
142 B2 **Malvern** U.S.A.
90 B1 **Malyn** Ukr.
83 L2 **Malyy Anyuy** *r.* Russia
83 K2 **Malyy Lyakhovskiy, Ostrov** *i.* Russia
123 C2 **Mamafubedu** S. Africa
119 C2 **Mambasa** Dem. Rep. Congo
118 B2 **Mambéré** *r.* C.A.R.
123 C2 **Mamelodi** S. Africa
150 B3 **Mamoré** *r.* Bol./Brazil
114 A3 **Mamou** Guinea
61 C2 **Mamuju** Indon.
114 B4 **Man** Côte d'Ivoire
98 A1 **Man, Isle of** *i.* Irish Sea
150 B2 **Manacapuru** Brazil
107 D2 **Manacor** Spain
59 C2 **Manado** Indon.
146 B3 **Managua** Nic.
121 □D3 **Manakara** Madag.
78 B3 **Manākhah** Yemen
79 C2 **Manama** Bahrain
59 D3 **Manam Island** P.N.G.
121 □D3 **Mananara** *r.* Madag.
121 □D2 **Mananara Avaratra** Madag.
121 □D3 **Mananjary** Madag.
114 A3 **Manantali, Lac de** *l.* Mali
77 E2 **Manas Hu** *l.* China
59 C3 **Manatuto** East Timor
62 A2 **Man-aung Kyun** *i.* Myanmar
150 B2 **Manaus** Brazil
80 B2 **Manavgat** Turkey
98 B2 **Manchester** U.K.
141 E2 **Manchester** U.S.A.
121 □D3 **Mandabe** Madag.
93 E4 **Mandal** Norway
59 D3 **Mandala, Puncak** *mt.* Indon.
62 A1 **Mandalay** Myanmar
69 D1 **Mandalgovĭ** Mongolia
136 C1 **Mandan** U.S.A.
118 B1 **Mandara Mountains** Cameroon/Nigeria
108 A3 **Mandas** *Sardinia* Italy
100 C2 **Manderscheid** Ger.
74 B1 **Mandi** India
114 B3 **Mandiana** Guinea
75 C2 **Mandla** India
121 □D2 **Mandritsara** Madag.
74 B2 **Mandsaur** India
50 A3 **Mandurah** Austr.
73 B3 **Mandya** India
108 B1 **Manerbio** Italy
90 B1 **Manevychi** Ukr.
109 C2 **Manfredonia** Italy
114 B3 **Manga** Burkina Faso
118 B3 **Mangai** Dem. Rep. Congo
54 C1 **Mangakino** N.Z.
110 C2 **Mangalia** Romania
Mangalore India *see* **Mangaluru**
73 B3 **Mangaluru** India
123 C2 **Mangaung** S. Africa
61 B2 **Manggar** Indon.
76 B2 **Mangistau** Kazakh.
121 C2 **Mangochi** Malawi
59 C3 **Mangole** *i.* Indon.
154 B3 **Mangueirinha** Brazil
69 E1 **Mangui** China
137 D3 **Manhattan** U.S.A.
121 C3 **Manhica** Moz.
155 D2 **Manhuaçu** Brazil
121 □D2 **Mania** *r.* Madag.
108 B1 **Maniago** Italy
150 B2 **Manicoré** Brazil
131 D2 **Manicouagan** *r.* Can.
131 D1 **Manicouagan, Petit Lac** *l.* Can.
131 D1 **Manicouagan, Réservoir** *resr* Can.
79 B2 **Manīfah** Saudi Arabia
64 B1 **Manila** Phil.
53 D2 **Manilla** Austr.
111 C3 **Manisa** Turkey
140 B2 **Manistee** U.S.A.
129 E2 **Manitoba** *prov.* Can.
129 E2 **Manitoba, Lake** Can.
140 B1 **Manitou Islands** U.S.A.
130 B2 **Manitoulin Island** Can.
140 B2 **Manitowoc** U.S.A.
130 C2 **Maniwaki** Can.
150 A1 **Manizales** Col.
137 E2 **Mankato** U.S.A.
114 B4 **Mankono** Côte d'Ivoire
129 D3 **Mankota** Can.
73 C4 **Mankulam** Sri Lanka
74 B2 **Manmad** India
73 B4 **Mannar, Gulf of** India/Sri Lanka
101 D3 **Mannheim** Ger.
128 C2 **Manning** Can.
52 A2 **Mannum** Austr.
129 C2 **Mannville** Can.
59 C3 **Manokwari** Indon.
119 C3 **Manono** Dem. Rep. Congo
63 A2 **Manoron** Myanmar
105 D3 **Manosque** France
131 C1 **Manouane, Lac** *l.* Can.
107 D1 **Manresa** Spain
121 B2 **Mansa** Zambia
127 F2 **Mansel Island** Can.
92 I2 **Mansel'kya** *ridge* Fin./Russia
53 C3 **Mansfield** Austr.
98 C2 **Mansfield** U.K.
142 B2 **Mansfield** *LA* U.S.A.
140 C2 **Mansfield** *OH* U.S.A.
141 D2 **Mansfield** *PA* U.S.A.
150 A2 **Manta** Ecuador
143 E1 **Manteo** U.S.A.
104 C2 **Mantes-la-Jolie** France
155 C2 **Mantiqueira, Serra da** *mts* Brazil
Mantova Italy *see* **Mantua**
108 B1 **Mantua** Italy
151 C2 **Manuelzinho** Brazil
59 C3 **Manui** *i.* Indon.
54 B1 **Manukau** N.Z.
142 B2 **Many** U.S.A.
119 D3 **Manyoni** Tanz.
106 C2 **Manzanares** Spain
144 B3 **Manzanillo** Mex.
69 D1 **Manzhouli** China
123 D2 **Manzini** Swaziland
115 D3 **Mao** Chad
107 D2 **Maó** Spain
59 D3 **Maoke, Pegunungan** *mts* Indon.
123 C2 **Maokeng** S. Africa
65 A1 **Maokui Shan** *mt.* China
71 B3 **Maoming** China
121 C3 **Mapai** Moz.
75 C1 **Mapam Yumco** *l.* China
145 C3 **Mapastepec** Mex.
144 B2 **Mapimí** Mex.
144 B2 **Mapimí, Bolsón de** *des.* Mex.
121 C3 **Mapinhane** Moz.
129 D3 **Maple Creek** Can.
121 C3 **Maputo** Moz.
123 D2 **Maputo** *r.* Moz./S. Africa
120 A1 **Maquela do Zombo** Angola
153 B5 **Maquinchao** Arg.
137 E2 **Maquoketa** U.S.A.
150 B2 **Maraã** Brazil
151 D2 **Marabá** Brazil
151 C1 **Maracá, Ilha de** *i.* Brazil
150 A1 **Maracaibo** Venez.
150 A1 **Maracaibo, Lake** Venez.
152 C3 **Maracaju, Serra de** *hills* Brazil
150 B1 **Maracay** Venez.
115 D2 **Marādah** Libya
115 C3 **Maradi** Niger
81 C2 **Marāgheh** Iran
151 D2 **Marajó, Baía de** *est.* Brazil
151 C2 **Marajó, Ilha de** *i.* Brazil
79 C2 **Marākī** Iran
119 D2 **Maralal** Kenya
50 C3 **Maralinga** Austr.
81 C2 **Marand** Iran
150 A2 **Marañón** *r.* Peru
130 B2 **Marathon** Can.
143 D4 **Marathon** U.S.A.
106 C2 **Marbella** Spain
50 A2 **Marble Bar** Austr.
123 C1 **Marble Hall** S. Africa
101 D2 **Marburg** Ger.
123 D3 **Marburg** S. Africa
99 D2 **March** U.K.
100 B2 **Marche-en-Famenne** Belgium
106 B2 **Marchena** Spain
152 B4 **Mar Chiquita, Laguna** *l.* Arg.
141 E2 **Marcy, Mount** U.S.A.
74 B1 **Mardan** Pak.
153 C4 **Mar del Plata** Arg.
81 C2 **Mardin** Turkey
96 B2 **Maree, Loch** *l.* U.K.
108 B3 **Marettimo, Isola** *i.* *Sicily* Italy
89 D2 **Marevo** Russia
138 C2 **Marfa** U.S.A.
73 B3 **Margao** India
50 A3 **Margaret River** Austr.
147 D3 **Margarita, Isla de** *i.* Venez.
123 D3 **Margate** S. Africa
99 D3 **Margate** U.K.
119 C2 **Margherita Peak** Dem. Rep. Congo/Uganda
76 C3 **Mārgō, Dasht-e** *des.* Afgh.
91 C2 **Marhanets'** Ukr.
62 A1 **Mari** Myanmar
152 B3 **María Elena** Chile

156 C5 **Mariana Trench** N. Pacific Ocean
142 B2 **Marianna** *AR* U.S.A.
143 C2 **Marianna** *FL* U.S.A.
102 C2 **Mariánské Lázně** Czech Rep.
144 B2 **Marías, Islas** *is* Mex.
78 B3 **Ma'rib** Yemen
109 C1 **Maribor** Slovenia
117 A4 **Maridi** *watercourse* South Sudan
55 M2 **Marie Byrd Land** *reg.* Antarctica
147 D3 **Marie-Galante** *i.* Guadeloupe
93 G3 **Mariehamn** Fin.
122 A1 **Mariental** Namibia
93 F4 **Mariestad** Sweden
143 D2 **Marietta** *GA* U.S.A.
140 C3 **Marietta** *OH* U.S.A.
105 D3 **Marignane** France
83 K3 **Marii, Mys** *pt* Russia
88 B3 **Marijampol** Lith.
154 C2 **Marília** Brazil
106 B1 **Marín** Spain
109 C3 **Marina di Gioiosa Ionica** Italy
88 C3 **Mar''ina Horka** Belarus
140 B1 **Marinette** U.S.A.
154 B2 **Maringá** Brazil
106 B2 **Marinha Grande** Port.
140 B2 **Marion** *IN* U.S.A.
140 C2 **Marion** *OH* U.S.A.
143 E2 **Marion** *SC* U.S.A.
140 C3 **Marion** *VA* U.S.A.
143 D2 **Marion, Lake** U.S.A.
52 A3 **Marion Bay** Austr.
152 B3 **Mariscal José Félix Estigarribia** Para.
110 C2 **Maritsa** *r.* Bulg.
91 D2 **Mariupol'** Ukr.
117 C4 **Marka** Somalia
123 C1 **Marken** S. Africa
100 B1 **Markermeer** *l.* Neth.
83 I2 **Markha** *r.* Russia
91 D2 **Markivka** Ukr.
142 B2 **Marksville** U.S.A.
101 D3 **Marktheidenfeld** Ger.
101 F2 **Marktredwitz** Ger.
100 C2 **Marl** Ger.
100 A3 **Marle** France
139 D2 **Marlin** U.S.A.
104 C3 **Marmande** France
111 C2 **Marmara, Sea of** *g.* Turkey
Marmara Denizi *g.* Turkey *see* **Marmara, Sea of**
111 C3 **Marmaris** Turkey
104 C2 **Marne-la-Vallée** France
121 □D2 **Maroantsetra** Madag.
101 E2 **Maroldsweisach** Ger.
121 □D2 **Maromokotro** *mt.* Madag.
121 C2 **Marondera** Zimbabwe
151 C1 **Maroni** *r.* Fr. Guiana
118 B1 **Maroua** Cameroon
121 □D2 **Marovoay** Madag.
49 I3 **Marquesas Islands** *is* Fr. Polynesia
140 B1 **Marquette** U.S.A.
117 A3 **Marra, Jebel** *mt.* Sudan
117 A3 **Marra, Jebel** *plat.* Sudan
123 D2 **Marracuene** Moz.
114 B1 **Marrakech** Morocco
52 A1 **Marree** Austr.
121 C2 **Marromeu** Moz.
121 C2 **Marrupa** Moz.
116 B2 **Marsá al 'Alam** Egypt
115 D1 **Marsá al Burayqah** Libya
119 D2 **Marsabit** Kenya
108 B3 **Marsala** *Sicily* Italy
116 A1 **Marsá Maṭrūḥ** Egypt
101 D2 **Marsberg** Ger.
108 B2 **Marsciano** Italy
53 C2 **Marsden** Austr.
105 D3 **Marseille** France
137 D2 **Marshall** *MN* U.S.A.
137 E3 **Marshall** *MO* U.S.A.
139 E2 **Marshall** *TX* U.S.A.
48 F2 **Marshall Islands** *country* N. Pacific Ocean
137 E2 **Marshalltown** U.S.A.
140 A2 **Marshfield** U.S.A.
143 E3 **Marsh Harbour** Bahamas
142 B3 **Marsh Island** U.S.A.
88 A2 **Märsta** Sweden
61 C2 **Martapura** *Kalimantan Selatan* Indon.
60 B2 **Martapura** *Sumatera Selatan* Indon.
141 E2 **Martha's Vineyard** *i.* U.S.A.
105 D2 **Martigny** Switz.
103 D2 **Martin** Slovakia
136 C2 **Martin** U.S.A.
145 C2 **Martínez** Mex.
147 D3 **Martinique** *terr.* West Indies
141 D3 **Martinsburg** U.S.A.
140 D3 **Martinsville** U.S.A.
76 B1 **Martok** Kazakh.
54 C2 **Marton** N.Z.
107 D1 **Martorell** Spain
106 C2 **Martos** Spain
105 C3 **Marvejols** France
76 C3 **Mary** Turkm.
51 E2 **Maryborough** Austr.
122 B2 **Marydale** S. Africa
141 D3 **Maryland** *state* U.S.A.
137 D3 **Marysville** U.S.A.
137 E2 **Maryville** *MO* U.S.A.
143 D1 **Maryville** *TN* U.S.A.
119 D3 **Masai Steppe** *plain* Tanz.
119 D3 **Masaka** Uganda
65 B2 **Masan** S. Korea
119 D4 **Masasi** Tanz.
64 B1 **Masbate** Phil.
64 B1 **Masbate** *i.* Phil.
123 C2 **Maseru** Lesotho
76 B3 **Mashhad** Iran
123 D2 **Mashishing** S. Africa
74 A2 **Mashkel, Hamun-i-** *salt flat* Pak.
123 C3 **Masibambane** S. Africa
79 C3 **Masīlah, Wādī al** *watercourse* Yemen
123 C2 **Masilo** S. Africa
119 D2 **Masindi** Uganda
79 C2 **Maṣīrah, Jazīrat** *i.* Oman
79 C3 **Maṣīrah, Khalīj** *b.* Oman
81 C2 **Masjed Soleymān** Iran
97 B2 **Mask, Lough** *l.* Ireland
121 □E2 **Masoala, Tanjona** *c.* Madag.
137 E2 **Mason City** U.S.A.
Masqaṭ Oman *see* **Muscat**
108 B2 **Massa** Italy
141 E2 **Massachusetts** *state* U.S.A.
141 E2 **Massachusetts Bay** U.S.A.
121 C3 **Massangena** Moz.
120 A1 **Massango** Angola
116 B3 **Massawa** Eritrea
141 E2 **Massena** U.S.A.
128 A2 **Masset** Can.
105 C2 **Massif Central** *mts* France
140 C2 **Massillon** U.S.A.
121 C3 **Massinga** Moz.
123 D1 **Massingir** Moz.
78 A2 **Mastābah** Saudi Arabia
54 C2 **Masterton** N.Z.
74 A2 **Mastung** Pak.
78 A2 **Mastūrah** Saudi Arabia
88 B3 **Masty** Belarus
67 B4 **Masuda** Japan
Masuku Gabon *see* **Franceville**
121 C3 **Masvingo** Zimbabwe
69 D1 **Matad** Mongolia
118 B3 **Matadi** Dem. Rep. Congo
146 B3 **Matagalpa** Nic.
130 C2 **Matagami** Can.
130 C2 **Matagami, Lac** *l.* Can.
139 D3 **Matagorda Island** U.S.A.
120 A2 **Matala** Angola
114 A3 **Matam** Senegal
144 B2 **Matamoros** *Coahuila* Mex.
145 C2 **Matamoros** *Tamaulipas* Mex.
119 D3 **Matandu** *r.* Tanz.
131 D2 **Matane** Can.
146 B2 **Matanzas** Cuba
73 C4 **Matara** Sri Lanka
61 C2 **Mataram** Indon.
50 C1 **Mataranka** Austr.
107 D1 **Mataró** Spain
123 C3 **Matatiele** S. Africa
54 A3 **Mataura** N.Z.
49 G3 **Matā'utu** Wallis and Futuna Is
54 C1 **Matawai** N.Z.
152 B2 **Mategua** Bol.
145 B2 **Matehuala** Mex.
109 C2 **Matera** Italy
139 D3 **Mathis** U.S.A.
75 B2 **Mathura** India
64 B2 **Mati** Phil.
98 C2 **Matlock** U.K.
123 D2 **Matola** Moz.
Matou China *see* **Pingguo**
67 B3 **Matsue** Japan
66 D2 **Matsumae** Japan
67 C4 **Matsusaka** Japan
67 B4 **Matsuyama** Japan
130 B1 **Mattagami** *r.* Can.
105 D2 **Matterhorn** *mt.* Italy/Switz.
134 C2 **Matterhorn** *mt.* U.S.A.
140 B3 **Mattoon** U.S.A.
150 B1 **Maturín** Venez.
123 C2 **Matwabeng** S. Africa
100 A2 **Maubeuge** France
104 C3 **Maubourguet** France
124 E5 **Maui** *i.* U.S.A.
140 C2 **Maumee** *r.* U.S.A.
120 B2 **Maun** Botswana
62 A1 **Maungdaw** Myanmar
50 C2 **Maurice, Lake** *imp. l.* Austr.
114 A3 **Mauritania** *country* Africa
113 K9 **Mauritius** *country* Indian Ocean
78 B2 **Māwān, Khashm** *mt.* Saudi Arabia
118 B3 **Mawanga** Dem. Rep. Congo
71 B3 **Mawei** China
62 A1 **Mawkmai** Myanmar
62 A1 **Mawlaik** Myanmar
Mawlamyine Myanmar *see* **Moulmein**
78 B2 **Mawqaq** Saudi Arabia
78 B3 **Mawza'** Yemen
108 A3 **Maxia, Punta** *mt. Sardinia* Italy
83 J2 **Maya** *r.* Russia
147 C2 **Mayaguana** *i.* Bahamas
96 B3 **Maybole** U.K.
104 B2 **Mayenne** France
104 B2 **Mayenne** *r.* France
128 C2 **Mayerthorpe** Can.
140 B3 **Mayfield** U.S.A.
87 D4 **Maykop** Russia
126 B2 **Mayo** Can.
118 B3 **Mayoko** Congo
121 D2 **Mayotte** *terr.* Africa
83 J3 **Mayskiy** Russia
140 C3 **Maysville** U.S.A.
118 B3 **Mayumba** Gabon

137	D1	Mayville U.S.A.
151	C2	Mazagão Brazil
104	C3	Mazamet France
77	C3	Mazar China
108	B3	Mazara del Vallo *Sicily* Italy
77	C3	Mazar-e Sharif Afgh.
144	A2	Mazatán Mex.
146	A3	Mazatenango Guat.
144	A2	Mazatlán Mex.
88	B2	Mažeikiai Lith.
88	B2	Mazirbe Latvia
71	C3	Mazu Dao *i.* Taiwan
121	B3	Mazunga Zimbabwe
88	C3	Mazyr Belarus
123	D2	Mbabane Swaziland
118	B2	Mbaïki C.A.R.
121	C1	Mbala Zambia
119	D2	Mbale Uganda
118	B2	Mbalmayo Cameroon
118	B3	Mbandaka Dem. Rep. Congo
118	A2	Mbanga Cameroon
120	A1	M'banza Congo Angola
119	D3	Mbeya Tanz.
119	D4	Mbinga Tanz.
123	D2	Mbombela S. Africa
118	B2	Mbomo Congo
118	B2	Mbouda Cameroon
114	A3	Mbour Senegal
114	A3	Mbout Maur.
118	C3	Mbuji-Mayi Dem. Rep. Congo
119	D3	Mbuyuni Tanz.
139	D2	McAlester U.S.A.
139	D3	McAllen U.S.A.
128	B2	McBride Can.
134	C2	McCall U.S.A.
126	E2	McClintock Channel Can.
126	E2	McClure Strait Can.
142	B2	McComb U.S.A.
136	C2	McConaughy, Lake U.S.A.
136	C2	McCook U.S.A.
134	C2	McDermitt U.S.A.
134	C2	McDonald Peak U.S.A.
134	D1	McGuire, Mount U.S.A.
		McKinley, Mount *mt* U.S.A. *see* Denali
128	C2	McLennan Can.
128	B2	McLeod Lake Can.
134	C2	McMinnville *OR* U.S.A.
142	C1	McMinnville *TN* U.S.A.
137	D3	McPherson U.S.A.
123	C3	Mdantsane S. Africa
135	D3	Mead, Lake *resr* U.S.A.
129	D2	Meadow Lake Can.
140	C2	Meadville U.S.A.
66	D2	Meaken-dake *vol.* Japan
106	B3	Mealhada Port.
131	E1	Mealy Mountains Can.
128	C2	Meander River Can.
78	A2	Mecca Saudi Arabia
100	B2	Mechelen Belgium
100	B2	Mechelen Neth.
100	C2	Mechernich Ger.
100	C2	Meckenheim Ger.
106	B1	Meda Port.
60	C1	Medan Indon.
153	B5	Medanosa, Punta *pt* Arg.
73	C4	Medawachchiya Sri Lanka
107	D2	Médéa Alg.
150	A1	Medellín Col.
115	D1	Medenine Tunisia
134	B2	Medford U.S.A.
110	C2	Medgidia Romania
110	B1	Medias Romania
136	B2	Medicine Bow Mountains U.S.A.
136	B2	Medicine Bow Peak U.S.A.
129	C2	Medicine Hat Can.
137	D3	Medicine Lodge U.S.A.
155	D1	Medina Brazil
78	A2	Medina Saudi Arabia
106	C1	Medinaceli Spain
106	C1	Medina del Campo Spain
106	B1	Medina de Rioseco Spain
84	G5	Mediterranean Sea
129	C2	Medley Can.
87	E3	Mednogorsk Russia
83	L2	Medvezh'i, Ostrova *is* Russia
86	C2	Medvezh'yegorsk Russia
50	A2	Meekatharra Austr.
136	B2	Meeker U.S.A.
75	B2	Meerut India
100	A2	Meetkerke Belgium
111	B3	Megalopoli Greece
75	C2	Meghasani *mt.* India
80	A2	Megisti *i.* Greece
92	I1	Mehamn Norway
50	A2	Meharry, Mount Austr.
79	C2	Mehrān *watercourse* Iran
76	C3	Mehrestān Iran
74	B1	Mehtar Lām Afgh.
154	C1	Meia Ponte *r.* Brazil
118	B2	Meiganga Cameroon
65	B1	Meihekou China
		Meijiang China *see* Ningdu
62	A1	Meiktila Myanmar
101	E2	Meiningen Ger.
102	C1	Meißen Ger.
71	B3	Meizhou China
152	B3	Mejicana *mt.* Arg.
152	A3	Mejillones Chile
117	B3	Mek'elē Eth.
114	C2	Mek'errhane, Sebkha *salt pan* Alg.
114	B1	Meknès Morocco
63	B2	Mekong *r.* Asia
63	B3	Mekong, Mouths of the Vietnam
60	B1	Melaka Malaysia
53	B3	Melbourne Austr.
143	D3	Melbourne U.S.A.
108	A2	Mele, Capo *c.* Italy
131	C1	Mélèzes, Rivière aux *r.* Can.
115	D3	Melfi Chad
109	C2	Melfi Italy
129	D2	Melfort Can.
92	F3	Melhus Norway
106	B1	Melide Spain
114	B1	Melilla N. Africa
91	D2	Melitopol' Ukr.
101	D1	Melle Ger.
93	F4	Mellerud Sweden
101	E2	Mellrichstadt Ger.
100	D1	Mellum *i.* Ger.
152	C4	Melo Uru.
115	C1	Melrhir, Chott *salt l.* Alg.
99	C2	Melton Mowbray U.K.
104	C2	Melun France
129	D2	Melville Can.
51	D1	Melville, Cape Austr.
131	E1	Melville, Lake Can.
50	C1	Melville Island Austr.
126	D1	Melville Island Can.
127	F2	Melville Peninsula Can.
102	C2	Memmingen Ger.
61	B1	Mempawah Indon.
80	B3	Memphis *tourist site* Egypt
142	B1	Memphis *TN* U.S.A.
139	C2	Memphis *TX* U.S.A.
91	C1	Mena Ukr.
142	B2	Mena U.S.A.
114	C3	Ménaka Mali
		Mènam Khong *r.* Laos/Thai. *see* Mekong
105	C3	Mende France
116	B3	Mendefera Eritrea
145	C2	Méndez Mex.
119	D2	Mendī Eth.
59	D3	Mendi P.N.G.
99	B3	Mendip Hills U.K.
153	B4	Mendoza Arg.
111	C3	Menemen Turkey
60	B2	Menggala Indon.
62	B1	Mengzi China
131	D1	Menihek Can.
52	B3	Menindee Austr.
52	B2	Menindee Lake Austr.
52	A3	Meningie Austr.
104	C2	Mennecy France
140	B1	Menominee U.S.A.
120	A2	Menongue Angola
		Menorca *i.* Spain *see* Minorca
60	A2	Mentawai, Kepulauan *is* Indon.
60	B2	Mentok Indon.
50	B2	Menzies Austr.
100	C1	Meppel Neth.
100	C1	Meppen Ger.
123	D1	Mepuze Moz.
123	C2	Meqheleng S. Africa
108	B1	Merano Italy
59	D3	Merauke Indon.
52	B2	Merbein Austr.
135	B3	Merced U.S.A.
152	C3	Mercedes Arg.
127	G2	Mercy, Cape Can.
139	C1	Meredith, Lake U.S.A.
91	D2	Merefa Ukr.
116	A3	Merga Oasis Sudan
63	A2	Mergui Archipelago *is* Myanmar
110	C2	Meriç *r.* Greece/Turkey
145	D2	Mérida Mex.
106	B2	Mérida Spain
150	A1	Mérida Venez.
142	C2	Meridian U.S.A.
53	C3	Merimbula Austr.
116	B3	Merowe Sudan
50	A3	Merredin Austr.
96	B3	Merrick *h.* U.K.
140	B1	Merrill U.S.A.
140	B2	Merrillville U.S.A.
128	B2	Merritt Can.
53	C2	Merrygoen Austr.
116	C3	Mersa Fatma Eritrea
100	C3	Mersch Lux.
101	E2	Merseburg Ger.
98	B2	Mersey *r.* U.K.
80	B2	Mersin Turkey
60	B1	Mersing Malaysia
99	D3	Mers-les-Bains France
74	B2	Merta India
99	B3	Merthyr Tydfil U.K.
106	B2	Mértola Port.
76	B2	Mertvyy Kultuk, Sor *dry lake* Kazakh.
119	D3	Meru *vol.* Tanz.
100	C3	Merzig Ger.
138	A2	Mesa U.S.A.
109	C2	Mesagne Italy
101	D2	Meschede Ger.
89	E3	Meshchovsk Russia
91	E2	Meshkovskaya Russia
111	B2	Mesimeri Greece
111	B3	Mesolongi Greece
121	D2	Messalo *r.* Moz.
109	C3	Messina *Sicily* Italy
109	C3	Messina, Strait of Italy
111	B3	Messiniakos Kolpos *g.* Greece
110	B2	Mesta *r.* Bulg.
150	B1	Meta *r.* Col./Venez.
127	G2	Meta Incognita Peninsula Can.
152	B3	Metán Arg.

111 B3 **Methoni** Greece
109 C2 **Metković** Croatia
60 B2 **Metro** Indon.
117 B4 **Metu** Eth.
105 D2 **Metz** France
100 B2 **Meuse** *r.* Belgium/France
139 D2 **Mexia** U.S.A.
144 A1 **Mexicali** Mex.
144 B2 **Mexico** *country* Central America
137 E3 **Mexico** U.S.A.
125 I5 **Mexico, Gulf of** Mex./U.S.A.
145 C3 **Mexico City** Mex.
101 F1 **Meyenburg** Ger.
86 D2 **Mezen'** Russia
86 D2 **Mezen'** *r.* Russia
86 E1 **Mezhdusharskiy, Ostrov** *i.* Russia
103 E2 **Meztúr** Hungary
144 B2 **Mezquitic** Mex.
88 C2 **Mežvidi** Latvia
121 C2 **Mfuwe** Zambia
123 D2 **Mhlume** Swaziland
74 B2 **Mhow** India
145 C3 **Miahuatlán** Mex.
106 B2 **Miajadas** Spain
143 D3 **Miami** *FL* U.S.A.
139 E1 **Miami** *OK* U.S.A.
143 D3 **Miami Beach** U.S.A.
81 C2 **Mīāndoāb** Iran
121 □D2 **Miandrivazo** Madag.
81 C2 **Mīāneh** Iran
74 B1 **Mianwali** Pak.
70 A2 **Mianyang** China
121 □D2 **Miarinarivo** Madag.
87 F3 **Miass** Russia
128 C2 **Mica Creek** Can.
103 E2 **Michalovce** Slovakia
140 B1 **Michigan** *state* U.S.A.
140 B2 **Michigan, Lake** U.S.A.
140 B2 **Michigan City** U.S.A.
130 B2 **Michipicoten Island** Can.
130 B2 **Michipicoten River** Can.
89 F3 **Michurinsk** Russia
48 D2 **Micronesia, Federated States of** *country* N. Pacific Ocean
100 A2 **Middelburg** Neth.
123 C3 **Middelburg** *E. Cape* S. Africa
123 C2 **Middelburg** *Mpumalanga* S. Africa
100 B2 **Middelharnis** Neth.
134 B2 **Middle Alkali Lake** U.S.A.
73 D3 **Middle Andaman** *i.* India
136 D2 **Middle Loup** *r.* U.S.A.
140 C3 **Middlesboro** U.S.A.
98 C1 **Middlesbrough** U.K.
141 E2 **Middletown** *NY* U.S.A.
140 C3 **Middletown** *OH* U.S.A.
78 B3 **Mīdī** Yemen
130 C2 **Midland** Can.
140 C2 **Midland** *MI* U.S.A.
139 C2 **Midland** *TX* U.S.A.
97 B3 **Midleton** Ireland
94 B1 **Miðvágur** Faroe Is
109 D2 **Midzhur** *mt.* Bulgaria/Serbia
103 E1 **Mielec** Pol.
110 C1 **Miercurea-Ciuc** Romania
106 B1 **Mieres del Camín** Spain
101 E1 **Mieste** Ger.
145 C3 **Miguel Alemán, Presa** *resr* Mex.
144 B2 **Miguel Auza** Mex.
144 B2 **Miguel Hidalgo, Presa** *resr* Mex.
63 A2 **Migyaunglaung** Myanmar
89 E3 **Mikhaylov** Russia
Mikhaylovgrad Bulg. *see* **Montana**
66 B2 **Mikhaylovka** Russia
77 D1 **Mikhaylovskoye** Russia
93 I3 **Mikkeli** Fin.
86 E2 **Mikun'** Russia
67 C3 **Mikuni-sanmyaku** *mts* Japan
108 A1 **Milan** Italy
121 C2 **Milange** Moz.
Milano Italy *see* **Milan**
111 C3 **Milas** Turkey
137 D1 **Milbank** U.S.A.
99 D2 **Mildenhall** U.K.
52 B2 **Mildura** Austr.
62 B1 **Mile** China
136 B1 **Miles City** U.S.A.
141 D3 **Milford** *DE* U.S.A.
135 D3 **Milford** *UT* U.S.A.
99 A3 **Milford Haven** U.K.
54 A2 **Milford Sound** N.Z.
107 D2 **Miliana** Alg.
116 B3 **Milk, Wadi el** *watercourse* Sudan
83 L3 **Mil'kovo** Russia
129 C3 **Milk River** Can.
105 C3 **Millau** France
143 D2 **Milledgeville** U.S.A.
137 E1 **Mille Lacs** *lakes* U.S.A.
130 A2 **Mille Lacs, Lac des** *l.* Can.
137 D2 **Miller** U.S.A.
91 E2 **Millerovo** Russia
52 A2 **Millers Creek** Austr.
52 B3 **Millicent** Austr.
141 F1 **Millinocket** U.S.A.
53 D1 **Millmerran** Austr.
98 B1 **Millom** U.K.
128 C1 **Mills Lake** Can.
111 B3 **Milos** *i.* Greece
89 E3 **Miloslavskoye** Russia
91 E2 **Milove** Ukr.
52 B1 **Milparinka** Austr.
54 A3 **Milton** N.Z.
99 C2 **Milton Keynes** U.K.
140 B2 **Milwaukee** U.S.A.
104 B3 **Mimizan** France
118 B3 **Mimongo** Gabon
79 C2 **Mīnāb** Iran
58 C2 **Minahasa, Semenanjung** *pen.* Indon.
79 C2 **Mīnā' Jabal 'Alī** U.A.E.
60 B1 **Minas** Indon.
153 C4 **Minas** Uru.
155 D1 **Minas Novas** Brazil
145 C3 **Minatitlán** Mex.
62 A1 **Minbu** Myanmar
64 B2 **Mindanao** *i.* Phil.
101 D1 **Minden** Ger.
142 B2 **Minden** *LA* U.S.A.
137 D2 **Minden** *NE* U.S.A.
64 B1 **Mindoro** *i.* Phil.
64 A1 **Mindoro Strait** Phil.
118 B3 **Mindouli** Congo
99 B3 **Minehead** U.K.
154 B1 **Mineiros** Brazil
139 D2 **Mineral Wells** U.S.A.
77 E3 **Minfeng** China
119 C4 **Minga** Dem. Rep. Congo
81 C1 **Mingäçevir** Azer.
131 D1 **Mingan** Can.
52 B2 **Mingary** Austr.
70 B2 **Mingguang** China
62 A1 **Mingin** Myanmar
107 C2 **Minglanilla** Spain
119 D4 **Mingoyo** Tanz.
69 E1 **Mingshui** China
71 B3 **Mingxi** China
73 B4 **Minicoy** *atoll* India
50 A2 **Minilya** Austr.
131 D1 **Minipi Lake** Can.
130 A1 **Miniss Lake** Can.
115 C4 **Minna** Nigeria
137 E2 **Minneapolis** U.S.A.
129 E2 **Minnedosa** Can.
137 E2 **Minnesota** *r.* U.S.A.
137 E1 **Minnesota** *state* U.S.A.
106 B1 **Miño** *r.* Port./Spain
107 D1 **Minorca** *i.* Spain
136 C1 **Minot** U.S.A.
88 C3 **Minsk** Belarus
103 E1 **Mińsk Mazowiecki** Pol.
131 D2 **Minto** Can.
130 C1 **Minto, Lac** *l.* Can.
70 A2 **Minxian** China
155 D1 **Mirabela** Brazil
155 D1 **Miralta** Brazil
131 D2 **Miramichi** Can.
111 C3 **Mirampellou, Kolpos** *b.* Greece
152 C3 **Miranda** Brazil
152 C2 **Miranda** *r.* Brazil
106 C1 **Miranda de Ebro** Spain
106 B1 **Mirandela** Port.
154 B2 **Mirandópolis** Brazil
79 C3 **Mirbāţ** Oman
61 C1 **Miri** *Sarawak* Malaysia
153 C4 **Mirim, Lagoa** *l.* Brazil/Uru.
79 D2 **Mīrjāveh** Iran
83 I2 **Mirnyy** Russia
101 F1 **Mirow** Ger.
74 A2 **Mirpur Khas** Pak.
111 B3 **Mirtoö Pelagos** *sea* Greece
65 B2 **Miryang** S. Korea
75 C2 **Mirzapur** India
66 B1 **Mishan** China
146 B3 **Miskitos, Cayos** *is* Nic.
103 E2 **Miskolc** Hungary
59 C3 **Misoöl** *i.* Indon.
115 D1 **Mişrātah** Libya
130 B1 **Missinaibi** *r.* Can.
130 B2 **Missinaibi Lake** Can.
128 B3 **Mission** Can.
130 B1 **Missisa Lake** Can.
142 C3 **Mississippi** *r.* U.S.A.
142 C2 **Mississippi** *state* U.S.A.
142 C3 **Mississippi Delta** U.S.A.
134 D1 **Missoula** U.S.A.
137 E3 **Missouri** *r.* U.S.A.
137 E3 **Missouri** *state* U.S.A.
131 C2 **Mistassibi** *r.* Can.
130 C1 **Mistassini, Lac** *l.* Can.
131 D1 **Mistastin Lake** Can.
103 D2 **Mistelbach** Austria
131 D1 **Mistinibi, Lac** *l.* Can.
130 C1 **Mistissini** Can.
51 D2 **Mitchell** Austr.
51 D1 **Mitchell** *r.* Austr.
137 D2 **Mitchell** U.S.A.
97 B2 **Mitchelstown** Ireland
74 A2 **Mithi** Pak.
67 D3 **Mito** Japan
119 D3 **Mitole** Tanz.
109 D2 **Mitrovicë** Kosovo
53 D2 **Mittagong** Austr.
101 E2 **Mittelhausen** Ger.
101 D1 **Mittellandkanal** *canal* Ger.
101 F3 **Mitterteich** Ger.
150 A1 **Mitú** Col.
119 C4 **Mitumba, Chaîne des** *mts* Dem. Rep. Congo
119 C3 **Mitumba, Monts** *mts* Dem. Rep. Congo
118 B2 **Mitzic** Gabon
78 B2 **Miyah, Wādī al** *watercourse* Saudi Arabia
67 C4 **Miyake-jima** *i.* Japan
66 D3 **Miyako** Japan
67 B4 **Miyakonojō** Japan
76 B2 **Miyaly** Kazakh.

Miyang China see Mile
67 B4 Miyazaki Japan
115 D1 Mizdah Libya
97 B3 Mizen Head Ireland
90 A2 Mizhhirr"ya Ukr.
93 G4 Mjölby Sweden
93 F3 Mjøsa l. Norway
119 D3 Mkomazi Tanz.
103 C1 Mladá Boleslav Czech Rep.
109 D2 Mladenovac Serbia
103 E1 Mława Pol.
123 C3 Mlungisi S. Africa
90 B1 Mlyniv Ukr.
123 C2 Mmabatho S. Africa
123 C2 Mmathethe Botswana
135 E3 Moab U.S.A.
123 D2 Moamba Moz.
119 C3 Moba Dem. Rep. Congo
118 C2 Mobayi-Mbongo Dem. Rep. Congo
137 E3 Moberly U.S.A.
142 C2 Mobile U.S.A.
142 C2 Mobile Bay U.S.A.
136 C1 Mobridge U.S.A.
121 D2 Moçambique Moz.
62 B1 Môc Châu Vietnam
78 B3 Mocha Yemen
123 C1 Mochudi Botswana
121 D2 Mocimboa da Praia Moz.
150 A1 Mocoa Col.
154 C2 Mococa Brazil
144 B2 Mocorito Mex.
144 B1 Moctezuma Chihuahua Mex.
145 B2 Moctezuma San Luis Potosí Mex.
144 B2 Moctezuma Sonora Mex.
121 C2 Mocuba Moz.
105 D2 Modane France
122 B2 Modder r. S. Africa
108 B2 Modena Italy
135 B3 Modesto U.S.A.
123 C1 Modimolle S. Africa
123 D1 Modjadjiskloof S. Africa
53 C3 Moe Austr.
100 C2 Moers Ger.
96 C3 Moffat U.K.
117 C4 Mogadishu Somalia
123 C1 Mogalakwena r. S. Africa
62 A1 Mogaung Myanmar
155 C2 Mogi das Cruzes Brazil
Mogilev Belarus see Mahilyow
154 C2 Mogi Mirim Brazil
83 I3 Mogocha Russia
62 A1 Mogok Myanmar
103 D2 Mohács Hungary
123 C3 Mohale's Hoek Lesotho
74 B1 Mohali India
79 C2 Moḩammadābād Iran
107 D2 Mohammadia Alg.
141 E2 Mohawk r. U.S.A.
119 D3 Mohoro Tanz.
90 B2 Mohyliv-Podil's'kyy Ukr.
110 C1 Moinești Romania
92 F2 Mo i Rana Norway
104 C3 Moissac France
135 C3 Mojave U.S.A.
135 C3 Mojave Desert U.S.A.
62 B1 Mojiang China
154 C2 Moji-Guaçu r. Brazil
54 B1 Mokau N.Z.
123 C2 Mokhotlong Lesotho
123 C1 Mokopane S. Africa
65 B3 Mokpo S. Korea
145 C2 Molango Mex.
Moldavia country Europe see Moldova
92 E3 Molde Norway
90 B2 Moldova country Europe
110 B1 Moldoveanu, Vârful mt. Romania
90 B2 Moldovei Centrale, Podișul plat. Moldova
123 C1 Molepolole Botswana
88 C2 Moltai Lith.
109 C2 Molfetta Italy
107 C1 Molina de Aragón Spain
150 A3 Mollendo Peru
53 C2 Molong Austr.
122 B2 Molopo watercourse Botswana/S. Africa
118 B2 Moloundou Cameroon
59 C3 Moluccas is Indon.
Molucca Sea Indon. see Maluku, Laut
52 B2 Momba Austr.
119 D3 Mombasa Kenya
154 B1 Mombuca, Serra da hills Brazil
93 F4 Møn i. Denmark
105 D3 Monaco country Europe
96 B2 Monadhliath Mountains U.K.
97 C1 Monaghan Ireland
89 D3 Monastyrshchina Russia
90 B2 Monastyryshche Ukr.
66 D2 Monbetsu Japan
108 A1 Moncalieri Italy
107 C1 Moncayo mt. Spain
86 C2 Monchegorsk Russia
100 C2 Mönchengladbach Ger.
144 B2 Monclova Mex.
131 D2 Moncton Can.
106 B1 Mondego r. Port.
108 A2 Mondovì Italy
111 B3 Monemvasia Greece
66 D1 Moneron, Ostrov i. Russia
137 E3 Monett U.S.A.
108 B1 Monfalcone Italy
106 B1 Monforte de Lemos Spain
62 B1 Mông Cai Vietnam
62 B1 Mong Lin Myanmar
68 C1 Mongolia country Asia
62 A1 Mong Pawk Myanmar
62 A1 Mong Ping Myanmar
120 B2 Mongu Zambia
135 C3 Monitor Range mts U.S.A.
114 C4 Mono r. Benin/Togo
135 C3 Mono Lake U.S.A.
109 C2 Monopoli Italy
107 C1 Monreal del Campo Spain
142 B2 Monroe LA U.S.A.
140 B2 Monroe WI U.S.A.
142 C2 Monroeville U.S.A.
114 A4 Monrovia Liberia
100 A2 Mons Belgium
122 B3 Montagu S. Africa
109 C3 Montalto mt. Italy
110 B2 Montana Bulg.
134 E1 Montana state U.S.A.
104 C2 Montargis France
104 C3 Montauban France
141 E2 Montauk Point U.S.A.
123 C2 Mont-aux-Sources mt. Lesotho
105 C2 Montbard France
105 C2 Montbrison France
100 B3 Montcornet France
104 B3 Mont-de-Marsan France
104 C2 Montdidier France
151 C2 Monte Alegre Brazil
105 D3 Monte-Carlo Monaco
152 C4 Monte Caseros Arg.
146 C3 Montego Bay Jamaica
105 C3 Montélimar France
109 C2 Montella Italy
145 C2 Montemorelos Mex.
104 B2 Montendre France
109 C2 Montenegro country Europe
121 C2 Montepuez Moz.
108 B2 Montepulciano Italy
135 B3 Monterey U.S.A.
135 B3 Monterey Bay U.S.A.
150 A1 Montería Col.
152 B2 Montero Bol.
145 B2 Monterrey Mex.
109 C2 Montesano sulla Marcellana Italy
109 C2 Monte Sant'Angelo Italy
151 E3 Monte Santo Brazil
108 A2 Monte Santu, Capo di c. Sardinia Italy
155 D1 Montes Claros Brazil
153 C4 Montevideo Uru.
137 D2 Montevideo U.S.A.
136 B3 Monte Vista U.S.A.
142 C2 Montgomery U.S.A.
105 D2 Monthey Switz.
142 B2 Monticello AR U.S.A.
135 E3 Monticello UT U.S.A.
104 C2 Montignac France
105 D2 Montigny-le-Roi France
106 C2 Montilla Spain
131 D2 Mont-Joli Can.
130 C2 Mont-Laurier Can.
104 C2 Montluçon France
131 C2 Montmagny Can.
104 C2 Montmorillon France
51 E2 Monto Austr.
134 D2 Montpelier ID U.S.A.
141 E2 Montpelier VT U.S.A.
105 C3 Montpellier France
130 C2 Montréal Can.
129 D2 Montreal Lake Can.
129 D2 Montreal Lake Can.
99 D3 Montreuil France
105 D2 Montreux Switz.
96 C2 Montrose U.K.
136 B3 Montrose U.S.A.
147 D3 Montserrat terr. West Indies
62 A1 Monywa Myanmar
108 A1 Monza Italy
107 D1 Monzón Spain
123 C1 Mookane Botswana
123 C1 Mookgophong S. Africa
53 D1 Moonie Austr.
53 C1 Moonie r. Austr.
52 A2 Moonta Austr.
50 A2 Moore, Lake imp. l. Austr.
137 D1 Moorhead U.S.A.
53 C3 Mooroopna Austr.
122 A3 Moorreesburg S. Africa
130 B1 Moose r. Can.
130 B1 Moose Factory Can.
141 F1 Moosehead Lake U.S.A.
129 D2 Moose Jaw Can.
137 E1 Moose Lake U.S.A.
129 D2 Moosomin Can.
130 B1 Moosonee Can.
52 B2 Mootwingee Austr.
123 C1 Mopane S. Africa
114 B3 Mopti Mali
150 A3 Moquegua Peru
93 F3 Mora Sweden
137 E1 Mora U.S.A.
75 B2 Moradabad India
121 □D2 Moramanga Madag.
103 D2 Morava r. Europe
96 B2 Moray Firth b. U.K.
100 C3 Morbach Ger.
74 B2 Morbi India
104 B3 Morcenx France
69 E1 Mordaga China
129 E3 Morden Can.
91 E1 Mordovo Russia

98 B1 **Morecambe** U.K.
98 B1 **Morecambe Bay** U.K.
53 C1 **Moree** Austr.
59 D3 **Morehead** P.N.G.
140 C3 **Morehead** U.S.A.
143 E2 **Morehead City** U.S.A.
145 B3 **Morelia** Mex.
107 C1 **Morella** Spain
106 B2 **Morena, Sierra** *mts* Spain
110 C2 **Moreni** Romania
128 A2 **Moresby, Mount** Can.
142 B3 **Morgan City** U.S.A.
143 D1 **Morganton** U.S.A.
140 D3 **Morgantown** U.S.A.
105 D2 **Morges** Switz.
77 C3 **Morghāb, Daryā-ye** *r.* Afgh.
66 D2 **Mori** Japan
128 B2 **Morice Lake** Can.
66 D3 **Morioka** Japan
53 D2 **Morisset** Austr.
104 B2 **Morlaix** France
51 C1 **Mornington Island** Austr.
59 D3 **Morobe** P.N.G.
114 B1 **Morocco** *country* Africa
119 D3 **Morogoro** Tanz.
64 B2 **Moro Gulf** Phil.
122 B2 **Morokweng** S. Africa
121 □D3 **Morombe** Madag.
68 C1 **Mörön** Mongolia
121 □D3 **Morondava** Madag.
121 D2 **Moroni** Comoros
59 C2 **Morotai** *i.* Indon.
119 D2 **Moroto** Uganda
98 C1 **Morpeth** U.K.
86 F2 **Morrasale** Russia
154 C1 **Morrinhos** Brazil
129 E3 **Morris** Can.
137 D1 **Morris** U.S.A.
143 D1 **Morristown** U.S.A.
87 D3 **Morshansk** Russia
154 B1 **Mortes, Rio das** *r.* Brazil
52 B3 **Mortlake** Austr.
53 D3 **Moruya** Austr.
96 B2 **Morvern** *reg.* U.K.
53 C3 **Morwell** Austr.
102 B2 **Mosbach** Ger.
89 E2 **Moscow** Russia
134 C1 **Moscow** U.S.A.
100 C2 **Mosel** *r.* Ger.
105 D2 **Moselle** *r.* France
134 C1 **Moses Lake** U.S.A.
92 □A3 **Mosfellsbær** Iceland
54 B3 **Mosgiel** N.Z.
89 D2 **Moshenskoye** Russia
119 D3 **Moshi** Tanz.
92 F2 **Mosjøen** Norway
Moskva Russia *see* **Moscow**
103 D2 **Mosonmagyaróvár** Hungary
146 B3 **Mosquitos, Costa de** *coastal area* Nic.
146 B4 **Mosquitos, Golfo de los** *b.* Panama
93 F4 **Moss** Norway
122 B3 **Mossel Bay** S. Africa
122 B3 **Mossel Bay** S. Africa
118 B3 **Mossendjo** Congo
52 B2 **Mossgiel** Austr.
51 D1 **Mossman** Austr.
151 E2 **Mossoró** Brazil
53 D2 **Moss Vale** Austr.
102 C1 **Most** Czech Rep.
114 C1 **Mostaganem** Alg.
109 C2 **Mostar** Bos. & Herz.
152 C4 **Mostardas** Brazil
81 C2 **Mosul** Iraq
93 G4 **Motala** Sweden
96 C3 **Motherwell** U.K.
107 C2 **Motilla del Palancar** Spain
122 B1 **Motokwe** Botswana
106 C2 **Motril** Spain
110 B2 **Motru** Romania
63 A2 **Mottama, Gulf of** Myanmar
145 D2 **Motul** Mex.
111 C3 **Moudros** Greece
118 B3 **Mouila** Gabon
52 B3 **Moulamein** Austr.
105 C2 **Moulins** France
63 A2 **Moulmein** Myanmar
143 D2 **Moultrie** U.S.A.
143 E2 **Moultrie, Lake** U.S.A.
140 B3 **Mound City** U.S.A.
115 D4 **Moundou** Chad
137 E3 **Mountain Grove** U.S.A.
142 B1 **Mountain Home** *AR* U.S.A.
134 C2 **Mountain Home** *ID* U.S.A.
143 D1 **Mount Airy** U.S.A.
52 A3 **Mount Barker** Austr.
53 C3 **Mount Beauty** Austr.
121 C2 **Mount Darwin** Zimbabwe
141 F2 **Mount Desert Island** U.S.A.
123 C3 **Mount Fletcher** S. Africa
123 C3 **Mount Frere** S. Africa
52 B3 **Mount Gambier** Austr.
59 D3 **Mount Hagen** P.N.G.
53 C2 **Mount Hope** Austr.
51 C2 **Mount Isa** Austr.
50 A2 **Mount Magnet** Austr.
52 B2 **Mount Manara** Austr.
54 C1 **Mount Maunganui** N.Z.
111 B2 **Mount Olympus** Greece
137 E2 **Mount Pleasant** *IA* U.S.A.
140 C2 **Mount Pleasant** *MI* U.S.A.
139 E2 **Mount Pleasant** *TX* U.S.A.
99 A3 **Mount's Bay** U.K.
134 B2 **Mount Shasta** U.S.A.
140 B3 **Mount Vernon** *IL* U.S.A.
140 C2 **Mount Vernon** *OH* U.S.A.
134 B1 **Mount Vernon** *WA* U.S.A.
51 D2 **Moura** Austr.
115 E3 **Mourdi, Dépression du** *depr.* Chad
97 C1 **Mourne Mountains** *hills* U.K.
100 A2 **Mouscron** Belgium
115 D3 **Moussoro** Chad
58 C2 **Moutong** Indon.
115 C2 **Mouydir, Monts du** *plat.* Alg.
100 B3 **Mouzon** France
97 B1 **Moy** *r.* Ireland
117 B4 **Moyale** Eth.
123 C3 **Moyeni** Lesotho
76 B2 **Mo'ynoq** Uzbek.
77 D2 **Moyynty** Kazakh.
121 C3 **Mozambique** *country* Africa
113 I9 **Mozambique Channel** Africa
89 E2 **Mozhaysk** Russia
119 D3 **Mpanda** Tanz.
121 C2 **Mpika** Zambia
121 C1 **Mporokoso** Zambia
123 C2 **Mpumalanga** *prov.* S. Africa
62 A1 **Mrauk-U** Myanmar
88 C2 **Mshinskaya** Russia
107 D2 **M'Sila** Alg.
89 D2 **Msta** *r.* Russia
89 D2 **Mstinskiy Most** Russia
89 D3 **Mstsislaw** Belarus
123 C3 **Mthatha** S. Africa
89 E3 **Mtsensk** Russia
119 E4 **Mtwara** Tanz.
118 B3 **Muanda** Dem. Rep. Congo
63 B2 **Muang Không** Laos
62 B1 **Muang Ngoy** Laos
62 B2 **Muang Pakbeng** Laos
62 B1 **Muang Sing** Laos
62 B2 **Muang Vangviang** Laos
60 B1 **Muar** Malaysia
60 B2 **Muarabungo** Indon.
60 B2 **Muaradua** Indon.
61 C2 **Muaralaung** Indon.
60 A2 **Muarasiberut** Indon.
60 B2 **Muaratembesi** Indon.
61 C2 **Muarateweh** Indon.
119 D2 **Mubende** Uganda
115 D3 **Mubi** Nigeria
120 B2 **Muconda** Angola
155 E1 **Mucuri** Brazil
155 E1 **Mucuri** *r.* Brazil
66 A2 **Mudanjiang** China
66 A1 **Mudan Jiang** *r.* China
111 C2 **Mudanya** Turkey
101 E1 **Müden (Örtze)** Ger.
53 C2 **Mudgee** Austr.
63 A2 **Mudon** Myanmar
80 B1 **Mudurnu** Turkey
121 C2 **Mueda** Moz.
121 B2 **Mufulira** Zambia
120 B2 **Mufumbwe** Zambia
111 C3 **Muğla** Turkey
116 B2 **Muhammad Qol** Sudan
101 F2 **Mühlberg/Elbe** Ger.
101 E2 **Mühlhausen/Thüringen** Ger.
121 C2 **Muite** Moz.
65 B2 **Muju** S. Korea
90 A2 **Mukacheve** Ukr.
61 C1 **Mukah** *Sarawak* Malaysia
79 B3 **Mukalla** Yemen
63 B2 **Mukdahan** Thai.
50 A3 **Mukinbudin** Austr.
60 B2 **Mukomuko** Indon.
121 C2 **Mulanje, Mount** Malawi
101 F2 **Mulde** *r.* Ger.
144 A2 **Mulegé** Mex.
139 C2 **Muleshoe** U.S.A.
106 C2 **Mulhacén** *mt.* Spain
100 C2 **Mülheim an der Ruhr** Ger.
105 D2 **Mulhouse** France
66 B2 **Muling** China
66 B1 **Muling He** *r.* China
96 B2 **Mull** *i.* U.K.
53 C2 **Mullaley** Austr.
136 C2 **Mullen** U.S.A.
61 C1 **Muller, Pegunungan** *mts* Indon.
50 A2 **Mullewa** Austr.
97 C2 **Mullingar** Ireland
96 B3 **Mull of Galloway** *c.* U.K.
96 B3 **Mull of Kintyre** *hd* U.K.
96 A3 **Mull of Oa** *hd* U.K.
120 B2 **Mulobezi** Zambia
74 B1 **Multan** Pak.
73 B3 **Mumbai** India
120 B2 **Mumbwa** Zambia
145 D2 **Muna** Mex.
101 E2 **Münchberg** Ger.
München Ger. *see* **Munich**
140 B2 **Muncie** U.S.A.
50 B3 **Mundrabilla** Austr.
119 C2 **Mungbere** Dem. Rep. Congo
75 C2 **Munger** India
52 A1 **Mungeranie** Austr.
53 C1 **Mungindi** Austr.
102 C2 **Munich** Ger.
155 D2 **Muniz Freire** Brazil
101 E1 **Munster** *Niedersachsen* Ger.
100 C2 **Münster** *Nordrhein-Westfalen* Ger.
97 B2 **Munster** *reg.* Ireland
100 C2 **Münsterland** *reg.* Ger.
62 B1 **Mương Nhe** Vietnam
92 H2 **Muonio** Fin.
92 H2 **Muonioälven** *r.* Fin./Sweden
Muqdisho Somalia *see* **Mogadishu**
103 D2 **Mur** *r.* Austria
119 C3 **Muramvya** Burundi

Murang'a

119 D3 Murang'a Kenya
81 B2 Murat r. Turkey
111 C2 Muratlı Turkey
50 A2 Murchison watercourse Austr.
107 C2 Murcia Spain
111 C2 Mürefte Turkey
110 B1 Mureşul r. Romania
104 C3 Muret France
142 C1 Murfreesboro U.S.A.
77 D3 Murghob Tajik.
155 D2 Muriaé Brazil
120 B1 Muriege Angola
101 F1 Müritz l. Ger.
86 C2 Murmansk Russia
87 D3 Murom Russia
66 D2 Muroran Japan
106 B1 Muros Spain
67 B4 Muroto Japan
67 B4 Muroto-zaki pt Japan
143 D1 Murphy U.S.A.
53 C1 Murra Murra Austr.
52 A3 Murray r. Austr.
128 B2 Murray r. Can.
140 B3 Murray r. U.S.A.
143 D2 Murray, Lake U.S.A.
52 A3 Murray Bridge Austr.
122 B3 Murraysburg S. Africa
52 B3 Murrayville Austr.
52 B2 Murrumbidgee r. Austr.
121 C2 Murrupula Moz.
53 D2 Murrurundi Austr.
109 C1 Murska Sobota Slovenia
54 C1 Murupara N.Z.
49 I4 Mururoa atoll Fr. Polynesia
Murwara India see Katni
53 D1 Murwillumbah Austr.
115 D2 Murziq Libya
115 D2 Murzuq, Idhän des. Libya
81 C2 Muş Turkey
110 B2 Musala mt. Bulg.
65 B1 Musan N. Korea
78 B3 Musaymir Yemen
79 C2 Muscat Oman
137 E2 Muscatine U.S.A.
50 C2 Musgrave Ranges mts
        Austr.
118 B3 Mushie Dem. Rep. Congo
60 B2 Musi r. Indon.
123 D1 Musina S. Africa
140 B2 Muskegon U.S.A.
139 D1 Muskogee U.S.A.
128 B2 Muskwa r. Can.
74 A1 Muslimbagh Pak.
78 A3 Musmar Sudan
119 D3 Musoma Tanz.
96 C3 Musselburgh U.K.
88 B2 Mustjala Estonia
53 D2 Muswellbrook Austr.
116 A2 Müt Egypt
121 C2 Mutare Zimbabwe
66 D2 Mutsu Japan
121 C2 Mutuali Moz.
92 I2 Muurola Fin.
70 A2 Mu Us Shadi des. China
120 A1 Muxaluando Angola
86 C2 Muyezerskiy Russia
119 D3 Muyinga Burundi
74 B1 Muzaffargarh Pak.
75 C2 Muzaffarpur India
144 B2 Múzquiz Mex.
75 C1 Muz Shan mt. China
119 C3 Mwanza Dem. Rep. Congo
119 D3 Mwanza Tanz.
118 C3 Mweka Dem. Rep. Congo
121 B2 Mwenda Zambia
118 C3 Mwene-Ditu Dem. Rep.
        Congo
121 C3 Mwenezi Zimbabwe

119 C3 Mweru, Lake Dem. Rep.
        Congo/Zambia
118 C3 Mwimba Dem. Rep. Congo
120 B2 Mwinilunga Zambia
88 C3 Myadzyel Belarus
62 A2 Myanaung Myanmar
62 A1 Myanmar country Asia
63 A2 Myaungmya Myanmar
63 C3 Myeik Myanmar
        Myeik Kyunzu is Myanmar see
        Mergui Archipelago
62 A1 Myingyan Myanmar
62 A1 Myitkyina Myanmar
91 C2 Mykolayiv Ukr.
111 C3 Mykonos Greece
111 C3 Mykonos i. Greece
86 E2 Myla Russia
75 D2 Mymensingh Bangl.
67 C3 Myōkō Japan
65 B1 Myŏnggan N. Korea
88 C2 Myory Belarus
92 □B3 Mýrdalsjökull Iceland
91 C2 Myrhorod Ukr.
90 C2 Myronivka Ukr.
143 E2 Myrtle Beach U.S.A.
53 C3 Myrtleford Austr.
134 B2 Myrtle Point U.S.A.
89 E2 Myshkin Russia
103 C1 Myślibórz Pol.
        Mysore India see Mysuru
83 N3 Mys Shmidta Russia
73 B3 Mysuru India
63 B2 My Tho Vietnam
111 C3 Mytilini Greece
89 E3 Mytishchi Russia
123 C3 Mzamomhle S. Africa
121 C2 Mzimba Malawi
121 C2 Mzuzu Malawi

N

97 C2 Naas Ireland
122 A2 Nababeep S. Africa
87 E3 Naberezhnyye Chelny Russia
59 D3 Nabire Indon.
80 B2 Näblus West Bank
121 D2 Nacala Moz.
63 A2 Nachuge India
139 E2 Nacogdoches U.S.A.
144 B1 Nacozari de García Mex.
74 B2 Nadiad India
90 A2 Nadvirna Ukr.
86 C2 Nadvoitsy Russia
86 G2 Nadym Russia
93 F4 Næstved Denmark
111 B3 Nafpaktos Greece
111 B3 Nafplio Greece
115 D1 Nafüsah, Jabal hills Libya
78 B2 Nafy Saudi Arabia
64 B1 Naga Phil.
130 B1 Nagagami r. Can.
67 C3 Nagano Japan
67 C3 Nagaoka Japan
75 D2 Nagaon India
74 B1 Nagar India
74 B2 Nagar Parkar Pak.
67 A4 Nagasaki Japan
67 B4 Nagato Japan
74 B2 Nagaur India
73 B4 Nagercoil India
74 A2 Nagha Kalat Pak.
72 B2 Nagina India
81 C1 Nagorno-Karabakh disp. terr.
        Azer.
67 C3 Nagoya Japan
75 B2 Nagpur India

68 C2 Nagqu China
103 D2 Nagyatád Hungary
103 D2 Nagykanizsa Hungary
128 B1 Nahanni Butte Can.
76 A3 Nahävand Iran
101 E1 Nahrendorf Ger.
153 A5 Nahuel Huapí, Lago l. Arg.
131 D1 Nain Can.
81 D2 Nä'īn Iran
96 C2 Nairn U.K.
119 D3 Nairobi Kenya
119 D3 Naivasha Kenya
81 D2 Najafäbäd Iran
78 B2 Najd reg. Saudi Arabia
106 C1 Nájera Spain
65 C1 Najin N. Korea
78 B3 Najrän Saudi Arabia
        Nakambé r. Burkina Faso/
        Ghana see White Volta
67 C3 Nakatsugawa Japan
78 A3 Nakfa Eritrea
66 B2 Nakhodka Russia
63 B2 Nakhon Pathom Thai.
63 B2 Nakhon Ratchasima Thai.
63 B2 Nakhon Sawan Thai.
63 A3 Nakhon Si Thammarat Thai.
130 B1 Nakina Can.
93 F4 Nakonde Zambia
119 D3 Nakuru Kenya
128 C2 Nakusp Can.
75 D2 Nalbari India
87 D4 Nal'chik Russia
115 D1 Nälūt Libya
123 C2 Namahadi S. Africa
77 D2 Namangan Uzbek.
122 A2 Namaqualand reg. S. Africa
51 E2 Nambour Austr.
53 D2 Nambucca Heads Austr.
75 D1 Nam Co salt l. China
62 B1 Nam Đinh Vietnam
120 A3 Namib Desert Namibia
120 A2 Namibe Angola
120 A3 Namibia country Africa
72 D2 Namjagbarwa Feng mt.
        China
59 C3 Namlea Indon.
53 C2 Namoi r. Austr.
134 C2 Nampa U.S.A.
114 B3 Nampala Mali
65 B2 Namp'o N. Korea
121 C2 Nampula Moz.
62 A1 Namrup India
62 A1 Namsang Myanmar
92 F3 Namsos Norway
63 A2 Nam Tok Thai.
83 J2 Namtsy Russia
62 A1 Namtu Myanmar
100 B2 Namur Belgium
120 B2 Namwala Zambia
65 B2 Namwon S. Korea
62 A1 Namya Ra Myanmar
62 B2 Nan Thai.
128 B3 Nanaimo Can.
71 B3 Nan'an China
122 A1 Nananib Plateau Namibia
67 C3 Nanao Japan
71 B3 Nanchang Jiangxi China
71 B3 Nanchang Jiangxi China
70 A2 Nanchong China
63 A3 Nancowry i. India
105 D2 Nancy France
75 C1 Nanda Devi mt. India
71 A3 Nandan China
73 B3 Nanded India
74 B2 Nandurbar India
73 B3 Nandyal India
71 B3 Nanfeng China

118 B2 Nanga Eboko Cameroon
61 C2 Nangahpinoh Indon.
74 B1 Nanga Parbat mt. Pak.
61 C2 Nangatayap Indon.
70 B2 Nangong China
119 D3 Nangulangwa Tanz.
70 C2 Nanhui China
70 B2 Nanjing China
Nanking China see Nanjing
120 A2 Nankova Angola
71 B3 Nan Ling mts China
71 A3 Nanning China
127 H2 Nanortalik Greenland
71 A3 Nanpan Jiang r. China
75 C2 Nanpara India
71 B3 Nanping China
Nansei-shotō is Japan see
Ryukyu Islands
104 B2 Nantes France
70 C2 Nantong China
141 F2 Nantucket Island U.S.A.
155 D1 Nanuque Brazil
64 B2 Nanusa, Kepulauan is Indon.
71 B3 Nanxiong China
70 B2 Nanyang China
70 B2 Nanzhang China
107 D2 Nao, Cabo de la c. Spain
131 C1 Naococane, Lac l. Can.
135 B3 Napa U.S.A.
126 D2 Napaktulik Lake Can.
127 H2 Napasoq Greenland
54 C1 Napier N.Z.
108 B2 Naples Italy
143 D3 Naples U.S.A.
150 A2 Napo r. Ecuador/Peru
Napoli Italy see Naples
114 B3 Nara Mali
93 I4 Narach Belarus
52 B3 Naracoorte Austr.
145 C2 Naranjos Mex.
63 B3 Narathiwat Thai.
105 C3 Narbonne France
63 A2 Narcondam Island India
127 G1 Nares Strait Can./Greenland
122 A1 Narib Namibia
87 D4 Narimanov Russia
67 D3 Narita Japan
74 B2 Narmada r. India
74 B2 Narnaul India
108 B2 Narni Italy
90 B1 Narodychi Ukr.
89 E2 Naro-Fominsk Russia
88 C3 Narooma Austr.
88 C3 Narowlya Belarus
53 C2 Narrabri Austr.
53 C2 Narrandera Austr.
53 C2 Narromine Austr.
88 C2 Narva Estonia
88 C2 Narva Bay Estonia/Russia
92 H2 Narvik Norway
88 C2 Narvskoye Vodokhranilishche
resr Estonia/Russia
86 E2 Nar'yan-Mar Russia
77 D2 Naryn Kyrg.
74 B2 Nashik India
141 E2 Nashua U.S.A.
142 C1 Nashville U.S.A.
49 I5 Nasinu Fiji
117 B4 Nasir South Sudan
32 Nass r. Can.
146 C2 Nassau Bahamas
116 B2 Nasser, Lake resr Egypt
93 F4 Nässjö Sweden
130 C1 Nastapoca r. Can.
130 C1 Nastapoka Islands Can.
67 D3 Nasushiobara Japan
120 B3 Nata Botswana
151 E2 Natal Brazil

131 D1 Natashquan Can.
131 D1 Natashquan r. Can.
142 B2 Natchez U.S.A.
142 B2 Natchitoches U.S.A.
53 C3 Nathalia Austr.
107 D1 Nati, Punta pt Spain
114 C3 Natitingou Benin
151 D3 Natividade Brazil
67 C3 Natori Japan
131 D1 Natuashish Can.
61 B1 Natuna, Kepulauan is Indon.
61 B1 Natuna Besar i. Indon.
120 A3 Nauchas Namibia
101 F1 Nauen Ger.
88 B2 Naujoji Akmen Lith.
74 A2 Naukot Pak.
101 E2 Naumburg (Saale) Ger.
48 F3 Nauru country S. Pacific Ocean
145 C2 Nautla Mex.
88 C3 Navahrudak Belarus
106 B2 Navalmoral de la Mata Spain
106 C2 Navalvillar de Pela Spain
97 C2 Navan Ireland
88 C2 Navapolatsk Belarus
83 M2 Navarin, Mys c. Russia
153 B6 Navarino, Isla i. Chile
96 B1 Naver r. U.K.
73 B3 Navi Mumbai India
89 D3 Navlya Russia
110 C2 Năvodari Romania
77 C2 Navoiy Uzbek.
144 B2 Navojoa Mex.
144 B2 Navolato Mex.
74 A2 Nawabshah Pak.
62 A1 Nawnghkio Myanmar
62 A1 Nawngleng Myanmar
81 C2 Naxçıvan Azer.
111 C3 Naxos i. Greece
144 B2 Nayar Mex.
66 D2 Nayoro Japan
144 B2 Nazas Mex.
144 B2 Nazas r. Mex.
150 A3 Nazca Peru
80 B2 Nazerat Israel
111 C3 Nazilli Turkey
117 B4 Nazrët Eth.
79 C2 Nazwá Oman
121 B1 Nchelenge Zambia
122 B1 Ncojane Botswana
120 A1 N'dalatando Angola
118 C2 Ndélé C.A.R.
118 B3 Ndendé Gabon
115 D3 Ndjamena Chad
121 B2 Ndola Zambia
97 C1 Neagh, Lough l. U.K.
50 C2 Neale, Lake imp. l. Austr.
111 B2 Nea Roda Greece
99 B3 Neath U.K.
53 C1 Nebine Creek r. Austr.
151 B4 Neblina, Pico da mt. Brazil
89 D2 Nebolchi Russia
136 C2 Nebraska state U.S.A.
137 D2 Nebraska City U.S.A.
108 B3 Nebrodi, Monti mts Sicily
Italy
153 C4 Necochea Arg.
131 D1 Nedlouc, Lac l. Can.
135 D4 Needles U.S.A.
74 B2 Neemuch India
129 E2 Neepawa Can.
87 D3 Neftekamsk Russia
82 F2 Neftegugansk Russia
120 A1 Negage Angola
117 B4 Negēlē Eth.
150 A2 Negra, Punta pt Peru
63 A2 Negrais, Cape Myanmar
153 B5 Negro r. Arg.
150 C2 Negro r. S. America

152 C4 Negro r. Uru.
106 B2 Negro, Cabo c. Morocco
64 B2 Negros i. Phil.
69 E1 Nehe China
70 A3 Neijiang China
129 D2 Neilburg Can.
150 A1 Neiva Col.
129 E2 Nejanilini Lake Can.
117 B4 Nek'emtē Eth.
89 F2 Nekrasovskoye Russia
89 D2 Nelidovo Russia
73 B3 Nellore India
128 C3 Nelson Can.
129 E2 Nelson r. Can.
54 B2 Nelson N.Z.
52 B3 Nelson, Cape Austr.
53 D2 Nelson Bay Austr.
129 E2 Nelson House Can.
134 E1 Nelson Reservoir U.S.A.
114 B3 Néma Maur.
88 B2 Neman Russia
104 C2 Nemours France
66 D2 Nemuro Japan
90 B2 Nemyriv Ukr.
97 B2 Nenagh Ireland
99 D2 Nene r. U.K.
69 E1 Nenjiang China
137 E3 Neosho U.S.A.
75 C2 Nepal country Asia
75 C2 Nepalganj Nepal
135 D3 Nephi U.S.A.
97 B1 Nephin h. Ireland
97 B1 Nephin Beg Range hills
Ireland
131 D1 Nepisiguit r. Can.
119 C2 Nepoko r. Dem. Rep. Congo
104 C3 Nérac France
53 D1 Nerang Austr.
69 D1 Nerchinsk Russia
89 F2 Nerekhta Russia
109 C2 Neretva r. Bos. & Herz./
Croatia
120 B2 Neriquinha Angola
88 B3 Neris r. Lith.
89 E2 Nerl' r. Russia
86 F2 Nerokhi Russia
154 C1 Nerópolis Brazil
83 J3 Neryungri Russia
92 □C2 Neskaupstaður Iceland
96 B2 Ness, Loch l. U.K.
136 D3 Ness City U.S.A.
111 B2 Nestos r. Greece
100 B1 Netherlands country Europe
127 G2 Nettilling Lake Can.
101 F1 Neubrandenburg Ger.
105 D2 Neuchâtel Switz.
100 C2 Neuerburg Ger.
100 B3 Neufchâteau Belgium
105 D2 Neufchâteau France
104 C2 Neufchâtel-en-Bray France
101 D2 Neuhof Ger.
102 B1 Neumünster Ger.
102 B2 Neunkirchen Ger.
153 B4 Neuquén Arg.
153 B4 Neuquén r. Arg.
101 F1 Neuruppin Ger.
100 C2 Neuss Ger.
101 D1 Neustadt am Rübenberge
Ger.
101 E3 Neustadt an der Aisch Ger.
101 F1 Neustrelitz Ger.
100 C2 Neuwied Ger.
137 E3 Nevada U.S.A.
135 C3 Nevada state U.S.A.
106 C2 Nevada, Sierra mts Spain
135 B2 Nevada, Sierra mts U.S.A.
88 C2 Nevel' Russia
105 C2 Nevers France

53 C2 Nevertire Austr.
109 C2 Nevesinje Bos. & Herz.
87 D4 Nevinnomyssk Russia
128 B2 New Aiyansh Can.
140 B3 New Albany U.S.A.
151 C1 New Amsterdam Guyana
141 E3 Newark NJ U.S.A.
140 C2 Newark OH U.S.A.
98 C2 Newark-on-Trent U.K.
141 E2 New Bedford U.S.A.
143 E1 New Bern U.S.A.
143 D2 Newberry U.S.A.
139 C2 New Boston U.S.A.
139 D3 New Braunfels U.S.A.
97 C2 Newbridge Ireland
48 E3 New Britain i. P.N.G.
131 D2 New Brunswick prov. Can.
99 C3 Newbury U.K.
64 A1 New Busuanga Phil.
48 E4 New Caledonia terr.
S. Pacific Ocean
53 C2 Newcastle Austr.
123 C2 Newcastle S. Africa
97 D1 Newcastle U.K.
140 C2 New Castle U.S.A.
136 C2 Newcastle U.S.A.
98 B2 Newcastle-under-Lyme U.K.
98 C1 Newcastle upon Tyne U.K.
97 C1 Newcastle West Ireland
74 B2 New Delhi India
128 C3 New Denver Can.
53 D2 New England Range mts
Austr.
131 E2 Newfoundland i. Can.
131 E1 Newfoundland and Labrador
prov. Can.
131 D2 New Glasgow Can.
59 D3 New Guinea i. Indon./P.N.G.
141 E2 New Hampshire state U.S.A.
141 E2 New Haven U.S.A.
128 B2 New Hazelton Can.
New Hebrides country
S. Pacific Ocean see Vanuatu
142 B3 New Iberia U.S.A.
48 E3 New Ireland i. P.N.G.
141 E3 New Jersey state U.S.A.
130 C2 New Liskeard Can.
50 A2 Newman Austr.
138 B2 New Mexico state U.S.A.
142 B3 New Orleans U.S.A.
140 C2 New Philadelphia U.S.A.
54 B1 New Plymouth N.Z.
99 C3 Newport England U.K.
99 B3 Newport Wales U.K.
142 B1 Newport AR U.S.A.
134 B2 Newport OR U.S.A.
141 E2 Newport RI U.S.A.
141 E2 Newport VT U.S.A.
134 C1 Newport WA U.S.A.
141 D3 Newport News U.S.A.
143 E3 New Providence i. Bahamas
99 A3 Newquay U.K.
142 B2 New Roads U.S.A.
97 C2 New Ross Ireland
97 C1 Newry U.K.
83 K1 New Siberia Islands Russia
52 B2 New South Wales state Austr.
137 E2 Newton IA U.S.A.
137 D3 Newton KS U.S.A.
99 B3 Newton Abbot U.K.
96 B3 Newton Stewart U.K.
97 B2 Newtown Ireland
99 B2 Newtown U.K.
136 C2 New Town U.S.A.
97 D1 Newtownabbey U.K.
97 D1 Newtownards U.K.
96 C3 Newtown St Boswells U.K.
97 C1 Newtownstewart U.K.

137 E2 New Ulm U.S.A.
141 E2 New York U.S.A.
141 D2 New York state U.S.A.
54 B2 New Zealand country Oceania
79 C2 Neyriz Iran
76 B3 Neyshābūr Iran
145 C3 Nezahualcóyotl, Presa resr
Mex.
61 B1 Ngabang Indon.
75 C2 Ngamring China
75 C1 Ngangla Ringco salt l. China
77 E3 Nganglong Kangri mt. China
75 C1 Nganglong Kangri mts China
75 C1 Ngangzê Co salt l. China
62 A2 Ngao Thai.
118 B2 Ngaoundéré Cameroon
54 C1 Ngaruawahia N.Z.
62 A2 Ngathainggyaung Myanmar
121 D2 Ngazidja i. Comoros
118 B3 Ngo Congo
115 C4 Ngol Bembo Nigeria
68 C2 Ngoring Hu l. China
115 D3 Ngourti Niger
115 D3 Nguigmi Niger
59 D2 Ngulu atoll Micronesia
115 D4 Nguru Nigeria
123 D2 Ngwelezana S. Africa
121 C2 Nhamalabué Moz.
63 B2 Nha Trang Vietnam
52 B3 Nhill Austr.
123 D2 Nhlangano Swaziland
51 C1 Nhulunbuy Austr.
141 D2 Niagara Falls Can.
114 C3 Niamey Niger
119 C2 Niangara Dem. Rep. Congo
114 B3 Niangay, Lac l. Mali
60 A1 Nias i. Indon.
146 B3 Nicaragua country
Central America
146 B3 Nicaragua, Lake Nic.
105 D3 Nice France
73 D4 Nicobar Islands India
80 B2 Nicosia Cyprus
146 B4 Nicoya, Golfo de b. Costa Rica
88 B2 Nida Lith.
103 E1 Nidzica Pol.
102 B1 Niebüll Ger.
101 D2 Niederaula Ger.
118 B2 Niefang Equat. Guinea
101 F1 Niemegk Ger.
101 D1 Nienburg (Weser) Ger.
100 B1 Nieuwe Niedorp Neth.
151 C1 Nieuw Nickerie Suriname
122 A3 Nieuwoudtville S. Africa
100 A2 Nieuwpoort Belgium
80 B2 Niğde Turkey
115 C3 Niger country Africa
115 C4 Niger r. Africa
115 C4 Niger, Mouths of the Nigeria
115 C4 Nigeria country Africa
130 B2 Nighthawk Lake Can.
111 B2 Nigrita Greece
67 C3 Niigata Japan
67 B4 Niihama Japan
67 C3 Nii-jima i. Japan
67 C3 Niitsu Japan
100 B2 Nijmegen Neth.
100 C1 Nijverdal Neth.
92 J2 Nik'el' Russia
83 M3 Nikol'skoye Russia
91 C2 Nikopol' Ukr.
80 B1 Niksar Turkey
79 D2 Nīkshahr Iran
109 C2 Nikšić Montenegro
116 B1 Nile r. Africa
140 B2 Niles U.S.A.
74 A1 Nīlī Afgh.
105 C3 Nîmes France

53 C3 Nimmitabel Austr.
117 B4 Nimule Sudan
53 C1 Nindigully Austr.
73 B4 Nine Degree Channel India
53 C3 Ninety Mile Beach Austr.
54 B1 Ninety Mile Beach N.Z.
70 C3 Ningbo China
71 B3 Ningde China
71 B3 Ningdu China
70 B2 Ningguo China
71 C3 Ninghai China
Ningjiang China see
Songyuan
68 C2 Ningjing Shan mts China
70 A2 Ningxia Huizu Zizhiqu
aut. reg. China
70 B2 Ningyang China
62 B1 Ninh Binh Vietnam
63 B2 Ninh Hoa Vietnam
66 C3 Ninohe Japan
137 D2 Niobrara r. U.S.A.
114 B3 Niono Mali
114 B3 Nioro Mali
104 B2 Niort France
129 D2 Nipawin Can.
130 B2 Nipigon Can.
130 B2 Nipigon, Lake Can.
131 D1 Nipishish Lake Can.
130 C2 Nipissing, Lake Can.
135 C3 Nipton U.S.A.
151 D3 Niquelândia Brazil
73 B3 Nirmal India
109 D2 Niš Serbia
109 D2 Nišava r. Serbia
108 B3 Niscemi Sicily Italy
67 B4 Nishino-omote Japan
155 D2 Niterói Brazil
96 C3 Nith r. U.K.
103 D2 Nitra Slovakia
49 G4 Niue terr. S. Pacific Ocean
92 H3 Nivala Fin.
100 B2 Nivelles Belgium
73 B3 Nizamabad India
87 E3 Nizhnekamsk Russia
83 H3 Nizhneudinsk Russia
82 G2 Nizhnevartovsk Russia
89 F3 Nizhniy Kislyay Russia
87 D3 Nizhniy Lomov Russia
87 D3 Nizhniy Novgorod Russia
86 E2 Nizhniy Odes Russia
86 E3 Nizhniy Tagil Russia
83 G2 Nizhnyaya Tunguska r.
Russia
91 C1 Nizhyn Ukr.
119 D3 Njinjo Tanz.
119 D3 Njombe Tanz.
121 C2 Nkhotakota Malawi
118 A2 Nkongsamba Cameroon
123 C3 Nkululeko S. Africa
123 C3 Nkwenkwezi S. Africa
67 B4 Nobeoka Japan
52 B1 Noccundra Austr.
144 A1 Nogales Mex.
138 A2 Nogales U.S.A.
104 C2 Nogent-le-Rotrou France
89 E2 Noginsk Russia
74 B2 Nohar India
100 C3 Nohfelden Ger.
104 B2 Noirmoutier, Île de i. France
104 B2 Noirmoutier-en-l'Île France
67 C4 Nojima-zaki c. Japan
74 B2 Nokha India
93 H3 Nokia Fin.
74 A2 Nok Kundi Pak.
118 B2 Nola C.A.R.
86 D3 Nolinsk Russia
123 C3 Nomonde S. Africa
123 D2 Nondweni S. Africa

62 B2 Nong Khai Thai.
52 A2 Nonning Austr.
144 B2 Nonoava Mex.
65 B2 Nonsan S. Korea
63 B2 Nonthaburi Thai.
122 B3 Nonzwakazi S. Africa
77 C3 Norak Tajik.
141 D1 Noranda Can.
82 C1 Nordaustlandet i. Svalbard
128 C2 Nordegg Can.
100 C1 Norden Ger.
83 H1 Nordenshel'da, Arkhipelag is
   Russia
100 C1 Norderney Ger.
100 C1 Norderney i. Ger.
101 E1 Norderstedt Ger.
93 E3 Nordfjordeid Norway
101 E2 Nordhausen Ger.
101 D1 Nordholz Ger.
100 C1 Nordhorn Ger.
   Nordkapp c. Norway see
   North Cape
92 F3 Nordli Norway
102 E3 Nördlingen Ger.
92 G3 Nordmaling Sweden
94 B1 Norðoyar is Faroe Is
97 C2 Nore r. Ireland
137 D2 Norfolk NE U.S.A.
141 D3 Norfolk VA U.S.A.
48 F4 Norfolk Island terr.
   S. Pacific Ocean
93 E3 Norheimsund Norway
82 G2 Noril'sk Russia
75 C2 Norkyung China
139 D1 Norman U.S.A.
   Normandes, Îles is English
   Chan. see Channel Islands
51 D1 Normanton Austr.
93 G4 Norrköping Sweden
93 G4 Norrtälje Sweden
50 B3 Norseman Austr.
92 G3 Norsjö Sweden
98 C1 Northallerton U.K.
50 A2 Northampton Austr.
99 C2 Northampton U.K.
73 D3 North Andaman i. India
129 D2 North Battleford Can.
130 C2 North Bay Can.
130 C1 North Belcher Islands Can.
96 C2 North Berwick U.K.
92 I1 North Cape Norway
54 B1 North Cape N.Z.
130 A1 North Caribou Lake Can.
143 E3 North Carolina state U.S.A.
130 B2 North Channel lake channel
   Can.
96 A3 North Channel U.K.
128 B3 North Cowichan Can.
136 C1 North Dakota state U.S.A.
99 C3 North Downs hills U.K.
143 E3 Northeast Providence
   Channel Bahamas
101 D2 Northeim Ger.
122 A2 Northern Cape prov. S. Africa
129 E2 Northern Indian Lake Can.
97 C1 Northern Ireland prov. U.K.
48 D1 Northern Mariana Islands
   terr. N. Pacific Ocean
50 C1 Northern Territory admin. div.
   Austr.
96 C2 North Esk r. U.K.
137 E2 Northfield U.S.A.
99 D3 North Foreland c. U.K.
54 B1 North Island N.Z.
129 E2 North Knife Lake Can.
65 B1 North Korea country Asia
62 A1 North Lakhimpur India
128 B1 North Nahanni r. Can.

136 C2 North Platte U.S.A.
136 C2 North Platte r. U.S.A.
96 C1 North Ronaldsay i. U.K.
94 D2 North Sea Europe
130 A1 North Spirit Lake Can.
53 D1 North Stradbroke Island
   Austr.
54 B1 North Taranaki Bight b. N.Z.
130 C1 North Twin Island Can.
96 B1 North Tyne r. U.K.
96 A2 North Uist i. U.K.
131 D2 Northumberland Strait Can.
122 C2 North West prov. S. Africa
50 A2 North West Cape Austr.
143 E3 Northwest Providence
   Channel Bahamas
131 E1 North West River Can.
128 B1 Northwest Territories
   admin. div. Can.
98 C1 North York Moors moorland
   U.K.
140 C3 Norton U.S.A.
121 C2 Norton Zimbabwe
140 C2 Norwalk U.S.A.
93 F3 Norway country Europe
129 E2 Norway House Can.
160 A3 Norwegian Sea N. Atlantic
   Ocean
99 D2 Norwich U.K.
141 E2 Norwich CT U.S.A.
141 E2 Norwich NY U.S.A.
66 D2 Noshiro Japan
91 C1 Nosivka Ukr.
122 B2 Nosop watercourse Africa
86 E2 Nosovaya Russia
79 C2 Noşratābād Iran
   Nossob watercourse Africa see
   Nosop
103 D1 Noteć r. Pol.
93 E4 Notodden Norway
67 C3 Noto-hantō pen. Japan
131 D2 Notre-Dame, Monts mts
   Can.
131 E2 Notre Dame Bay Can.
130 C1 Nottaway r. Can.
99 C2 Nottingham U.K.
114 A2 Nouâdhibou Maur.
114 A3 Nouakchott Maur.
114 A3 Nouâmghâr Maur.
48 F4 Nouméa New Caledonia
114 B3 Nouna Burkina Faso
154 B2 Nova Esperança Brazil
155 D2 Nova Friburgo Brazil
109 C1 Nova Gradiška Croatia
154 B2 Nova Granada Brazil
155 D2 Nova Iguaçu Brazil
91 C2 Nova Kakhovka Ukr.
155 D1 Nova Lima Brazil
154 B2 Nova Londrina Brazil
91 C2 Nova Odesa Ukr.
131 D2 Nova Scotia prov. Can.
154 C1 Nova Venécia Brazil
83 K1 Novaya Sibir', Ostrov i.
   Russia
86 E1 Novaya Zemlya is Russia
103 D2 Nové Zámky Slovakia
91 C1 Novhorod-Sivers'kyy Ukr.
110 B2 Novi Iskar Bulg.
108 A2 Novi Ligure Italy
109 D2 Novi Pazar Serbia
109 C1 Novi Sad Serbia
87 D3 Novoanninskiy Russia
150 B2 Novo Aripuanã Brazil
91 D2 Novoazovs'k Ukr.
86 D2 Novodvinsk Russia
152 C3 Novo Hamburgo Brazil
154 C2 Novo Horizonte Brazil

90 B1 Novohrad-Volyns'kyy Ukr.
103 D2 Novo mesto Slovenia
91 D3 Novomikhaylovskiy Russia
89 E3 Novomoskovsk Russia
91 D2 Novomoskovs'k Ukr.
91 C2 Novomyrhorod Ukr.
91 C2 Novooleksiyivka Ukr.
150 B1 Novo Paraíso Brazil
91 D2 Novopskov Ukr.
91 D3 Novorossiysk Russia
88 C2 Novorzhev Russia
87 E3 Novosergiyevka Russia
91 D2 Novoshakhtinsk Russia
82 G3 Novosibirsk Russia
   Novosibirskiye Ostrova is
   Russia see New Siberia Islands
89 E3 Novosil' Russia
89 D2 Novosokol'niki Russia
91 C2 Novotroyits'ke Ukr.
91 C2 Novoukrayinka Ukr.
90 A1 Novovolyns'k Ukr.
89 D3 Novozybkov Russia
103 D2 Nový Jičín Czech Rep.
86 E2 Nyyy Bor Russia
91 D1 Novyy Oskol Russia
86 G2 Novyy Port Russia
86 G2 Novyy Urengoy Russia
69 E1 Novyy Urgal Russia
103 D1 Nowogard Pol.
53 D2 Nowra Austr.
81 D2 Nowshahr Iran
74 B1 Nowshera Pak.
103 E2 Nowy Sącz Pol.
103 E2 Nowy Targ Pol.
87 G2 Noyabr'sk Russia
105 C2 Noyon France
68 C2 Noyon Mongolia
121 C2 Nsanje Malawi
118 B3 Ntandembele
   Dem. Rep. Congo
111 B3 Ntoro, Kavo pt Greece
119 D3 Ntungamo Uganda
79 C2 Nu'aym reg. Oman
116 B2 Nubian Desert Sudan
129 E1 Nueltin Lake Can.
153 A5 Nueva Lubecka Arg.
145 B2 Nueva Rosita Mex.
144 B1 Nuevo Casas Grandes Mex.
144 B2 Nuevo Ideal Mex.
145 C2 Nuevo Laredo Mex.
117 C4 Nugaal watercourse Somalia
117 C4 Nugaaleed, Dooxo val.
   Somalia
105 C2 Nuits-St-Georges France
   Nu Jiang r. China/Myanmar
   see Salween
49 C4 Nuku'alofa Tonga
49 I3 Nuku Hiva i. Fr. Polynesia
76 B2 Nukus Uzbek.
50 B2 Nullagine Austr.
50 B3 Nullarbor Plain Austr.
115 D4 Numan Nigeria
67 C3 Numazu Japan
93 E3 Numedal val. Norway
59 C3 Numfoor i. Indon.
53 C3 Numurkah Austr.
   Nunap Isua c. Greenland see
   Farewell, Cape
127 G3 Nunavik reg. Can.
129 E1 Nunavut admin. div. Can.
99 C2 Nuneaton U.K.
106 B1 Nuñomoral Spain
108 A2 Nuoro Sardinia Italy
78 B2 Nuqrah Saudi Arabia
101 E3 Nuremberg Ger.
52 A2 Nuriootpa Austr.
92 I3 Nurmes Fin.
   Nürnberg Ger. see Nuremberg

62 A1 Nu Shan *mts* China
74 A2 Nushki Pak.
127 H2 Nuuk Greenland
127 H2 Nuussuaq Greenland
127 H2 Nuussuaq *pen.* Greenland
78 A2 Nuwaybi' al Muzayyinah Egypt
122 B3 Nuweveldberge *mts* S. Africa
86 F2 Nyagan' Russia
75 D1 Nyainqêntanglha Feng *mt.* China
75 D2 Nyainqêntanglha Shan *mts* China
117 A3 Nyala Sudan
86 D2 Nyandoma Russia
118 B3 Nyanga *r.* Gabon
121 C2 Nyanga Zimbabwe
Nyang'oma Kenya *see* Kogelo
121 C1 Nyasa, Lake Africa
93 F4 Nyborg Denmark
92 I1 Nyborg Norway
93 G4 Nybro Sweden
119 D3 Nyeri Kenya
68 C3 Nyingchi China
103 E2 Nyíregyháza Hungary
93 F4 Nykøbing Falster Denmark
93 G4 Nyköping Sweden
53 C2 Nymagee Austr.
93 G4 Nynäshamn Sweden
53 C2 Nyngan Austr.
88 B3 Nyoman *r.* Belarus/Lith.
105 D3 Nyons France
86 E2 Nyrob Russia
134 C2 Nyssa U.S.A.
119 C3 Nyunzu Dem. Rep. Congo
91 C2 Nyzhn'ohirs'kyy Ukr.
119 D3 Nzega Tanz.
114 B4 Nzérékoré Guinea
120 A1 N'zeto Angola

O

136 C2 Oahe, Lake U.S.A.
124 E5 O'ahu *i.* U.S.A.
52 B2 Oakbank Austr.
142 B2 Oakdale U.S.A.
53 D1 Oakey Austr.
134 B1 Oak Harbor U.S.A.
140 C3 Oak Hill U.S.A.
135 B3 Oakland U.S.A.
50 B2 Oakover *r.* Austr.
134 B2 Oakridge U.S.A.
143 D1 Oak Ridge U.S.A.
54 B3 Oamaru N.Z.
64 B1 Oas Phil.
145 C3 Oaxaca Mex.
86 F2 Ob' *r.* Russia
118 B2 Obala Cameroon
96 B2 Oban U.K.
106 B1 O Barco Spain
53 C2 Oberon Austr.
101 F3 Oberpfälzer Wald *mts* Ger.
101 F3 Oberviechtach Ger.
59 C3 Obi *i.* Indon.
151 C2 Óbidos Brazil
66 D2 Obihiro Japan
69 E1 Obluch'ye Russia
89 E2 Obninsk Russia
119 C2 Obo C.A.R.
117 C3 Obock Djibouti
118 B3 Obouya Congo
89 E3 Oboyan' Russia
144 B2 Obregón, Presa *resr* Mex.
109 D2 Obrenovac Serbia
87 E3 Obshchiy Syrt *hills* Kazakh./Russia

86 G2 Obskaya Guba *sea chan.* Russia
114 B4 Obuasi Ghana
90 C1 Obukhiv Ukr.
86 C2 Ob"yachevo Russia
143 D3 Ocala U.S.A.
144 B2 Ocampo Mex.
106 C2 Ocaña Spain
150 A1 Occidental, Cordillera *mts* Col.
150 A2 Occidental, Cordillera *mts* Peru
141 D3 Ocean City U.S.A.
128 B2 Ocean Falls Can.
135 C4 Oceanside U.S.A.
91 C2 Ochakiv Ukr.
88 E3 Ocher Russia
101 E3 Ochsenfurt Ger.
143 D2 Oconee *r.* U.S.A.
145 C3 Ocosingo Mex.
58 C3 Ocussi *enclave* East Timor
116 B2 Oda, Jebel *mt.* Sudan
66 D2 Ōdate Japan
67 C3 Odawara Japan
93 E3 Odda Norway
106 B2 Odemira Port.
111 C3 Ödemiş Turkey
93 F4 Odense Denmark
101 D3 Odenwald *reg.* Ger.
102 C1 Oderbucht *b.* Ger.
Odesa Ukr. *see* Odessa
90 C2 Odessa Ukr.
139 C2 Odessa U.S.A.
114 B4 Odienné Côte d'Ivoire
103 D2 Odra *r.* Ger./Pol.
151 D2 Oeiras Brazil
100 D2 Oelde Ger.
136 C2 Oelrichs U.S.A.
101 F2 Oelsnitz/Vogtland Ger.
100 B1 Oenkerk Neth.
108 E2 Ofanto *r.* Italy
101 D2 Offenbach am Main Ger.
102 B2 Offenburg Ger.
117 C4 Ogadēn *reg.* Eth.
66 C3 Oga-hantō *pen.* Japan
67 C3 Ōgaki Japan
136 C2 Ogallala U.S.A.
115 C4 Ogbomoso Nigeria
134 D2 Ogden U.S.A.
141 D2 Ogdensburg U.S.A.
126 B2 Ogilvie *r.* Can.
126 B2 Ogilvie Mountains Can.
143 D2 Oglethorpe, Mount U.S.A.
130 B1 Ogoki *r.* Can.
130 B1 Ogoki Reservoir Can.
88 B2 Ogre Latvia
109 C1 Ogulin Croatia
140 B3 Ohio *r.* U.S.A.
140 C2 Ohio *state* U.S.A.
101 E2 Ohrdruf Ger.
109 D2 Ohrid Macedonia
111 B2 Ohrid, Lake *l.* Albania/Macedonia

151 C1 Oiapoque Brazil
141 D2 Oil City U.S.A.
100 A3 Oise *r.* France
67 B4 Ōita Japan
144 B2 Ojinaga Mex.
152 B3 Ojos del Salado, Nevado *mt.* Arg./Chile
89 F2 Oka *r.* Russia
120 A3 Okahandja Namibia
120 A3 Okakarara Namibia
128 C3 Okanagan Falls Can.
128 C3 Okanagan Lake Can.
134 C1 Okanogan U.S.A.
134 C1 Okanogan *r.* U.S.A.
74 B1 Okara Pak.

120 B2 Okavango *r.* Africa
120 B2 Okavango Delta *swamp* Botswana
67 C3 Okaya Japan
67 B4 Okayama Japan
67 C4 Okazaki Japan
143 D3 Okeechobee, Lake U.S.A.
143 D2 Okefenokee Swamp U.S.A.
99 A3 Okehampton U.K.
74 A2 Okha India
83 K3 Okha Russia
75 C2 Okhaldhunga Nepal
83 K3 Okhota *r.* Russia
83 K3 Okhotsk Russia
156 C2 Okhotsk, Sea of Japan/Russia
91 C1 Okhtyrka Ukr.
67 B3 Okinawa *i.* Japan
67 B3 Oki-shotō *is* Japan
139 D1 Oklahoma *state* U.S.A.
139 D1 Oklahoma City U.S.A.
139 D1 Okmulgee U.S.A.
78 A2 Oko, Wadi *watercourse* Sudan
118 B3 Okondja Gabon
128 C2 Okotoks Can.
89 D3 Okovskiy Les *for.* Russia
118 B3 Okoyo Congo
92 H1 Øksfjord Norway
62 A2 Oktwin Myanmar
86 D2 Oktyabr'skiy Arkhangel'skaya Oblast' Russia
83 L3 Oktyabr'skiy Kamchatskiy Kray Russia
87 E3 Oktyabr'skiy Respublika Bashkortostan Russia
86 F2 Oktyabr'skoye Russia
83 H1 Oktyabr'skoy Revolyutsii, Ostrov *i.* Russia
89 D2 Okulovka Russia
66 C2 Okushiri-tō *i.* Japan
92 □A3 Ólafsvík Iceland
93 G4 Öland *i.* Sweden
52 B2 Olary Austr.
153 B4 Olavarría Arg.
103 D1 Oława Pol.
108 A2 Olbia Sardinia Italy
126 B2 Old Crow Can.
101 D1 Oldenburg Ger.
102 C1 Oldenburg in Holstein Ger.
100 C1 Oldenzaal Neth.
97 B3 Old Head of Kinsale Ireland
128 C2 Olds Can.
129 D2 Old Wives Lake Can.
141 D2 Olean U.S.A.
103 E1 Olecko Pol.
83 J2 Olekminsk Russia
91 C2 Oleksandriya Ukr.
86 C2 Olenegorsk Russia
83 I2 Olenëk Russia
83 I2 Olenëk *r.* Russia
89 D2 Olenino Russia
90 B1 Olevs'k Ukr.
106 B2 Olhão Port.
122 A2 Olifants *watercourse* Namibia
123 D1 Olifants S. Africa
123 D1 Olifants *r. Northern* S. Africa
122 A3 Olifants *r. W. Cape* S. Africa
122 B2 Olifantshoek S. Africa
154 C2 Olímpia Brazil
151 E2 Olinda Brazil
123 C1 Oliphants Drift Botswana
107 C2 Oliva Spain
155 D2 Oliveira Brazil
106 B2 Olivenza Spain
152 B3 Ollagüe Chile
150 A2 Olmos Peru
140 B3 Olney U.S.A.
103 D2 Olomouc Czech Rep.
64 B1 Olongapo Phil.

104 B3 **Oloron-Ste-Marie** France
107 D1 **Olot** Spain
69 D1 **Olovyannaya** Russia
83 L2 **Oloy** *r.* Russia
100 C2 **Olpe** Ger.
103 E1 **Olsztyn** Pol.
110 B2 **Olt** *r.* Romania
81 C1 **Oltu** Turkey
111 B3 **Olympia** *tourist site* Greece
134 B1 **Olympia** U.S.A.
134 B1 **Olympus, Mount** U.S.A.
83 M3 **Olyutorskiy, Mys** *c.* Russia
97 C1 **Omagh** U.K.
137 D2 **Omaha** U.S.A.
79 C2 **Oman** *country* Asia
79 C2 **Oman, Gulf of** Asia
120 A3 **Omaruru** Namibia
120 B2 **Omatako** *watercourse* Namibia
116 B3 **Omdurman** Sudan
53 C3 **Omeo** Austr.
145 C3 **Ometepec** Mex.
78 A3 **Om Hajēr** Eritrea
128 B2 **Omineca Mountains** Can.
67 C3 **Ōmiya** Japan
100 C1 **Ommen** Neth.
83 L2 **Omolon** *r.* Russia
100 B3 **Omont** France
82 F3 **Omsk** Russia
83 L2 **Omsukchan** Russia
110 C1 **Omu, Vârful** *mt.* Romania
67 A4 **Ōmura** Japan
141 D3 **Onancock** U.S.A.
140 C1 **Onaping Lake** Can.
131 C2 **Onatchiway, Lac** *l.* Can.
120 A2 **Oncócua** Angola
122 B3 **Onderstedorings** S. Africa
120 A2 **Ondjiva** Angola
86 C2 **Onega** Russia
86 C2 **Onega** *r.* Russia
141 D2 **Oneida Lake** U.S.A.
137 D2 **O'Neill** U.S.A.
141 D2 **Oneonta** U.S.A.
110 C1 **Onești** Romania
Onezhskoye, Ozero *l.* Russia *see* **Onezhskoye Ozero**
86 C2 **Onezhskoye Ozero** *l.* Russia
122 B2 **Ongers** *watercourse* S. Africa
65 B2 **Ongjin** N. Korea
73 C3 **Ongole** India
121 □D3 **Onilahy** *r.* Madag.
115 C4 **Onitsha** Nigeria
122 A2 **Onseepkans** S. Africa
50 A2 **Onslow** Austr.
143 E2 **Onslow Bay** U.S.A.
130 A1 **Ontario** *prov.* Can.
134 C2 **Ontario** U.S.A.
141 D2 **Ontario, Lake** Can./U.S.A.
51 C2 **Oodnadatta** Austr.
Oostende Belgium *see* **Ostend**
100 B2 **Oosterhout** Neth.
100 A2 **Oosterschelde** *est.* Neth.
100 B1 **Oost-Vlieland** Neth.
128 B2 **Ootsa Lake** Can.
128 B2 **Ootsa Lake** Can.
118 C3 **Opala** Dem. Rep. Congo
130 C1 **Opataca, Lac** *l.* Can.
103 D2 **Opava** Czech Rep.
143 C2 **Opelika** U.S.A.
142 B2 **Opelousas** U.S.A.
130 C1 **Opinaca, Réservoir** *resr* Can.
131 D1 **Opiscotéo, Lac** *l.* Can.
88 C2 **Opochka** Russia
144 A2 **Opodepe** Mex.
103 D1 **Opole** Pol.
106 B1 **Oporto** Port.
54 C1 **Opotiki** N.Z.
93 E3 **Oppdal** Norway
54 B1 **Opunake** N.Z.
120 A2 **Opuwo** Namibia
110 B1 **Oradea** Romania
Orahovac Kosovo *see* **Rahovec**
114 B1 **Oran** Alg.
152 B3 **Orán** Arg.
65 B1 **Ŏrang** N. Korea
53 C2 **Orange** Austr.
105 C3 **Orange** France
122 A2 **Orange** *r.* Namibia/S. Africa
139 E2 **Orange** U.S.A.
143 D2 **Orangeburg** U.S.A.
140 C2 **Orangeville** Can.
101 F1 **Oranienburg** Ger.
122 A2 **Oranjemund** Namibia
120 B3 **Orapa** Botswana
110 B1 **Orăștie** Romania
108 B2 **Orbetello** Italy
53 C3 **Orbost** Austr.
50 B1 **Ord, Mount** *h.* Austr.
106 B1 **Ordes** Spain
Ordos China *see* **Dongsheng**
80 B1 **Ordu** Turkey
Ordzhonikidze Russia *see* **Vladikavkaz**
91 C2 **Ordzhonikidze** Ukr.
93 G4 **Örebro** Sweden
134 B2 **Oregon** *state* U.S.A.
134 B1 **Oregon City** U.S.A.
87 C3 **Orekhovo-Zuyevo** Russia
89 E3 **Orel** Russia
83 K3 **Orel', Ozero** *l.* Russia
135 D2 **Orem** U.S.A.
111 C3 **Ören** Turkey
87 E3 **Orenburg** Russia
54 A3 **Orepuki** N.Z.
93 F4 **Öresund** *str.* Denmark/Sweden
99 D2 **Orford Ness** *hd* U.K.
111 C3 **Orhaneli** Turkey
111 C2 **Orhangazi** Turkey
83 I3 **Orhon Gol** *r.* Mongolia
152 B2 **Oriental, Cordillera** *mts* Bol.
150 A1 **Oriental, Cordillera** *mts* Col.
150 A3 **Oriental, Cordillera** *mts* Peru
107 C2 **Orihuela** Spain
91 D2 **Orikhiv** Ukr.
130 C2 **Orillia** Can.
150 B1 **Orinoco** *r.* Col./Venez.
150 B1 **Orinoco, Delta del** Venez.
93 H4 **Orissaare** Estonia
108 A3 **Oristano** *Sardinia* Italy
93 I3 **Orivesi** *l.* Fin.
151 C2 **Oriximiná** Brazil
145 C3 **Orizaba** Mex.
145 C3 **Orizaba, Pico de** *vol.* Mex.
92 E3 **Orkanger** Norway
93 F4 **Örkelljunga** Sweden
92 E3 **Orkla** *r.* Norway
96 C1 **Orkney Islands** U.K.
154 C2 **Orlândia** Brazil
143 D3 **Orlando** U.S.A.
104 C2 **Orléans** France
74 A2 **Ormara** Pak.
64 B1 **Ormoc** Phil.
98 B2 **Ormskirk** U.K.
92 G3 **Örnsköldsvik** Sweden
114 B3 **Orodara** Burkina Faso
134 C1 **Orofino** U.S.A.
Oroqen Zizhiqi China *see* **Alihe**
64 B2 **Oroquieta** Phil.
108 A2 **Orosei** *Sardinia* Italy
108 A2 **Orosei, Golfo di** *b. Sardinia* Italy
103 E2 **Orosháza** Hungary
135 B3 **Oroville** U.S.A.
52 A2 **Orroroo** Austr.
89 D3 **Orsha** Belarus
87 E3 **Orsk** Russia
93 E3 **Ørsta** Norway
106 B1 **Ortegal, Cabo** *c.* Spain
104 B3 **Orthez** France
106 B1 **Ortigueira** Spain
137 D1 **Ortonville** U.S.A.
83 J2 **Orulgan, Khrebet** *mts* Russia
Orūmīyeh, Daryācheh-ye *salt l.* Iran *see* **Urmia, Lake**
152 B2 **Oruro** Bol.
108 B2 **Orvieto** Italy
146 B4 **Osa, Península de** *pen.* Costa Rica
137 E3 **Osage** *r.* U.S.A.
67 C4 **Ōsaka** Japan
66 D3 **Ōsaki** Japan
110 B2 **Osam** *r.* Bulg.
101 E1 **Oschersleben (Bode)** Ger.
108 A2 **Oschiri** *Sardinia* Italy
89 E3 **Osetr** *r.* Russia
77 D2 **Osh** Kyrg.
120 A2 **Oshakati** Namibia
130 C2 **Oshawa** Can.
66 C2 **Ō-shima** *i.* Japan
67 C4 **Ō-shima** *i.* Japan
140 B2 **Oshkosh** U.S.A.
81 C2 **Oshnavīyeh** Iran
118 B3 **Oshwe** Dem. Rep. Congo
109 C1 **Osijek** Croatia
128 B2 **Osilinka** *r.* Can.
108 B2 **Osimo** Italy
123 D2 **oSizweni** S. Africa
137 E2 **Oskaloosa** U.S.A.
93 G4 **Oskarshamn** Sweden
89 E3 **Oskol** *r.* Russia
93 F4 **Oslo** Norway
93 F4 **Oslofjorden** *sea chan.* Norway
80 B1 **Osmancık** Turkey
111 C2 **Osmaneli** Turkey
80 B2 **Osmaniye** Turkey
88 C2 **Os'mino** Russia
100 D1 **Osnabrück** Ger.
115 C4 **Osogbo** Nigeria
153 A5 **Osorno** Chile
106 C1 **Osorno** Spain
128 C3 **Osoyoos** Can.
100 B2 **Oss** Neth.
51 D4 **Ossa, Mount** Austr.
83 L3 **Ossora** Russia
89 D2 **Ostashkov** Russia
101 D1 **Oste** *r.* Ger.
100 A2 **Ostend** Belgium
101 E1 **Osterburg (Altmark)** Ger.
93 F3 **Österdalälven** *r.* Sweden
101 D1 **Osterholz-Scharmbeck** Ger.
101 E2 **Osterode am Harz** Ger.
92 F3 **Östersund** Sweden
Ostfriesische Inseln *is* Ger. *see* **East Frisian Islands**
100 C1 **Ostfriesland** *reg.* Ger.
93 G3 **Östhammar** Sweden
103 D2 **Ostrava** Czech Rep.
103 D1 **Ostróda** Pol.
89 E3 **Ostrogozhsk** Russia
103 E1 **Ostrołęka** Pol.
101 F2 **Ostrov** Czech Rep.
88 C2 **Ostrov** Russia
103 E1 **Ostrowiec Świętokrzyski** Pol.
103 E1 **Ostrów Mazowiecka** Pol.
103 D1 **Ostrów Wielkopolski** Pol.
67 B4 **Ōsumi-kaikyō** *sea chan.* Japan
67 B4 **Ōsumi-shotō** *is* Japan
106 B2 **Osuna** Spain
141 D2 **Oswego** U.S.A.
99 B2 **Oswestry** U.K.

Ōta

67 C3 Ōta Japan
54 B3 Otago Peninsula N.Z.
54 C1 Otaki N.Z.
66 D2 Otaru Japan
120 A2 Otavi Namibia
134 C1 Othello U.S.A.
120 A3 Otjiwarongo Namibia
117 B3 Otoro, Jebel *mt.* Sudan
93 E4 Otra *r.* Norway
109 C2 Otranto, Strait of Albania/ Italy
67 C3 Ōtsu Japan
93 E3 Otta Norway
130 C2 Ottawa Can.
130 C2 Ottawa *r.* Can.
140 B2 Ottawa *IL* U.S.A.
137 D3 Ottawa *KS* U.S.A.
130 B1 Otter Rapids Can.
100 B2 Ottignies Belgium
137 E2 Ottumwa U.S.A.
150 A2 Otuzco Peru
52 B3 Otway, Cape Austr.
142 B2 Ouachita *r.* U.S.A.
142 B2 Ouachita, Lake U.S.A.
142 B2 Ouachita Mountains U.S.A.
118 C2 Ouadda C.A.R.
115 D3 Ouaddaï *reg.* Chad
114 B3 Ouagadougou Burkina Faso
114 B3 Ouahigouya Burkina Faso
114 B3 Oualâta Maur.
118 C2 Ouanda Djallé C.A.R.
114 B3 Ouarâne *reg.* Maur.
115 C1 Ouargla Alg.
114 B1 Ouarzazate Morocco
100 A2 Oudenaarde Belgium
122 B3 Oudtshoorn S. Africa
107 C2 Oued Tlélat Alg.
104 A2 Ouessant, Île d' *i.* France
118 B2 Ouesso Congo
114 B1 Oujda Morocco
107 D2 Ouled Farès Alg.
92 I2 Oulu Fin.
92 I3 Oulujärvi *l.* Fin.
108 A1 Oulx Italy
115 D3 Oum-Chalouba Chad
115 D3 Oum-Hadjer Chad
115 E3 Ounianga Kébir Chad
100 B2 Oupeye Belgium
106 B1 Ourense Spain
154 C2 Ourinhos Brazil
155 D2 Ouro Preto Brazil
100 B2 Ourthe *r.* Belgium
98 C2 Ouse *r.* U.K.
131 D2 Outardes, Rivière aux *r.* Can.
131 D2 Outardes Quatre, Réservoir *resr* Can.
96 A2 Outer Hebrides *is* U.K.
120 A3 Outjo Namibia
129 D2 Outlook Can.
92 I3 Outokumpu Fin.
52 B3 Ouyen Austr.
106 B1 Ovar Port.
92 H2 Överkalix Sweden
135 D3 Overton U.S.A.
92 H2 Övertorneå Sweden
106 B1 Oviedo Spain
93 E3 Øvre Årdal Norway
93 F3 Øvre Rendal Norway
90 B1 Ovruch Ukr.
118 B3 Owando Congo
67 C4 Owase Japan
137 E2 Owatonna U.S.A.
140 B3 Owensboro U.S.A.
135 C3 Owens Lake U.S.A.
130 B2 Owen Sound Can.
115 C4 Owerri Nigeria

140 C2 Owosso U.S.A.
134 C2 Owyhee U.S.A.
134 C2 Owyhee *r.* U.S.A.
54 B2 Oxford N.Z.
99 C3 Oxford U.K.
142 C2 Oxford *KY* U.S.A.
129 E2 Oxford Lake Can.
52 B2 Oxley Austr.
97 B1 Ox Mountains *hills* Ireland
135 C4 Oxnard U.S.A.
67 C3 Oyama Japan
118 B2 Oyem Gabon
129 C2 Oyen Can.
105 D2 Oyonnax France
64 B2 Ozamis Phil.
142 C2 Ozark U.S.A.
137 E3 Ozark Plateau U.S.A.
137 E3 Ozarks, Lake of the U.S.A.
83 L3 Ozernovskiy Russia
88 B3 Ozersk Russia
89 E3 Ozery Russia
87 D3 Ozinki Russia

P

127 H2 Paamiut Greenland
122 A3 Paarl S. Africa
103 D1 Pabianice Pol.
75 C2 Pabna Bangl.
74 A2 Pab Range *mts* Pak.
109 C3 Pachino *Sicily* Italy
145 C2 Pachuca Mex.
156 Pacific Ocean
103 D1 Paczków Pol.
60 B2 Padang Indon.
60 B1 Padang Endau Malaysia
60 B2 Padangpanjang Indon.
60 A1 Padangsidimpuan Indon.
101 D2 Paderborn Ger.
Padova Italy *see* Padua
139 D3 Padre Island U.S.A.
52 B3 Padthaway Austr.
108 B1 Padua Italy
140 B3 Paducah *KY* U.S.A.
139 C2 Paducah *TX* U.S.A.
65 B1 Paegam N. Korea
Paengnyŏng-do *i.* S. Korea *see* Baengnyeong-do
54 C1 Paeroa N.Z.
Pafos Cyprus *see* Paphos
109 C2 Pag Croatia
64 B2 Pagadian Phil.
60 B2 Pagai Selatan *i.* Indon.
60 B2 Pagai Utara *i.* Indon.
59 D1 Pagan *i.* N. Mariana Is
61 C2 Pagatan Indon.
138 A1 Page U.S.A.
88 B2 Paggiai Lith.
136 B3 Pagosa Springs U.S.A.
93 I3 Paide Estonia
93 I3 Päijänne *l.* Fin.
75 C2 Paikü Co *l.* China
138 A1 Painted Desert U.S.A.
96 B3 Paisley U.K.
92 H2 Pajala Sweden
150 B1 Pakaraima Mountains S. America
150 C1 Pakaraima Mountains S. America
74 A2 Pakistan *country* Asia
62 A1 Pakokku Myanmar
88 B2 Pakruojis Lith.
103 D2 Paks Hungary
130 A1 Pakwash Lake Can.
62 B2 Pakxan Laos

63 B2 Pakxé Laos
115 D4 Pala Chad
60 B2 Palabuhanratu, Teluk *b.* Indon.
111 B3 Palaikastro Greece
111 B3 Palaiochora Greece
122 B1 Palamakoloi Botswana
107 D1 Palamós Spain
83 L3 Palana Russia
64 B1 Palanan Phil.
61 C2 Palangkaraya Indon.
74 B2 Palanpur India
83 L3 Palapye Botswana
83 L2 Palatka Russia
143 D3 Palatka U.S.A.
59 C2 Palau *country* N. Pacific Ocean
63 A2 Palaw Myanmar
64 A2 Palawan *i.* Phil.
93 H4 Paldiski Estonia
60 B2 Palembang Indon.
106 C1 Palencia Spain
145 C3 Palenque Mex.
108 B3 Palermo *Sicily* Italy
139 D2 Palestine U.S.A.
62 A1 Paletwa Myanmar
74 B2 Pali India
48 E2 Palikir Micronesia
109 C2 Palinuro, Capo *c.* Italy
111 B3 Paliouri, Akrotirio *pt* Greece
100 B3 Paliseul Belgium
92 I3 Paljakka *h.* Fin.
88 C2 Palkino Russia
73 B4 Palk Strait India/Sri Lanka
54 C2 Palliser, Cape N.Z.
49 I3 Palliser, Îles *is* Fr. Polynesia
106 B2 Palma del Río Spain
107 D2 Palma de Mallorca Spain
154 B3 Palmas *Paraná* Brazil
151 D3 Palmas Tocantins Brazil
114 B4 Palmas, Cape Liberia
154 C3 Palmeira Brazil
151 D2 Palmeirais Brazil
55 P2 Palmer Land *reg.* Antarctica
54 C2 Palmerston North N.Z.
109 C3 Palmi Italy
145 C2 Palmillas Mex.
150 A1 Palmira Col.
135 C4 Palm Springs U.S.A.
49 H2 Palmyra Atoll N. Pacific Ocean
117 B3 Paloich South Sudan
145 C3 Palomares Mex.
107 C2 Palos, Cabo de *c.* Spain
92 I3 Paltamo Fin.
58 B3 Palu Indon.
83 M2 Palyavaam *r.* Russia
104 C3 Pamiers France
77 D3 Pamir *mts* Asia
143 E1 Pamlico Sound *sea chan.* U.S.A.
152 B2 Pampa Grande Bol.
153 B4 Pampas *reg.* Arg.
150 A1 Pamplona Col.
107 C1 Pamplona Spain
111 D2 Pamukova Turkey
60 B2 Panaan *i.* Indon.
73 B3 Panaji India
146 B4 Panama *country* Central America
146 C4 Panamá, Canal de Panama
146 C4 Panama, Gulf of Panama
146 C4 Panama City Panama
142 C2 Panama City U.S.A.
135 C3 Panamint Range *mts* U.S.A.
61 B1 Panarik Indon.
64 B1 Panay *i.* Phil.
109 D2 Pančevo Serbia
64 B1 Pandan Phil.

224

75 C2 Pandaria India
73 B3 Pandharpur India
88 B2 Panevžys Lith.
61 C2 Pangkalanbuun Indon.
60 A1 Pangkalansusu Indon.
60 B2 Pangkalpinang Indon.
127 G2 Pangnirtung Can.
86 G2 Pangody Russia
89 F3 Panino Russia
74 B2 Panipat India
74 A2 Panjgur Pak.
65 C1 Pan Ling *mts* China
75 C2 Panna India
50 A2 Pannawonica Austr.
154 B2 Panorama Brazil
65 B1 Panshi China
145 C2 Pánuco Mex.
62 B1 Panzhihua China
109 C3 Paola Italy
63 B2 Paôy Pêt Cambodia
103 D2 Pápa Hungary
54 B1 Papakura N.Z.
145 C2 Papantla Mex.
49 I4 Papeete Fr. Polynesia
100 C1 Papenburg Ger.
80 B2 Paphos Cyprus
59 D3 Papua, Gulf of P.N.G.
59 D3 Papua New Guinea *country* Oceania
50 A2 Paraburdoo Austr.
154 C1 Paracatu Brazil
155 C1 Paracatu *r.* Brazil
52 A2 Parachilna Austr.
109 D2 Paraćin Serbia
155 D1 Pará de Minas Brazil
135 B3 Paradise U.S.A.
142 B1 Paragould U.S.A.
151 C3 Paraguai *r.* Brazil
152 C3 Paraguay *r.* Arg./Para.
152 C3 Paraguay *country* S. America
155 D2 Paraíba do Sul *r.* Brazil
154 B1 Paraíso Brazil
145 C3 Paraíso Mex.
114 C4 Parakou Benin
151 C1 Paramaribo Suriname
83 L3 Paramushir, Ostrov *i.* Russia
152 B4 Paraná Arg.
154 A3 Paraná *r.* S. America
154 C3 Paranaguá Brazil
154 B1 Paranaíba Brazil
154 B2 Paranaíba *r.* Brazil
154 B2 Paranapanema *r.* Brazil
154 B2 Paranapiacaba, Serra *mts* Brazil
154 B2 Paranavaí Brazil
54 B2 Paraparaumu N.Z.
155 D2 Parati Brazil
154 B1 Paraúna Brazil
105 C2 Paray-le-Monial France
73 B3 Parbhani India
101 E1 Parchim Ger.
155 E1 Pardo *r. Bahia* Brazil
154 B2 Pardo *r. Mato Grosso do Sul* Brazil
154 C2 Pardo *r. São Paulo* Brazil
103 D1 Pardubice Czech Rep.
150 B3 Parecis, Serra dos *hills* Brazil
130 C2 Parent, Lac *l.* Can.
58 B3 Parepare Indon.
89 D2 Parfino Russia
111 B3 Parga Greece
109 C3 Parghelia Italy
147 D3 Paria, Gulf of Trin. and Tob./Venez.
150 B1 Parima, Serra *mts* Brazil
151 C2 Parintins Brazil
104 C2 Paris France
142 C1 Paris *TN* U.S.A.
139 D2 Paris *TX* U.S.A.
93 H3 Parkano Fin.
138 A2 Parker U.S.A.
140 C3 Parkersburg U.S.A.
53 C2 Parkes Austr.
140 A1 Park Falls U.S.A.
137 D1 Park Rapids U.S.A.
108 B2 Parma Italy
151 D2 Parnaíba Brazil
151 D2 Parnaíba *r.* Brazil
54 B2 Parnassus N.Z.
111 B3 Parnonas *mts* Greece
88 B2 Pärnu Estonia
65 B2 Paro-ho *l.* S. Korea
52 B2 Paroo *watercourse* Austr.
Paropamisus *mts* Afgh. *see* Safēd Kōh, Silsilah-ye
111 C3 Paros *i.* Greece
144 B2 Parras Mex.
126 C2 Parry, Cape Can.
126 D1 Parry Islands Can.
130 B2 Parry Sound Can.
79 C2 Pārsīān Iran
137 D3 Parsons U.S.A.
101 D2 Partenstein Ger.
104 B2 Parthenay France
97 B2 Partry Mountains *hills* Ireland
151 C2 Paru *r.* Brazil
135 C4 Pasadena U.S.A.
62 A2 Pasawng Myanmar
142 C2 Pascagoula U.S.A.
110 C1 Paşcani Romania
134 C1 Pasco U.S.A.
102 C1 Pasewalk Ger.
129 D2 Pasfield Lake Can.
64 B1 Pasig Phil.
60 B1 Pasir Putih Malaysia
74 A2 Pasni Pak.
153 A5 Paso Río Mayo Arg.
135 B3 Paso Robles U.S.A.
155 D2 Passa Tempo Brazil
102 C2 Passau Ger.
152 C3 Passo Fundo Brazil
155 C2 Passos Brazil
88 C2 Pastavy Belarus
150 A2 Pastaza *r.* Peru
150 A1 Pasto Col.
61 C2 Pasuruan Indon.
88 B2 Pasvalys Lith.
153 A6 Patagonia *reg.* Arg.
75 C2 Patan Nepal
54 B1 Patea N.Z.
141 E2 Paterson U.S.A.
Pathein Myanmar *see* Bassein
136 B2 Pathfinder Reservoir U.S.A.
61 C2 Pati Indon.
62 A1 Patkai Bum *mts* India/Myanmar
111 C3 Patmos *i.* Greece
75 C2 Patna India
81 C2 Patnos Turkey
154 B3 Pato Branco Brazil
152 C4 Patos, Lagoa dos *l.* Brazil
155 C1 Patos de Minas Brazil
152 B4 Patquía Arg.
111 B3 Patras Greece
75 C2 Patratu India
154 C1 Patrocínio Brazil
63 B3 Pattani Thai.
63 B2 Pattaya Thai.
128 B2 Pattullo, Mount Can.
129 D2 Patuanak Can.
144 B3 Pátzcuaro Mex.
104 B3 Pau France
104 B2 Pauillac France
62 A1 Pauk Myanmar
151 D2 Paulistana Brazil
151 E2 Paulo Afonso Brazil
123 D2 Paulpietersburg S. Africa
139 D2 Pauls Valley U.S.A.
62 A2 Paungde Myanmar
155 D1 Pavão Brazil
108 A1 Pavia Italy
88 B2 Pāvilosta Latvia
110 C2 Pavlikeni Bulg.
77 D1 Pavlodar Kazakh.
91 D2 Pavlohrad Ukr.
91 E1 Pavlovsk Russia
91 D2 Pavlovskaya Russia
60 B2 Payakumbuh Indon.
134 C2 Payette U.S.A.
134 C2 Payette *r.* U.S.A.
86 F2 Pay-Khoy, Khrebet *hills* Russia
152 C4 Paysandú Uru.
81 C1 Pazar Turkey
110 B2 Pazardzhik Bulg.
111 C3 Pazarköy Turkey
108 B1 Pazin Croatia
63 A2 Pe Myanmar
128 C2 Peace *r.* Can.
128 C2 Peace River Can.
135 E3 Peale, Mount U.S.A.
142 C2 Pearl *r.* U.S.A.
139 D3 Pearsall U.S.A.
121 C2 Pebane Moz.
Peć Kosovo *see* Pejë
155 D1 Peçanha Brazil
86 E2 Pechora Russia
86 E2 Pechora *r.* Russia
88 C2 Pechory Russia
139 C2 Pecos U.S.A.
139 C3 Pecos *r.* U.S.A.
103 D2 Pécs Hungary
155 D1 Pedra Azul Brazil
154 C2 Pedregulho Brazil
151 D2 Pedreiras Brazil
73 C4 Pedro, Point Sri Lanka
151 D2 Pedro Afonso Brazil
154 B1 Pedro Gomes Brazil
152 C3 Pedro Juan Caballero Para.
96 C3 Peebles U.K.
143 E2 Pee Dee *r.* U.S.A.
126 C2 Peel *r.* Can.
98 A1 Peel Isle of Man
128 C2 Peerless Lake Can.
54 B2 Pegasus Bay N.Z.
101 E3 Pegnitz Ger.
62 A2 Pegu Myanmar
62 A2 Pegu Yoma *mts* Myanmar
153 B4 Pehuajó Arg.
101 E1 Peine Ger.
88 C2 Peipus, Lake Estonia/Russia
154 B2 Peixe *r.* Brazil
109 D2 Pejë Kosovo
61 B2 Pekalongan Indon.
60 B1 Pekan Malaysia
60 B1 Pekanbaru Indon.
Peking China *see* Beijing
130 B2 Pelee Island Can.
59 C3 Peleng *i.* Indon.
92 I2 Pelkosenniemi Fin.
122 A2 Pella S. Africa
59 D3 Pelleluhu Islands P.N.G.
92 H2 Pello Fin.
128 A1 Pelly *r.* Can.
128 A1 Pelly Mountains Can.
152 C4 Pelotas Brazil
152 C3 Pelotas, Rio das *r.* Brazil
141 F1 Pemadumcook Lake U.S.A.
61 B1 Pemangkat Indon.
60 A1 Pematangsiantar Indon.
121 D2 Pemba Moz.
120 B2 Pemba Zambia
119 D3 Pemba Island Tanz.
128 B2 Pemberton Can.
137 D1 Pembina *r.* Can./U.S.A.

130 C2 **Pembroke** Can.
99 A3 **Pembroke** U.K.
145 C2 **Peña Nevada, Cerro** *mt.* Mex.
154 B2 **Penápolis** Brazil
106 B1 **Peñaranda de Bracamonte** Spain
107 C1 **Peñarroya** *mt.* Spain
106 B2 **Peñarroya-Pueblonuevo** Spain
106 B1 **Peñas, Cabo de** *c.* Spain
153 A5 **Penas, Golfo de** *g.* Chile
134 C1 **Pendleton** U.S.A.
128 B2 **Pendleton Bay** Can.
134 C1 **Pend Oreille Lake** U.S.A.
74 B3 **Penganga** *r.* India
123 D1 **Penge** S. Africa
70 C2 **Penglai** China
71 A3 **Pengshui** China
106 B2 **Peniche** Port.
98 B1 **Penicuik** U.K.
108 B2 **Penne** Italy
98 B1 **Pennines** *hills* U.K.
141 D2 **Pennsylvania** *state* U.S.A.
127 G2 **Penny Icecap** Can.
141 F2 **Penobscot** *r.* U.S.A.
52 B3 **Penola** Austr.
50 C3 **Penong** Austr.
49 H3 **Penrhyn** *atoll* Cook Is
98 B1 **Penrith** U.K.
142 C2 **Pensacola** U.S.A.
55 Q1 **Pensacola Mountains** Antarctica
61 C1 **Pensiangan** *Sabah* Malaysia
128 C3 **Penticton** Can.
96 C1 **Pentland Firth** *sea chan.* U.K.
99 B2 **Penygadair** *h.* U.K.
87 D3 **Penza** Russia
99 A3 **Penzance** U.K.
140 B2 **Peoria** U.S.A.
107 C1 **Perales del Alfambra** Spain
51 E2 **Percy Isles** Austr.
86 F2 **Peregrebnoye** Russia
150 A1 **Pereira** Col.
154 B2 **Pereira Barreto** Brazil
90 A2 **Peremyshlyany** Ukr.
89 E2 **Pereslavl'-Zalesskiy** Russia
91 C1 **Pereyaslav-Khmel'nyts'kyy** Ukr.
153 B4 **Pergamino** Arg.
92 H3 **Perhonjoki** *r.* Fin.
131 C1 **Péribonka, Lac** *l.* Can.
152 B3 **Perico** Arg.
144 B2 **Pericos** Mex.
104 C2 **Périgueux** France
147 C4 **Perijá, Sierra de** *mts* Venez.
153 A5 **Perito Moreno** Arg.
101 E1 **Perleberg** Ger.
86 E3 **Perm'** Russia
52 A2 **Pernatty Lagoon** *imp. l.* Austr.
105 C2 **Péronne** France
145 C3 **Perote** Mex.
104 C3 **Perpignan** France
99 A3 **Perranporth** U.K.
143 D2 **Perry** *FL* U.S.A.
143 D2 **Perry** *GA* U.S.A.
137 E2 **Perry** *IA* U.S.A.
139 C1 **Perryton** U.S.A.
137 F3 **Perryville** U.S.A.
99 B2 **Pershore** U.K.
91 D2 **Pershotravens'k** Ukr.
Persian Gulf Asia *see* **The Gulf**
50 A3 **Perth** Austr.
96 C2 **Perth** U.K.
105 D3 **Pertuis** France
108 A2 **Pertusato, Capo** *c. Corsica* France
150 A2 **Peru** *country* S. America
157 H4 **Peru-Chile Trench** S. Pacific Ocean
108 B2 **Perugia** Italy
154 C2 **Peruíbe** Brazil
100 A2 **Péruwelz** Belgium
90 C2 **Pervomays'k** Ukr.
91 C2 **Pervomays'ke** Ukr.
89 F3 **Pervomayskiy** Russia
91 D2 **Pervomays'kyy** Ukr.
108 B2 **Pesaro** Italy
108 B2 **Pescara** Italy
108 B2 **Pescara** *r.* Italy
74 B1 **Peshawar** Pak.
109 D2 **Peshkopi** Albania
109 C1 **Pesnica** Slovenia
89 E2 **Pestovo** Russia
144 B3 **Petatlán** Mex.
140 B2 **Petenwell Lake** U.S.A.
52 A2 **Peterborough** Austr.
130 C2 **Peterborough** Can.
99 C2 **Peterborough** U.K.
96 D2 **Peterhead** U.K.
50 B2 **Petermann Ranges** *mts* Austr.
129 D2 **Peter Pond Lake** Can.
128 A2 **Petersburg** *AK* U.S.A.
141 D3 **Petersburg** *VA* U.S.A.
101 D1 **Petershagen** Ger.
131 E1 **Petit Mécatina** *r.* Can.
145 D2 **Peto** Mex.
140 C1 **Petoskey** U.S.A.
80 B2 **Petra** *tourist site* Jordan
66 B2 **Petra Velikogo, Zaliv** *b.* Russia
111 B2 **Petrich** Bulg.
151 D2 **Petrolina** Brazil
83 L3 **Petropavlovsk-Kamchatskiy** Russia
77 C1 **Petropavlovskoye** Kazakh.
110 B1 **Petroșani** Romania
89 F3 **Petrovskoye** Russia
86 C2 **Petrozavodsk** Russia
123 C2 **Petrusburg** S. Africa
123 C2 **Petrus Steyn** S. Africa
122 B3 **Petrusville** S. Africa
89 E2 **Petushki** Russia
60 A1 **Peureula** Indon.
83 M2 **Pevek** Russia
102 B2 **Pforzheim** Ger.
102 C2 **Pfunds** Austria
101 D3 **Pfungstadt** Ger.
123 C2 **Phahameng** S. Africa
123 D1 **Phalaborwa** S. Africa
74 B2 **Phalodi** India
63 B2 **Phang Hoei, San Khao** *mts* Thai.
63 A3 **Phangnga** Thai.
62 B1 **Phăng Xi Păng** *mt.* Vietnam
63 B2 **Phan Rang-Thap Cham** Vietnam
63 B2 **Phan Thiêt** Vietnam
63 B3 **Phatthalung** Thai.
62 A2 **Phayao** Thai.
129 D2 **Phelps Lake** Can.
143 C2 **Phenix City** U.S.A.
63 A2 **Phet Buri** Thai.
63 B2 **Phichit** Thai.
141 D3 **Philadelphia** U.S.A.
136 C2 **Philip** U.S.A.
100 B2 **Philippeville** Belgium
100 A2 **Philippine** Neth.
64 B1 **Philippines** *country* Asia
64 B1 **Philippine Sea** N. Pacific Ocean
126 B2 **Philip Smith Mountains** U.S.A.
122 B3 **Philipstown** S. Africa
53 C3 **Phillip Island** Austr.
137 D3 **Phillipsburg** U.S.A.
123 C2 **Phiritona** S. Africa
63 B2 **Phitsanulok** Thai.
63 B2 **Phnom Penh** Cambodia
Phnum Pénh Cambodia *see* **Phnom Penh**
138 A2 **Phoenix** U.S.A.
49 G3 **Phoenix Islands** Kiribati
63 B2 **Phon** Thai.
62 B1 **Phôngsali** Laos
62 B2 **Phônsavan** Laos
62 B2 **Phônsavan** *mt.* Laos
62 B2 **Phrae** Thai.
63 A3 **Phuket** Thai.
62 A2 **Phumiphon, Khuan** Thai.
123 C2 **Phuthaditjhaba** S. Africa
62 A2 **Phyu** Myanmar
108 A1 **Piacenza** Italy
110 C1 **Piatra Neamț** Romania
151 D2 **Piauí** *r.* Brazil
108 B1 **Piave** *r.* Italy
117 B4 **Pibor** *r.* South Sudan
117 B4 **Pibor Post** South Sudan
142 C2 **Picayune** U.S.A.
152 B3 **Pichanal** Arg.
144 A2 **Pichilingue** Mex.
98 C1 **Pickering** U.K.
130 A1 **Pickle Lake** Can.
151 D2 **Picos** Brazil
153 B5 **Pico Truncado** Arg.
53 D2 **Picton** Austr.
150 B3 **Piedras, Río de las** *r.* Peru
145 C3 **Piedras Negras** Guat.
145 B2 **Piedras Negras** Mex.
93 I3 **Pieksämäki** Fin.
92 I3 **Pielinen** *l.* Fin.
136 C2 **Pierre** U.S.A.
105 C3 **Pierrelatte** France
123 D2 **Pietermaritzburg** S. Africa
110 B1 **Pietrosa** *mt.* Romania
128 C2 **Pigeon Lake** Can.
153 B4 **Pigüé** Arg.
92 I3 **Pihtipudas** Fin.
145 C3 **Pijijiapan** Mex.
136 C3 **Pikes Peak** U.S.A.
122 A3 **Piketberg** S. Africa
140 C3 **Pikeville** U.S.A.
103 D1 **Piła** Pol.
153 C4 **Pilar** Arg.
152 C3 **Pilar** Para.
50 A2 **Pilbara** *reg.* Austr.
53 C2 **Pilliga** Austr.
154 C1 **Pilões, Serra dos** *mts* Brazil
150 B3 **Pimenta Bueno** Brazil
153 C4 **Pinamar** Arg.
80 B2 **Pınarbaşı** Turkey
146 B2 **Pinar del Río** Cuba
103 E1 **Pińczów** Pol.
151 D2 **Pindaré** *r.* Brazil
Pindos *mts* Greece *see* **Pindus Mountains**
111 B2 **Pindus Mountains** Greece
142 B2 **Pine Bluff** U.S.A.
50 C1 **Pine Creek** Austr.
136 B2 **Pinedale** U.S.A.
86 D2 **Pinega** Russia
129 D2 **Pinehouse Lake** Can.
111 B3 **Pineios** *r.* Greece
128 C1 **Pine Point (abandoned)** Can.
136 C2 **Pine Ridge** U.S.A.
123 D2 **Pinetown** S. Africa
70 B2 **Pingdingshan** China
71 A3 **Pingguo** China
71 B3 **Pingjiang** China
70 A2 **Pingliang** China
71 C3 **P'ingtung** Taiwan
71 A3 **Pingxiang** *Guangxi* China
71 B3 **Pingxiang** *Jiangxi* China
70 B2 **Pingyin** China
151 D2 **Pinheiro** Brazil
101 D1 **Pinneberg** Ger.
145 C3 **Pinotepa Nacional** Mex.

88 C3 Pinsk Belarus
135 D3 Pioche U.S.A.
119 C3 Piodi Dem. Rep. Congo
103 E1 Pionki Pol.
103 D1 Piotrków Trybunalski Pol.
137 D2 Pipestone U.S.A.
131 C2 Pipmuacan, Réservoir resr Can.
154 B2 Piquiri r. Brazil
154 C1 Piracanjuba Brazil
154 C2 Piracicaba Brazil
155 D1 Piracicaba r. Brazil
111 B3 Piraeus Greece
154 C2 Piraí do Sul Brazil
154 C2 Pirajuí Brazil
154 B1 Piranhas Brazil
151 E2 Piranhas r. Brazil
155 D1 Pirapora Brazil
154 C2 Pirassununga Brazil
154 C1 Pires do Rio Brazil
151 D2 Piripiri Brazil
59 C3 Piru Indon.
108 B2 Pisa Italy
152 A2 Pisagua Chile
150 A3 Pisco Peru
102 C2 Písek Czech Rep.
79 D2 Pishin Iran
145 D2 Pisté Mex.
109 C2 Pisticci Italy
108 B2 Pistoia Italy
106 C1 Pisuerga r. Spain
134 B2 Pit r. U.S.A.
154 B2 Pitanga Brazil
155 D1 Pitangui Brazil
49 J4 Pitcairn Island i. Pitcairn Is
92 H2 Piteå Sweden
92 H2 Piteälven r. Sweden
110 B2 Pitești Romania
75 C2 Pithoragarh India
96 C2 Pitlochry U.K.
128 B2 Pitt Island Can.
137 E3 Pittsburg U.S.A.
141 E2 Pittsburgh U.S.A.
141 E2 Pittsfield U.S.A.
53 D1 Pittsworth Austr.
155 C2 Piumhi Brazil
150 A2 Piura Peru
90 C2 Pivdennyy Buh r. Ukr.
131 E2 Placentia Can.
135 B3 Placerville U.S.A.
146 C2 Placetas Cuba
139 C2 Plainview U.S.A.
154 C1 Planaltina Brazil
142 B2 Plaquemine U.S.A.
106 B1 Plasencia Spain
147 C4 Plato Col.
137 D2 Platte r. U.S.A.
101 E2 Plattsburgh U.S.A.
101 F1 Plau am See Ger.
101 F2 Plauen Ger.
101 F1 Plauer See l. Ger.
89 E3 Plavsk Russia
129 E2 Playgreen Lake Can.
63 B2 Plây Ku Vietnam
153 B4 Plaza Huincul Arg.
139 D3 Pleasanton U.S.A.
54 B2 Pleasant Point N.Z.
104 B2 Pleaux France
130 B1 Pledger Lake Can.
54 C1 Plenty, Bay of g. N.Z.
136 C1 Plentywood U.S.A.
86 D2 Plesetsk Russia
131 C1 Plétipi, Lac l. Can.
100 C2 Plettenberg Ger.
122 B3 Plettenberg Bay S. Africa
110 B2 Pleven Bulg.
109 C2 Pljevlja Montenegro
109 C2 Ploče Croatia

103 D1 Płock Pol.
109 C2 Pločno mt. Bos. & Herz.
104 B2 Ploemeur France
110 C2 Ploiești Romania
89 F2 Ploskoye Russia
104 B2 Plouzané France
110 B2 Plovdiv Bulg.
88 B2 Plung Lith.
88 C3 Plyeshchanitsy Belarus
99 A3 Plymouth U.K.
140 B2 Plymouth U.S.A.
147 D3 Plymouth (abandoned)
          Montserrat
88 C2 Plyussa Russia
102 C2 Plzeň Czech Rep.
114 B3 Pô Burkina Faso
108 B1 Po r. Italy
77 E2 Pobeda Peak China/Kyrg.
142 B1 Pocahontas U.S.A.
134 D2 Pocatello U.S.A.
90 B1 Pochayiv Ukr.
89 D3 Pochep Russia
89 D3 Pochinok Russia
145 C3 Pochutla Mex.
141 D3 Pocomoke City U.S.A.
155 C2 Poços de Caldas Brazil
89 D2 Poddor'ye Russia
91 D1 Podgorenskiy Russia
109 C2 Podgorica Montenegro
82 G3 Podgornoye Russia
83 H2 Podkamennaya Tunguska r.
          Russia
89 E2 Podol'sk Russia
109 D2 Podujevo Kosovo
          Podujevo Kosovo see
          Podujevë
122 A2 Pofadder S. Africa
89 D3 Pogar Russia
109 D2 Pogradec Albania
65 B2 Pohang S. Korea
48 E2 Pohnpei atoll Micronesia
110 B2 Poiana Mare Romania
118 C3 Poie Dem. Rep. Congo
118 B3 Pointe-Noire Congo
126 D2 Point Lake Can.
140 C2 Point Pleasant U.S.A.
104 C2 Poitiers France
74 B2 Pokaran India
75 C2 Pokhara Nepal
83 J2 Pokrovsk Russia
91 D2 Pokrovskoye Russia
138 A1 Polacca U.S.A.
103 D1 Poland country Europe
88 C2 Polatsk Belarus
61 C2 Polewali Indon.
118 B2 Poli Cameroon
102 C1 Police Pol.
109 C2 Policoro Italy
105 D2 Poligny France
64 B1 Polillo Islands Phil.
90 B1 Polis'ke (abandoned) Ukr.
103 D1 Polkowice Pol.
109 C3 Pollino, Monte mt. Italy
91 D2 Polohy Ukr.
123 C1 Polokwane S. Africa
123 D1 Polokwane r. S. Africa
90 B1 Polonne Ukr.
134 D1 Polson U.S.A.
91 C2 Poltava Ukr.
66 B2 Poltavka Russia
91 D2 Poltavskaya Russia
88 C2 Põlva Estonia
111 B2 Polygyros Greece
106 B2 Pombal Port.
102 C1 Pomeranian Bay Ger./Pol.
108 B2 Pomezia Italy
92 I2 Pomokaira reg. Fin.
110 C2 Pomorie Bulg.

Pomorska, Zatoka b. Ger./Pol.
   see Pomeranian Bay
155 D1 Pompéu Brazil
139 D1 Ponca City U.S.A.
147 D3 Ponce Puerto Rico
127 F2 Pond Inlet Can.
106 B1 Ponferrada Spain
117 A4 Pongo watercourse
          South Sudan
123 D2 Pongola r. S. Africa
128 C2 Ponoka Can.
84 D5 Ponta Delgada
          Arquipélago dos Açores
154 B3 Ponta Grossa Brazil
154 C1 Pontalina Brazil
105 D2 Pont-à-Mousson France
105 D2 Pontarlier France
102 B2 Pontcharra France
142 B2 Pontchartrain, Lake U.S.A.
106 B2 Ponte de Sor Port.
129 D3 Ponteix Can.
150 A2 Pontes e Lacerda Brazil
106 B1 Pontevedra Spain
140 B2 Pontiac IL U.S.A.
140 C2 Pontiac MI U.S.A.
61 B2 Pontianak Indon.
104 B2 Pontivy France
151 C1 Pontoetoe Suriname
129 E2 Ponton Can.
99 B3 Pontypool U.K.
108 B2 Ponziane, Isole is Italy
99 C3 Poole U.K.
          Poona India see Pune
52 B2 Pooncarie Austr.
152 B2 Poopó, Lago de l. Bol.
150 A1 Popayán Col.
83 I2 Popigay r. Russia
52 B2 Popilta Austr.
129 E2 Poplar r. Can.
137 E3 Poplar Bluff U.S.A.
118 B3 Popokabaka Dem. Rep. Congo
110 C2 Popovo Bulg.
103 E2 Poprad Slovakia
151 D3 Porangatu Brazil
74 A2 Porbandar India
126 B2 Porcupine r. Can./U.S.A.
108 B1 Poreč Croatia
54 B2 Porirua N.Z.
88 C2 Porkhov Russia
104 B2 Pornic France
83 K3 Poronaysk Russia
111 B3 Poros Greece
93 E4 Porsgrunn Norway
97 C1 Portadown U.K.
97 D1 Portaferry U.K.
140 B2 Portage U.S.A.
129 E3 Portage la Prairie Can.
128 B3 Port Alberni Can.
106 B2 Portalegre Port.
139 C2 Portales U.S.A.
128 A2 Port Alexander U.S.A.
128 B2 Port Alice Can.
134 B1 Port Angeles U.S.A.
51 D4 Port Arthur Austr.
53 D2 Port Arthur U.S.A.
96 A3 Port Askaig U.K.
52 A2 Port Augusta Austr.
147 C3 Port-au-Prince Haiti
131 E1 Port aux Choix Can.
73 D3 Port Blair India
52 B3 Port Campbell Austr.
54 B3 Port Chalmers N.Z.
143 D3 Port Charlotte U.S.A.
147 C3 Port-de-Paix Haiti
128 A2 Port Edward Can.
155 D1 Porteirinha Brazil
151 C2 Portel Brazil
130 B2 Port Elgin Can.

Port Elizabeth

123 C3 Port Elizabeth S. Africa
96 A3 Port Ellen U.K.
98 A1 Port Erin Isle of Man
122 A3 Porterville S. Africa
135 E2 Porterville U.S.A.
52 B3 Port Fairy Austr.
54 C1 Port Fitzroy N.Z.
118 A3 Port-Gentil Gabon
115 C4 Port Harcourt Nigeria
128 B2 Port Hardy Can.
Port Harrison Can. see
Inukjuak
131 D2 Port Hawkesbury Can.
50 A2 Port Hedland Austr.
99 A2 Porthmadog U.K.
131 E1 Port Hope Simpson Can.
140 C2 Port Huron U.S.A.
106 B2 Portimão Port.
53 C2 Portland N.S.W. Austr.
52 B3 Portland Vic. Austr.
141 E2 Portland ME U.S.A.
134 B1 Portland OR U.S.A.
128 A2 Portland Canal inlet Can.
97 C2 Portlaoise Ireland
139 D3 Port Lavaca U.S.A.
52 A2 Port Lincoln Austr.
114 A4 Port Loko Sierra Leone
113 K9 Port Louis Mauritius
53 D2 Port Macquarie Austr.
128 B2 Port McNeill Can.
131 D2 Port-Menier Can.
59 D3 Port Moresby P.N.G.
122 A2 Port Nolloth S. Africa
Port-Nouveau-Québec Can.
see Kangiqsualujjuaq
Porto Port. see Oporto
150 B2 Porto Acre Brazil
152 C4 Porto Alegre Brazil
151 C3 Porto Artur Brazil
151 C3 Porto dos Gaúchos Óbidos
Brazil
150 C2 Porto Esperidião Brazil
108 B2 Portoferraio Italy
96 A1 Port of Ness U.K.
151 D2 Porto Franco Brazil
147 D3 Port of Spain Trin. and Tob.
108 B1 Portogruaro Italy
108 B2 Portomaggiore Italy
154 B2 Porto Mendes Brazil
152 C3 Porto Murtinho Brazil
151 D3 Porto Nacional Brazil
114 C4 Porto-Novo Benin
154 B2 Porto Primavera, Represa resr
Brazil
155 E1 Porto Seguro Brazil
108 B2 Porto Tolle Italy
108 A2 Porto Torres Sardinia Italy
105 D3 Porto-Vecchio Corsica France
150 B2 Porto Velho Brazil
150 A2 Portoviejo Ecuador
52 A2 Port Pirie Austr.
96 A2 Portree U.K.
128 B3 Port Renfrew Can.
97 C1 Portrush U.K.
116 B1 Port Said Egypt
143 C3 Port St Joe U.S.A.
123 C3 Port St Johns S. Africa
123 D3 Port Shepstone S. Africa
99 C3 Portsmouth U.K.
141 E2 Portsmouth NH U.S.A.
140 C3 Portsmouth OH U.S.A.
141 D3 Portsmouth VA U.S.A.
97 C1 Portstewart U.K.
117 B3 Port Sudan Sudan
142 C3 Port Sulphur U.S.A.
99 B3 Port Talbot U.K.
106 B2 Portugal country Europe

97 B2 Portumna Ireland
105 C3 Port-Vendres France
48 F4 Port Vila Vanuatu
93 I3 Porvoo Fin.
152 C3 Posadas Arg.
89 E2 Poshekhon'ye Russia
92 I2 Posio Fin.
58 C3 Poso Indon.
151 D3 Posse Brazil
101 E2 Pößneck Ger.
139 C2 Post U.S.A.
Poste-de-la-Baleine Can. see
Kuujjuarapik
122 B2 Postmasburg S. Africa
109 C2 Posušje Bos. & Herz.
139 E1 Poteau U.S.A.
109 C2 Potenza Italy
151 D2 Poti r. Brazil
81 C1 Poti Georgia
115 D3 Potiskum Nigeria
141 D3 Potomac, South Branch r.
U.S.A.
152 B2 Potosí Bol.
64 B1 Pototan Phil.
138 C3 Potrero del Llano Mex.
101 F1 Potsdam Ger.
141 E2 Potsdam U.S.A.
141 D2 Pottstown U.S.A.
141 D2 Pottsville U.S.A.
131 E2 Pouch Cove Can.
141 E2 Poughkeepsie U.S.A.
98 B2 Poulton-le-Fylde U.K.
155 C2 Pouso Alegre Brazil
63 B2 Poŭthĭsăt Cambodia
103 D2 Považská Bystrica Slovakia
109 C2 Povlen mt. Serbia
106 B1 Póvoa de Varzim Port.
136 B2 Powell U.S.A.
135 D3 Powell, Lake resr U.S.A.
128 B3 Powell River Can.
154 B1 Poxoréo Brazil
71 B3 Poyang China
71 B3 Poyang Hu l. China
109 D2 Požarevac Serbia
145 C2 Poza Rica Mex.
109 C1 Požega Croatia
109 D2 Požega Serbia
79 D2 Pozm Tīāb Iran
103 D1 Poznań Pol.
106 C2 Pozoblanco Spain
108 B2 Pozzuoli Italy
60 B2 Prabumulih Indon.
102 C2 Prachatice Czech Rep.
63 A2 Prachuap Khiri Khan Thai.
155 E1 Prado Brazil
102 C1 Prague Czech Rep.
Praha Czech Rep. see
Prague
112 C5 Praia Cape Verde
139 C2 Prairie Dog Town Fork r.
U.S.A.
140 A2 Prairie du Chien U.S.A.
60 B1 Prapat Indon.
154 C1 Prata Brazil
105 E3 Prato Italy
137 D3 Pratt U.S.A.
61 C2 Praya Indon.
63 B2 Preăh Vihéar Cambodia
89 F2 Prechistoye Russia
129 D2 Preeceville Can.
88 C2 Preiļi Latvia
53 C2 Premer Austr.
105 C2 Prémery France
101 F1 Premnitz Ger.
102 C1 Prenzlau Ger.
63 A2 Preparis Island Cocos Is
63 A2 Preparis North Channel
Cocos Is

63 A2 Preparis South Channel
Cocos Is
103 D2 Přerov Czech Rep.
138 A2 Prescott U.S.A.
109 D2 Preševo Serbia
152 B3 Presidencia Roque Sáenz
Peña Arg.
151 D2 Presidente Dutra Brazil
154 B2 Presidente Epitácio Brazil
64 B2 Presidente Manuel A Roxas
Phil.
154 B2 Presidente Prudente Brazil
138 C3 Presidio U.S.A.
103 E2 Prešov Slovakia
141 F1 Presque Isle U.S.A.
101 F2 Pressel Ger.
98 B2 Preston U.K.
134 D2 Preston U.S.A.
96 B3 Prestwick U.K.
155 C1 Preto r. Brazil
123 C2 Pretoria S. Africa
111 B3 Preveza Greece
63 B2 Prey Vêng Cambodia
109 C2 Priboj Serbia
135 D3 Price U.S.A.
88 B3 Prienai Lith.
122 B2 Prieska S. Africa
103 D2 Prievidza Slovakia
109 C2 Prijedor Bos. & Herz.
Prikaspiyskaya Nizmennost'
lowland Kazakh./Russia see
Caspian Lowland
109 D2 Prilep Macedonia
91 D2 Primorsko-Akhtarsk Russia
129 D2 Primrose Lake Can.
129 D2 Prince Albert Can.
122 B3 Prince Albert S. Africa
126 D2 Prince Albert Peninsula Can.
122 B3 Prince Albert Road S. Africa
126 C2 Prince Alfred, Cape Can.
122 A3 Prince Alfred Hamlet
S. Africa
127 F2 Prince Charles Island Can.
55 E2 Prince Charles Mountains
Antarctica
131 D2 Prince Edward Island prov.
Can.
113 I11 Prince Edward Islands
Indian Ocean
128 B2 Prince George Can.
126 E2 Prince of Wales Island Can.
128 A2 Prince of Wales Island U.S.A.
126 D2 Prince of Wales Strait Can.
126 D1 Prince Patrick Island Can.
126 E2 Prince Regent Inlet sea chan.
Can.
128 A2 Prince Rupert Can.
51 D1 Princess Charlotte Bay Austr.
128 B2 Princess Royal Island Can.
128 B3 Princeton Can.
134 B2 Prineville U.S.A.
90 B1 Pripet r. Belarus/Ukr.
90 A1 Pripet Marshes Belarus/Ukr.
109 D2 Prishtinë Kosovo
Priština Kosovo see Prishtinë
101 F1 Pritzwalk Ger.
109 C2 Privlaka Croatia
89 F2 Privolzhsk Russia
109 D2 Prizren Kosovo
151 C1 Professor van Blommestein
Meer resr Suriname
145 D2 Progreso Mex.
81 C1 Prokhladnyy Russia
109 D2 Prokuplje Serbia
87 D4 Proletarskoye
Vodokhranilishche l. Russia
126 C3 Prophet r. Can.
128 B2 Prophet River Can.

51 D2 **Proserpine** Austr.
89 E3 **Protvino** Russia
110 C2 **Provadia** Bulg.
141 E2 **Providence** U.S.A.
146 B3 **Providencia, Isla de** *i.* Caribbean Sea
83 N2 **Provideniya** Russia
135 D2 **Provo** U.S.A.
129 C2 **Provost** Can.
154 B3 **Prudentópolis** Brazil
126 B2 **Prudhoe Bay** U.S.A.
100 C2 **Prüm** Ger.
105 D3 **Prunelli-di-Fiumorbo** *Corsica* France
103 E1 **Pruszków** Pol.
110 C1 **Prut** *r.* Europe
91 C1 **Pryluky** Ukr.
91 D2 **Prymors'k** Ukr.
139 D1 **Pryor** U.S.A.
Prypyats' *r.* Belarus *see* **Pripet**
103 E2 **Przemyśl** Pol.
111 C3 **Psara** *i.* Greece
81 C1 **Psebay** Russia
91 D3 **Pshekha** *r.* Russia
88 C2 **Pskov** Russia
88 C2 **Pskov, Lake** Estonia/Russia
111 B2 **Ptolemaïda** Greece
109 C1 **Ptuj** Slovenia
150 A2 **Pucallpa** Peru
71 B3 **Pucheng** China
103 D1 **Puck** Pol.
92 I2 **Pudasjärvi** Fin.
86 C2 **Pudozh** Russia
Puduchcheri India *see* **Puducherry**
73 B3 **Puducherry** India
145 C3 **Puebla** Mex.
136 C3 **Pueblo** U.S.A.
153 B4 **Puelén** Arg.
106 C2 **Puente Genil** Spain
153 A5 **Puerto Aisén** Chile
152 B2 **Puerto Alegre** Bol.
145 C3 **Puerto Ángel** Mex.
146 B4 **Puerto Armuelles** Panama
150 B1 **Puerto Ayacucho** Venez.
146 B3 **Puerto Cabezas** Nic.
150 B1 **Puerto Carreño** Col.
153 A5 **Puerto Cisnes** Chile
144 A2 **Puerto Cortés** Mex.
145 C3 **Puerto Escondido** Mex.
152 B2 **Puerto Frey** Bol.
150 B1 **Puerto Inírida** Col.
152 C2 **Puerto Isabel** Bol.
150 A2 **Puerto Leguizamo** Col.
144 A2 **Puerto Libertad** Mex.
146 B3 **Puerto Limón** Costa Rica
106 C2 **Puertollano** Spain
153 B5 **Puerto Madryn** Arg.
150 B3 **Puerto Maldonado** Peru
153 A5 **Puerto Montt** Chile
153 A6 **Puerto Natales** Chile
150 B1 **Puerto Nuevo** Col.
144 A1 **Puerto Peñasco** Mex.
152 C3 **Puerto Pinasco** Para.
147 C3 **Puerto Plata** Dom. Rep.
150 A2 **Puerto Portillo** Peru
64 A2 **Puerto Princesa** Phil.
147 D3 **Puerto Rico** *terr.* West Indies
158 B3 **Puerto Rico Trench** Caribbean Sea
146 A3 **Puerto San José** Guat.
153 B6 **Puerto Santa Cruz** Arg.
144 B2 **Puerto Vallarta** Mex.
87 D3 **Pugachev** Russia
74 B2 **Pugal** India
54 B2 **Pukaki, Lake** N.Z.
129 D2 **Pukatawagan** Can.
65 B1 **Pukchin** N. Korea
65 B1 **Pukch'ŏng** N. Korea
54 B1 **Pukekohe** N.Z.
65 B1 **Puksubaek-san** *mt.* N. Korea
108 B2 **Pula** Croatia
108 A3 **Pula** *Sardinia* Italy
152 B3 **Pulacayo** Bol.
77 C3 **Pul-e Khumrī** Afgh.
92 I3 **Pulkkila** Fin.
134 C1 **Pullman** U.S.A.
64 B1 **Pulog, Mount** Phil.
150 A2 **Puná, Isla** *i.* Ecuador
54 B2 **Punakaiki** N.Z.
123 D1 **Punda Maria** S. Africa
73 B3 **Pune** India
65 B1 **P'ungsan** N. Korea
121 C2 **Púnguè** *r.* Moz.
153 B4 **Punta Alta** Arg.
153 A6 **Punta Arenas** Chile
146 B3 **Puntarenas** Costa Rica
117 C4 **Puntland** *reg.* Somalia
150 A1 **Punto Fijo** Venez.
92 I3 **Puolanka** Fin.
82 G2 **Pur** *r.* Russia
75 C3 **Puri** India
100 B1 **Purmerend** Neth.
75 C2 **Purnia** India
75 C2 **Puruliya** India
150 B2 **Purus** *r.* Brazil/Peru
61 B2 **Purwakarta** Indon.
65 B1 **Puryŏng** N. Korea
Pusan S. Korea *see* **Busan**
88 C2 **Pushkinskiye Gory** Russia
88 C2 **Pustoshka** Russia
62 A1 **Putao** Myanmar
71 B3 **Putian** China
Puting China *see* **De'an**
61 C2 **Puting, Tanjung** *pt* Indon.
101 F1 **Putlitz** Ger.
58 A2 **Putrajaya** Malaysia
102 C1 **Puttgarden** Ger.
150 A2 **Putumayo** *r.* Col.
61 C1 **Putusibau** Indon.
91 C1 **Putyvl'** Ukr.
127 F2 **Puvirnituq** Can.
70 B2 **Puyang** China
104 C3 **Puylaurens** France
99 A2 **Pwllheli** U.K.
63 A2 **Pyapon** Myanmar
82 G2 **Pyasina** *r.* Russia
87 D4 **Pyatigorsk** Russia
91 C2 **P"yatykhatky** Ukr.
62 A2 **Pyè** Myanmar
88 C3 **Pyetrykaw** Belarus
93 H3 **Pyhäjärvi** *l.* Fin.
92 H3 **Pyhäjoki** *r.* Fin.
92 I3 **Pyhäsalmi** Fin.
62 A2 **Pyinmana** Myanmar
65 B2 **Pyŏksŏng** N. Korea
65 B2 **P'yŏnggang** N. Korea
65 B2 **P'yŏngsan** N. Korea
65 B2 **P'yŏngyang** N. Korea
135 C2 **Pyramid Lake** U.S.A.
80 B3 **Pyramids of Giza** *tourist site* Egypt
107 D1 **Pyrenees** *mts* Europe
111 B3 **Pyrgos** Greece
91 C1 **Pyryatyn** Ukr.
103 C1 **Pyrzyce** Pol.
88 C2 **Pytalovo** Russia
111 B3 **Pyxaria** *mt.* Greece

Q

127 G1 **Qaanaaq** Greenland
123 C3 **Qacha's Nek** Lesotho
68 C2 **Qaidam Pendi** *basin* China
74 A1 **Qalāt** Afgh.
78 B2 **Qal'at Bīshah** Saudi Arabia
129 E1 **Qamanirjuaq Lake** Can.
Qamanittuaq Can. *see* **Baker Lake**
79 C3 **Qamar, Ghubbat al** *b.* Yemen
78 B3 **Qam Hadīl** Saudi Arabia
77 C3 **Qarshi** Uzbek.
78 B2 **Qaryat al Ulyā** Saudi Arabia
127 H2 **Qasigiannguit** Greenland
79 D2 **Qaṣr-e Qand** Iran
81 C2 **Qaṣr-e Shīrīn** Iran
127 H2 **Qassimiut** Greenland
78 B3 **Qa'tabah** Yemen
79 C2 **Qatar** *country* Asia
116 A2 **Qattara Depression** Egypt
81 C1 **Qazax** Azer.
81 C2 **Qazvīn** Iran
127 H2 **Qeqertarsuaq** *i.* Greenland
127 H2 **Qeqertarsuatsiaat** Greenland
127 H2 **Qeqertarsuup Tunua** *b.* Greenland
79 C2 **Qeshm** Iran
70 A3 **Qianjiang** *Chongqing* China
70 B2 **Qianjiang** *Hubei* China
65 A1 **Qian Shan** *mts* China
70 C2 **Qidong** China
68 B2 **Qiemo** China
71 A3 **Qijiang** China
68 C2 **Qijiaojing** China
127 G2 **Qikiqtarjuaq** Can.
74 A2 **Qila Ladgasht** Pak.
68 C2 **Qilian Shan** *mts* China
127 I2 **Qillak** *i.* Greenland
70 B3 **Qimen** China
127 G1 **Qimusseriarsuaq** *b.* Greenland
116 B2 **Qinā** Egypt
70 C2 **Qingdao** China
68 C2 **Qinghai Hu** *salt l.* China
68 C2 **Qinghai Nanshan** *mts* China
70 A2 **Qingyang** China
71 B3 **Qingyuan** *Guangdong* China
65 A1 **Qingyuan** *Liaoning* China
Qingzang Gaoyuan *plat.* China *see* **Tibet, Plateau of**
70 B2 **Qingzhou** China
70 B2 **Qinhuangdao** China
70 A2 **Qin Ling** *mts* China
Qinting China *see* **Lianhua**
70 B2 **Qinyang** China
71 A3 **Qinzhou** China
71 B4 **Qionghai** China
68 C2 **Qionglai Shan** *mts* China
71 B4 **Qiongshan** China
69 E1 **Qiqihar** China
79 C2 **Qīr** Iran
66 B1 **Qitaihe** China
70 B2 **Qixian** China
Qogir Feng *mt.* China/Pak. *see* **K2**
81 D2 **Qom** Iran
Qomolangma Feng *mt.* China/Nepal *see* **Everest, Mount**
76 B2 **Qo'ng'irot** Uzbek.
Qoqek China *see* **Tacheng**
77 D2 **Qo'qon** Uzbek.
76 B2 **Qoraqalpog'iston** Uzbek.
100 C1 **Quakenbrück** Ger.
63 B2 **Quang Ngai** Vietnam
63 B2 **Quang Tri** Vietnam
71 B3 **Quanzhou** *Fujian* China
71 B3 **Quanzhou** *Guangxi* China
108 A3 **Quartu Sant'Elena** *Sardinia* Italy
138 A2 **Quartzsite** U.S.A.
81 C1 **Quba** Azer.
53 C3 **Queanbeyan** Austr.

131 C2 **Québec** Can.
130 C1 **Québec** *prov.* Can.
101 E2 **Quedlinburg** Ger.
128 A2 **Queen Charlotte** Can.
128 B2 **Queen Charlotte Sound** *sea chan.* Can.
128 B2 **Queen Charlotte Strait** Can.
126 D1 **Queen Elizabeth Islands** Can.
55 A1 **Queen Elizabeth Land** *reg.* Antarctica
126 E2 **Queen Maud Gulf** Can.
55 B2 **Queen Maud Land** *reg.* Antarctica
55 N1 **Queen Maud Mountains** Antarctica
52 B1 **Queensland** *state* Austr.
51 D4 **Queenstown** Austr.
54 A3 **Queenstown** N.Z.
123 C3 **Queenstown** S. Africa
121 C2 **Quelimane** Moz.
138 B2 **Quemado** U.S.A.
154 B2 **Querência do Norte** Brazil
145 B2 **Querétaro** Mex.
101 E2 **Querfurt** Ger.
128 B2 **Quesnel** Can.
128 B2 **Quesnel Lake** Can.
74 A1 **Quetta** Pak.
64 A2 **Quezon** Phil.
64 B1 **Quezon City** Phil.
120 A2 **Quibala** Angola
150 A1 **Quibdó** Col.
104 B2 **Quiberon** France
104 C3 **Quillan** France
153 C4 **Quilmes** Arg.
Quilon India *see* **Kollam**
51 D2 **Quilpie** Austr.
153 A4 **Quilpué** Chile
120 A1 **Quimbele** Angola
152 B3 **Quimilí** Arg.
104 B2 **Quimper** France
104 B2 **Quimperlé** France
137 E3 **Quincy** *IL* U.S.A.
141 E2 **Quincy** *MA* U.S.A.
107 C1 **Quinto** Spain
121 D2 **Quionga** Moz.
53 D2 **Quirindi** Austr.
120 A2 **Quitapa** Angola
150 A2 **Quito** Ecuador
151 E2 **Quixadá** Brazil
71 A3 **Qujing** China
52 A2 **Quorn** Austr.
79 C2 **Qurayat** Oman
77 C3 **Qŭrghonteppa** Tajik.
63 B2 **Quy Nhơn** Vietnam
71 B3 **Quzhou** China

R

103 D2 **Raab** *r.* Austria
92 H3 **Raahe** Fin.
100 C1 **Raalte** Neth.
61 C2 **Raas** *i.* Indon.
61 C2 **Raba** Indon.
114 B1 **Rabat** Morocco
48 E3 **Rabaul** P.N.G.
78 A2 **Rābigh** Saudi Arabia
131 E2 **Race, Cape** Can.
142 B3 **Raceland** U.S.A.
63 B3 **Rach Gia** Vietnam
140 B2 **Racine** U.S.A.
78 B3 **Radā'** Yemen
110 C1 **Rădăuți** Romania
140 B3 **Radcliff** U.S.A.
74 B2 **Radhanpur** India
130 C1 **Radisson** Can.
103 E1 **Radom** Pol.
103 D1 **Radomsko** Pol.
90 B1 **Radomyshl'** Ukr.
109 D2 **Radoviš** Macedonia
88 B2 **Radviliškis** Lith.
78 A2 **Raḍwá, Jabal** *mt.* Saudi Arabia
90 B1 **Radyvyliv** Ukr.
75 C2 **Rae Bareli** India
100 C2 **Raeren** Belgium
54 C1 **Raetihi** N.Z.
152 B4 **Rafaela** Arg.
118 C2 **Rafaï** C.A.R.
78 B2 **Rafḥā'** Saudi Arabia
79 C1 **Rafsanjān** Iran
64 B2 **Ragang, Mount** *vol.* Phil.
109 B3 **Ragusa** *Sicily* Italy
59 C3 **Raha** Indon.
88 D3 **Rahachow** Belarus
74 B2 **Rahimyar Khan** Pak.
109 D2 **Rahovec** Kosovo
73 B3 **Raichur** India
75 C2 **Raigarh** India
128 C2 **Rainbow Lake** Can.
134 B1 **Rainier, Mount** *vol.* U.S.A.
130 A2 **Rainy Lake** Can./U.S.A.
129 E3 **Rainy River** Can.
75 C2 **Raipur** India
93 H3 **Raisio** Fin.
73 C3 **Rajahmundry** India
61 C1 **Rajang** *r. Sarawak* Malaysia
74 B2 **Rajanpur** Pak.
73 B4 **Rajapalayam** India
74 B2 **Rajasthan Canal** India
74 B2 **Rajgarh** India
74 B2 **Rajkot** India
74 B2 **Rajpur** India
75 C2 **Rajshahi** Bangl.
54 B2 **Rakaia** *r.* N.Z.
90 A2 **Rakhiv** Ukr.
91 D1 **Rakitnoye** Russia
88 C2 **Rakke** Estonia
88 C2 **Rakvere** Estonia
143 E1 **Raleigh** U.S.A.
48 F2 **Ralik Chain** *is* Marshall Is
117 B4 **Ramciel** S. Sudan
74 B2 **Ramgarh** India
81 C2 **Rāmhormoz** Iran
Ramlat Rabyānah *des.* Libya *see* **Rebiana Sand Sea**
110 C1 **Râmnicu Sărat** Romania
110 B1 **Râmnicu Vâlcea** Romania
89 E3 **Ramon'** Russia
123 C1 **Ramotswa** Botswana
75 B2 **Rampur** India
62 A2 **Ramree Island** Myanmar
98 A1 **Ramsey** Isle of Man
99 D3 **Ramsgate** U.K.
75 C2 **Ranaghat** India
61 C1 **Ranau** *Sabah* Malaysia
153 A4 **Rancagua** Chile
75 C2 **Ranchi** India
93 F4 **Randers** Denmark
54 B2 **Rangiora** N.Z.
54 C1 **Rangitaiki** *r.* N.Z.
62 A2 **Rangoon** Myanmar
75 C2 **Rangpur** Bangl.
129 E1 **Rankin Inlet** Can.
53 C2 **Rankin's Springs** Austr.
96 B2 **Rannoch Moor** *moorland* U.K.
63 A3 **Ranong** Thai.
59 C3 **Ransiki** Indon.
61 C2 **Rantaupanjang** Indon.
60 A1 **Rantauprapat** Indon.
92 I2 **Ranua** Fin.
78 B2 **Ranyah, Wādī** *watercourse* Saudi Arabia
49 I4 **Rapa** *i.* Fr. Polynesia
136 C2 **Rapid City** U.S.A.
88 B2 **Rapla** Estonia
74 A2 **Rapur** India
49 H4 **Rarotonga** *i.* Cook Is
153 B5 **Rasa, Punta** *pt* Arg.
79 C2 **Ra's al Khaymah** U.A.E.
117 B3 **Ras Dejen** *mt.* Eth.
88 B2 **Raseiniai** Lith.
81 C2 **Rasht** Iran
74 A2 **Ras Koh** *mt.* Pak.
88 C2 **Rasony** Belarus
87 D3 **Rasskazovo** Russia
79 C2 **Ras Tannūrah** Saudi Arabia
101 D1 **Rastede** Ger.
93 F3 **Rätan** Sweden
74 B2 **Ratangarh** India
63 A2 **Rat Buri** Thai.
62 A1 **Rathedaung** Myanmar
101 F1 **Rathenow** Ger.
97 C1 **Rathlin Island** U.K.
74 B2 **Ratlam** India
73 B3 **Ratnagiri** India
73 C4 **Ratnapura** Sri Lanka
90 A1 **Ratne** Ukr.
138 C1 **Raton** U.S.A.
96 D2 **Rattray Head** U.K.
101 E1 **Ratzeburg** Ger.
92 □B2 **Raufarhöfn** Iceland
54 C1 **Raukumara Range** *mts* N.Z.
93 H3 **Rauma** Fin.
61 C2 **Raung, Gunung** *vol.* Indon.
75 C2 **Raurkela** India
134 D1 **Ravalli** U.S.A.
81 C2 **Ravānsar** Iran
108 B2 **Ravenna** Italy
102 B2 **Ravensburg** Ger.
74 B1 **Ravi** *r.* Pak.
74 B1 **Rawalpindi** Pak.
103 D1 **Rawicz** Pol.
50 B3 **Rawlinna** Austr.
136 B2 **Rawlins** U.S.A.
153 B5 **Rawson** Arg.
73 C3 **Rayagada** India
69 E1 **Raychikhinsk** Russia
78 B3 **Raydah** Yemen
87 E3 **Rayevskiy** Russia
134 B1 **Raymond** U.S.A.
53 D2 **Raymond Terrace** Austr.
139 D3 **Raymondville** U.S.A.
145 C2 **Rayón** Mex.
63 B2 **Rayong** Thai.
78 A2 **Rayyis** Saudi Arabia
104 B2 **Raz, Pointe du** *pt* France
81 C2 **Razāzah, Buḥayrat ar** *l.* Iraq
110 C2 **Razgrad** Bulg.
110 C2 **Razim, Lacul** *lag.* Romania
110 B2 **Razlog** Bulg.
104 B2 **Ré, Île de** *i.* France
99 C3 **Reading** U.K.
141 D2 **Reading** U.S.A.
115 E2 **Rebiana Sand Sea** *des.* Libya
66 D1 **Rebun-tō** *i.* Japan
50 B3 **Recherche, Archipelago of the** *is* Austr.
89 D3 **Rechytsa** Belarus
151 E2 **Recife** Brazil
123 C3 **Recife, Cape** S. Africa
100 C2 **Recklinghausen** Ger.
152 C3 **Reconquista** Arg.
142 B2 **Red** *r.* U.S.A.
131 E1 **Red Bay** Can.
135 B2 **Red Bluff** U.S.A.
98 C1 **Redcar** U.K.
129 C2 **Redcliff** Can.
52 B2 **Red Cliffs** Austr.
128 C2 **Red Deer** Can.
126 D3 **Red Deer** *r.* Can.
129 D2 **Red Deer Lake** Can.
134 B2 **Redding** U.S.A.
99 C2 **Redditch** U.K.

137 D2 Redfield U.S.A.
130 A1 Red Lake Can.
137 E1 Red Lakes U.S.A.
134 E1 Red Lodge U.S.A.
134 B2 Redmond U.S.A.
137 D2 Red Oak U.S.A.
106 B2 Redondo Port.
159 B2 Red Sea Africa/Asia
128 B1 Redstone r. Can.
100 B1 Reduzum Neth.
137 E2 Red Wing U.S.A.
137 D2 Redwood Falls U.S.A.
97 C2 Ree, Lough l. Ireland
134 B2 Reedsport U.S.A.
54 B2 Reefton N.Z.
102 C2 Regen Ger.
155 E1 Regência Brazil
102 C2 Regensburg Ger.
114 C2 Reggane Alg.
109 C3 Reggio di Calabria Italy
108 B2 Reggio nell'Emilia Italy
110 B1 Reghin Romania
129 D2 Regina Can.
122 A1 Rehoboth Namibia
101 F2 Reichenbach im Vogtland Ger.
143 E1 Reidsville U.S.A.
99 C3 Reigate U.K.
105 C2 Reims France
101 E1 Reinbek Ger.
129 D2 Reindeer r. Can.
129 E2 Reindeer Island Can.
129 D2 Reindeer Lake Can.
92 F2 Reine Norway
100 C3 Reinsfeld Ger.
123 C2 Reitz S. Africa
122 B2 Reivilo S. Africa
129 D1 Reliance Can.
107 D2 Relizane Alg.
79 C2 Remeshk Iran
105 D2 Remiremont France
100 C2 Remscheid Ger.
102 B1 Rendsburg Ger.
141 D1 Renfrew Can.
60 B1 Rengat Indon.
90 B1 Reni Ukr.
52 B2 Renmark Austr.
100 D2 Rennerod Ger.
104 B2 Rennes France
129 D1 Rennie Lake Can.
108 B2 Reno r. Italy
135 C3 Reno U.S.A.
70 A3 Renshou China
140 B2 Rensselaer U.S.A.
75 C2 Renukut India
54 B2 Renwick N.Z.
53 C2 Reo Indon.
136 D3 Republican r. U.S.A.
127 F2 Repulse Bay Can.
150 A2 Requena Peru
107 C2 Requena Spain
154 B2 Reserva Brazil
152 C3 Resistencia Arg.
110 B1 Reşiţa Romania
126 E2 Resolute Can.
127 G2 Resolution Island Can.
105 C2 Rethel France
111 B3 Rethymno Greece
113 K9 Réunion terr. Indian Ocean
107 D1 Reus Spain
102 B2 Reutlingen Ger.
128 C3 Revelstoke Can.
144 A3 Revillagigedo, Islas is Mex.
128 A2 Revillagigedo Island U.S.A.
75 C2 Rewa India
134 D2 Rexburg U.S.A.
92 □A3 Reykjanestá pt Iceland
92 □A3 Reykjavík Iceland

145 C2 Reynosa Mex.
88 C2 Rēzekne Latvia
        Rhein r. Ger. see Rhine
100 C3 Rheine Ger.
101 F1 Rheinsberg Ger.
        Rhin r. France see Rhine
100 C2 Rhine r. Europe
140 B1 Rhinelander U.S.A.
101 F1 Rhinluch marsh Ger.
101 F1 Rhinow Ger.
141 E2 Rhode Island state U.S.A.
111 C3 Rhodes Greece
111 C3 Rhodes i. Greece
110 B2 Rhodope Mountains Bulg./Greece
105 C3 Rhône r. France/Switz.
        Rhuthun U.K. see Ruthin
98 B2 Rhyl U.K.
154 C1 Rianápolis Brazil
60 B1 Riau, Kepulauan is Indon.
106 B1 Ribadeo Spain
106 B1 Ribadesella/Ribeseya Spain
154 B2 Ribas do Rio Pardo Brazil
121 C2 Ribáuè Moz.
98 B2 Ribble r. U.K.
154 B2 Ribeira r. Brazil
154 C2 Ribeirão Preto Brazil
104 C2 Ribérac France
152 B2 Riberalta Bol.
90 B2 Rîbniţa Moldova
102 C1 Ribnitz-Damgarten Ger.
140 A1 Rice Lake U.S.A.
123 D2 Richards Bay S. Africa
126 B2 Richardson Mountains Can.
135 D3 Richfield U.S.A.
134 C1 Richland U.S.A.
140 A2 Richland Center U.S.A.
53 D2 Richmond N.S.W. Austr.
51 D2 Richmond Qld Austr.
54 B2 Richmond N.Z.
122 B3 Richmond S. Africa
98 C1 Richmond U.K.
140 C3 Richmond IN U.S.A.
140 C3 Richmond KY U.S.A.
141 D3 Richmond VA U.S.A.
130 C2 Rideau Lakes Can.
135 C3 Ridgecrest U.S.A.
102 C1 Riesa Ger.
100 D1 Rieste Ger.
123 B2 Riet r. S. Africa
101 D2 Rietberg Ger.
108 B2 Rieti Italy
136 B3 Rifle U.S.A.
88 B2 Riga Latvia
88 B2 Riga, Gulf of Estonia/Latvia
134 D2 Rigby U.S.A.
131 E1 Rigolet Can.
93 H3 Riihimäki Fin.
108 B1 Rijeka Croatia
134 C2 Riley U.S.A.
78 B2 Rimah, Wādī ar watercourse Saudi Arabia
103 D2 Rimavská Sobota Slovakia
128 C2 Rimbey Can.
108 B2 Rimini Italy
131 D2 Rimouski Can.
93 F3 Ringebu Norway
93 E4 Ringkøbing Denmark
92 G2 Ringvassøya i. Norway
101 D1 Rinteln Ger.
150 A2 Riobamba Ecuador
150 B3 Rio Branco Brazil
154 C3 Rio Branco do Sul Brazil
154 B2 Rio Brilhante Brazil
153 B4 Río Claro Brazil
153 B4 Río Colorado Arg.
153 B4 Río Cuarto Arg.
155 D2 Rio de Janeiro Brazil

153 B6 Río Gallegos Arg.
153 B6 Río Grande Arg.
152 C4 Rio Grande Brazil
144 B2 Río Grande Mex.
139 D3 Rio Grande r. Mex./U.S.A.
139 D3 Rio Grande City U.S.A.
150 A1 Ríohacha Col.
150 A2 Rioja Peru
145 C2 Río Lagartos Mex.
105 C2 Riom France
152 B2 Río Mulatos Bol.
154 C3 Rio Negro Brazil
155 D1 Rio Pardo de Minas Brazil
138 B1 Rio Rancho U.S.A.
150 A2 Río Tigre Ecuador
154 B1 Rio Verde Brazil
145 C2 Río Verde Mex.
154 B1 Rio Verde de Mato Grosso Brazil
90 C1 Ripky Ukr.
98 C2 Ripley U.K.
107 D1 Ripoll Spain
98 C1 Ripon U.K.
66 D1 Rishiri-tō i. Japan
93 E4 Risør Norway
122 B2 Ritchie S. Africa
63 A2 Ritchie's Archipelago is India
134 C1 Ritzville U.S.A.
152 B3 Rivadavia Arg.
108 B1 Riva del Garda Italy
146 B3 Rivas Nic.
152 C4 Rivera Uru.
129 D2 Riverhurst Can.
52 B2 Riverina reg. Austr.
122 B3 Riversdale S. Africa
135 C4 Riverside U.S.A.
136 B2 Riverton U.S.A.
131 D2 Riverview Can.
104 C3 Rivesaltes France
131 D2 Rivière-du-Loup Can.
90 B1 Rivne Ukr.
54 B2 Riwaka N.Z.
78 B2 Riyadh Saudi Arabia
81 C1 Rize Turkey
70 B2 Rizhao China
105 C2 Roanne France
141 D3 Roanoke U.S.A.
143 E1 Roanoke r. U.S.A.
143 E1 Roanoke Rapids U.S.A.
52 A3 Robe Austr.
101 F1 Röbel/Müritz Ger.
130 C1 Robert-Bourassa, Réservoir resr Can.
92 H3 Robertsfors Sweden
122 A3 Robertson S. Africa
114 A4 Robertsport Liberia
131 C2 Roberval Can.
50 A2 Robinson Ranges hills Austr.
52 B2 Robinvale Austr.
129 D2 Roblin Can.
128 C2 Robson, Mount Can.
153 C4 Rocha Uru.
98 B2 Rochdale U.K.
154 B1 Rochedo Brazil
100 B2 Rochefort Belgium
104 B2 Rochefort France
137 E2 Rochester MN U.S.A.
141 E2 Rochester NH U.S.A.
141 D2 Rochester NY U.S.A.
140 B2 Rockford U.S.A.
51 E2 Rockhampton Austr.
143 D2 Rock Hill U.S.A.
50 A3 Rockingham Austr.
140 A2 Rock Island U.S.A.
136 B1 Rock Springs MT U.S.A.
139 C3 Rocksprings U.S.A.
136 B2 Rock Springs WY U.S.A.

136 C3 **Rocky Ford** U.S.A.
143 E1 **Rocky Mount** U.S.A.
128 C2 **Rocky Mountain House** Can.
124 G3 **Rocky Mountains** Can./U.S.A.
100 B3 **Rocroi** France
102 C1 **Rødbyhavn** Denmark
131 E1 **Roddickton** Can.
104 C3 **Rodez** France
89 F2 **Rodniki** Russia
Rodos Greece *see* **Rhodes**
Rodos *i.* Greece *see* **Rhodes**
50 A2 **Roebourne** Austr.
50 B1 **Roebuck Bay** Austr.
123 C1 **Roedtan** S. Africa
100 B2 **Roermond** Neth.
100 A2 **Roeselare** Belgium
127 F2 **Roes Welcome Sound** *sea chan.* Can.
142 B1 **Rogers** U.S.A.
122 B3 **Roggeveldberge** *esc.* S. Africa
92 G2 **Rognan** Norway
134 B2 **Rogue** *r.* U.S.A.
88 B2 **Roja** Latvia
60 B1 **Rokan** *r.* Indon.
88 C2 **Rokiškis** Lith.
90 B1 **Rokytne** Ukr.
154 B2 **Rolândia** Brazil
137 E3 **Rolla** U.S.A.
51 D2 **Roma** Austr.
Roma Italy *see* **Rome**
123 C2 **Roma** Lesotho
139 D3 **Roma** U.S.A.
143 E2 **Romain, Cape** U.S.A.
110 C1 **Roman** Romania
59 C3 **Romang, Pulau** *i.* Indon.
110 B1 **Romania** *country* Europe
91 C3 **Roman-Kosh** *mt.* Ukr.
69 D1 **Romanovka** Russia
105 D2 **Rombas** France
64 B1 **Romblon** Phil.
108 B2 **Rome** Italy
143 C2 **Rome** *GA* U.S.A.
141 D2 **Rome** *NY* U.S.A.
99 D3 **Romford** U.K.
105 C2 **Romilly-sur-Seine** France
91 C1 **Romny** Ukr.
104 C2 **Romorantin-Lanthenay** France
99 C3 **Romsey** U.K.
96 □ **Ronas Hill** U.K.
151 C3 **Roncador, Serra do** *hills* Brazil
106 B2 **Ronda** Spain
154 B2 **Rondon** Brazil
154 B1 **Rondonópolis** Brazil
74 B1 **Rondu** Pak.
71 A3 **Rong'an** China
71 A3 **Rongjiang** China
62 A1 **Rongklang Range** *mts* Myanmar
Rongmei China *see* **Hefeng**
93 F4 **Rønne** Denmark
55 P2 **Ronne Ice Shelf** Antarctica
101 D1 **Ronnenberg** Ger.
100 A2 **Ronse** Belgium
75 B2 **Roorkee** India
100 B2 **Roosendaal** Neth.
135 E2 **Roosevelt** U.S.A.
128 B2 **Roosevelt, Mount** Can.
55 K2 **Roosevelt Sub-glacial Island** Antarctica
104 B3 **Roquefort** France
150 B1 **Roraima, Mount** Guyana
93 F3 **Røros** Norway
153 B4 **Rosario** Arg.
144 A1 **Rosario** *Baja California* Mex.
144 B2 **Rosario** *Sinaloa* Mex.
144 B2 **Rosario** *Sonora* Mex.
151 C3 **Rosário Oeste** Brazil
144 A2 **Rosarito** Mex.
109 C3 **Rosarno** Italy
104 B2 **Roscoff** France
97 B2 **Roscommon** Ireland
97 C2 **Roscrea** Ireland
147 D3 **Roseau** Dominica
137 D1 **Roseau** U.S.A.
134 B2 **Roseburg** U.S.A.
139 D3 **Rosenberg** U.S.A.
101 D1 **Rosengarten** Ger.
102 C2 **Rosenheim** Ger.
129 D2 **Rosetown** Can.
122 A2 **Rosh Pinah** Namibia
110 C2 **Roșiori de Vede** Romania
93 F4 **Roskilde** Denmark
89 D3 **Roslavl'** Russia
109 C3 **Rossano** Italy
97 B1 **Rossan Point** Ireland
55 K1 **Ross Ice Shelf** Antarctica
131 D2 **Rossignol, Lake** Can.
128 C3 **Rossland** Can.
97 C2 **Rosslare** Ireland
114 A3 **Rosso** Maur.
105 D3 **Rosso, Capo** *c. Corsica* France
99 B3 **Ross-on-Wye** U.K.
91 D1 **Rossosh'** Russia
128 A1 **Ross River** Can.
55 K2 **Ross Sea** Antarctica
92 F2 **Røssvatnet** *l.* Norway
79 C2 **Rostāq** Iran
129 D2 **Rosthern** Can.
102 C1 **Rostock** Ger.
89 E2 **Rostov** Russia
91 D2 **Rostov-na-Donu** Russia
92 H2 **Rosvik** Sweden
138 C2 **Roswell** U.S.A.
59 D2 **Rota** *i.* N. Mariana Is
59 C3 **Rote** *i.* Indon.
101 D1 **Rotenburg (Wümme)** Ger.
102 C2 **Roth** Ger.
98 C1 **Rothbury** U.K.
98 C2 **Rotherham** U.K.
96 B3 **Rothesay** U.K.
53 C2 **Roto** Austr.
105 D3 **Rotondo, Monte** *mt. Corsica* France
54 C1 **Rotorua** N.Z.
54 C1 **Rotorua, Lake** N.Z.
101 E2 **Rottenbach** Ger.
102 C2 **Rottenmann** Austria
100 B2 **Rotterdam** Neth.
102 B2 **Rottweil** Ger.
49 F3 **Rotuma** *i.* Fiji
105 C1 **Roubaix** France
104 C2 **Rouen** France
Roulers Belgium *see* **Roeselare**
53 D2 **Round Mountain** Austr.
131 E2 **Round Pond** *l.* Can.
139 D2 **Round Rock** U.S.A.
134 E1 **Roundup** U.S.A.
96 C1 **Rousay** *i.* U.K.
130 C2 **Rouyn-Noranda** Can.
92 I2 **Rovaniemi** Fin.
91 D2 **Roven'ki** Russia
91 D2 **Roven'ky** Ukr.
108 B1 **Rovereto** Italy
63 B2 **Rôviěng Tbong** Cambodia
108 B1 **Rovigo** Italy
108 B1 **Rovinj** Croatia
53 C1 **Rowena** Austr.
64 B1 **Roxas** *Capiz* Phil.
64 B1 **Roxas** *Oriental Mindoro* Phil.
64 A1 **Roxas** *Palawan* Phil.
52 A2 **Roxby Downs** Austr.
138 C1 **Roy** U.S.A.
140 B1 **Royale, Isle** *i.* U.S.A.
99 D3 **Royal Tunbridge Wells** U.K.
104 B2 **Royan** France
99 C2 **Royston** U.K.
90 C2 **Rozdil'na** Ukr.
91 C2 **Rozdol'ne** Ukr.
100 B3 **Rozoy-sur-Serre** France
87 D3 **Rtishchevo** Russia
54 C1 **Ruapehu, Mount** *vol.* N.Z.
54 A3 **Ruapuke Island** N.Z.
78 B3 **Rub' al Khālī** *des.* Saudi Arabi
91 D2 **Rubizhne** Ukr.
77 E1 **Rubtsovsk** Russia
135 C2 **Ruby Mountains** U.S.A.
66 C2 **Rudnaya Pristan'** Russia
89 D3 **Rudnya** Russia
76 C1 **Rudnyy** Kazakh.
82 E1 **Rudol'fa, Ostrov** *i.* Russia
101 E2 **Rudolstadt** Ger.
119 D3 **Rufiji** *r.* Tanz.
153 B4 **Rufino** Arg.
121 B2 **Rufunsa** Zambia
70 C2 **Rugao** China
99 C2 **Rugby** U.K.
136 C1 **Rugby** U.S.A.
102 C1 **Rügen** *i.* Ger.
101 E2 **Ruhla** Ger.
88 B2 **Ruhnu** *i.* Estonia
100 C2 **Ruhr** *r.* Ger.
71 C3 **Rui'an** China
138 B2 **Ruidoso** U.S.A.
144 B2 **Ruiz** Mex.
119 D3 **Rukwa, Lake** Tanz.
96 A2 **Rum** *i.* U.K.
109 C1 **Ruma** Serbia
78 B2 **Rumāh** Saudi Arabia
117 A4 **Rumbek** South Sudan
105 D2 **Rumilly** France
50 C1 **Rum Jungle (abandoned)** Austr.
54 B2 **Runanga** N.Z.
98 B2 **Runcorn** U.K.
120 A2 **Rundu** Namibia
68 B2 **Ruoqiang** China
130 C1 **Rupert** *r.* Can.
130 C1 **Rupert Bay** Can.
121 C2 **Rusape** Zimbabwe
110 C2 **Ruse** Bulg.
77 D3 **Rŭshon** Tajik.
136 C2 **Rushville** U.S.A.
53 C3 **Rushworth** Austr.
129 D2 **Russell** Can.
54 B1 **Russell** N.Z.
142 C2 **Russellville** *AL* U.S.A.
142 B1 **Russellville** *AR* U.S.A.
140 B3 **Russellville** *KY* U.S.A.
101 D2 **Rüsselsheim** Ger.
82 F2 **Russia** *country* Asia/Europe
123 C2 **Rustenburg** S. Africa
142 B2 **Ruston** U.S.A.
98 B2 **Ruthin** U.K.
141 E2 **Rutland** U.S.A.
119 E4 **Ruvuma** *r.* Moz./Tanz.
79 C2 **Ruweis** U.A.E.
77 C1 **Ruzayevka** Kazakh.
87 D3 **Ruzayevka** Russia
119 C3 **Rwanda** *country* Africa
89 E3 **Ryazan'** Russia
89 F3 **Ryazhsk** Russia
86 C2 **Rybachiy, Poluostrov** *pen.* Russia
89 E2 **Rybinsk** Russia
89 E2 **Rybinskoye Vodokhranilishche** *resr* Russia
103 D1 **Rybnik** Pol.
89 E3 **Rybnoye** Russia
99 D3 **Rye** U.K.
89 D3 **Ryl'sk** Russia
67 C3 **Ryōtsu** Japan

69 E3 Ryukyu Islands Japan
103 E1 Rzeszów Pol.
89 D2 Rzhev Russia

S

79 C2 Sa'ādatābād Iran
101 E2 Saale r. Ger.
101 E2 Saalfeld/Saale Ger.
102 B2 Saarbrücken Ger.
88 B2 Saaremaa i. Estonia
92 I2 Saarenkylä Fin.
93 I3 Saarijärvi Fin.
102 B2 Saarlouis Ger.
80 B2 Sab' Ābār Syria
107 D1 Sabadell Spain
61 C1 Sabah state Malaysia
61 C2 Sabalana i. Indon.
155 D1 Sabará Brazil
108 B2 Sabaudia Italy
115 D2 Sabhā Libya
123 D2 Sabie r. Moz./S. Africa
145 B2 Sabinas Mex.
145 B2 Sabinas Hidalgo Mex.
131 C2 Sable, Cape Can.
143 D3 Sable, Cape U.S.A.
131 E2 Sable Island Can.
106 B1 Sabugal Port.
78 B3 Şabyā Saudi Arabia
76 B3 Sabzevār Iran
120 A2 Sachanga Angola
65 B3 Sacheon S. Korea
130 A1 Sachigo Lake Can.
126 C2 Sachs Harbour Can.
135 B3 Sacramento U.S.A.
138 B2 Sacramento Mountains U.S.A.
135 B3 Sacramento Valley U.S.A.
123 C3 Sada S. Africa
107 C1 Sádaba Spain
78 B3 Şa'dah Yemen
62 A1 Sadiya India
67 C3 Sadoga-shima i. Japan
107 D2 Sa Dragonera i. Spain
76 C3 Safēd Kōh, Silsilah-ye mts
        Afgh.
93 F4 Säffle Sweden
138 B2 Safford U.S.A.
99 D2 Saffron Walden U.K.
114 B1 Safi Morocco
86 D2 Safonovo Arkhangel'skaya
        Oblast' Russia
89 D2 Safonovo Smolenskaya Oblast'
        Russia
75 C2 Saga China
67 B4 Saga Japan
62 A1 Sagaing Myanmar
67 C3 Sagamihara Japan
62 B2 Sagar India
140 C2 Saginaw U.S.A.
140 C2 Saginaw Bay U.S.A.
106 B2 Sagres Port.
146 B2 Sagua la Grande Cuba
141 F1 Saguenay r. Can.
107 C2 Sagunto Spain
76 B2 Sagyndyk, Mys pt Kazakh.
106 B1 Sahagún Spain
114 C3 Sahara des. Africa
        Saharan Atlas mts Alg. see
        Atlas Saharien
75 B3 Saharanpur India
75 C2 Saharsa India
114 B3 Sahel reg. Africa
144 B2 Sahuayo Mex.
78 B3 Şāḥūq reg. Saudi Arabia
        Saïda Lebanon see Sidon
75 C2 Saidpur Bangl.

        Saigon Vietnam see
        Ho Chi Minh City
62 A1 Saiha India
69 D2 Saihan Tal China
67 B4 Saiki Japan
93 I3 Saimaa l. Fin.
144 B2 Sain Alto Mex.
96 C3 St Abb's Head U.K.
99 C3 St Albans U.K.
99 B3 St Aldhelm's Head U.K.
96 C2 St Andrews U.K.
131 E1 St Anthony Can.
134 D2 St Anthony U.S.A.
52 B3 St Arnaud Austr.
131 E1 St-Augustin Can.
131 E1 St Augustin r. Can.
143 D3 St Augustine U.S.A.
99 A3 St Austell U.K.
104 C2 St-Avertin France
98 B1 St Bees Head U.K.
105 D3 St-Bonnet-en-Champsaur
        France
99 A3 St Bride's Bay U.K.
104 B2 St-Brieuc France
130 C2 St Catharines Can.
99 C3 St Catherine's Point U.K.
137 E3 St Charles U.S.A.
140 C2 St Clair, Lake Can./U.S.A.
105 D2 St-Claude France
137 E1 St Cloud U.S.A.
140 A1 St Croix r. U.S.A.
147 D3 St Croix i. Virgin Is (U.S.A.)
105 C2 St-Dizier France
131 D2 Ste-Anne-des-Monts Can.
105 D2 St-Égrève France
126 B2 St Elias Mountains Can.
131 D1 Ste-Marguerite r. Can.
129 E2 Sainte Rose du Lac Can.
104 B2 Saintes France
105 C2 St-Étienne France
108 A2 St-Florent Corsica France
105 C2 St-Flour France
136 C3 St Francis U.S.A.
104 C3 St-Gaudens France
53 C1 St George Austr.
135 D3 St George U.S.A.
143 D3 St George Island U.S.A.
131 C2 St-Georges Can.
147 D3 St George's Grenada
97 C3 St George's Channel Ireland/
        U.K.
105 D2 St Gotthard Pass Switz.
113 E8 St Helena i. S. Atlantic
        Ocean
113 E8 St Helena, Ascension and
        Tristan da Cunha terr.
        S. Atlantic Ocean
122 A3 St Helena Bay S. Africa
122 A3 St Helena Bay S. Africa
98 B2 St Helens U.K.
134 B1 St Helens, Mount vol. U.S.A.
95 C4 St Helier Channel Is
100 B2 St-Hubert Belgium
141 E1 St-Hyacinthe Can.
140 C1 St Ignace U.S.A.
130 B2 St Ignace Island Can.
99 A3 St Ives U.K.
128 A2 St James, Cape Can.
104 B2 St-Jean-d'Angély France
104 B2 St-Jean-de-Monts France
130 C2 St-Jean-sur-Richelieu Can.
141 E1 St-Jérôme Can.
134 C1 Saint John Can.
141 F1 St John r. U.S.A.
147 D3 St John's Antigua
131 E2 St John's Can.

138 B2 St Johns U.S.A.
141 E2 St Johnsbury U.S.A.
137 E3 St Joseph U.S.A.
130 A1 St Joseph, Lake Can.
130 B2 St Joseph Island Can.
104 C2 St-Junien France
94 B2 St Kilda i. U.K.
147 D3 St Kitts and Nevis country
        West Indies
        St-Laurent, Golfe du g. Can.
        see St Lawrence, Gulf of
151 C1 St-Laurent-du-Maroni
        Fr. Guiana
131 D2 St Lawrence inlet Can.
131 D2 St Lawrence, Gulf of Can.
83 N2 St Lawrence Island U.S.A.
104 B2 St-Lô France
114 A3 St-Louis Senegal
137 E3 St Louis U.S.A.
137 E3 St Louis r. U.S.A.
147 D3 St Lucia country West Indies
123 D2 St Lucia Estuary S. Africa
104 B2 St-Malo France
104 B2 St-Malo, Golfe de g. France
122 A3 St Martin, Cape S. Africa
141 D2 St Marys U.S.A.
130 C2 St-Maurice r. Can.
104 B2 St-Nazaire France
104 C1 St-Omer France
129 C2 St Paul Can.
137 E2 St Paul U.S.A.
137 E2 St Peter U.S.A.
95 C4 St Peter Port Channel Is
89 D2 St Petersburg Russia
143 D3 St Petersburg U.S.A.
131 E2 St-Pierre
        St Pierre and Miquelon
131 E2 St Pierre and Miquelon terr.
        N. America
104 B2 St-Pierre-d'Oléron France
105 C2 St-Pourçain-sur-Sioule
        France
131 D2 St Quentin Can.
105 C2 St-Quentin France
131 D2 St-Siméon Can.
129 E2 St Theresa Point Can.
130 B2 St Thomas Can.
105 D3 St-Tropez France
105 D3 St-Tropez, Cap de c. France
52 A3 St Vincent, Gulf Austr.
147 D3 St Vincent and the
        Grenadines country
        West Indies
100 C2 St-Vith Belgium
129 C2 St Walburg Can.
104 C2 St-Yrieix-la-Perche France
59 D1 Saipan i. N. Mariana Is
152 B2 Sajama, Nevado mt. Bol.
122 B2 Sak watercourse S. Africa
67 C4 Sakai Japan
67 B4 Sakaide Japan
78 B2 Sakākah Saudi Arabia
136 C1 Sakakawea, Lake U.S.A.
111 D2 Sakarya r. Turkey
66 C3 Sakata Japan
65 B1 Sakchu N. Korea
83 K3 Sakhalin i. Russia
123 C2 Sakhile S. Africa
81 B1 Şäki Azer.
88 B3 Sakiai Lith.
69 E3 Sakishima-shotō is Japan
62 B2 Sakon Nakhon Thai.
122 B3 Sakrivier S. Africa
67 D3 Sakura Japan
91 C2 Saky Ukr.
93 G4 Sala Sweden
130 C2 Salaberry-de-Valleyfield
        Can.

88 B2 **Salacgrīva** Latvia
109 C2 **Sala Consilina** Italy
135 C4 **Salada, Laguna** *salt l.* Mex.
152 B4 **Salado** *r.* Arg.
145 C2 **Salado** *r.* Mex.
114 B4 **Salaga** Ghana
122 B1 **Salajwe** Botswana
115 D3 **Salal** Chad
78 A2 **Salāla** Sudan
79 C3 **Şalālah** Oman
106 B1 **Salamanca** Mex.
106 C1 **Salamanca** Spain
106 B1 **Salas** Spain
63 B2 **Salavan** Laos
59 C3 **Salawati** *i.* Indon.
104 C2 **Salbris** France
88 C3 **Šalčininkai** Lith.
106 C1 **Saldaña** Spain
122 A3 **Saldanha** S. Africa
88 B2 **Saldus** Latvia
53 C3 **Sale** Austr.
86 F2 **Salekhard** Russia
73 B3 **Salem** India
137 E3 **Salem** MO U.S.A.
134 B2 **Salem** OR U.S.A.
96 B2 **Salem** U.K.
109 B2 **Salerno** Italy
98 B2 **Salford** U.K.
151 E2 **Salgado** *r.* Brazil
103 D2 **Salgótarján** Hungary
151 E2 **Salgueiro** Brazil
136 B3 **Salida** U.S.A.
111 C3 **Salihli** Turkey
88 C3 **Salihorsk** Belarus
121 C2 **Salima** Malawi
122 C2 **Salimo** Moz.
137 D3 **Salina** U.S.A.
108 B3 **Salina, Isola** *i.* Italy
145 C3 **Salina Cruz** Mex.
155 D1 **Salinas** Brazil
144 B2 **Salinas** Mex.
135 B3 **Salinas** U.S.A.
107 D2 **Salines, Cap de ses** *c.* Spain
151 D2 **Salinópolis** Brazil
99 C3 **Salisbury** U.K.
141 D3 **Salisbury** MD U.S.A.
143 D1 **Salisbury** NC U.S.A.
99 C3 **Salisbury Plain** U.K.
151 D2 **Salitre** *r.* Brazil
92 I2 **Salla** Fin.
127 F2 **Salluit** Can.
75 C2 **Sallyana** Nepal
81 C2 **Salmās** Iran
134 D1 **Salmon** U.S.A.
134 C1 **Salmon** *r.* U.S.A.
128 C2 **Salmon Arm** Can.
134 C2 **Salmon River Mountains** U.S.A.
100 C3 **Salmtal** Ger.
93 H3 **Salo** Fin.
87 D4 **Sal'sk** Russia
122 B3 **Salt** *watercourse* S. Africa
138 A2 **Salt** *r.* U.S.A.
152 B3 **Salta** Arg.
145 B2 **Saltillo** Mex.
134 D2 **Salt Lake City** U.S.A.
154 C2 **Salto** Brazil
152 C4 **Salto** Uru.
155 E1 **Salto da Divisa** Brazil
154 B2 **Salto do Guairá** Para.
135 C4 **Salton Sea** *salt l.* U.S.A.
143 D2 **Saluda** U.S.A.
108 A2 **Saluzzo** Italy
151 E3 **Salvador** Brazil
79 C2 **Salwah** Saudi Arabia
62 A1 **Salween** *r.* China/Myanmar
81 C2 **Salyan** Azer.

102 C2 **Salzburg** Austria
101 E1 **Salzgitter** Ger.
101 D2 **Salzkotten** Ger.
101 E1 **Salzwedel** Ger.
144 B1 **Salzwedel** Ger.
144 B1 **Samalayuca** Mex.
66 D2 **Samani** Japan
64 B1 **Samar** *i.* Phil.
87 E3 **Samara** Russia
61 C3 **Samarinda** Indon.
77 C3 **Samarqand** Uzbek.
81 C2 **Sāmarrā'** Iraq
81 C1 **Şamaxı** Azer.
119 C3 **Samba** Dem. Rep. Congo
61 C1 **Sambaliung** *mts* Indon.
75 C2 **Sambalpur** India
61 C2 **Sambar, Tanjung** *pt* Indon.
61 B1 **Sambas** Indon.
121 DE2 **Sambava** Madag.
90 A2 **Sambir** Ukr.
153 C4 **Samborombón, Bahía** *b.* Arg.
65 B2 **Samcheok** S. Korea
81 C2 **Samdi Dag** *mt.* Turkey
119 D3 **Same** Tanz.
78 B2 **Samīrah** Saudi Arabia
65 B1 **Samjiyŏn** N. Korea
49 G3 **Samoa** *country* S. Pacific Ocean
109 C1 **Samobor** Croatia
111 C3 **Samos** *i.* Greece
111 C2 **Samothraki** Greece
111 C2 **Samothraki** *i.* Greece
61 C2 **Sampit** Indon.
119 C3 **Sampwe** Dem. Rep. Congo
63 B2 **Sâmraông** Cambodia
139 E2 **Sam Rayburn Reservoir** U.S.A.
80 B1 **Samsun** Turkey
81 C1 **Samt'redia** Georgia
63 B3 **Samui, Ko** *i.* Thai.
63 B2 **Samut Songkhram** Thai.
114 B3 **Sam** Mali
78 B3 **Şan'ā'** Yemen
118 A2 **Sanaga** *r.* Cameroon
81 C2 **Sanandaj** Iran
146 B3 **San Andrés, Isla de** *i.* Caribbean Sea
138 B2 **San Andres Mountains** U.S.A.
145 C3 **San Andrés Tuxtla** Mex.
139 C2 **San Angelo** U.S.A.
139 D3 **San Antonio** U.S.A.
135 C4 **San Antonio, Mount** U.S.A.
152 B3 **San Antonio de los Cobres** Arg.
153 B5 **San Antonio Oeste** Arg.
108 B2 **San Benedetto del Tronto** Italy
144 A3 **San Benedicto, Isla** *i.* Mex.
135 C4 **San Bernardino** U.S.A.
135 C4 **San Bernardino Mountains** U.S.A.
143 D3 **San Blas, Cape** U.S.A.
152 B2 **San Borja** Bol.
144 B2 **San Buenaventura** Mex.
64 B1 **San Carlos** Phil.
147 D4 **San Carlos** Venez.
153 A5 **San Carlos de Bariloche** Arg.
147 C4 **San Carlos del Zulia** Venez.
135 C4 **San Clemente Island** U.S.A.
104 C2 **Sancoins** France
150 A1 **San Cristóbal** Venez.
145 C3 **San Cristóbal de las Casas** Mex.
146 C2 **Sancti Spíritus** Cuba
61 C1 **Sandakan** *Sabah* Malaysia
93 E3 **Sandane** Norway
111 B2 **Sandanski** Bulg.
96 C1 **Sanday** *i.* U.K.
139 C2 **Sanderson** U.S.A.
150 B3 **Sandia** Peru

135 C4 **San Diego** U.S.A.
80 B2 **Sandıklı** Turkey
93 E4 **Sandnes** Norway
92 F2 **Sandnessjøen** Norway
118 C3 **Sandoa** Dem. Rep. Congo
103 E1 **Sandomierz** Pol.
89 E2 **Sandovo** Russia
94 B1 **Sandoy** *i.* Faroe Is
134 C1 **Sandpoint** U.S.A.
71 B3 **Sandu** China
94 B1 **Sandur** Faroe Is
140 C2 **Sandusky** U.S.A.
122 A3 **Sandveld** *mts* S. Africa
93 F4 **Sandvika** Norway
93 G3 **Sandviken** Sweden
131 E1 **Sandwich Bay** Can.
129 D2 **Sandy Bay** Can.
51 E2 **Sandy Cape** Austr.
130 A1 **Sandy Lake** Can.
130 A1 **Sandy Lake** Can.
144 A1 **San Felipe** *Baja California* Mex.
145 B2 **San Felipe** *Guanajuato* Mex.
150 B1 **San Felipe** Venez.
144 A2 **San Fernando** *Baja California* Mex.
145 C2 **San Fernando** *Tamaulipas* Mex.
64 B1 **San Fernando** *La Union* Phil.
64 B1 **San Fernando** *Pampanga* Phil.
106 B2 **San Fernando** Spain
147 D3 **San Fernando** Trin. and Tob.
150 B1 **San Fernando de Apure** Venez.
143 D3 **Sanford** FL U.S.A.
141 E2 **Sanford** ME U.S.A.
152 B4 **San Francisco** Arg.
135 B3 **San Francisco** U.S.A.
74 B3 **Sangamner** India
83 J2 **Sangar** Russia
108 A3 **San Gavino Monreale** *Sardinia* Italy
101 E2 **Sangerhausen** Ger.
118 B3 **Sangha** *r.* Congo
109 C3 **Sanggau** Indon.
118 B3 **Sangha** *r.* Congo
109 C3 **San Giovanni in Fiore** Italy
64 B2 **Sangir** *i.* Indon.
59 C2 **Sangir, Kepulauan** *is* Indon.
65 B2 **Sangju** S. Korea
61 C1 **Sangkulirang** Indon.
73 B3 **Sangli** India
118 B2 **Sangmélima** Cameroon
121 C3 **Sango** Zimbabwe
136 B3 **Sangre de Cristo Range** *mts* U.S.A.
75 C2 **Sangsang** China
144 A2 **San Hipólito, Punta** *pt* Mex.
144 A2 **San Ignacio** Mex.
130 C1 **Sanikiluaq** Can.
71 A3 **Sanjiang** China
135 B3 **San Joaquin** *r.* U.S.A.
153 B5 **San Jorge, Golfo de** *g.* Arg.
146 B4 **San José** Costa Rica
64 B1 **San Jose** *Nueva Ecija* Phil.
64 B1 **San Jose** *Occidental Mindoro* Phil.
135 B3 **San Jose** U.S.A.
144 A2 **San José, Isla** *i.* Mex.
144 B2 **San José de Bavicora** Mex.
144 A2 **San José de Buenavista** Phil.
144 A2 **San José de Comondú** Mex.
144 B2 **San José del Cabo** Mex.
150 A1 **San José del Guaviare** Col.
152 B4 **San José de Mayo** ...
146 B3 **San Juan** *r.* Costa Rica/Nic.
147 D3 **San Juan** Puerto Rico
135 D3 **San Juan** *r.* U.S.A.

145 C3 **San Juan Bautista Tuxtepec** Mex.
134 B1 **San Juan Islands** U.S.A.
144 B2 **San Juanito** Mex.
136 B3 **San Juan Mountains** U.S.A.
153 B5 **San Julián** Arg.
75 C2 **Sankh** *r.* India
105 D2 **Sankt Gallen** Switz.
105 D2 **Sankt Moritz** Switz.
Sankt-Peterburg Russia *see* **St Petersburg**
102 C2 **Sankt Veit an der Glan** Austria
100 C3 **Sankt Wendel** Ger.
80 B2 **Şanlıurfa** Turkey
138 B3 **San Lorenzo** Mex.
106 B2 **Sanlúcar de Barrameda** Spain
153 B4 **San Luis** Arg.
145 B2 **San Luis de la Paz** Mex.
138 A2 **San Luisito** Mex.
135 B3 **San Luis Obispo** U.S.A.
145 B2 **San Luis Potosí** Mex.
144 A1 **San Luis Río Colorado** Mex.
139 D3 **San Marcos** U.S.A.
108 B2 **San Marino** *country* Europe
108 B2 **San Marino** San Marino
144 B2 **San Martín de Bolaños** Mex.
153 A5 **San Martín de los Andes** Arg.
153 B5 **San Matías, Golfo** *g.* Arg.
70 B2 **Sanmenxia** China
146 B3 **San Miguel** El Salvador
152 B3 **San Miguel de Tucumán** Arg.
145 C3 **San Miguel Sola de Vega** Mex.
71 B3 **Sanming** China
153 B4 **San Nicolás de los Arroyos** Arg.
135 C4 **San Nicolas Island** U.S.A.
123 C2 **Sannieshof** S. Africa
103 E2 **Sanok** Pol.
64 B1 **San Pablo** Phil.
144 B2 **San Pablo Balleza** Mex.
152 B3 **San Pedro** Arg.
152 B2 **San Pedro** Bol.
114 B4 **San-Pédro** Côte d'Ivoire
144 A2 **San Pedro** Mex.
138 A2 **San Pedro** *watercourse* U.S.A.
106 B2 **San Pedro, Sierra de** *mts* Spain
144 B2 **San Pedro de las Colonias** Mex.
146 B3 **San Pedro Sula** Hond.
108 A3 **San Pietro, Isola di** *i. Sardinia* Italy
144 A1 **San Quintín, Cabo** *c.* Mex.
153 B4 **San Rafael** Arg.
108 A2 **Sanremo** Italy
146 B3 **San Salvador** El Salvador
152 B3 **San Salvador de Jujuy** Arg.
107 C1 **San Sebastián** Spain
107 C1 **San Sebastián** Spain
109 C2 **San Severo** Italy
109 C2 **Sanski Most** Bos. & Herz.
152 B2 **Santa Ana** Bol.
146 B3 **Santa Ana** El Salvador
144 A1 **Santa Ana** Mex.
135 C4 **Santa Ana** U.S.A.
144 B2 **Santa Bárbara** Mex.
135 C4 **Santa Barbara** U.S.A.
154 B2 **Santa Bárbara, Serra de** *hills* Brazil
152 B3 **Santa Catalina** Chile
150 B2 **Santa Clara** Col.
146 C2 **Santa Clara** Cuba
135 C4 **Santa Clarita** U.S.A.
109 C3 **Santa Croce, Capo** *c. Sicily* Italy
153 B6 **Santa Cruz** *r.* Arg.
152 B2 **Santa Cruz** Bol.
64 B1 **Santa Cruz** Phil.
135 B3 **Santa Cruz** U.S.A.
145 C3 **Santa Cruz Barillas** Guat.
155 E1 **Santa Cruz Cabrália** Brazil
107 C2 **Santa Cruz de Moya** Spain
114 A2 **Santa Cruz de Tenerife** Canary Islands
135 C4 **Santa Cruz Island** U.S.A.
48 F3 **Santa Cruz Islands** Solomon Is
152 B4 **Santa Fe** Arg.
138 B1 **Santa Fe** U.S.A.
154 B1 **Santa Helena de Goiás** Brazil
153 B4 **Santa Isabel** Arg.
154 B1 **Santa Luisa, Serra de** *hills* Brazil
152 C3 **Santa Maria** Brazil
144 B1 **Santa María** *r.* Mex.
135 B4 **Santa Maria** U.S.A.
123 D2 **Santa Maria, Cabo de** *c.* Moz.
106 B2 **Santa Maria, Cabo de** *c.* Port.
151 D2 **Santa Maria das Barreiras** Brazil
109 C3 **Santa Maria di Leuca, Capo** *c.* Italy
150 A1 **Santa Marta** Col.
135 C4 **Santa Monica** U.S.A.
151 C2 **Santana** *Amapá* Brazil
151 D3 **Santana** *Bahia* Brazil
106 C1 **Santander** Spain
108 A3 **Sant'Antioco** *Sardinia* Italy
108 A3 **Sant'Antioco, Isola di** *i. Sardinia* Italy
107 D2 **Sant Antoni de Portmany** Spain
151 C2 **Santarém** Brazil
106 B2 **Santarém** Port.
154 B1 **Santa Rita do Araguaia** Brazil
153 B4 **Santa Rosa** Arg.
152 C3 **Santa Rosa** Brazil
135 B3 **Santa Rosa** *CA* U.S.A.
138 C2 **Santa Rosa** *NM* U.S.A.
146 B3 **Santa Rosa de Copán** Hond.
135 B4 **Santa Rosa Island** U.S.A.
144 A2 **Santa Rosalía** Mex.
107 D2 **Sant Francesc de Formentera** Spain
152 C3 **Santiago** Brazil
153 A4 **Santiago** Chile
147 C3 **Santiago** Dom. Rep.
144 B2 **Santiago** Mex.
146 B4 **Santiago** Panama
64 B1 **Santiago** Phil.
106 B1 **Santiago de Compostela** Spain
144 B2 **Santiago Ixcuintla** Mex.
144 B2 **Santiago Papasquiaro** Mex.
107 D2 **Sant Joan de Labritja** Spain
107 D1 **Sant Jordi, Golf de** *g.* Spain
155 D2 **Santo Amaro de Campos** Brazil
155 C2 **Santo André** Brazil
152 C3 **Santo Ângelo** Brazil
154 B2 **Santo Antônio da Platina** Brazil
151 E3 **Santo Antônio de Jesus** Brazil
150 B2 **Santo Antônio do Içá** Brazil
155 C2 **Santo Antônio do Monte** Brazil
147 D3 **Santo Domingo** Dom. Rep.
138 B1 **Santo Domingo Pueblo** U.S.A.
111 C3 **Santorini** *i.* Greece
155 C2 **Santos** Brazil
152 C3 **Santo Tomé** Arg.
153 A5 **San Valentín, Cerro** *mt.* Chile
146 B3 **San Vicente** El Salvador
144 A1 **San Vicente** Mex.
150 A3 **San Vicente de Cañete** Peru
108 B2 **San Vincenzo** Italy
108 B3 **San Vito, Capo** *c. Sicily* Italy
71 A4 **Sanya** China
155 C2 **São Bernardo do Campo** Brazil
152 C3 **São Borja** Brazil
154 C2 **São Carlos** Brazil
151 C3 **São Félix do Araguaia** Brazil
151 C2 **São Félix do Xingu** Brazil
155 D2 **São Fidélis** Brazil
155 D1 **São Francisco** Brazil
151 E3 **São Francisco** *r.* Brazil
154 C3 **São Francisco, Ilha de** *i.* Brazil
154 C3 **São Francisco do Sul** Brazil
152 C4 **São Gabriel** Brazil
150 B2 **São Gabriel da Cachoeira** Brazil
155 D2 **São Gonçalo** Brazil
155 C1 **São Gotardo** Brazil
155 D2 **São João da Barra** Brazil
155 C2 **São João da Boa Vista** Brazil
106 B1 **São João da Madeira** Port.
155 D1 **São João do Paraíso** Brazil
155 D2 **São João Nepomuceno** Brazil
154 C2 **São Joaquim da Barra** Brazil
154 C2 **São José do Rio Preto** Brazil
155 C2 **São José dos Campos** Brazil
154 C3 **São José dos Pinhais** Brazil
155 C2 **São Lourenço** Brazil
151 D2 **São Luís** Brazil
154 C2 **São Manuel** Brazil
154 C1 **São Marcos** *r.* Brazil
151 D2 **São Marcos, Baía de** *b.* Brazil
155 E1 **São Mateus** Brazil
155 C2 **São Paulo** Brazil
151 D2 **São Raimundo Nonato** Brazil
155 C1 **São Romão** Brazil
155 C2 **São Sebastião, Ilha do** *i.* Brazil
154 C2 **São Sebastião do Paraíso** Brazil
154 B1 **São Simão** Brazil
154 B1 **São Simão, Barragem de** *resr* Brazil
59 C2 **Sao-Siu** Indon.
113 F6 **São Tomé** São Tomé and Príncipe
155 D2 **São Tomé, Cabo de** *c.* Brazil
113 F6 **São Tomé and Príncipe** *country* Africa
155 C2 **São Vicente** Brazil
106 B2 **São Vicente, Cabo de** *c.* Port.
59 C3 **Saparua** Indon.
89 F3 **Sapozhok** Russia
66 D2 **Sapporo** Japan
109 C2 **Sapri** Italy
81 C2 **Saqqez** Iran
81 C2 **Sarāb** Iran
63 B2 **Sara Buri** Thai.
109 C2 **Sarajevo** Bos. & Herz.
76 B1 **Saraktash** Russia
141 E2 **Saranac Lake** U.S.A.
109 D3 **Sarandë** Albania
64 B2 **Sarangani Islands** Phil.
87 D3 **Saransk** Russia
87 E3 **Sarapul** Russia
143 D3 **Sarasota** U.S.A.
90 B2 **Sarata** Ukr.
136 B2 **Saratoga** U.S.A.
141 E2 **Saratoga Springs** U.S.A.
61 C1 **Saratok** *Sarawak* Malaysia
87 D3 **Saratov** Russia
79 D2 **Sarāvān** Iran
61 C1 **Sarawak** *state* Malaysia
110 C2 **Saray** Turkey
111 C3 **Sarayköy** Turkey
79 D2 **Sarbāz** Iran

74 B2 Sardarshahr India
Sardegna i. Italy see Sardinia
108 A2 Sardinia i. Italy
92 G2 Sarektjåkkå mt. Sweden
77 C3 Sar-e Pul Afgh.
158 B3 Sargasso Sea N. Atlantic Ocean
74 B1 Sargodha Pak.
115 D4 Sarh Chad
79 D2 Sarhad reg. Iran
81 D2 Sārī Iran
111 C3 Sarıgöl Turkey
81 C1 Sarıkamış Turkey
61 C1 Sarikei Sarawak Malaysia
51 D2 Sarina Austr.
65 B2 Sariwŏn N. Korea
111 C2 Sarıyer Turkey
78 B2 Sark, Safrā' as esc. Saudi Arabia
77 D2 Sarkand Kazakh.
111 C2 Şarköy Turkey
59 D3 Sarmi Indon.
140 C2 Sarnia Can.
90 B1 Sarny Ukr.
60 B2 Sarolangun Indon.
111 B3 Saronikos Kolpos g. Greece
111 C2 Saros Körfezi b. Turkey
87 D3 Sarov Russia
105 D2 Sarrebourg France
106 B1 Sarria Spain
107 C1 Sarrión Spain
105 D3 Sartène Corsica France
103 D2 Sárvár Hungary
77 D1 Saryarka plain Kazakh.
76 B2 Sarykamyshskoye Ozero salt l. Turkm./Uzbek.
77 D2 Saryozek Kazakh.
77 D2 Saryshagan Kazakh.
82 F3 Sarysu watercourse Kazakh.
77 D3 Sary-Tash Kyrg.
75 C2 Sasaram India
67 A4 Sasebo Japan
129 D2 Saskatchewan prov. Can.
129 D2 Saskatchewan r. Can.
129 D2 Saskatoon Can.
83 I2 Saskylakh Russia
123 C2 Sasolburg S. Africa
87 D3 Sasovo Russia
114 B4 Sassandra Côte d'Ivoire
108 A2 Sassari Sardinia Italy
102 C1 Sassnitz Ger.
114 A3 Satadougou Mali
123 D1 Satara S. Africa
75 C2 Satna India
74 B2 Satpura Range mts India
67 B4 Satsuma-Sendai Japan
63 B2 Sattahip Thai.
110 B1 Satu Mare Romania
63 B3 Satun Thai.
144 B2 Saucillo Mex.
93 E4 Sauda Norway
92 □B2 Sauðárkrókur Iceland
78 B2 Saudi Arabia country Asia
105 C3 Saugues France
105 C2 Saulieu France
130 B2 Sault Sainte Marie Can.
140 C1 Sault Sainte Marie U.S.A.
77 C1 Saumalkol' Kazakh.
59 C3 Saumlakki Indon.
104 B2 Saumur France
120 B1 Saurimo Angola
109 D2 Sava r. Europe
49 G3 Savai'i i. Samoa
91 E1 Savala r. Russia
143 D2 Savannah GA U.S.A.
142 C1 Savannah TN U.S.A.
143 D2 Savannah r. U.S.A.
63 B2 Savannakhét Laos
130 A1 Savant Lake Can.
111 C3 Savaştepe Turkey
89 F2 Savino Russia
86 D2 Savinskiy Russia
108 A2 Savona Italy
93 I3 Savonlinna Fin.
92 I2 Savukoski Fin.
Savu Sea Indon. see Sawu, Laut
74 B2 Sawai Madhopur India
62 A2 Sawankhalok Thai.
136 B3 Sawatch Range mts U.S.A.
53 D2 Sawtell Austr.
59 C3 Sawu, Laut sea Indon.
79 C3 Sayḩūt Yemen
69 D2 Saynshand Mongolia
141 D2 Sayre U.S.A.
144 B3 Sayula Jalisco Mex.
145 C3 Sayula Veracruz Mex.
128 B2 Sayward Can.
89 E2 Sazonovo Russia
114 B2 Sbaa Alg.
98 B1 Scafell Pike h. U.K.
109 C3 Scalea Italy
96 C1 Scapa Flow inlet U.K.
130 C2 Scarborough Can.
147 D3 Scarborough Trin. and Tob.
98 C1 Scarborough U.K.
64 A1 Scarborough Reef sea feature S. China Sea
96 A2 Scarinish U.K.
Scarpanto i. Greece see Karpathos
105 D2 Schaffhausen Switz.
100 B1 Schagen Neth.
102 C2 Schärding Austria
100 A2 Scharendijke Neth.
101 D1 Scharhörn i. Ger.
101 D1 Scheeßel Ger.
131 D1 Schefferville Can.
135 D3 Schell Creek Range mts U.S.A.
141 E2 Schenectady U.S.A.
101 E3 Scheßlitz Ger.
100 C1 Schiermonnikoog i. Neth.
108 B1 Schio Italy
101 F2 Schkeuditz Ger.
101 E1 Schladen Ger.
101 E2 Schleiz Ger.
102 B1 Schleswig Ger.
101 D2 Schloß Holte-Stukenbrock Ger.
101 D2 Schlüchtern Ger.
101 E3 Schlüsselfeld Ger.
101 D2 Schmallenberg Ger.
101 D1 Schneverdingen Ger.
101 E1 Schönebeck (Elbe) Ger.
101 E1 Schöningen Ger.
100 B2 Schoonhoven Neth.
59 D3 Schouten Islands P.N.G.
102 B2 Schwäbische Alb mts Ger.
102 C2 Schwandorf Ger.
61 C2 Schwaner, Pegunungan mts Indon.
101 E1 Schwarzenbek Ger.
101 F2 Schwarzenberg/Erzgebirge Ger.
122 A2 Schwarzrand mts Namibia
Schwarzwald mts Ger. see Black Forest
102 C2 Schwaz Austria
102 C1 Schwedt/Oder Ger.
101 E2 Schweinfurt Ger.
101 E1 Schwerin Ger.
101 E1 Schweriner See l. Ger.
105 D2 Schwyz Switz.
108 B3 Sciacca Sicily Italy
95 B4 Scilly, Isles of U.K.
140 C3 Scioto r. U.S.A.
136 B1 Scobey U.S.A.
53 D2 Scone Austr.
55 R3 Scotia Ridge S. Atlantic Ocean
158 C8 Scotia Sea S. Atlantic Ocean
96 C2 Scotland admin. div. U.K.
128 B2 Scott, Cape Can.
123 D3 Scottburgh S. Africa
136 C3 Scott City U.S.A.
136 C2 Scottsbluff U.S.A.
142 C2 Scottsboro U.S.A.
96 B1 Scourie U.K.
141 D2 Scranton U.S.A.
98 C2 Scunthorpe U.K.
105 E2 Scuol Switz.
129 E2 Seal r. Can.
122 B3 Seal, Cape S. Africa
52 B3 Sea Lake Austr.
139 D3 Sealy U.S.A.
142 B1 Searcy U.S.A.
98 B1 Seascale U.K.
134 B1 Seattle U.S.A.
141 E2 Sebago Lake U.S.A.
144 A2 Sebastián Vizcaíno, Bahía b. Mex.
110 B1 Sebeş Romania
60 B2 Sebesi i. Indon.
88 C2 Sebezh Russia
80 B1 Şebinkarahisar Turkey
143 D3 Sebring U.S.A.
128 B3 Sechelt Can.
150 A2 Sechura Peru
73 B3 Secunderabad India
137 E3 Sedalia U.S.A.
105 C2 Sedan France
54 B2 Seddon N.Z.
138 A2 Sedona U.S.A.
101 E2 Seeburg Ger.
101 E1 Seehausen (Altmark) Ger.
122 A2 Seeheim Namibia
104 C2 Sées France
101 E2 Seesen Ger.
101 E1 Seevetal Ger.
123 C1 Sefare Botswana
60 B1 Segamat Malaysia
86 C2 Segezha Russia
114 B3 Ségou Mali
106 C1 Segovia Spain
115 D2 Séguédine Niger
114 B4 Séguéla Côte d'Ivoire
139 D3 Seguin U.S.A.
107 C2 Segura r. Spain
106 C2 Segura, Sierra de mts Spain
120 B3 Sehithwa Botswana
93 H3 Seinäjoki Fin.
104 C2 Seine r. France
104 B2 Seine, Baie de b. France
103 E1 Sejny Pol.
60 B2 Sekayu Indon.
114 B4 Sekondi Ghana
59 C3 Selaru i. Indon.
61 C2 Selatan, Tanjung pt Indon.
126 A2 Selawik U.S.A.
58 C3 Selayar, Pulau i. Indon.
98 C2 Selby U.K.
120 B3 Selebi-Phikwe Botswana
105 D2 Sélestat France
92 □A3 Selfoss Iceland
114 A3 Sélibabi Maur.
138 A1 Seligman U.S.A.
116 A2 Selîma Oasis Sudan
111 C3 Selimiye Turkey
114 B3 Sélingué, Lac de l. Mali
89 D2 Selizharovo Russia
93 E4 Seljord Norway
129 E2 Selkirk Can.
96 C3 Selkirk U.K.
128 C2 Selkirk Mountains Can.

142 C2 **Selma** *AL* U.S.A.
135 C3 **Selma** *CA* U.S.A.
89 D3 **Sel'tso** Russia
150 A2 **Selvas** *reg.* Brazil
134 C1 **Selway** *r.* U.S.A.
129 D1 **Selwyn Lake** Can.
128 A1 **Selwyn Mountains** Can.
51 C2 **Selwyn Range** *hills* Austr.
60 B2 **Semangka, Teluk** *b.* Indon.
61 C2 **Semarang** Indon.
61 B1 **Sematan** *Sarawak* Malaysia
118 B2 **Sembé** Congo
81 C2 **Şemdinli** Turkey
91 C1 **Semenivka** Ukr.
86 D3 **Semenov** Russia
61 C2 **Semeru, Gunung** *vol.* Indon.
77 E1 **Semey** Kazakh.
89 E3 **Semiluki** Russia
136 B2 **Seminoe Reservoir** U.S.A.
139 C2 **Seminole** U.S.A.
143 D2 **Seminole, Lake** U.S.A.
61 C1 **Semitau** Indon.
81 D2 **Semnān** Iran
61 C1 **Semporna** *Sabah* Malaysia
105 C2 **Semur-en-Auxois** France
150 B2 **Sena Madureira** Brazil
120 B2 **Senanga** Zambia
67 D3 **Sendai** Japan
114 A3 **Senegal** *country* Africa
114 A3 **Sénégal** *r.* Maur./Senegal
102 C1 **Senftenberg** Ger.
119 D3 **Sengerema** Tanz.
151 D3 **Senhor do Bonfim** Brazil
108 B2 **Senigallia** Italy
109 B2 **Senj** Croatia
92 G2 **Senja** *i.* Norway
122 B2 **Senlac** S. Africa
104 C2 **Senlis** France
63 B2 **Sênmônoŭrôm** Cambodia
117 B3 **Sennar** Sudan
130 C2 **Senneterre** Can.
123 C3 **Senqu** *r.* Lesotho
105 C2 **Sens** France
109 D1 **Senta** Serbia
128 B2 **Sentinel Peak** Can.
123 C1 **Senwabarwana** S. Africa
65 B2 **Seocheon** S. Korea
65 B2 **Seongnam** S. Korea
75 B2 **Seoni** India
65 B2 **Seosan** S. Korea
65 B2 **Seoul** S. Korea
155 D2 **Sepetiba, Baía de** *b.* Brazil
59 D3 **Sepik** *r.* P.N.G.
61 C1 **Sepinang** Indon.
131 D1 **Sept-Îles** Can.
59 C3 **Seram** *i.* Indon.
59 C3 **Seram, Laut** *sea* Indon.
60 B2 **Serang** Indon.
109 D2 **Serbia** *country* Europe
76 B3 **Serdar** Turkm.
89 E3 **Serebryanyye Prudy** Russia
60 B1 **Seremban** Malaysia
119 D3 **Serengeti Plain** Tanz.
87 D3 **Sergach** Russia
89 E2 **Sergiyev Posad** Russia
74 A1 **Serhetabat** Turkm.
61 C1 **Seria** Brunei
61 C1 **Serian** *Sarawak* Malaysia
111 B3 **Serifos** *i.* Greece
80 B2 **Serik** Turkey
59 C3 **Sermata, Kepulauan** *is* Indon.
86 F3 **Serov** Russia
120 B3 **Serowe** Botswana
106 B2 **Serpa** Port.
89 E3 **Serpukhov** Russia
151 D3 **Serra da Mesa, Represa** *resr* Brazil
154 B1 **Serranópolis** Brazil
100 A3 **Serre** *r.* France
111 B2 **Serres** Greece
151 E3 **Serrinha** Brazil
155 D1 **Serro** Brazil
154 C2 **Sertãozinho** Brazil
61 C2 **Seruyan** *r.* Indon.
68 C2 **Sêrxü** China
120 A2 **Sesfontein** Namibia
108 B2 **Sessa Aurunca** Italy
108 A2 **Sestri Levante** Italy
105 C3 **Sète** France
155 D1 **Sete Lagoas** Brazil
92 G2 **Setermoen** Norway
93 E4 **Setesdal** *val.* Norway
115 C1 **Sétif** Alg.
67 B4 **Seto-naikai** *sea* Japan
114 B1 **Settat** Morocco
98 B1 **Settle** U.K.
106 B2 **Setúbal** Port.
106 B2 **Setúbal, Baía de** *b.* Port.
130 A1 **Seul, Lac** *l.* Can.
76 A2 **Sevan, Lake** Armenia
Sevana Lich *l.* Armenia *see* **Sevan, Lake**
91 C3 **Sevastopol'** Ukr.
105 C3 **Sévérac-le-Château** France
130 B1 **Severn** *r.* Can.
122 B2 **Severn** S. Africa
99 B3 **Severn** *r.* U.K.
86 D2 **Severnaya Dvina** *r.* Russia
83 H1 **Severnaya Zemlya** *is* Russia
86 D2 **Severnyy** *Nenetskiy Avtonomnyy Okrug* Russia
86 F2 **Severnyy** *Respublika Komi* Russia
82 E1 **Severnyy, Ostrov** *i.* Russia
86 C2 **Severodvinsk** Russia
86 C2 **Severomorsk** Russia
91 D3 **Severskaya** Russia
135 D3 **Sevier** *r.* U.S.A.
135 D3 **Sevier Lake** U.S.A.
Sevilla Spain *see* **Seville**
106 B2 **Seville** Spain
128 A2 **Sewell Inlet** Can.
128 C2 **Sexsmith** Can.
144 B2 **Sextín** *r.* Mex.
86 G1 **Seyakha** Russia
113 J7 **Seychelles** *country* Indian Ocean
92 □C2 **Seyðisfjörður** Iceland
91 C1 **Seym** *r.* Russia/Ukr.
83 L2 **Seymchan** Russia
53 C3 **Seymour** Austr.
140 B3 **Seymour** *IN* U.S.A.
139 D2 **Seymour** *TX* U.S.A.
105 C2 **Sézanne** France
110 C1 **Sfântu Gheorghe** Romania
115 D1 **Sfax** Tunisia
's-Gravenhage Neth. *see* **The Hague**
96 A2 **Sgurr Alasdair** *h.* U.K.
70 A2 **Shaanxi** *prov.* China
91 D2 **Shabel'skoye** Russia
77 D3 **Shache** China
55 R1 **Shackleton Range** *mts* Antarctica
99 B3 **Shaftesbury** U.K.
75 C2 **Shahdol** India
77 C3 **Shāh Fōlād** *mt.* Afgh.
75 B2 **Shahjahanpur** India
81 D2 **Shahr-e Kord** Iran
81 D2 **Shahreẓā** Iran
77 C3 **Shahrisabz** Uzbek.
81 D2 **Shāhrūd** Iran
89 E2 **Shakhovskaya** Russia
91 E2 **Shakhty** Russia
86 D3 **Shakhun'ya** Russia
66 D2 **Shakotan-misaki** *c.* Japan
76 B2 **Shalkar** Kazakh.
68 C2 **Shaluli Shan** *mts* China
129 E2 **Shamattawa** Can.
139 C1 **Shamrock** U.S.A.
70 B2 **Shancheng** China
68 C2 **Shandan** China
70 B2 **Shandong** *prov.* China
70 C2 **Shandong Bandao** *pen.* China
121 B2 **Shangani** *r.* Zimbabwe
70 C2 **Shanghai** China
70 C2 **Shanghai** *mun.* China
71 B3 **Shanghang** China
70 A2 **Shangluo** China
71 B3 **Shangrao** China
70 B2 **Shangshui** China
70 C2 **Shangyu** China
69 E1 **Shangzhi** China
97 B2 **Shannon** *r.* Ireland
97 B2 **Shannon, Mouth of the** Ireland
71 B3 **Shantou** China
70 B2 **Shanxi** *prov.* China
71 B3 **Shaoguan** China
71 B3 **Shaowu** China
70 C2 **Shaoxing** China
71 B3 **Shaoyang** China
116 C2 **Shaqrā'** Saudi Arabia
90 B2 **Sharhorod** Ukr.
79 C2 **Sharjah** U.A.E.
88 C2 **Sharkawshchyna** Belarus
50 A2 **Shark Bay** Austr.
78 A2 **Sharm ash Shaykh** Egypt
140 C2 **Sharon** U.S.A.
86 D3 **Shar'ya** Russia
121 B3 **Shashe** *r.* Botswana/Zimbabwe
117 B4 **Shashemenē** Eth.
134 B2 **Shasta, Mount** *vol.* U.S.A.
134 B2 **Shasta Lake** U.S.A.
156 D3 **Shatsky Rise** *sea feature* N. Pacific Ocean
89 E2 **Shatura** Russia
129 D3 **Shaunavon** Can.
140 B2 **Shawano** U.S.A.
131 C2 **Shawinigan** Can.
139 D1 **Shawnee** U.S.A.
50 B2 **Shay Gap (abandoned)** Austr.
89 E3 **Shchekino** Russia
89 E2 **Shchelkovo** Russia
89 E3 **Shchigry** Russia
91 C1 **Shchors** Ukr.
88 B3 **Shchuchyn** Belarus
91 D1 **Shebekino** Russia
117 C4 **Shebelē Wenz, Wabē** *r.* Eth./Somalia
140 B2 **Sheboygan** U.S.A.
91 D3 **Shebsh** *r.* Russia
97 C2 **Sheelin, Lough** *l.* Ireland
98 C2 **Sheffield** U.K.
89 E2 **Sheksna** Russia
89 E2 **Sheksninskoye Vodokhranilishche** *resr* Russia
83 M2 **Shelagskiy, Mys** *pt* Russia
131 D2 **Shelburne** Can.
134 D1 **Shelby** U.S.A.
140 B3 **Shelbyville** *IN* U.S.A.
142 C1 **Shelbyville** *TN* U.S.A.
83 L2 **Shelikhova, Zaliv** *g.* Russia
129 D2 **Shellbrook** Can.
134 B1 **Shelton** U.S.A.
137 D2 **Shenandoah** U.S.A.
141 D3 **Shenandoah** *r.* U.S.A.
141 D3 **Shenandoah Mountains** U.S.A.
118 A2 **Shendam** Nigeria
86 D2 **Shenkursk** Russia
65 A1 **Shenyang** China

71 B3 **Shenzhen** China
90 B1 **Shepetivka** Ukr.
53 C3 **Shepparton** Austr.
99 D3 **Sheppey, Isle of** *i.* U.K.
131 D2 **Sherbrooke** *N.S.* Can.
131 C2 **Sherbrooke** *Que.* Can.
116 B3 **Shereiq** Sudan
136 B2 **Sheridan** U.S.A.
139 D2 **Sherman** U.S.A.
100 B2 **'s-Hertogenbosch** Neth.
96 □ **Shetland Islands** U.K.
76 B2 **Shetpe** Kazakh.
Shevchenko Kazakh. *see* **Aktau**
137 D1 **Sheyenne** *r.* U.S.A.
79 B3 **Shibām** Yemen
66 D2 **Shibetsu** Japan
77 C3 **Shibirghān** Afgh.
71 B3 **Shicheng** China
96 B2 **Shiel, Loch** *l.* U.K.
77 E2 **Shihezi** China
Shijiao China *see* **Fogang**
70 B2 **Shijiazhuang** China
74 A2 **Shikarpur** Pak.
67 B4 **Shikoku** *i.* Japan
66 D2 **Shikotsu-ko** *l.* Japan
86 D2 **Shilega** Russia
75 C2 **Shiliguri** India
75 D2 **Shillong** India
89 F3 **Shilovo** Russia
117 C3 **Shimbiris** *mt.* Somalia
67 C3 **Shimizu** Japan
Shimoga India *see* **Shivamogga**
67 B4 **Shimonoseki** Japan
96 B1 **Shin, Loch** *l.* U.K.
67 C4 **Shingū** Japan
123 D1 **Shingwedzi** S. Africa
123 D1 **Shingwedzi** *r.* S. Africa
119 D3 **Shinyanga** Tanz.
67 C4 **Shiono-misaki** *c.* Japan
138 B1 **Shiprock** U.S.A.
71 A3 **Shiqian** China
70 A2 **Shiquan** China
Shiquanhe China *see* **Gar**
Shiquan He *r.* China/Pak. *see* **Indus**
67 C3 **Shirane-san** *mt.* Japan
81 D3 **Shīrāz** Iran
66 D2 **Shiretoko-misaki** *c.* Japan
66 D2 **Shiriya-zaki** *c.* Japan
74 B2 **Shiv** India
73 B3 **Shivamogga** India
69 D1 **Shiveegovī** Mongolia
75 B2 **Shivpuri** India
70 B2 **Shiyan** China
70 B2 **Shizhong** China
70 A2 **Shizuishan** China
67 C4 **Shizuoka** Japan
89 D3 **Shklow** Belarus
109 C2 **Shkodër** Albania
83 H1 **Shmidta, Ostrov** *i.* Russia
135 C3 **Shoshone** U.S.A.
135 C3 **Shoshone Mountains** U.S.A.
123 C1 **Shoshong** Botswana
91 C1 **Shostka** Ukr.
70 B2 **Shouxian** China
138 A2 **Show Low** U.S.A.
91 C2 **Shpola** Ukr.
142 B2 **Shreveport** U.S.A.
99 B2 **Shrewsbury** U.K.
62 A1 **Shuangjiang** China
87 E4 **Shubarkudyk** Kazakh.
116 B1 **Shubrā al Khaymah** Egypt
89 D2 **Shugozero** Russia
120 B2 **Shumba** Zimbabwe
110 C2 **Shumen** Bulg.
88 C2 **Shumilina** Belarus
89 D3 **Shumyachi** Russia
126 A2 **Shungnak** U.S.A.
78 B3 **Shuqrah** Yemen
89 F2 **Shushkodom** Russia
81 C2 **Shūshtar** Iran
128 C2 **Shuswap Lake** Can.
89 F2 **Shuya** Russia
89 F2 **Shuyskoye** Russia
62 A1 **Shwebo** Myanmar
62 A1 **Shwedwin** Myanmar
62 A2 **Shwegyin** Myanmar
77 D2 **Shyganak** Kazakh.
77 C2 **Shymkent** Kazakh.
91 C2 **Shyroke** Ukr.
59 C3 **Sia** Indon.
74 A2 **Siahan Range** *mts* Pak.
74 B1 **Sialkot** Pak.
64 B2 **Siargao** *i.* Phil.
88 B2 **Šiauliai** Lith.
109 C2 **Šibenik** Croatia
83 I2 **Siberia** *reg.* Russia
60 A2 **Siberut** *i.* Indon.
74 A2 **Sibi** Pak.
110 B1 **Sibiu** Romania
60 A1 **Sibolga** Indon.
61 C1 **Sibu** *Sarawak* Malaysia
118 B2 **Sibut** C.A.R.
64 B1 **Sibuyan** *i.* Phil.
64 B1 **Sibuyan Sea** Phil.
128 C2 **Sicamous** Can.
63 A3 **Sichon** Thai.
70 A2 **Sichuan** *prov.* China
70 A3 **Sichuan Pendi** *basin* China
105 D3 **Sicié, Cap** *c.* France
Sicilia *i.* Italy *see* **Sicily**
108 B3 **Sicilian Channel** Italy/Tunisia
108 B3 **Sicily** *i.* Italy
150 A3 **Sicuani** Peru
111 C3 **Sideros, Akrotirio** *pt* Greece
74 B2 **Sidhpur** India
107 D2 **Sidi Aïssa** Alg.
107 D2 **Sidi Ali** Alg.
114 B1 **Sidi Bel Abbès** Alg.
114 A2 **Sidi Ifni** Morocco
60 A1 **Sidikalang** Indon.
96 C2 **Sidlaw Hills** U.K.
99 B3 **Sidmouth** U.K.
134 B1 **Sidney** Can.
136 C1 **Sidney** *MT* U.S.A.
136 C2 **Sidney** *NE* U.S.A.
140 C2 **Sidney** *OH* U.S.A.
143 D2 **Sidney Lanier, Lake** U.S.A.
80 B2 **Sidon** Lebanon
154 B2 **Sidrolândia** Brazil
103 E1 **Siedlce** Pol.
100 C2 **Sieg** *r.* Ger.
100 D2 **Siegen** Ger.
63 B2 **Siĕm Réab** Cambodia
108 B2 **Siena** Italy
103 D1 **Sieradz** Pol.
153 B5 **Sierra Grande** Arg.
114 A4 **Sierra Leone** *country* Africa
144 B2 **Sierra Mojada** Mex.
138 A2 **Sierra Vista** U.S.A.
105 D2 **Sierre** Switz.
111 B3 **Sifnos** *i.* Greece
107 C2 **Sig** Alg.
127 H2 **Sigguup Nunaa** *pen.* Greenland
103 E2 **Sighetu Marmației** Romania
110 B1 **Sighișoara** Romania
60 A1 **Sigli** Indon.
92 □B2 **Siglufjörður** Iceland
102 B2 **Sigmaringen** Ger.
100 B3 **Signy-l'Abbaye** France
106 C1 **Sigüenza** Spain
114 B3 **Siguiri** Guinea
88 B2 **Sigulda** Latvia
63 B2 **Sihanoukville** Cambodia
92 I3 **Siilinjärvi** Fin.
81 C2 **Siirt** Turkey
60 B2 **Sijunjung** Indon.
74 B2 **Sikar** India
114 B3 **Sikasso** Mali
137 F3 **Sikeston** U.S.A.
66 B2 **Sikhote-Alin'** *mts* Russia
111 C3 **Sikinos** *i.* Greece
75 C2 **Sikkim** *state* India
144 B2 **Silao** Mex.
75 D2 **Silchar** India
77 D1 **Siletyteniz, Ozero** *salt l.* Kazakh.
75 C2 **Silgarhi** Nepal
80 B2 **Silifke** Turkey
75 C1 **Siling Co** *salt l.* China
110 C2 **Silistra** Bulg.
80 A1 **Silivri** Turkey
93 F3 **Siljan** *l.* Sweden
93 E4 **Silkeborg** Denmark
88 C2 **Sillamäe** Estonia
142 B1 **Siloam Springs** U.S.A.
123 D2 **Silobela** S. Africa
88 B2 **Šilut** Lith.
81 C2 **Silvan** Turkey
138 B2 **Silver City** U.S.A.
136 B3 **Silverton** U.S.A.
62 B1 **Simao** China
130 C2 **Simard, Lac** *l.* Can.
111 C3 **Simav** Turkey
111 C3 **Simav Dağları** *mts* Turkey
118 C2 **Simba** Dem. Rep. Congo
141 D2 **Simcoe, Lake** Can.
60 A1 **Simeulue** *i.* Indon.
91 C3 **Simferopol'** Ukr.
110 B1 **Șimleu Silvaniei** Romania
100 C3 **Simmern/Hunsrück** Ger.
129 D2 **Simonhouse** Can.
51 C2 **Simpson Desert** Austr.
93 F4 **Simrishamn** Sweden
60 A1 **Sinabang** Indon.
116 B2 **Sinai** *pen.* Egypt
71 A3 **Sinan** China
65 B2 **Sinanju** N. Korea
62 A1 **Sinbo** Myanmar
150 A1 **Sincelejo** Col.
111 C3 **Sındırgı** Turkey
86 E2 **Sindor** Russia
111 C2 **Sinekçi** Turkey
106 B2 **Sines** Port.
106 B2 **Sines, Cabo de** *c.* Port.
75 C2 **Singahi** India
60 B1 **Singapore** *country* Asia
61 C2 **Singaraja** Indon.
119 D3 **Singida** Tanz.
62 A1 **Singkaling Hkamti** Myanmar
61 B1 **Singkawang** Indon.
60 A1 **Singkil** Indon.
53 D2 **Singleton** Austr.
62 A1 **Singu** Myanmar
108 A2 **Siniscola** *Sardinia* Italy
109 C2 **Sinj** Croatia
116 B3 **Sinkat** Sudan
Sinkiang Uighur Autonomous Region *aut. reg.* China *see* **Xinjiang Uygur Zizhiqu**
80 B1 **Sinop** Turkey
65 B1 **Sinp'o** N. Korea
61 C1 **Sintang** Indon.
100 B2 **Sint Anthonis** Neth.
100 A2 **Sint-Laureins** Belgium
147 D3 **Sint Maarten** *i.* West Indies
147 D3 **Sint Maarten** *terr.* West Indies
100 B2 **Sint-Niklaas** Belgium
139 D3 **Sinton** U.S.A.
65 A1 **Sinŭiju** N. Korea
103 D2 **Siófok** Hungary

105 D2 **Sion** Switz.
137 D2 **Sioux Center** U.S.A.
137 D2 **Sioux City** U.S.A.
137 D2 **Sioux Falls** U.S.A.
130 A1 **Sioux Lookout** Can.
65 A1 **Siping** China
129 E2 **Sipiwesk Lake** Can.
60 A2 **Sipura** *i.* Indon.
93 E4 **Sira** *r.* Norway
Siracusa Italy *see* **Syracuse**
51 C1 **Sir Edward Pellew Group** *is* Austr.
116 B1 **Sirhān, Wādī an** *watercourse* Saudi Arabia
79 C2 **Sīrīk** Iran
62 B2 **Siri Kit, Khuan** Thai.
128 B1 **Sir James MacBrien, Mount** Can.
79 C2 **Sīrjān** Iran
81 C2 **Şırnak** Turkey
74 B2 **Sirohi** India
74 B2 **Sirsa** India
115 D1 **Sirte** Libya
115 D1 **Sirte, Gulf of** Libya
88 B2 **Širvintos** Lith.
145 C2 **Sisal** Mex.
122 B2 **Sishen** S. Africa
81 C2 **Sisian** Armenia
129 D2 **Sisipuk Lake** Can.
63 B2 **Sisŏphŏn** Cambodia
105 D3 **Sisteron** France
75 C2 **Sitapur** India
123 D2 **Siteki** Swaziland
128 A2 **Sitka** U.S.A.
100 B2 **Sittard** Neth.
62 A2 **Sittaung** *r.* Myanmar
Sittwe Myanmar *see* **Akyab**
61 C2 **Situbondo** Indon.
80 B2 **Sivas** Turkey
62 A1 **Sivasagar** India
111 C3 **Sivaslı** Turkey
80 B2 **Siverek** Turkey
80 B2 **Sivrihisar** Turkey
116 A2 **Sīwah** Egypt
74 B1 **Siwalik Range** *mts* India/ Nepal
105 D3 **Six-Fours-les-Plages** France
70 B2 **Sixian** China
123 C2 **Siyabuswa** S. Africa
Sjælland *i.* Denmark *see* **Zealand**
109 D2 **Sjenica** Serbia
92 G2 **Sjøvegan** Norway
91 C2 **Skadovs'k** Ukr.
93 F4 **Skagen** Denmark
93 E4 **Skagerrak** *str.* Denmark/ Norway
134 B1 **Skagit** *r.* U.S.A.
128 A2 **Skagway** U.S.A.
74 B1 **Skardu** Pak.
103 E1 **Skarżysko-Kamienna** Pol.
103 D2 **Skawina** Pol.
128 B2 **Skeena** *r.* Can.
128 B2 **Skeena Mountains** Can.
98 D2 **Skegness** U.K.
92 H3 **Skellefteå** Sweden
92 H3 **Skellefteälven** *r.* Sweden
97 C2 **Skerries** Ireland
111 B3 **Skiathos** *i.* Greece
97 B3 **Skibbereen** Ireland
92 □B2 **Skíðadalsjökull** *glacier* Iceland
98 B1 **Skiddaw** *h.* U.K.
93 E4 **Skien** Norway
103 E1 **Skierniewice** Pol.
115 C1 **Skikda** Alg.
52 B3 **Skipton** Austr.
98 B2 **Skipton** U.K.
93 E4 **Skive** Denmark
92 H1 **Skjervøy** Norway
111 B3 **Skopelos** *i.* Greece
89 E3 **Skopin** Russia
109 D2 **Skopje** Macedonia
93 F4 **Skövde** Sweden
141 F2 **Skowhegan** U.S.A.
88 B2 **Skrunda** Latvia
128 A1 **Skukum, Mount** Can.
123 D1 **Skukuza** S. Africa
88 B2 **Skuodas** Lith.
96 A2 **Skye** *i.* U.K.
111 B3 **Skyros** Greece
111 B3 **Skyros** *i.* Greece
93 F4 **Slagelse** Denmark
97 C2 **Slaney** *r.* Ireland
88 C2 **Slantsy** Russia
109 C1 **Slatina** Croatia
110 B2 **Slatina** Romania
129 C1 **Slave** *r.* Can.
114 C4 **Slave Coast** Africa
128 C2 **Slave Lake** Can.
82 G3 **Slavgorod** Russia
109 C1 **Slavonski Brod** Croatia
90 B1 **Slavuta** Ukr.
90 C1 **Slavutych** Ukr.
91 D2 **Slavyansk-na-Kubani** Russia
89 D3 **Slawharad** Belarus
103 D1 **Sławno** Pol.
97 A2 **Slea Head** Ireland
130 C1 **Sleeper Islands** Can.
97 D1 **Slieve Donard** *h.* U.K.
Slieve Gamph *hills* Ireland *see* **Ox Mountains**
97 B1 **Sligo** Ireland
97 B1 **Sligo Bay** Ireland
93 G4 **Slite** Sweden
110 C2 **Sliven** Bulg.
110 C2 **Slobozia** Romania
128 C3 **Slocan** Can.
88 C3 **Slonim** Belarus
100 B1 **Sloten** Neth.
99 C3 **Slough** U.K.
103 D2 **Slovakia** *country* Europe
108 B1 **Slovenia** *country* Europe
91 D2 **Slov"yans'k** Ukr.
90 B1 **Sluch** *r.* Ukr.
103 D1 **Słupsk** Pol.
88 C3 **Slutsk** Belarus
97 A2 **Slyne Head** Ireland
131 D1 **Smallwood Reservoir** Can.
88 C3 **Smalyavichy** Belarus
88 C3 **Smarhon'** Belarus
129 D2 **Smeaton** Can.
109 D2 **Smederevo** Serbia
109 D2 **Smederevska Palanka** Serbia
91 C2 **Smila** Ukr.
88 C2 **Smiltene** Latvia
128 B2 **Smithers** Can.
143 E1 **Smithfield** U.S.A.
141 D3 **Smith Mountain Lake** U.S.A.
130 C2 **Smiths Falls** Can.
137 D3 **Smoky Hills** U.S.A.
92 E3 **Smøla** *i.* Norway
89 D3 **Smolensk** Russia
110 B2 **Smolyan** Bulg.
130 B2 **Smooth Rock Falls** Can.
Smyrna Turkey *see* **İzmir**
92 □B3 **Snæfell** *mt.* Iceland
98 A1 **Snaefell** *h.* Isle of Man
132 B2 **Snake** *r.* U.S.A.
134 D2 **Snake River Plain** U.S.A.
100 B1 **Sneek** Neth.
97 B3 **Sneem** Ireland
122 B3 **Sneeuberge** *mts* S. Africa
108 B1 **Snežnik** *mt.* Slovenia
91 C2 **Snihurivka** Ukr.
93 E3 **Snøhetta** *mt.* Norway
129 D1 **Snowbird Lake** Can.
98 A2 **Snowdon** *mt.* U.K.
Snowdrift Can. *see* **Łutselk'e**
129 C1 **Snowdrift** *r.* Can.
138 A2 **Snowflake** U.S.A.
129 D2 **Snow Lake** Can.
52 A2 **Snowtown** Austr.
53 C3 **Snowy** *r.* Austr.
53 C3 **Snowy Mountains** Austr.
139 C2 **Snyder** U.S.A.
121 □D2 **Soalala** Madag.
117 B4 **Sobat** *r.* South Sudan
89 F2 **Sobinka** Russia
151 D3 **Sobradinho, Barragem de** *resr* Brazil
151 D2 **Sobral** Brazil
87 C4 **Sochi** Russia
49 H4 **Society Islands** Fr. Polynesia
150 A1 **Socorro** Col.
138 B2 **Socorro** U.S.A.
144 A3 **Socorro, Isla** *i.* Mex.
56 F6 **Socotra** *i.* Yemen
63 B3 **Soc Trăng** Vietnam
106 C2 **Socuéllamos** Spain
92 I2 **Sodankylä** Fin.
134 D2 **Soda Springs** U.S.A.
93 G3 **Söderhamn** Sweden
93 G4 **Södertälje** Sweden
116 A3 **Sodiri** Sudan
117 B4 **Sodo** Eth.
93 G3 **Södra Kvarken** *str.* Fin./ Sweden
100 D2 **Soest** Ger.
110 B2 **Sofia** Bulg.
Sofiya Bulg. *see* **Sofia**
68 C2 **Sog** China
93 E3 **Sognefjorden** *inlet* Norway
111 D2 **Söğüt** Turkey
100 B2 **Soignies** Belgium
105 C2 **Soissons** France
90 A1 **Sokal'** Ukr.
111 C3 **Söke** Turkey
81 C1 **Sokhumi** Georgia
114 C4 **Sokodé** Togo
89 F2 **Sokol** Russia
101 F2 **Sokolov** Czech Rep.
115 C3 **Sokoto** Nigeria
115 C3 **Sokoto** *r.* Nigeria
90 B2 **Sokyryany** Ukr.
73 B3 **Solapur** India
135 B3 **Soledad** U.S.A.
89 F2 **Soligalich** Russia
99 C2 **Solihull** U.K.
86 E3 **Solikamsk** Russia
87 E3 **Sol'-Iletsk** Russia
100 C2 **Solingen** Ger.
92 G3 **Sollefteå** Sweden
93 G4 **Sollentuna** Sweden
101 D2 **Solling** *hills* Ger.
89 E2 **Solnechnogorsk** Russia
60 B2 **Solok** Indon.
48 E3 **Solomon Islands** *country* S. Pacific Ocean
48 E3 **Solomon Sea** S. Pacific Ocean
105 D2 **Solothurn** Switz.
101 D1 **Soltau** Ger.
89 D2 **Sol'tsy** Russia
96 C3 **Solway Firth** *est.* U.K.
120 B2 **Solwezi** Zambia
111 C3 **Soma** Turkey
117 C4 **Somalia** *country* Africa
117 A4 **Somaliland** *disp. terr.* Somalia
120 B1 **Sombo** Angola
109 C1 **Sombor** Serbia
144 B2 **Sombrerete** Mex.
140 C3 **Somerset** U.S.A.
123 C3 **Somerset East** S. Africa

Somerset Island

126 E2 **Somerset Island** Can.
122 A3 **Somerset West** S. Africa
101 C2 **Sömmerda** Ger.
75 C2 **Son** r. India
65 C1 **Sönbong** N. Korea
93 E4 **Sønderborg** Denmark
101 C2 **Sondershausen** Ger.
Søndre Strømfjord inlet
Greenland see **Kangerlussuaq**
62 B1 **Sông Đa, Hô** resr Vietnam
119 D4 **Songea** Tanz.
65 B1 **Sŏnggan** N. Korea
65 B1 **Songhua Hu** resr China
Sŏngjin N. Korea see
Kimch'aek
63 B3 **Songkhla** Thai.
65 B2 **Songnim** N. Korea
120 A1 **Songo** Angola
121 C2 **Songo** Moz.
69 E1 **Songyuan** China
Sonid Youqi China see
Saihan Tal
89 E2 **Sonkovo** Russia
62 B1 **Sơn La** Vietnam
74 A2 **Sonmiani** Pak.
101 E2 **Sonneberg** Ger.
138 A2 **Sonoita** Mex.
144 A2 **Sonora** r. Mex.
135 B3 **Sonora** CA U.S.A.
139 C2 **Sonora** TX U.S.A.
117 A4 **Sopo** watercourse
South Sudan
103 D2 **Sopron** Hungary
108 B2 **Sora** Italy
130 C2 **Sorel** Can.
51 D4 **Sorell** Austr.
106 C1 **Soria** Spain
90 B2 **Soroca** Moldova
154 C2 **Sorocaba** Brazil
87 E3 **Sorochinsk** Russia
59 D2 **Sorol** atoll Micronesia
59 C3 **Sorong** Indon.
119 D2 **Soroti** Uganda
92 H1 **Sørøya** i. Norway
92 G2 **Sorsele** Sweden
64 B1 **Sorsogon** Phil.
86 C2 **Sortavala** Russia
92 G2 **Sortland** Norway
123 C2 **Soshanguve** S. Africa
89 E3 **Sosna** r. Russia
153 B4 **Sosneado** mt. Arg.
86 E2 **Sosnogorsk** Russia
86 D2 **Sosnovka** Russia
88 C2 **Sosnovyy Bor** Russia
103 D1 **Sosnowiec** Pol.
91 D2 **Sosyka** r. Russia
145 C2 **Soto la Marina** Mex.
118 B2 **Souanké** Congo
104 C3 **Souillac** France
Sŏul S. Korea see **Seoul**
104 B2 **Soulac-sur-Mer** France
104 B3 **Soulom** France
Soûr Lebanon see **Tyre**
107 D2 **Sour el Ghozlane** Alg.
129 D3 **Souris** Man. Can.
131 D2 **Souris** P.E.I. Can.
129 E3 **Souris** r. Can.
151 E2 **Sousa** Brazil
115 D1 **Sousse** Tunisia
104 B3 **Soustons** France
122 B3 **South Africa** country Africa
99 C3 **Southampton** U.K.
127 F2 **Southampton Island** Can.
73 D3 **South Andaman** i. India
52 A1 **South Australia** state Austr.
130 B2 **South Baymouth** Can.
140 B2 **South Bend** U.S.A.
143 D2 **South Carolina** state U.S.A.

156 B4 **South China Sea**
N. Pacific Ocean
136 C2 **South Dakota** state U.S.A.
99 C3 **South Downs** hills U.K.
129 D2 **Southend** Can.
99 D3 **Southend-on-Sea** U.K.
54 B2 **Southern Alps** mts N.Z.
50 A3 **Southern Cross** Austr.
129 E2 **Southern Indian Lake** Can.
159 **Southern Ocean**
143 E1 **Southern Pines** U.S.A.
'96 B3 **Southern Uplands** hills U.K.
149 F6 **South Georgia and the
South Sandwich Islands** terr.
S. Atlantic Ocean
96 A2 **South Harris** pen. U.K.
129 E1 **South Henik Lake** Can.
54 B2 **South Island** N.Z.
65 B2 **South Korea** country Asia
135 B3 **South Lake Tahoe** U.S.A.
55 Q4 **South Orkney Islands**
S. Atlantic Ocean
81 C1 **South Ossetia** disp. terr.
Georgia
136 C2 **South Platte** r. U.S.A.
98 B2 **Southport** U.K.
143 E2 **Southport** U.S.A.
96 C1 **South Ronaldsay** i. U.K.
123 D3 **South Sand Bluff** pt S. Africa
129 D2 **South Saskatchewan** r. Can.
129 E2 **South Seal** r. Can.
55 P3 **South Shetland Islands**
Antarctica
98 C1 **South Shields** U.K.
117 A4 **South Sudan** country Africa
54 B1 **South Taranaki Bight** b. N.Z.
130 C1 **South Twin Island** Can.
96 A2 **South Uist** i. U.K.
53 D2 **South West Rocks** Austr.
99 D2 **Southwold** U.K.
109 C3 **Soverato** Italy
88 B2 **Sovetsk** Russia
86 F2 **Sovetskiy** Russia
91 C2 **Sovyets'kyy** Ukr.
123 C2 **Soweto** S. Africa
66 D1 **Sōya-misaki** c. Japan
90 C1 **Sozh** r. Europe
110 C2 **Sozopol** Bulg.
106 C1 **Spain** country Europe
99 C2 **Spalding** U.K.
135 D2 **Spanish Fork** U.S.A.
97 B2 **Spanish Point** Ireland
108 B3 **Sparagio, Monte** mt. Sicily
Italy
135 C3 **Sparks** U.S.A.
140 A2 **Sparta** U.S.A.
143 D2 **Spartanburg** U.S.A.
111 B3 **Sparti** Greece
109 C3 **Spartivento, Capo** c. Italy
89 D3 **Spas-Demensk** Russia
89 F2 **Spas-Klepiki** Russia
66 B2 **Spassk-Dal'niy** Russia
89 F3 **Spassk-Ryazanskiy** Russia
111 B3 **Spatha, Akrotirio** pt Greece
136 C2 **Spearfish** U.S.A.
139 C1 **Spearman** U.S.A.
137 D2 **Spencer** U.S.A.
52 A2 **Spencer Gulf** est. Austr.
98 C1 **Spennymoor** U.K.
54 B2 **Spenser Mountains** N.Z.
101 D3 **Spessart** reg. Ger.
99 C3 **Spey** r. U.K.
102 B2 **Speyer** Ger.
100 C1 **Spiekeroog** i. Ger.
100 B2 **Spijkenisse** Neth.
128 C2 **Spirit River** Can.
103 E2 **Spišská Nová Ves** Slovakia
82 C1 **Spitsbergen** i. Svalbard

102 C2 **Spittal an der Drau** Austria
109 C2 **Split** Croatia
129 E2 **Split Lake** Can.
129 E2 **Split Lake** Can.
134 C1 **Spokane** U.S.A.
140 A1 **Spooner** U.S.A.
102 C1 **Spree** r. Ger.
122 A2 **Springbok** S. Africa
131 E2 **Springdale** Can.
142 B1 **Springdale** U.S.A.
101 D1 **Springe** Ger.
138 C1 **Springer** U.S.A.
138 B2 **Springerville** U.S.A.
136 C3 **Springfield** CO U.S.A.
140 B3 **Springfield** IL U.S.A.
141 E2 **Springfield** MA U.S.A.
137 E3 **Springfield** MO U.S.A.
140 C3 **Springfield** OH U.S.A.
134 B2 **Springfield** OR U.S.A.
123 C3 **Springfontein** S. Africa
131 D2 **Springhill** Can.
143 D3 **Spring Hill** U.S.A.
54 B2 **Springs Junction** N.Z.
98 D2 **Spurn Head** U.K.
128 B3 **Squamish** Can.
109 C3 **Squillace, Golfo di** g. Italy
Srbija country Europe see
Serbia
109 C2 **Srebrenica** Bos. & Herz.
110 C2 **Sredets** Bulg.
83 L3 **Sredinnyy Khrebet** mts Russia
83 L2 **Srednekolymsk** Russia
**Sredne-Russkaya
Vozvyshennost'** hills Russia see
Central Russian Upland
**Sredne-Sibirskoye
Ploskogor'ye** plat. Russia see
Central Siberian Plateau
69 D1 **Sretensk** Russia
61 C1 **Sri Aman** Sarawak Malaysia
73 B4 **Sri Jayewardenepura Kotte**
Sri Lanka
73 C3 **Srikakulam** India
73 C4 **Sri Lanka** country Asia
74 B1 **Srinagar** India
73 B3 **Srivardhan** India
101 D1 **Stade** Ger.
101 E1 **Stadensen** Ger.
100 C1 **Stadskanaal** Neth.
101 D2 **Stadtallendorf** Ger.
101 D2 **Stadthagen** Ger.
99 B2 **Stafford** U.K.
99 C3 **Staines-upon-Thames** U.K.
91 D2 **Stakhanov** Ukr.
Stalingrad Russia see
Volgograd
103 E1 **Stalowa Wola** Pol.
141 E2 **Stamford** U.K.
141 E2 **Stamford** CT U.S.A.
139 D2 **Stamford** TX U.S.A.
122 A1 **Stampriet** Namibia
92 F2 **Stamsund** Norway
123 C2 **Standerton** S. Africa
136 C1 **Standish** U.S.A.
153 C6 **Stanley** Falkland Is
136 C1 **Stanley** U.S.A.
83 I3 **Stanovoye Nagor'ye** mts
Russia
83 J3 **Stanovoy Khrebet** mts Russia
53 D1 **Stanthorpe** Austr.
103 E1 **Staphorst** Neth.
Stara Planina mts Bulgaria/
Serbia see **Balkan Mountains**
89 D2 **Staraya Russa** Russia
89 D2 **Staraya Toropa** Russia
110 C2 **Stara Zagora** Bulg.
103 D1 **Stargard Szczeciński** Pol.
142 C2 **Starkville** U.S.A.

91 D2 Starobil's'k Ukr.
89 D3 Starodub Russia
103 D1 Starogard Gdański Pol.
90 B2 Starokostyantyniv Ukr.
91 D2 Starominskaya Russia
91 D2 Staroshcherbinovskaya Russia
91 D2 Starotitarovskaya Russia
89 F3 Staroyur'yevo Russia
99 B3 Start Point U.K.
88 C3 Staryya Darohi Belarus
89 E3 Staryy Oskol Russia
101 E2 Staßfurt Ger.
141 D2 State College U.S.A.
143 D2 Statesboro U.S.A.
143 D1 Statesville U.S.A.
93 E4 Stavanger Norway
87 D4 Stavropol' Russia
87 D4 Stavropol'skaya Vozvyshennost' hills Russia
52 B3 Stawell Austr.
123 C2 Steadville S. Africa
136 B2 Steamboat Springs U.S.A.
101 E2 Stedten Ger.
128 C2 Steen River Can.
134 C2 Steens Mountain U.S.A.
100 C1 Steenwijk Neth.
126 D2 Stefansson Island Can.
101 E3 Steigerwald mts Ger.
129 E3 Steinbach Can.
100 C1 Steinfurt Ger.
120 A3 Steinhausen Namibia
92 F3 Steinkjer Norway
122 A2 Steinkopf S. Africa
122 B2 Stella S. Africa
105 D2 Stenay France
101 E1 Stendal Ger.
Stepanakert Azer. see Xankändi
131 E2 Stephenville Can.
139 D2 Stephenville U.S.A.
122 B3 Sterling S. Africa
136 C2 Sterling CO U.S.A.
140 B2 Sterling IL U.S.A.
87 E3 Sterlitamak Russia
101 E1 Sternberg Ger.
128 C2 Stettler Can.
140 C2 Steubenville U.S.A.
99 C3 Stevenage U.K.
129 E2 Stevenson Lake Can.
126 B2 Stevens Village U.S.A.
128 B2 Stewart Can.
126 B2 Stewart r. Can.
54 A3 Stewart Island N.Z.
127 F2 Stewart Lake Can.
102 C2 Steyr Austria
122 B3 Steytlerville S. Africa
128 A2 Stikine r. Can.
128 A2 Stikine Plateau Can.
122 B3 Stilbaai S. Africa
139 D1 Stillwater U.S.A.
109 D2 Štip Macedonia
96 C2 Stirling U.K.
92 F3 Stjørdalshalsen Norway
103 D2 Stockerau Austria
93 G4 Stockholm Sweden
98 B2 Stockport U.K.
135 B3 Stockton U.S.A.
98 C1 Stockton-on-Tees U.K.
63 B2 Stœng Trêng Cambodia
96 B1 Stoer, Point of U.K.
98 B2 Stoke-on-Trent U.K.
109 D2 Stol mt. Serbia
109 C2 Stolac Bos. & Herz.
100 C2 Stolberg (Rheinland) Ger.
88 C3 Stolin Belarus
101 F2 Stollberg Ger.
101 D1 Stolzenau Ger.

96 C2 Stonehaven U.K.
129 E2 Stonewall Can.
129 D2 Stony Rapids Can.
92 G2 Storavan l. Sweden
Store Bælt sea chan. Denmark see Great Belt
92 F3 Støren Norway
92 F2 Storforshei Norway
126 D2 Storkerson Peninsula Can.
137 D2 Storm Lake U.S.A.
93 E3 Stornosa mt. Norway
96 A1 Stornoway U.K.
86 C2 Storozhevsk Russia
90 B2 Storozhynets' Ukr.
92 F3 Storsjön l. Sweden
92 H2 Storslett Norway
92 G2 Storuman Sweden
99 C2 Stour r. England U.K.
99 C3 Stour r. England U.K.
130 A1 Stout Lake Can.
88 C3 Stowbtsy Belarus
97 C3 Strabane U.K.
102 C2 Strakonice Czech Rep.
102 C1 Stralsund Ger.
122 A3 Strand S. Africa
93 E3 Stranda Norway
97 D1 Strangford Lough inlet U.K.
96 C3 Stranraer U.K.
105 D2 Strasbourg France
140 C2 Stratford Can.
54 B1 Stratford N.Z.
139 C1 Stratford U.S.A.
99 C2 Stratford-upon-Avon U.K.
128 C2 Strathmore Can.
96 C2 Strathspey val. U.K.
102 C2 Straubing Ger.
51 C3 Streaky Bay Austr.
140 B2 Streator U.S.A.
110 B2 Strehaia Romania
153 B5 Stroeder Arg.
101 D1 Ströhen Ger.
109 C3 Stromboli, Isola i. Italy
96 C1 Stromness U.K.
92 G3 Strömsund Sweden
96 C1 Stronsay i. U.K.
53 D2 Stroud Austr.
99 B3 Stroud U.K.
100 C1 Strücklingen (Saterland) Ger.
109 D2 Struga Macedonia
88 C2 Strugi-Krasnyye Russia
122 B3 Struis Bay S. Africa
109 D2 Strumica Macedonia
122 B2 Strydenburg S. Africa
111 B2 Strymonas r. Greece
90 A2 Stryy Ukr.
128 B2 Stuart Lake Can.
130 A1 Stull Lake Can.
89 E3 Stupino Russia
140 B2 Sturgeon Bay U.S.A.
130 C2 Sturgeon Falls Can.
130 A2 Sturgeon Lake Can.
140 B2 Sturgis MI U.S.A.
136 C2 Sturgis SD U.S.A.
50 B1 Sturt Creek watercourse Austr.
52 B1 Sturt Stony Desert Austr.
123 C3 Stutterheim S. Africa
102 B2 Stuttgart Ger.
142 B2 Stuttgart U.S.A.
92 □A2 Stykkishólmur Iceland
90 B1 Styr r. Belarus/Ukr.
155 D1 Suaçuí Grande r. Brazil
116 B3 Suakin Sudan
78 A3 Suara Eritrea
109 C1 Subotica Serbia
110 C1 Suceava Romania
97 B2 Suck r. Ireland
152 B2 Sucre Bol.
154 B2 Sucuriú r. Brazil

89 E2 Suda Russia
91 C3 Sudak Ukr.
117 A3 Sudan country Africa
130 B2 Sudbury Can.
117 A4 Sudd swamp South Sudan
89 F2 Sudislavl' Russia
89 F2 Sudogda Russia
94 B1 Suðuroy i. Faroe Is
107 C2 Sueca Spain
116 B2 Suez Egypt
116 B2 Suez, Gulf of Egypt
80 B2 Suez Canal Egypt
141 D3 Suffolk U.S.A.
116 B2 Sühäj Egypt
79 C2 Şuḩār Oman
69 D1 Sühbaatar Mongolia
101 E2 Suhl Ger.
109 C1 Suhopolje Croatia
70 B2 Suide China
66 B2 Suifenhe China
69 E1 Suihua China
70 A2 Suining China
70 B2 Suiping China
70 B2 Suiyang China
70 B2 Suizhou China
74 B2 Sujangarh India
74 B1 Sujanpur India
74 A2 Sujawal Pak.
60 B2 Sukabumi Indon.
61 B2 Sukadana Indon.
61 C2 Sukaraja Indon.
89 E3 Sukhinichi Russia
89 F2 Sukhona r. Russia
62 A2 Sukhothai Thai.
74 A2 Sukkur Pak.
89 E2 Sukromny Russia
59 C3 Sula, Kepulauan is Indon.
74 A1 Sulaiman Range mts Pak.
Sulawesi i. Indon. see Celebes
150 A2 Sullana Peru
137 E3 Sullivan U.S.A.
139 D2 Sulphur Springs U.S.A.
64 B2 Sulu Archipelago is Phil.
64 A2 Sulu Sea N. Pacific Ocean
101 E3 Sulzbach-Rosenberg Ger.
79 C2 Sumāil Oman
Sumatera i. Indon. see Sumatra
60 A1 Sumatra i. Indon.
58 C3 Sumba i. Indon.
61 C2 Sumbawa i. Indon.
61 C2 Sumbawabesar Indon.
119 D3 Sumbawanga Tanz.
120 A2 Sumbe Angola
96 □ Sumburgh U.K.
96 □ Sumburgh Head U.K.
61 C2 Sumenep Indon.
67 D4 Sumisu-jima i. Japan
131 D2 Summerside Can.
140 C3 Summersville U.S.A.
128 B3 Summit Lake Can.
103 D2 Šumperk Czech Rep.
81 C1 Sumqayıt Azer.
143 D2 Sumter U.S.A.
91 C1 Sumy Ukr.
75 D2 Sunamganj Bangl.
65 B2 Sunan N. Korea
79 C2 Şunaynah Oman
52 B3 Sunbury Austr.
141 D2 Sunbury U.S.A.
65 B3 Suncheon S. Korea
65 B2 Sunch'ŏn N. Korea
123 C2 Sun City S. Africa
60 B2 Sunda, Selat str. Indon.
136 C2 Sundance U.S.A.
75 C2 Sundarbans coastal area Bangl./India

Sunderland

98 C1 Sunderland U.K.
128 C2 Sundre Can.
93 G3 Sundsvall Sweden
123 D2 Sundumbili S. Africa
60 B2 Sungailiat Indon.
60 B2 Sungaipenuh Indon.
60 B1 Sungai Petani Malaysia
80 B1 Sungurlu Turkey
93 E3 Sunndalsøra Norway
134 C1 Sunnyside U.S.A.
135 B3 Sunnyvale U.S.A.
83 I2 Suntar Pak.
74 A2 Suntsar Pak.
114 B4 Sunyani Ghana
82 D2 Suoyarvi Russia
138 A2 Superior AZ U.S.A.
137 D2 Superior NE U.S.A.
140 A1 Superior WI U.S.A.
140 B1 Superior, Lake Can./U.S.A.
89 D3 Suponevo Russia
81 C2 Sūq ash Shuyūkh Iraq
70 B2 Suqian China
78 A2 Sūq Suwayq Saudi Arabia
79 C2 Şūr Oman
74 A2 Surab Pak.
61 C2 Surabaya Indon.
61 C2 Surakarta Indon.
74 B2 Surat India
74 B2 Suratgarh India
63 A3 Surat Thani Thai.
89 D3 Surazh Russia
109 D2 Surdulica Serbia
74 B2 Surendranagar India
82 F2 Surgut Russia
64 B2 Surigao Phil.
63 B2 Surin Thai.
151 C1 Suriname country S. America
Surt Libya see Sirte
Surt, Khalīj g. Libya see
    Sirte, Gulf of
60 B2 Surulangun Indon.
89 E2 Susanino Russia
135 B2 Susanville U.S.A.
80 B1 Suşehri Turkey
131 D2 Sussex Can.
101 D1 Süstedt Ger.
100 C1 Sustrum Ger.
82 C2 Susuman Russia
111 C3 Susurluk Turkey
75 B1 Sutak India
122 B3 Sutherland S. Africa
99 C2 Sutton Coldfield U.K.
66 D2 Suttsu Japan
49 F4 Suva Fiji
89 E3 Suvorov Russia
103 E1 Suwałki Pol.
63 B2 Suwannaphum Thai.
143 D3 Suwannee r. U.S.A.
Suweis, Qanâ el canal Egypt
    see Suez Canal
65 B2 Suwon S. Korea
79 C2 Sūzā Iran
89 F2 Suzdal' Russia
70 B2 Suzhou Anhui China
70 C2 Suzhou Jiangsu China
67 C3 Suzu Japan
67 C3 Suzu-misaki pt Japan
82 B1 Svalbard terr. Arctic Ocean
91 D2 Svatove Ukr.
63 B2 Svay Riĕng Cambodia
93 F3 Sveg Sweden
88 C2 Švenčionys Lith.
93 F4 Svendborg Denmark
Sverdlovsk Russia see
    Yekaterinburg
109 D2 Sveti Nikole Macedonia
88 B3 Svetlogorsk Russia
88 B3 Svetlyy Russia

93 I3 Svetogorsk Russia
110 C2 Svilengrad Bulg.
110 B2 Svinecea Mare, Vârful mt. Romania
110 C2 Svishtov Bulg.
103 D2 Svitavy Czech Rep.
91 C2 Svitlovods'k Ukr.
69 E1 Svobodnyy Russia
92 F2 Svolvær Norway
88 C3 Svyetlahorsk Belarus
143 D2 Swainsboro U.S.A.
120 A3 Swakopmund Namibia
52 B3 Swan Hill Austr.
128 C2 Swan Hills Can.
129 D2 Swan Lake Can.
129 D2 Swan River Can.
53 D2 Swansea Austr.
99 B3 Swansea U.K.
123 C2 Swartruggens S. Africa
Swatow China see Shantou
123 D2 Swaziland country Africa
93 G3 Sweden country Europe
139 C2 Sweetwater U.S.A.
136 B2 Sweetwater r. U.S.A.
122 B3 Swellendam S. Africa
103 D1 Świdnica Pol.
103 D1 Świdwin Pol.
103 D1 Świebodzin Pol.
103 D1 Świecie Pol.
129 D2 Swift Current Can.
97 C1 Swilly, Lough inlet Ireland
99 C3 Swindon U.K.
102 C1 Świnoujście Pol.
105 D2 Switzerland country Europe
97 C2 Swords Ireland
88 C3 Syanno Belarus
89 D2 Sychevka Russia
53 D2 Sydney Austr.
131 D2 Sydney Can.
131 D2 Sydney Mines Can.
91 D2 Syeverodonets'k Ukr.
86 E2 Syktyvkar Russia
142 C2 Sylacauga U.S.A.
75 D2 Sylhet Bangl.
102 B1 Sylt i. Ger.
111 C3 Symi i. Greece
91 D2 Synel'nykove Ukr.
109 C3 Syracuse Sicily Italy
136 C3 Syracuse KS U.S.A.
141 D2 Syracuse NY U.S.A.
77 C2 Syrdar'ya r. Asia
80 B2 Syria country Asia
80 B2 Syrian Desert Asia
111 B3 Syros i. Greece
87 D3 Syzran' Russia
102 C1 Szczecin Pol.
103 D1 Szczecinek Pol.
103 E1 Szczytno Pol.
103 E2 Szeged Hungary
103 D2 Székesfehérvár Hungary
103 D2 Szekszárd Hungary
103 E2 Szentes Hungary
103 D2 Szentgotthárd Hungary
103 D2 Szigetvár Hungary
103 E2 Szolnok Pol.
103 D2 Szombathely Hungary

T

78 B2 Tābah Saudi Arabia
76 B3 Tabas Iran
81 D3 Tābask, Kūh-e mt. Iran
150 B2 Tabatinga Brazil
114 B2 Tabelbala Alg.
129 C2 Taber Can.
102 C2 Tábor Czech Rep.

119 D3 Tabora Tanz.
114 B4 Tabou Côte d'Ivoire
81 C2 Tabriz Iran
78 A2 Tabūk Saudi Arabia
77 E2 Tacheng China
102 C2 Tachov Czech Rep.
64 B1 Tacloban Phil.
150 A3 Tacna Peru
134 B1 Tacoma U.S.A.
152 C4 Tacuarembó Uru.
138 B3 Tacupeto Mex.
117 C3 Tadjourah Djibouti
80 B2 Tadmur Syria
129 E2 Tadoule Lake Can.
65 B2 Taebaek S. Korea
Taegu S. Korea see Daegu
Taejŏn S. Korea see Daejeon
107 C1 Tafalla Spain
152 B3 Tafí Viejo Arg.
79 D2 Taftān, Kūh-e mt. Iran
91 D2 Taganrog Russia
91 D2 Taganrog, Gulf of Russia/Ukr.
62 A1 Tagaung Myanmar
64 B1 Tagaytay City Phil.
64 B2 Tagbilaran Phil.
64 B1 Tagudin Phil.
64 B1 Tagum Phil.
106 B2 Tagus r. Port./Spain
60 B1 Tahan, Gunung mt. Malaysia
115 C2 Tahat, Mont mt. Alg.
69 E1 Tahe China
49 I4 Tahiti i. Fr. Polynesia
139 E1 Tahlequah U.S.A.
135 B3 Tahoe, Lake U.S.A.
135 B3 Tahoe City U.S.A.
126 D2 Tahoe Lake Can.
115 C3 Tahoua Niger
79 C2 Tahrūd Iran
128 B3 Tahsis Can.
70 B2 Tai'an China
71 C3 Taibei Taiwan
Taibus Qi China see Baochang
54 C1 Taihape N.Z.
70 C2 Tai Hu l. China
52 A3 Tailem Bend Austr.
71 C3 Tainan Taiwan
155 D1 Taiobeiras Brazil
61 B1 Taiping Malaysia
71 B3 Taishan China
153 A5 Taitao, Península de pen. Chile
71 C3 T'aitung Taiwan
92 I2 Taivalkoski Fin.
92 H2 Taivaskero h. Fin.
71 C3 Taiwan country Asia
71 B3 Taiwan Strait China/Taiwan
70 B2 Taiyuan China
71 C3 Taizhong Taiwan
71 B3 Taizhou Jiangsu China
71 C3 Taizhou Zhejiang China
78 B3 Ta'izz Yemen
77 D3 Tajikistan country Asia
75 B2 Taj Mahal tourist site India
Tajo r. Port. see Tagus
58 A1 Tak Thai.
54 B2 Takaka N.Z.
67 B4 Takamatsu Japan
67 C3 Takaoka Japan
54 B1 Takapuna N.Z.
67 C3 Takasaki Japan
122 B1 Takatokwane Botswana
67 C3 Takayama Japan
Takefu Japan see Echizen
60 A1 Takengon Indon.
63 B2 Ta Khmau Cambodia
66 D2 Takikawa Japan
128 B2 Takla Lake Can.

128 B2 **Takla Landing** Can.
77 E3 **Taklimakan Desert** China
Taklimakan Shamo *des.* China *see* **Taklimakan Desert**
128 A2 **Taku** *r.* Can./U.S.A.
63 A3 **Takua Pa** Thai.
115 C4 **Takum** Nigeria
88 C3 **Talachyn** Belarus
74 B1 **Talagang** Pak.
150 A2 **Talara** Peru
59 C2 **Talaud, Kepulauan** *is* Indon.
106 C2 **Talavera de la Reina** Spain
153 A4 **Talca** Chile
153 A4 **Talcahuano** Chile
89 E2 **Taldom** Russia
77 D2 **Taldykorgan** Kazakh.
59 C3 **Taliabu** *i.* Indon.
64 B1 **Talisay** Phil.
61 C2 **Taliwang** Indon.
81 C2 **Tall 'Afar** Iraq
143 D2 **Tallahassee** U.S.A.
88 B2 **Tallinn** Estonia
142 B2 **Tallulah** U.S.A.
104 B2 **Talmont-St-Hilaire** France
90 C2 **Tal'ne** Ukr.
117 B3 **Talodi** Sudan
91 E1 **Talovaya** Russia
126 E2 **Taloyoak** Can.
88 B2 **Talsi** Latvia
152 A3 **Taltal** Chile
129 C1 **Taltson** *r.* Can.
60 A1 **Talu** Indon.
53 C1 **Talwood** Austr.
114 B4 **Tamale** Ghana
115 C2 **Tamanrasset** Alg.
99 A3 **Tamar** *r.* U.K.
145 C2 **Tamazunchale** Mex.
114 A3 **Tambacounda** Senegal
61 B1 **Tambelan, Kepulauan** *is* Indon.
91 E1 **Tambov** Russia
145 C2 **Tamiahua, Laguna de** *lag.* Mex.
143 D3 **Tampa** U.S.A.
143 D3 **Tampa Bay** U.S.A.
93 H3 **Tampere** Fin.
145 C2 **Tampico** Mex.
69 D1 **Tamsagbulag** Mongolia
102 C2 **Tamsweg** Austria
53 D2 **Tamworth** Austr.
99 C2 **Tamworth** U.K.
119 E3 **Tana** *r.* Kenya
117 B3 **Tana, Lake** Eth.
67 C4 **Tanabe** Japan
92 I1 **Tana Bru** Norway
61 C2 **Tanahgrogot** Indon.
58 C3 **Tanahjampea** *i.* Indon.
50 C1 **Tanami Desert** Austr.
126 A2 **Tanana** U.S.A.
108 A1 **Tanaro** *r.* Italy
65 B1 **Tanch'ŏn** N. Korea
64 B2 **Tandag** Phil.
110 C2 **Ţăndărei** Romania
153 C4 **Tandil** Arg.
74 A2 **Tando Adam** Pak.
74 A2 **Tando Muhammad Khan** Pak.
114 B2 **Tanezrouft** *reg.* Alg./Mali
119 D3 **Tanga** Tanz.
119 C3 **Tanganyika, Lake** Africa
Tanger Morocco *see* **Tangier**
101 E1 **Tangermünde** Ger.
75 C1 **Tanggula Shan** *mts* China
114 B1 **Tangier** Morocco
75 C1 **Tangra Yumco** *salt l.* China
70 B2 **Tangshan** China
68 C2 **Taniantaweng Shan** *mts* China
59 C3 **Tanimbar, Kepulauan** *is* Indon.
64 B2 **Tanjay** Phil.
60 A1 **Tanjungbalai** Indon.
61 B2 **Tanjungpandan** Indon.
60 B1 **Tanjungpinang** Indon.
61 C1 **Tanjungredeb** Indon.
61 C1 **Tanjungselor** Indon.
74 B1 **Tank** Pak.
115 C3 **Tanout** Niger
75 C2 **Tansen** Nepal
116 B1 **Ţanţā** Egypt
119 D3 **Tanzania** *country* Africa
69 E1 **Taonan** China
138 B1 **Taos** U.S.A.
114 B2 **Taoudenni** Mali
88 C2 **Tapa** Estonia
145 C3 **Tapachula** Mex.
151 C2 **Tapajós** *r.* Brazil
60 A1 **Tapaktuan** Indon.
145 C3 **Tapanatepec** Mex.
150 B2 **Tapauá** Brazil
114 B4 **Tapeta** Liberia
141 D3 **Tappahannock** U.S.A.
74 B2 **Tapti** *r.* India
150 B2 **Tapurucuara** Brazil
154 A1 **Taquarí** *r.* Brazil
154 B1 **Taquari, Serra do** *hills* Brazil
154 C2 **Taquaritinga** Brazil
115 D4 **Taraba** *r.* Nigeria
Ţarābulus Libya *see* **Tripoli**
61 C1 **Tarakan** Indon.
88 A3 **Taran, Mys** *pt* Russia
54 B1 **Taranaki, Mount** *vol.* N.Z.
106 C1 **Tarancón** Spain
109 C2 **Taranto** Italy
109 C2 **Taranto, Golfo di** *g.* Italy
150 A2 **Tarapoto** Peru
91 E2 **Tarasovskiy** Russia
150 A2 **Tarauacá** Brazil
150 B2 **Tarauacá** *r.* Brazil
48 F2 **Tarawa** *atoll* Kiribati
77 D2 **Taraz** Kazakh.
107 C1 **Tarazona** Spain
77 E2 **Tarbagatay, Khrebet** *mts* Kazakh.
96 A2 **Tarbert** *Scotland* U.K.
96 B3 **Tarbert** *Scotland* U.K.
104 C3 **Tarbes** France
96 B2 **Tarbet** U.K.
51 C3 **Tarcoola** Austr.
53 D2 **Taree** Austr.
110 C2 **Târgovişte** Romania
110 B1 **Târgu Jiu** Romania
110 B1 **Târgu Mureş** Romania
110 C1 **Târgu Neamţ** Romania
79 C2 **Tarif** U.A.E.
152 B3 **Tarija** Bol.
79 B3 **Tarīm** Yemen
77 E3 **Tarim Basin** China
77 E2 **Tarim He** *r.* China
Tarim Pendi *basin* China *see* **Tarim Basin**
59 D3 **Taritatu** *r.* Indon.
82 G2 **Tarko-Sale** Russia
114 B4 **Tarkwa** Ghana
64 B1 **Tarlac** Phil.
68 C2 **Tarlag** China
92 G2 **Tärnaby** Sweden
77 C3 **Tarnak Rōd** *r.* Afgh.
110 B1 **Târnăveni** Romania
103 E1 **Tarnobrzeg** Pol.
103 E1 **Tarnów** Pol.
75 C1 **Taro Co** *salt l.* China
114 B1 **Taroudannt** Morocco
69 E1 **Tarqi** China
108 B2 **Tarquinia** Italy
107 D1 **Tarragona** Spain
107 D1 **Tàrrega** Spain
80 B2 **Tarsus** Turkey
152 B3 **Tartagal** Arg.
104 B3 **Tartas** France
88 C2 **Tartu** Estonia
80 B2 **Ţarţūs** Syria
155 D1 **Tarumirim** Brazil
89 E3 **Tarusa** Russia
108 B1 **Tarvisio** Italy
81 D3 **Tashk, Daryācheh-ye** *l.* Iran
Tashkent Uzbek. *see* **Toshkent**
127 I2 **Tasiilaq** Greenland
131 D1 **Tasiujaq** Can.
76 B1 **Taskala** Kazakh.
77 E2 **Taskesken** Kazakh.
54 B2 **Tasman Bay** N.Z.
51 D4 **Tasmania** *state* Austr.
54 B2 **Tasman Mountains** N.Z.
156 D8 **Tasman Sea** S. Pacific Ocean
103 D2 **Tatabánya** Hungary
90 B2 **Tatarbunary** Ukr.
83 K3 **Tatarskiy Proliv** *str.* Russia
67 C4 **Tateyama** Japan
128 C1 **Tathlina Lake** Can.
78 B3 **Tathlīth** Saudi Arabia
78 B2 **Tathlīth, Wādī** *watercourse* Saudi Arabia
53 C3 **Tathra** Austr.
62 A1 **Tatkon** Myanmar
128 B2 **Tatla Lake** Can.
103 D2 **Tatra Mountains** *mts* Pol./Slovakia
Tatry *mts* Pol./Slovakia *see* **Tatra Mountains**
154 C2 **Tatuí** Brazil
139 C2 **Tatum** U.S.A.
81 C2 **Tatvan** Turkey
151 D2 **Tauá** Brazil
155 C2 **Taubaté** Brazil
101 D3 **Tauberbischofsheim** Ger.
54 C1 **Taumarunui** N.Z.
62 A1 **Taunggyi** Myanmar
62 A2 **Taung-ngu** Myanmar
62 A2 **Taungup** Myanmar
99 B3 **Taunton** U.K.
100 C2 **Taunus** *hills* Ger.
54 C1 **Taupo** N.Z.
54 C1 **Taupo, Lake** N.Z.
88 B2 **Taurag** Lith.
54 C1 **Tauranga** N.Z.
80 B2 **Taurus Mountains** Turkey
111 C3 **Tavas** Turkey
86 F3 **Tavda** Russia
106 B2 **Tavira** Port.
99 A3 **Tavistock** U.K.
63 A2 **Tavoy** Myanmar
111 C3 **Tavşanlı** Turkey
99 A3 **Taw** *r.* U.K.
140 C2 **Tawas City** U.S.A.
61 C1 **Tawau** *Sabah* Malaysia
64 A2 **Tawi-Tawi** *i.* Phil.
145 C3 **Taxco** Mex.
76 B2 **Taxiatosh** Uzbek.
77 D3 **Taxkorgan** China
96 C2 **Tay** *r.* U.K.
96 C2 **Tay, Firth of** *est.* U.K.
96 B2 **Tay, Loch** *l.* U.K.
128 B2 **Taylor** Can.
139 D2 **Taylor** U.S.A.
140 B3 **Taylorville** U.S.A.
78 A2 **Taymā'** Saudi Arabia
83 H2 **Taymura** *r.* Russia
83 H2 **Taymyr, Ozero** *l.* Russia
Taymyr, Poluostrov *pen.* Russia *see* **Taymyr Peninsula**
83 G2 **Taymyr Peninsula** Russia
63 B2 **Tây Ninh** Vietnam
64 A1 **Taytay** Phil.

Taz

82 G2 Taz r. Russia
114 B1 Taza Morocco
129 D2 Tazin Lake Can.
86 G2 Tazovskaya Guba sea chan. Russia
81 C1 Tbilisi Georgia
91 E2 Tbilisskaya Russia
118 B3 Tchibanga Gabon
118 B2 Tcholliré Cameroon
103 D1 Tczew Pol.
144 B2 Teacapán Mex.
54 A1 Te Anau N.Z.
54 A3 Te Anau, Lake N.Z.
145 C3 Teapa Mex.
54 C1 Te Awamutu N.Z.
115 C1 Tébessa Alg.
60 B2 Tebingtinggi Sumatera Selatan Indon.
60 A1 Tebingtinggi Sumatera Utara Indon.
114 B4 Techiman Ghana
144 B3 Tecomán Mex.
144 B2 Tecoripa Mex.
144 B2 Técpan Mex.
144 B2 Tecuala Mex.
110 C1 Tecuci Romania
68 C1 Teeli Russia
98 C1 Tees r. U.K.
111 C3 Tefenni Turkey
61 B2 Tegal Indon.
146 B3 Tegucigalpa Hond.
115 C3 Teguidda-n-Tessoumt Niger
114 B4 Téhini Côte d'Ivoire
81 D2 Tehrān Iran
145 C3 Tehuacán Mex.
145 C3 Tehuantepec, Gulf of Mex.
145 C3 Tehuantepec, Istmo de isth. Mex.
99 A2 Teifi r. U.K.
Te Ika-a-Māui i. N.Z. see North Island
76 C3 Tejen Turkm.
76 C3 Tejen r. Turkm.
Tejo r. Spain see Tagus
145 B3 Tejupilco Mex.
54 B2 Tekapo, Lake N.Z.
145 D2 Tekax Mex.
116 B3 Tekezē Wenz r. Eritrea/Eth.
111 C2 Tekirdağ Turkey
54 C1 Te Kuiti N.Z.
75 C2 Tel r. India
81 C1 Telavi Georgia
80 B2 Tel Aviv-Yafo Israel
145 D2 Telchac Puerto Mex.
128 A2 Telegraph Creek Can.
154 B2 Telêmaco Borba Brazil
99 B2 Telford U.K.
60 A2 Telo Indon.
86 E2 Telpoziz, Gora mt. Russia
88 B2 Telšiai Lith.
61 B2 Telukbatang Indon.
60 A1 Telukdalam Indon.
60 B1 Teluk Intan Malaysia
130 C2 Temagami Lake Can.
61 C2 Temanggung Indon.
123 C3 Temba S. Africa
60 B2 Tembilahan Indon.
120 A1 Tembo Aluma Angola
99 B2 Teme r. U.K.
60 B1 Temerluh Malaysia
77 D1 Temirtau Kazakh.
53 C2 Temora Austr.
139 D2 Temple U.S.A.
97 C2 Templemore Ireland
145 C2 Tempoal Mex.
91 D2 Temryuk Russia
153 A4 Temuco Chile
54 B2 Temuka N.Z.

145 C2 Tenabo Mex.
73 C3 Tenali India
63 A2 Tenasserim Myanmar
99 A3 Tenby U.K.
117 C3 Tendaho Eth.
105 D3 Tende France
108 A2 Tende, Col de pass France/Italy
73 D4 Ten Degree Channel India
67 D3 Tendō Japan
114 B3 Ténenkou Mali
115 D3 Ténéré, Erg du des. Niger
115 D2 Ténéré du Tafassâsset des. Niger
114 A2 Tenerife i. Canary Islands
107 D2 Ténès Alg.
61 C2 Tengah, Kepulauan is Indon.
62 A1 Tengchong China
61 C2 Tenggarong Indon.
70 A2 Tengger Shamo des. China
71 B3 Tengxian China
119 C4 Tenke Dem. Rep. Congo
114 B3 Tenkodogo Burkina Faso
51 C1 Tennant Creek Austr.
142 C1 Tennessee r. U.S.A.
142 C1 Tennessee state U.S.A.
145 C3 Tenosique Mex.
53 D1 Tenterfield Austr.
154 B2 Teodoro Sampaio Brazil
155 D1 Teófilo Otoni Brazil
145 C3 Teopisca Mex.
144 B2 Tepache Mex.
54 B1 Te Paki N.Z.
144 B3 Tepalcatepec Mex.
144 B2 Tepatitlán Mex.
144 B2 Tepehuanes Mex.
109 D2 Tepelenë Albania
144 B2 Tepic Mex.
102 C1 Teplice Czech Rep.
89 E3 Teploye Russia
144 B2 Tequila Mex.
108 B2 Teramo Italy
89 E3 Terbuny Russia
90 B2 Terebovlya Ukr.
87 D4 Terek r. Russia
151 D2 Teresina Brazil
155 D2 Teresópolis Brazil
63 A3 Teressa Island India
80 B1 Terme Turkey
108 B3 Termini Imerese Sicily Italy
145 C3 Términos, Laguna de lag. Mex.
77 C3 Termiz Uzbek.
109 B2 Termoli Italy
59 C2 Ternate Indon.
100 A2 Terneuzen Neth.
108 B2 Terni Italy
90 B2 Ternopil' Ukr.
128 B2 Terrace Can.
130 B2 Terrace Bay Can.
122 B2 Terra Firma S. Africa
140 B3 Terre Haute U.S.A.
131 E2 Terrenceville Can.
100 B1 Terschelling i. Neth.
108 A3 Tertenia Sardinia Italy
107 C1 Teruel Spain
92 H2 Tervola Fin.
109 C2 Tešanj Bos. & Herz.
116 B3 Teseney Eritrea
66 D2 Teshio-gawa r. Japan
128 A1 Teslin Can.
128 A1 Teslin Lake Can.
154 B1 Tesouro Brazil
115 C3 Tessaoua Niger
Tete Moz.
90 C1 Teteriv r. Ukr.
101 F1 Teterow Ger.
90 B2 Tetiyiv Ukr.

114 B1 Tétouan Morocco
109 D2 Tetovo Macedonia
152 B3 Teuco r. Arg.
144 B2 Teul de González Ortega Mex.
101 D1 Teutoburger Wald hills Ger.
Tevere r. Italy see Tiber
54 A3 Teviot N.Z.
Te Waipounamu i. N.Z. see South Island
51 E2 Tewantin Austr.
54 C2 Te Wharau N.Z.
139 E2 Texarkana U.S.A.
53 D1 Texas Austr.
139 D2 Texas state U.S.A.
139 E3 Texas City U.S.A.
100 B1 Texel i. Neth.
139 D2 Texoma, Lake U.S.A.
123 C2 Teyateyaneng Lesotho
89 F2 Teykovo Russia
89 F2 Teza r. Russia
75 D2 Tezpur India
62 A1 Tezu India
129 E1 Tha-anne r. Can.
123 C2 Thabana-Ntlenyana mt. Lesotho
123 C2 Thaba Putsoa mt. Lesotho
123 C1 Thabazimbi S. Africa
123 C2 Thabong S. Africa
63 A2 Thagyettaw Myanmar
63 B2 Thai Binh Vietnam
63 B2 Thailand country Asia
63 B2 Thailand, Gulf of Asia
62 B1 Thai Nguyên Vietnam
62 B2 Thakhèk Laos
63 A3 Thalang Thai.
74 B1 Thal Desert Pak.
101 E2 Thale Ger.
62 B2 Tha Li Thai.
123 C1 Thamaga Botswana
116 C3 Thamar, Jabal mt. Yemen
79 C3 Thamarīt Oman
130 B2 Thames r. Can.
54 C1 Thames N.Z.
99 D3 Thames est. U.K.
99 D3 Thames r. U.K.
63 A2 Thanbyuzayat Myanmar
62 A1 Thandwè Myanmar
62 B2 Thanh Hoa Vietnam
73 B3 Thanjavur India
63 A2 Thanlyin Myanmar
74 A2 Thano Bula Khan Pak.
62 B1 Than Uyên Vietnam
74 A2 Thar Desert India/Pak.
52 B1 Thargomindah Austr.
81 C2 Tharthār, Buḩayrat ath l. Iraq
111 B3 Thasos i. Greece
62 B1 Thât Khê Vietnam
62 A2 Thaton Myanmar
74 A2 Thatta Pak.
62 A1 Thaungdut Myanmar
62 A1 Thayawadi Myanmar
62 A1 Thayetmyo Myanmar
62 A1 Thazi Myanmar
146 C2 The Bahamas country West Indies
98 B1 The Cheviot h. U.K.
134 B1 The Dalles U.S.A.
99 C2 The Fens reg. U.K.
114 A3 The Gambia country Africa
The Great Oasis Egypt see Khārijah, Wāḩāt al
79 C2 The Gulf Asia
100 B1 The Hague Neth.
129 E1 Thelon r. Can.
101 E2 Themar Ger.
96 A1 The Minch sea chan. U.K.
150 B2 Theodore Roosevelt r. Brazil
129 D2 The Pas Can.

244

111 B2 **Thermaïkos Kolpos** *g.* Greece
136 B2 **Thermopolis** U.S.A.
53 C3 **The Rock** Austr.
130 B2 **Thessalon** Can.
111 B2 **Thessaloniki** Greece
99 D2 **Thetford** U.K.
131 C2 **Thetford Mines** Can.
62 A1 **The Triangle** *mts* Myanmar
131 D1 **Thévenet, Lac** *l.* Can.
99 D2 **The Wash** *b.* U.K.
139 D2 **The Woodlands** U.S.A.
142 B3 **Thibodaux** U.S.A.
129 E2 **Thicket Portage** Can.
137 D1 **Thief River Falls** U.S.A.
105 C2 **Thiers** France
119 D3 **Thika** Kenya
73 B4 **Thiladhunmathi** Maldives
75 C2 **Thimphu** Bhutan
105 D2 **Thionville** France
Thira *i.* Greece *see* **Santorini**
98 C1 **Thirsk** U.K.
73 B4 **Thiruvananthapuram** India
93 E4 **Thisted** Denmark
129 E1 **Thlewiaza** *r.* Can.
62 A2 **Thoen** Thai.
123 D1 **Thohoyandou** S. Africa
101 E1 **Thomasburg** Ger.
97 C2 **Thomastown** Ireland
143 D2 **Thomasville** U.S.A.
100 C2 **Thommen** Belgium
129 E2 **Thompson** Can.
128 E3 **Thompson** *r.* U.S.A.
134 C1 **Thompson Falls** U.S.A.
128 B2 **Thompson Sound** Can.
Thoothukudi India *see* **Tuticorin**
96 C3 **Thornhill** U.K.
55 C2 **Thorshavnheiane** *reg.* Antarctica
104 B2 **Thouars** France
128 C2 **Three Hills** Can.
63 A2 **Three Pagodas Pass** Myanmar/Thai.
114 B4 **Three Points, Cape** Ghana
73 B3 **Thrissur** India
63 B2 **Thu Dâu Môt** Vietnam
100 B2 **Thuin** Belgium
127 G1 **Thule Air Base** Greenland
121 B3 **Thuli** Zimbabwe
130 B2 **Thunder Bay** Can.
63 A3 **Thung Song** Thai.
101 E2 **Thüringer Becken** *reg.* Ger.
101 E2 **Thüringer Wald** *mts* Ger.
97 C2 **Thurles** Ireland
96 C1 **Thurso** U.K.
96 C1 **Thurso** *r.* U.K.
151 D2 **Tianguá** Brazil
70 B2 **Tianjin** China
70 B2 **Tianjin** *mun.* China
71 A3 **Tianlin** China
70 B2 **Tianmen** China
70 A2 **Tianshui** China
154 B2 **Tibagi** *r.* Brazil
118 B2 **Tibati** Cameroon
108 B2 **Tiber** *r.* Italy
Tiberias, Lake Israel *see* **Galilee, Sea of**
115 D2 **Tibesti** *mts* Chad
Tibet *aut. reg.* China *see* **Xizang Zizhiqu**
68 B2 **Tibet, Plateau of** China
Tibet Autonomous Region *aut. reg.* China *see* **Xizang Zizhiqu**
115 D2 **Tībistī, Sarīr** *des.* Libya
52 B1 **Tibooburra** Austr.
144 A2 **Tiburón, Isla** *i.* Mex.
114 B3 **Tîchît** Maur.
114 A2 **Tichla** Western Sahara
145 D2 **Ticul** Mex.
114 A3 **Tidjikja** Maur.
100 B2 **Tiel** Neth.
65 A1 **Tieling** China
100 A2 **Tielt** Belgium
100 B2 **Tienen** Belgium
68 B2 **Tien Shan** *mts* China/Kyrg.
Tientsin China *see* **Tianjin**
93 G3 **Tierp** Sweden
145 C3 **Tierra Blanca** Mex.
145 C3 **Tierra Colorada** Mex.
153 B6 **Tierra del Fuego, Isla Grande de** *i.* Arg./Chile
154 C2 **Tietê** Brazil
Tiflis Georgia *see* **Tbilisi**
143 D2 **Tifton** U.S.A.
Tighina Moldova *see* **Bender**
131 D2 **Tignish** Can.
150 A2 **Tigre** *r.* Ecuador/Peru
81 C2 **Tigris** *r.* Asia
145 C2 **Tihuatlán** Mex.
144 A1 **Tijuana** Mex.
91 E2 **Tikhoretsk** Russia
89 D2 **Tikhvin** Russia
89 D2 **Tikhvinskaya Gryada** *ridge* Russia
54 C1 **Tikokino** N.Z.
81 C2 **Tikrīt** Iraq
83 J2 **Tiksi** Russia
100 B2 **Tilburg** Neth.
152 B3 **Tilcara** Arg.
52 B1 **Tilcha (abandoned)** Austr.
114 C3 **Tillabéri** Niger
134 B1 **Tillamook** U.S.A.
63 A3 **Tillanchong Island** India
111 C3 **Tilos** *i.* Greece
52 B2 **Tilpa** Austr.
86 F2 **Til'tim** Russia
89 E3 **Tim** Russia
86 D2 **Timanskiy Kryazh** *ridge* Russia
54 B2 **Timaru** N.Z.
91 D2 **Timashevsk** Russia
114 B3 **Timbedgha** Maur.
50 C1 **Timber Creek** Austr.
114 B3 **Timbuktu** Mali
114 C2 **Timimoun** Alg.
111 B2 **Timiou Prodromou, Akrotirio** *pt* Greece
110 B1 **Timiș** *r.* Romania
110 B1 **Timișoara** Romania
130 B2 **Timmins** Can.
89 E2 **Timokhino** Russia
151 D2 **Timon** Brazil
59 C3 **Timor** *i.* East Timor/Indonesia
Timor-Leste *country* Asia *see* **East Timor**
58 C3 **Timor Sea** Austr./Indon.
93 G3 **Timrå** Sweden
114 B2 **Tindouf** Alg.
75 C2 **Tingri** China
93 F4 **Tingsryd** Sweden
59 D2 **Tinian** *i.* N. Mariana Is
152 B3 **Tinogasta** Arg.
111 C3 **Tinos** Greece
111 C3 **Tinos** *i.* Greece
115 C2 **Tinrhert, Hamada de** Alg.
62 A1 **Tinsukia** India
107 D2 **Tipasa** Alg.
97 B2 **Tipperary** Ireland
109 C2 **Tirana** Albania
Tiranë Albania *see* **Tirana**
108 B1 **Tirano** Italy
90 B2 **Tiraspol** Moldova
111 C3 **Tire** Turkey
96 A2 **Tiree** *i.* U.K.
74 B1 **Tirich Mir** *mt.* Pak.
73 B3 **Tiruchchirappalli** India
73 B4 **Tirunelveli** India
73 B3 **Tirupati** India
73 B3 **Tiruppattur** India
73 B3 **Tiruppur** India
109 D1 **Tisa** *r.* Serbia
129 D2 **Tisdale** Can.
107 D2 **Tissemsilt** Alg.
152 B2 **Titicaca, Lake** Bol./Peru
75 C2 **Titlagarh** India
110 C2 **Titu** Romania
143 D3 **Titusville** U.S.A.
99 B3 **Tiverton** U.K.
108 B2 **Tivoli** Italy
79 C2 **Ţīwī** Oman
145 D2 **Tizimín** Mex.
107 D2 **Tizi Ouzou** Alg.
114 B2 **Tiznit** Morocco
145 C3 **Tlacotalpán** Mex.
144 B2 **Tlahualilo** Mex.
145 C3 **Tlapa** Mex.
145 C3 **Tlaxcala** Mex.
145 C3 **Tlaxiaco** Mex.
114 B1 **Tlemcen** Alg.
123 C1 **Tlokweng** Botswana
128 B2 **Toad River** Can.
121 □D2 **Toamasina** Madag.
60 A1 **Toba, Danau** *l.* Indon.
74 A1 **Toba and Kakar Ranges** *mts* Pak.
147 D3 **Tobago** *i.* Trin. and Tob.
59 C2 **Tobelo** Indon.
130 B2 **Tobermory** Can.
96 A2 **Tobermory** U.K.
129 D2 **Tobin Lake** Can.
60 B2 **Toboali** Indon.
86 F3 **Tobol'sk** Russia
76 C1 **Tobyl** *r.* Kazakh./Russia
151 D2 **Tocantinópolis** Brazil
151 D2 **Tocantins** *r.* Brazil
143 D2 **Toccoa** U.S.A.
108 A1 **Toce** *r.* Italy
152 A3 **Tocopilla** Chile
53 C3 **Tocumwal** Austr.
144 A2 **Todos Santos** Mex.
128 B3 **Tofino** Can.
96 □ **Toft** U.K.
49 G4 **Tofua** *i.* Tonga
59 C3 **Togian, Kepulauan** *is* Indon.
114 C4 **Togo** *country* Africa
74 B2 **Tohana** India
126 B2 **Tok** U.S.A.
116 B3 **Tokar** Sudan
69 E3 **Tokara-rettō** *is* Japan
91 E1 **Tokarevka** Russia
49 G3 **Tokelau** *terr.* S. Pacific Ocean
91 D2 **Tokmak** Ukr.
77 D2 **Tokmok** Kyrg.
54 C1 **Tokoroa** N.Z.
68 B2 **Toksun** China
67 B4 **Tokushima** Japan
67 C3 **Tōkyō** Japan
121 □D3 **Tôlañaro** Madag.
154 B2 **Toledo** Brazil
106 C2 **Toledo** Spain
140 C2 **Toledo** U.S.A.
106 C2 **Toledo, Montes de** *mts* Spain
142 B2 **Toledo Bend Reservoir** U.S.A.
121 □D3 **Toliara** Madag.
58 C2 **Tolitoli** Indon.
108 B1 **Tolmezzo** Italy
104 B3 **Tolosa** Spain
145 C3 **Toluca** Mex.
87 D3 **Tol'yatti** Russia
140 A2 **Tomah** U.S.A.
66 D2 **Tomakomai** Japan
106 B2 **Tomar** Port.
103 E1 **Tomaszów Lubelski** Pol.
103 E1 **Tomaszów Mazowiecki** Pol.

144 B3 **Tomatlán** Mex.
142 C2 **Tombigbee** *r.* U.S.A.
155 D2 **Tombos** Brazil
Tombouctou Mali *see* **Timbuktu**
120 A2 **Tombua** Angola
123 C1 **Tom Burke** S. Africa
53 C2 **Tomingley** Austr.
58 C3 **Tomini, Teluk** *g.* Indon.
109 C2 **Tomislavgrad** Bos. & Herz.
50 A2 **Tom Price** Austr.
82 G3 **Tomsk** Russia
93 F4 **Tomtabacken** *h.* Sweden
145 C3 **Tonalá** Mex.
150 B2 **Tonantins** Brazil
59 C2 **Tondano** Indon.
49 G4 **Tonga** *country* S. Pacific Ocean
49 G4 **Tongatapu Group** *is* Tonga
71 B3 **Tongcheng** China
70 A2 **Tongchuan** China
71 A3 **Tongdao** China
100 B2 **Tongeren** Belgium
62 B1 **Tonghai** China
65 B1 **Tonghua** China
65 B2 **Tongjosŏn-man** *b.* N. Korea
62 B1 **Tongking, Gulf of** China/Vietnam
69 E2 **Tongliao** China
70 B2 **Tongling** China
52 B2 **Tongo** Austr.
71 A3 **Tongren** China
Tongshan China *see* **Xuzhou**
Tongtian He *r.* China *see* **Yangtze**
96 B1 **Tongue** U.K.
65 B3 **Tongyeong** S. Korea
69 E2 **Tongyu** China
65 A1 **Tongyuanpu** China
74 B2 **Tonk** India
63 B2 **Tonle Sap** *l.* Cambodia
135 C3 **Tonopah** U.S.A.
93 F4 **Tønsberg** Norway
135 D2 **Tooele** U.S.A.
52 B3 **Tooleybuc** Austr.
53 D1 **Toowoomba** Austr.
137 D3 **Topeka** U.S.A.
144 B2 **Topia** Mex.
144 B2 **Topolobampo** Mex.
86 C2 **Topozero, Ozero** *l.* Russia
134 B1 **Toppenish** U.S.A.
111 C3 **Torbalı** Turkey
76 B3 **Torbat-e Ḩeydarīyeh** Iran
76 C3 **Torbat-e Jām** Iran
106 C1 **Tordesillas** Spain
107 C1 **Tordesilos** Spain
107 D1 **Torelló** Spain
101 F2 **Torgau** Ger.
76 C2 **Torgay** Kazakh.
100 A2 **Torhout** Belgium
Torino Italy *see* **Turin**
154 B1 **Torixoréu** Brazil
106 B1 **Tormes** *r.* Spain
92 H2 **Torneälven** *r.* Sweden
127 G3 **Torngat Mountains** Can.
92 H2 **Tornio** Fin.
106 B1 **Toro** Spain
130 C2 **Toronto** Can.
119 D2 **Tororo** Uganda
Toros Dağları *mts* Turkey *see* **Taurus Mountains**
99 B3 **Torquay** U.K.
106 B2 **Torrão** Port.
106 B1 **Torre** *mt.* Port.
107 D1 **Torreblanca** Spain
106 C1 **Torrecerredo** *mt.* Spain
106 B1 **Torre de Moncorvo** Port.
106 C1 **Torrelavega** Spain
106 C2 **Torremolinos** Spain
52 A2 **Torrens, Lake** *imp. l.* Austr.
107 C2 **Torrent** Spain
144 B2 **Torreón** Mex.
106 B2 **Torres Novas** Port.
106 B2 **Torres Vedras** Port.
107 C2 **Torrevieja** Spain
96 B2 **Torridon** U.K.
106 C2 **Torrijos** Spain
141 E2 **Torrington** *CT* U.S.A.
136 C2 **Torrington** *WY* U.S.A.
107 D1 **Torroella de Montgrí** Spain
94 B1 **Tórshavn** Faroe Is
108 A3 **Tortolì** *Sardinia* Italy
107 D1 **Tortosa** Spain
103 D1 **Toruń** Pol.
97 B1 **Tory Island** Ireland
89 D2 **Torzhok** Russia
122 B2 **Tosca** S. Africa
108 B2 **Toscano, Arcipelago** *is* Italy
77 C2 **Toshkent** Uzbek.
89 D2 **Tosno** Russia
152 B3 **Tostado** Arg.
101 D1 **Tostedt** Ger.
80 B1 **Tosya** Turkey
86 D3 **Tot'ma** Russia
67 B3 **Tottori** Japan
114 B4 **Touba** Côte d'Ivoire
114 B1 **Toubkal, Jebel** *mt.* Morocco
114 B3 **Tougan** Burkina Faso
115 C1 **Touggourt** Alg.
105 D2 **Toul** France
105 D3 **Toulon** France
104 C3 **Toulouse** France
100 A2 **Tournai** Belgium
105 C2 **Tournus** France
151 E2 **Touros** Brazil
104 C2 **Tours** France
115 D2 **Tousside, Pic** *mt.* Chad
122 B3 **Touwsrivier** S. Africa
66 D2 **Towada** Japan
134 D1 **Townsend** U.S.A.
51 D1 **Townsville** Austr.
77 E2 **Toxkan He** *r.* China
66 D2 **Tōya-ko** *l.* Japan
67 C3 **Toyama** Japan
67 C3 **Toyota** Japan
115 C1 **Tozeur** Tunisia
81 C1 **T'q'varcheli** Georgia
Trâblous Lebanon *see* **Tripoli**
80 B1 **Trabzon** Turkey
106 B2 **Trafalgar, Cabo** *c.* Spain
128 C3 **Trail** Can.
88 B3 **Trakai** Lith.
97 B2 **Tralee** Ireland
97 C2 **Tramore** Ireland
63 A3 **Trang** Thai.
59 C3 **Trangan** *i.* Indon.
55 I2 **Transantarctic Mountains** Antarctica
90 B2 **Transnistria** *disp. terr.* Moldova
110 B1 **Transylvanian Alps** *mts* Romania
108 B3 **Trapani** *Sicily* Italy
53 C3 **Traralgon** Austr.
63 B2 **Trat** Thai.
102 C2 **Traunstein** Ger.
140 B2 **Traverse City** U.S.A.
103 D2 **Třebíč** Czech Rep.
109 C2 **Trebinje** Bos. & Herz.
103 E2 **Trebišov** Slovakia
109 C1 **Trebnje** Slovenia
153 C4 **Treinta y Tres** Uru.
153 B5 **Trelew** Arg.
93 F4 **Trelleborg** Sweden
130 C2 **Tremblant, Mont** *h.* Can.
109 C2 **Tremiti, Isole** *is* Italy
134 D2 **Tremonton** U.S.A.
107 D1 **Tremp** Spain
103 D2 **Trenčín** Slovakia
153 B4 **Trenque Lauquén** Arg.
98 C2 **Trent** *r.* U.K.
108 B1 **Trento** Italy
141 D2 **Trenton** Can.
137 E2 **Trenton** *MO* U.S.A.
141 E2 **Trenton** *NJ* U.S.A.
131 E2 **Trepassey** Can.
153 B4 **Tres Arroyos** Arg.
155 C2 **Três Corações** Brazil
154 B2 **Três Lagoas** Brazil
153 A5 **Tres Lagos** Arg.
155 C1 **Três Marias, Represa** *resr* Brazil
155 C2 **Três Pontas** Brazil
153 B5 **Tres Puntas, Cabo** *c.* Arg.
155 D2 **Três Rios** Brazil
101 F1 **Treuenbrietzen** Ger.
108 A1 **Treviglio** Italy
108 B1 **Treviso** Italy
99 A3 **Trevose Head** U.K.
109 C3 **Tricase** Italy
53 C2 **Trida** Austr.
100 C3 **Trier** Ger.
108 B1 **Trieste** Italy
108 B1 **Triglav** *mt.* Slovenia
59 D3 **Trikora, Puncak** *mt.* Indon.
97 C2 **Trim** Ireland
73 C4 **Trincomalee** Sri Lanka
154 C1 **Trindade** Brazil
152 B2 **Trinidad** Bol.
147 D3 **Trinidad** *i.* Trin. and Tob.
136 C3 **Trinidad** U.S.A.
147 D3 **Trinidad and Tobago** *country* West Indies
131 E2 **Trinity Bay** Can.
111 B3 **Tripoli** Greece
80 B2 **Tripoli** Lebanon
115 D1 **Tripoli** Libya
113 D10 **Tristan da Cunha** *i.* S. Atlantic Ocean
Trivandrum India *see* **Thiruvananthapuram**
108 B2 **Trivento** Italy
103 D2 **Trnava** Slovakia
109 C2 **Trogir** Croatia
109 C2 **Troia** Italy
100 C2 **Troisdorf** Ger.
131 C2 **Trois-Rivières** Can.
86 E2 **Troitsko-Pechorsk** Russia
151 C2 **Trombetas** *r.* Brazil
123 C3 **Trompsburg** S. Africa
92 G2 **Tromsø** Norway
92 F3 **Trondheim** Norway
128 C2 **Trout Lake** Can.
128 B1 **Trout Lake** *N.W.T.* Can.
130 A1 **Trout Lake** *Ont.* Can.
99 B3 **Trowbridge** U.K.
142 C2 **Troy** *AL* U.S.A.
141 E2 **Troy** *NY* U.S.A.
105 C2 **Troyes** France
109 D2 **Trstenik** Serbia
89 D3 **Trubchevsk** Russia
106 B1 **Truchas** Spain
79 C2 **Trucial Coast** U.A.E.
146 B3 **Trujillo** Hond.
150 A2 **Trujillo** Peru
106 B2 **Trujillo** Spain
147 C4 **Trujillo** Venez.
142 B1 **Trumann** U.S.A.
131 D2 **Truro** Can.
99 A3 **Truro** U.K.
138 B2 **Truth or Consequences** U.S.A.
103 D1 **Trutnov** Czech Rep.
93 F3 **Trysil** Norway
103 D1 **Trzebiatów** Pol.

68 B1 **Tsagaannuur** Mongolia
121 ◻D2 **Tsaratanana, Massif du** *mts* Madag.
110 C2 **Tsarevo** Bulg.
122 A2 **Tses** Namibia
122 B1 **Tsetseng** Botswana
68 C1 **Tsetserleg** *Arhangay* Mongolia
68 C1 **Tsetserleg** *Hövsgöl* Mongolia
122 B2 **Tshabong** Botswana
122 B1 **Tshane** Botswana
118 B3 **Tshela** Dem. Rep. Congo
118 C3 **Tshikapa** Dem. Rep. Congo
118 C3 **Tshikapa** *r.* Dem. Rep. Congo
123 D1 **Tshipise** S. Africa
118 C3 **Tshitanzu** Dem. Rep. Congo
118 C3 **Tshuapa** *r.* Dem. Rep. Congo
Tshwane S. Africa *see* Pretoria
87 D4 **Tsimlyanskoye Vodokhranilishche** *resr* Russia
Tsingtao China *see* Qingdao
Tsining China *see* Ulan Qab
121 ◻D2 **Tsiroanomandidy** Madag.
123 C3 **Tsomo** S. Africa
67 C4 **Tsu** Japan
67 D3 **Tsuchiura** Japan
66 D2 **Tsugarū-kaikyō** *str.* Japan
120 A2 **Tsumeb** Namibia
122 A1 **Tsumis Park** Namibia
120 B2 **Tsumkwe** Namibia
67 C3 **Tsuruga** Japan
66 C3 **Tsuruoka** Japan
67 A4 **Tsushima** *is* Japan
Tsushima-kaikyō *str.* Japan/ S. Korea *see* Korea Strait
67 B3 **Tsuyama** Japan
123 C2 **Tswelelang** S. Africa
91 C2 **Tsyurupyns'k** Ukr.
Tthenaagoo Can. *see* Nahanni Butte
59 C3 **Tual** Indon.
97 B2 **Tuam** Ireland
91 D3 **Tuapse** Russia
54 A3 **Tuatapere** N.Z.
96 A1 **Tuath, Loch a'** *b.* U.K.
138 A1 **Tuba City** U.S.A.
61 C2 **Tuban** Indon.
152 D3 **Tubarão** Brazil
102 B2 **Tübingen** Ger.
115 E1 **Tubruq** Libya
49 H4 **Tubuai** Fr. Polynesia
49 I4 **Tubuai** i. Fr. Polynesia
144 A1 **Tubutama** Mex.
152 C2 **Tucavaca** Bol.
128 B1 **Tuchitua** Can.
138 A2 **Tucson** U.S.A.
139 C1 **Tucumcari** U.S.A.
150 B1 **Tucupita** Venez.
151 D2 **Tucuruí** Brazil
151 D2 **Tucuruí, Represa de** *resr* Brazil
107 C1 **Tudela** Spain
106 B1 **Tuela** *r.* Port.
62 A1 **Tuensang** India
64 B1 **Tuguegarao** Phil.
106 B1 **Tui** Spain
59 C3 **Tukangbesi, Kepulauan** *is* Indon.
126 C2 **Tuktoyaktuk** Can.
88 B2 **Tukums** Latvia
145 C2 **Tula** Mex.
89 E3 **Tula** Russia
145 C2 **Tulancingo** Mex.
135 C3 **Tulare** U.S.A.
138 B2 **Tularosa** U.S.A.
110 C1 **Tulcea** Romania
90 B2 **Tul'chyn** Ukr.
129 E1 **Tulemalu Lake** Can.
139 C1 **Tulia** U.S.A.
142 C1 **Tullahoma** U.S.A.

97 C2 **Tullamore** Ireland
104 C2 **Tulle** France
51 D1 **Tully** Austr.
139 D1 **Tulsa** U.S.A.
150 A1 **Tumaco** Col.
123 C2 **Tumahole** S. Africa
73 B3 **Tumakuru** India
93 G4 **Tumba** Sweden
118 B3 **Tumba, Lac** l. Dem. Rep. Congo
53 C3 **Tumbarumba** Austr.
150 A2 **Tumbes** Peru
128 B2 **Tumbler Ridge** Can.
52 A2 **Tumby Bay** Austr.
65 B1 **Tumen** China
150 B1 **Tumereng** Guyana
64 A2 **Tumindao** i. Phil.
Tumkur India *see* Tumakuru
74 A2 **Tump** Pak.
151 C1 **Tumucumaque, Serra** *hills* Brazil
53 C3 **Tumut** Austr.
80 B2 **Tunceli** Turkey
53 D2 **Tuncurry** Austr.
119 D4 **Tunduru** Tanz.
110 C2 **Tundzha** *r.* Bulg.
128 B1 **Tungsten (abandoned)** Can.
115 D1 **Tunis** Tunisia
115 C1 **Tunisia** *country* Africa
150 A1 **Tunja** Col.
93 F3 **Tunnsjøen** l. Norway
Tunxi China *see* Huangshan
154 B2 **Tupã** Brazil
154 C1 **Tupaciguara** Brazil
142 C2 **Tupelo** U.S.A.
152 B3 **Tupiza** Bol.
76 B2 **Tupkaragan, Mys** *pt* Kazakh.
83 H2 **Tura** Russia
86 F3 **Tura** *r.* Russia
78 B2 **Turabah** Saudi Arabia
83 J3 **Turana, Khrebet** *mts* Russia
54 C1 **Turangi** N.Z.
76 B2 **Turan Lowland** Asia
78 A1 **Ţurayf** Saudi Arabia
88 B2 **Turba** Estonia
74 A2 **Turbat** Pak.
150 A1 **Turbo** Col.
110 B1 **Turda** Romania
Turfan China *see* Turpan
111 C3 **Turgutlu** Turkey
111 C2 **Turhal** Turkey
107 C2 **Turia** *r.* Spain
108 A1 **Turin** Italy
86 F3 **Turinsk** Russia
90 A1 **Turiys'k** Ukr.
119 D2 **Turkana, Lake** *salt l.* Eth./ Kenya
80 B2 **Turkey** *country* Asia/Europe
77 C2 **Turkistan** Kazakh.
76 C3 **Türkmenabat** Turkm.
76 B2 **Türkmenbaşy** Turkm.
76 B2 **Turkmenistan** *country* Asia
147 C2 **Turks and Caicos Islands** *terr.* West Indies
93 H3 **Turku** Fin.
135 B3 **Turlock** U.S.A.
54 C2 **Turnagain, Cape** N.Z.
100 B2 **Turnhout** Belgium
129 D2 **Turnor Lake** Can.
110 B2 **Turnu Măgurele** Romania
68 B2 **Turpan** China
77 D2 **Turugart Pass** China/Kyrg.
142 C2 **Tuscaloosa** U.S.A.
142 C2 **Tuskegee** U.S.A.
73 B4 **Tuticorin** India
120 B3 **Tutume** Botswana
49 F3 **Tuvalu** *country* S. Pacific Ocean

78 B2 **Tuwayq, Jabal** *hills* Saudi Arabia
78 B2 **Tuwayq, Jabal** *mts* Saudi Arabia
78 A2 **Tuwwal** Saudi Arabia
144 B2 **Tuxpan** *Nayarit* Mex.
145 C2 **Tuxpan** *Veracruz* Mex.
145 C3 **Tuxtla Gutiérrez** Mex.
62 B1 **Tuyên Quang** Vietnam
63 B2 **Tuy Hoa** Vietnam
80 B2 **Tuz, Lake** *salt l.* Turkey
Tuz Gölü *salt l.* Turkey *see* Tuz, Lake
81 C2 **Tuz Khurmātū** Iraq
109 C2 **Tuzla** Bos. & Herz.
91 E2 **Tuzlov** *r.* Russia
89 E2 **Tver'** Russia
98 B1 **Tweed** *r.* U.K.
122 A2 **Twee Rivier** Namibia
135 C4 **Twentynine Palms** U.S.A.
131 E2 **Twillingate** Can.
134 D2 **Twin Falls** U.S.A.
137 E1 **Two Harbors** U.S.A.
129 C2 **Two Hills** Can.
139 D2 **Tyler** U.S.A.
83 J3 **Tynda** Russia
93 F3 **Tynset** Norway
80 B2 **Tyre** Lebanon
111 B3 **Tyrnavos** Greece
52 B3 **Tyrrell, Lake** *dry lake* Austr.
108 B2 **Tyrrhenian Sea** France/Italy
87 E3 **Tyul'gan** Russia
86 F3 **Tyumen'** Russia
83 J2 **Tyung** *r.* Russia
99 A3 **Tywi** *r.* U.K.
122 D1 **Tzaneen** S. Africa

U

120 B2 **Uamanda** Angola
155 D2 **Ubá** Brazil
155 D1 **Ubaí** Brazil
151 E3 **Ubaitaba** Brazil
118 B3 **Ubangi** *r.* C.A.R./ Dem. Rep. Congo
67 B4 **Ube** Japan
106 C2 **Úbeda** Spain
154 C1 **Uberaba** Brazil
154 C1 **Uberlândia** Brazil
106 B1 **Ubiña, Peña** *mt.* Spain
123 D2 **Ubombo** S. Africa
63 B2 **Ubon Ratchathani** Thai.
119 C3 **Ubundu** Dem. Rep. Congo
150 A2 **Ucayali** *r.* Peru
74 B2 **Uch** Pak.
66 D2 **Uchiura-wan** *b.* Japan
83 J3 **Uchur** *r.* Russia
128 B3 **Ucluelet** Can.
74 B2 **Udaipur** India
91 C1 **Uday** *r.* Ukr.
93 F4 **Uddevalla** Sweden
92 G2 **Uddjaure** l. Sweden
100 B2 **Uden** Neth.
74 B1 **Udhampur** India
108 B1 **Udine** Italy
89 E2 **Udomlya** Russia
62 B2 **Udon Thani** Thai.
73 B3 **Udupi** India
83 K3 **Udyl', Ozero** l. Russia
67 C3 **Ueda** Japan
58 C3 **Uekuli** Indon.
118 C2 **Uele** *r.* Dem. Rep. Congo
101 E1 **Uelzen** Ger.
119 C2 **Uere** *r.* Dem. Rep. Congo
87 E3 **Ufa** Russia
119 D3 **Ugalla** *r.* Tanz.

119 D2 **Uganda** *country* Africa
89 E2 **Uglich** Russia
89 D2 **Uglovka** Russia
89 D3 **Ugra** Russia
103 D2 **Uherské Hradiště** Czech Rep.
101 E2 **Uichteritz** Ger.
96 A2 **Uig** U.K.
120 A1 **Uíge** Angola
65 B2 **Uijeongbu** S. Korea
135 D2 **Uinta Mountains** U.S.A.
65 B2 **Uiseong** S. Korea
120 A3 **Uis Mine** Namibia
123 C3 **Uitenhage** S. Africa
100 C1 **Uithuizen** Neth.
74 B2 **Ujjain** India
89 F3 **Ukholovo** Russia
62 A1 **Ukhrul** India
86 E2 **Ukhta** Russia
135 B3 **Ukiah** U.S.A.
127 H2 **Ukkusissat** Greenland
88 B2 **Ukmerg** Lith.
90 C2 **Ukraine** *country* Europe
Ulaanbaatar Mongolia *see* **Ulan Bator**
68 C1 **Ulaangom** Mongolia
69 D1 **Ulan Bator** Mongolia
Ulanhad China *see* **Chifeng**
69 E1 **Ulanhot** China
87 D4 **Ulan-Khol** Russia
70 B1 **Ulan Qab** China
69 D1 **Ulan-Ude** Russia
75 D1 **Ulan Ul Hu** *l.* China
Uleåborg Fin. *see* **Oulu**
88 C2 **Ülenurme** Estonia
69 D1 **Uliastai** China
68 C1 **Uliastay** Mongolia
59 D2 **Ulithi** *atoll* Micronesia
65 B2 **Uljin** S. Korea
53 D3 **Ulladulla** Austr.
96 B2 **Ullapool** U.K.
65 C2 **Ulleung-do** *i.* S. Korea
98 B1 **Ullswater** *l.* U.K.
102 B2 **Ulm** Ger.
65 B2 **Ulsan** S. Korea
96 □ **Ulsta** U.K.
97 C1 **Ulster** *reg.* Ireland/U.K.
52 B3 **Ultima** Austr.
111 C3 **Ulubey** Turkey
111 C2 **Uludağ** *mt.* Turkey
126 D2 **Ulukhaktok** Can.
123 D2 **Ulundi** S. Africa
77 E2 **Ulungur Hu** *l.* China
50 C2 **Uluṟu** *h.* Austr.
98 B1 **Ulverston** U.K.
87 D3 **Ul'yanovsk** Russia
136 C3 **Ulysses** U.S.A.
90 C2 **Uman'** Ukr.
86 C2 **Umba** Russia
59 D3 **Umboi** *i.* P.N.G.
92 H3 **Umeå** Sweden
92 H3 **Umeälven** *r.* Sweden
127 I2 **Umiiviip Kangertiva** *inlet* Greenland
126 D2 **Umingmaktok** Can.
123 D2 **Umlazi** S. Africa
117 A3 **Umm Keddada** Sudan
78 A2 **Umm Lajj** Saudi Arabia
117 B3 **Umm Ruwaba** Sudan
115 E1 **Umm Sa'ad** Libya
134 B2 **Umpqua** *r.* U.S.A.
120 A2 **Umpulo** Angola
154 B2 **Umuarama** Brazil
109 C1 **Una** *r.* Bos. & Herz./Croatia
155 E1 **Una** Brazil
154 C1 **Unaí** Brazil
78 B2 **'Unayzah** Saudi Arabia
136 C1 **Underwood** U.S.A.
89 D3 **Unecha** Russia
53 C2 **Ungarie** Austr.
52 A2 **Ungarra** Austr.
127 G2 **Ungava, Péninsule d'** *pen.* Can.
127 G3 **Ungava Bay** Can.
90 B2 **Ungheni** Moldova
Unguja *i.* Tanz. *see* **Zanzibar Island**
154 B3 **União da Vitória** Brazil
150 B2 **Unini** *r.* Brazil
142 C1 **Union City** U.S.A.
122 B3 **Uniondale** S. Africa
141 D3 **Uniontown** U.S.A.
79 C2 **United Arab Emirates** *country* Asia
95 C3 **United Kingdom** *country* Europe
132 D3 **United States of America** *country* N. America
129 D2 **Unity** Can.
96 □ **Unst** *i.* U.K.
101 E2 **Unstrut** *r.* Ger.
89 E3 **Upa** *r.* Russia
119 C3 **Upemba, Lac** *l.* Dem. Rep. Congo
122 B2 **Upington** S. Africa
134 B2 **Upper Alkali Lake** U.S.A.
128 C2 **Upper Arrow Lake** Can.
134 B2 **Upper Klamath Lake** U.S.A.
128 B1 **Upper Liard** Can.
97 C1 **Upper Lough Erne** *l.* U.K.
93 G4 **Uppsala** Sweden
78 B2 **'Uqlat aş Şuqūr** Saudi Arabia
Urad Qianqi China *see* **Xishanzui**
76 B2 **Ural** *r.* Kazakh./Russia
53 D2 **Uralla** Austr.
87 E3 **Ural Mountains** Russia
76 B1 **Ural'sk** Kazakh.
Ural'skiy Khrebet *mts* Russia *see* **Ural Mountains**
119 D3 **Urambo** Tanz.
53 C3 **Urana** Austr.
129 D2 **Uranium City** Can.
86 F2 **Uray** Russia
98 C1 **Ure** *r.* U.K.
86 D3 **Uren'** Russia
144 A2 **Ures** Mex.
76 C2 **Urganch** Uzbek.
100 B1 **Urk** Neth.
111 C3 **Urla** Turkey
81 C2 **Urmia, Lake** *salt l.* Iran
Uroševac Kosovo *see* **Ferizaj**
144 B2 **Uruáchic** Mex.
151 D3 **Uruaçu** Brazil
144 B3 **Uruapan** Mex.
150 A3 **Urubamba** *r.* Peru
151 C2 **Urucará** Brazil
151 D2 **Uruçuí** Brazil
151 C2 **Urucurituba** Brazil
152 C3 **Uruguaiana** Brazil
153 C4 **Uruguay** *country* S. America
Urumchi China *see* **Ürümqi**
68 B2 **Ürümqi** China
53 D2 **Urunga** Austr.
110 C2 **Urziceni** Romania
67 B4 **Usa** Japan
82 E2 **Usa** *r.* Russia
111 C3 **Uşak** Turkey
82 G1 **Ushakova, Ostrov** *i.* Russia
77 E2 **Usharal** Kazakh.
77 D2 **Ushtobe** Kazakh.
153 B6 **Ushuaia** Arg.
86 E2 **Usinsk** Russia
88 C3 **Uskhodni** Belarus
89 E3 **Usman'** Russia
86 D2 **Usogorsk** Russia
104 C2 **Ussel** France
66 B2 **Ussuriysk** Russia
108 B3 **Ustica, Isola di** *i. Sicily* Italy
83 H3 **Ust'-Ilimsk** Russia
86 E2 **Ust'-Ilych** Russia
103 D1 **Ustka** Pol.
83 L3 **Ust'-Kamchatsk** Russia
77 E2 **Ust'-Kamenogorsk** Kazakh.
86 F2 **Ust'-Kara** Russia
86 E2 **Ust'-Kulom** Russia
83 I3 **Ust'-Kut** Russia
91 D2 **Ust'-Labinsk** Russia
88 C2 **Ust'-Luga** Russia
86 E2 **Ust'-Nem** Russia
83 K2 **Ust'-Nera** Russia
83 K2 **Ust'-Omchug** Russia
83 H3 **Ust'-Ordynskiy** Russia
86 E2 **Ust'-Tsil'ma** Russia
86 D2 **Ust'-Ura** Russia
76 B2 **Ustyurt Plateau** Kazakh./Uzbek.
89 E2 **Ustyuzhna** Russia
89 D2 **Usvyaty** Russia
135 D3 **Utah** *state* U.S.A.
135 D2 **Utah Lake** U.S.A.
93 I4 **Utena** Lith.
123 D2 **uThukela** *r.* S. Africa
141 D2 **Utica** U.S.A.
107 C2 **Utiel** Spain
128 C2 **Utikuma Lake** Can.
100 B1 **Utrecht** Neth.
106 B2 **Utrera** Spain
92 I2 **Utsjoki** Fin.
67 C3 **Utsunomiya** Japan
87 D4 **Utta** Russia
62 B2 **Uttaradit** Thai.
127 H2 **Uummannaq Fjord** *inlet* Greenland
93 H3 **Uusikaupunki** Fin.
139 D3 **Uvalde** U.S.A.
119 D3 **Uvinza** Tanz.
68 C1 **Uvs Nuur** *salt l.* Mongolia
67 B4 **Uwajima** Japan
78 A2 **'Uwayriḍ, Ḥarrat al** *lava field* Saudi Arabia
116 A2 **Uweinat, Jebel** *mt.* Sudan
83 H3 **Uyar** Russia
115 C4 **Uyo** Nigeria
152 B3 **Uyuni, Salar de** *salt flat* Bol.
76 C2 **Uzbekistan** *country* Asia
104 C2 **Uzerche** France
105 C3 **Uzès** France
90 C1 **Uzh** *r.* Ukr.
90 A2 **Uzhhorod** Ukr.
109 C2 **Užice** Serbia
89 E3 **Uzlovaya** Russia
111 C3 **Üzümlü** Turkey
111 C2 **Uzunköprü** Turkey

V

123 B2 **Vaal** *r.* S. Africa
123 C2 **Vaal Dam** S. Africa
123 C1 **Vaalwater** S. Africa
92 H3 **Vaasa** Fin.
103 D2 **Vác** Hungary
152 C3 **Vacaria** Brazil
154 B2 **Vacaria, Serra** *hills* Brazil
135 B3 **Vacaville** U.S.A.
74 B2 **Vadodara** India
92 I1 **Vadsø** Norway
105 D2 **Vaduz** Liechtenstein
94 B1 **Vágar** *i.* Faroe Is
94 B1 **Vágur** Faroe Is
103 D2 **Váh** *r.* Slovakia
49 F3 **Vaiaku** Tuvalu
88 B2 **Vaida** Estonia

136 B3 **Vail** U.S.A.
79 C2 **Vakīlābād** Iran
108 B1 **Valdagno** Italy
89 D2 **Valday** Russia
89 D2 **Valdayskaya Vozvyshennost'** *hills* Russia
106 B2 **Valdecañas, Embalse de** *resr* Spain
93 G4 **Valdemarsvik** Sweden
106 C2 **Valdepeñas** Spain
153 B5 **Valdés, Península** *pen.* Arg.
153 A4 **Valdivia** Chile
130 C2 **Val-d'Or** Can.
143 D2 **Valdosta** U.S.A.
128 C2 **Valemount** Can.
105 C3 **Valence** France
107 C2 **Valencia** Spain
150 B1 **Valencia** Venez.
107 D2 **Valencia, Golfo de** *g.* Spain
100 A2 **Valenciennes** France
136 C2 **Valentine** U.S.A.
64 B1 **Valenzuela** Phil.
150 A1 **Valera** Venez.
109 C2 **Valjevo** Serbia
88 C2 **Valka** Latvia
93 H3 **Valkeakoski** Fin.
100 B2 **Valkenswaard** Neth.
91 D2 **Valky** Ukr.
55 D2 **Valkyrie Dome** Antarctica
145 D2 **Valladolid** Mex.
106 C1 **Valladolid** Spain
93 E4 **Valle** Norway
145 C2 **Vallecillos** Mex.
150 B1 **Valle de la Pascua** Venez.
150 A1 **Valledupar** Col.
145 C2 **Valle Hermoso** Mex.
135 B3 **Vallejo** U.S.A.
152 A3 **Vallenar** Chile
85 H5 **Valletta** Malta
137 D1 **Valley City** U.S.A.
134 B2 **Valley Falls** U.S.A.
128 C2 **Valleyview** Can.
107 D1 **Valls** Spain
129 D3 **Val Marie** Can.
88 C2 **Valmiera** Latvia
88 C3 **Valozhyn** Belarus
154 B2 **Valparaíso** Brazil
153 A4 **Valparaíso** Chile
105 C3 **Valréas** France
59 D3 **Vals, Tanjung** *c.* Indon.
74 B2 **Valsad** India
122 B2 **Valspan** S. Africa
91 D1 **Valuyki** Russia
106 B2 **Valverde del Camino** Spain
81 C2 **Van** Turkey
81 C2 **Van, Lake** *salt l.* Turkey
141 F1 **Van Buren** U.S.A.
128 B3 **Vancouver** Can.
134 B1 **Vancouver** U.S.A.
128 B3 **Vancouver Island** Can.
140 B3 **Vandalia** U.S.A.
123 C2 **Vanderbijlpark** S. Africa
128 B2 **Vanderhoof** Can.
88 C2 **Vändra** Estonia
93 F4 **Vänern** *l.* Sweden
93 F4 **Vänersborg** Sweden
121 □D3 **Vangaindrano** Madag.
Van Gölü *salt l.* Turkey *see* **Van, Lake**
138 C2 **Van Horn** U.S.A.
59 D3 **Vanimo** P.N.G.
83 K3 **Vanino** Russia
104 B2 **Vannes** France
59 D3 **Van Rees, Pegunungan** *mts* Indon.
122 A3 **Vanrhynsdorp** S. Africa
93 H3 **Vantaa** Fin.
49 F4 **Vanua Levu** *i.* Fiji
48 E3 **Vanuatu** *country* S. Pacific Ocean
140 C2 **Van Wert** U.S.A.
122 B3 **Van Wyksvlei** S. Africa
122 B2 **Van Zylsrus** S. Africa
75 C2 **Varanasi** India
92 I1 **Varangerfjorden** *sea chan.* Norway
92 I1 **Varangerhalvøya** *pen.* Norway
109 C1 **Varaždin** Croatia
93 F4 **Varberg** Sweden
109 D2 **Vardar** *r.* Macedonia
93 E4 **Varde** Denmark
92 J1 **Vardø** Norway
100 D1 **Varel** Ger.
88 B3 **Varna** Lith.
108 A1 **Varese** Italy
155 C2 **Varginha** Brazil
93 I3 **Varkaus** Fin.
110 C2 **Varna** Bulg.
93 F4 **Värnamo** Sweden
155 D1 **Várzea da Palma** Brazil
86 C2 **Varzino** Russia
Vasa Fin. *see* **Vaasa**
88 C2 **Vasknarva** Estonia
110 C1 **Vaslui** Romania
93 G4 **Västerås** Sweden
93 G3 **Västerdalälven** *r.* Sweden
88 A2 **Västerhaninge** Sweden
93 G4 **Västervik** Sweden
108 B2 **Vasto** Italy
90 C1 **Vasyl'kiv** Ukr.
104 C2 **Vatan** France
108 B2 **Vatican City** Europe
92 □B3 **Vatnajökull** Iceland
110 C1 **Vatra Dornei** Romania
93 F4 **Vättern** *l.* Sweden
138 B2 **Vaughn** U.S.A.
105 C3 **Vauvert** France
49 G4 **Vava'u Group** *is* Tonga
88 B3 **Vawkavysk** Belarus
93 F4 **Växjö** Sweden
86 E1 **Vaygach, Ostrov** *i.* Russia
101 D1 **Vechta** Ger.
110 C2 **Vedea** *r.* Romania
100 C1 **Veendam** Neth.
100 B1 **Veenendaal** Neth.
129 C2 **Vegreville** Can.
106 B2 **Vejer de la Frontera** Spain
93 E4 **Vejle** Denmark
109 D2 **Velbŭzhdki Prokhod** *pass* Bulg./Macedonia
100 B2 **Veldhoven** Neth.
109 B2 **Velebit** *mts* Croatia
100 C2 **Velen** Ger.
109 C1 **Velenje** Slovenia
109 D2 **Veles** Macedonia
106 C2 **Vélez-Málaga** Spain
155 D1 **Velhas** *r.* Brazil
109 D2 **Velika Plana** Serbia
88 C2 **Velikaya** *r.* Russia
89 D2 **Velikiye Luki** Russia
89 D2 **Velikiy Novgorod** Russia
86 D2 **Velikiy Ustyug** Russia
110 C2 **Veliko Tarnovo** Bulg.
108 B2 **Veli Lošinj** Croatia
89 D2 **Velizh** Russia
86 D2 **Vel'sk** Russia
101 F1 **Velten** Ger.
91 D1 **Velykyy Burluk** Ukr.
108 B2 **Venafro** Italy
104 C2 **Vendôme** France
89 E3 **Venev** Russia
Venezia Italy *see* **Venice**
150 B1 **Venezuela** *country* S. America
150 A1 **Venezuela, Golfo de** *g.* Venez.
108 B1 **Venice** Italy
143 D3 **Venice** U.S.A.
108 B1 **Venice, Gulf of** Europe
100 C2 **Venlo** Neth.
100 B2 **Venray** Neth.
88 B2 **Venta** *r.* Latvia/Lith.
88 B2 **Venta** Lith.
123 C3 **Venterstad** S. Africa
99 C3 **Ventnor** U.K.
88 B2 **Ventspils** Latvia
135 C4 **Ventura** U.S.A.
139 C3 **Venustiano Carranza, Presa** *resr* Mex.
107 C2 **Vera** Spain
145 C3 **Veracruz** Mex.
74 B2 **Veraval** India
108 A1 **Verbania** Italy
108 A1 **Vercelli** Italy
105 D3 **Vercors** *reg.* France
92 F3 **Verdalsøra** Norway
154 B1 **Verde** *r. Goiás* Brazil
154 B2 **Verde** *r. Mato Grosso do Sul* Brazil
144 B2 **Verde** *r.* Mex.
138 A2 **Verde** *r.* U.S.A.
155 D1 **Verde Grande** *r.* Brazil
101 D1 **Verden (Aller)** Ger.
154 B1 **Verdinho, Serra do** *mts* Brazil
108 A2 **Verdon** *r.* France
105 D2 **Verdun** France
123 C2 **Vereeniging** S. Africa
106 B1 **Verín** Spain
91 D3 **Verkhnebakanskiy** Russia
92 J2 **Verkhnetulomskiy** Russia
91 E1 **Verkhniy Mamon** Russia
89 E3 **Verkhov'ye** Russia
90 A2 **Verkhovyna** Ukr.
83 J2 **Verkhoyanskiy Khrebet** *mts* Russia
129 C2 **Vermilion** Can.
137 D2 **Vermillion** U.S.A.
130 A2 **Vermillion Bay** Can.
141 E2 **Vermont** *state* U.S.A.
135 E2 **Vernal** U.S.A.
128 C2 **Vernon** Can.
139 D2 **Vernon** U.S.A.
143 D3 **Vero Beach** U.S.A.
111 B2 **Veroia** Greece
108 B1 **Verona** Italy
104 C2 **Versailles** France
104 B2 **Vertou** France
123 D2 **Verulam** S. Africa
100 B2 **Verviers** Belgium
105 C2 **Vervins** France
105 D3 **Vescovato** *Corsica* France
87 E3 **Veselaya, Gora** *mt.* Russia
91 C2 **Vesele** Ukr.
105 D2 **Vesoul** France
92 F2 **Vesterålen** *is* Norway
92 F2 **Vestfjorden** *sea chan.* Norway
94 B1 **Vestmanna** Faroe Is
93 E3 **Vestnes** Norway
108 B2 **Vesuvius** *vol.* Italy
89 E2 **Ves'yegonsk** Russia
93 G4 **Vetlanda** Sweden
86 D3 **Vetluga** Russia
100 A2 **Veurne** Belgium
105 D2 **Vevey** Switz.
91 D1 **Veydelevka** Russia
80 B1 **Vezirköprü** Turkey
151 D2 **Viana** Brazil
106 B1 **Viana do Castelo** Port.
Viangchan Laos *see* **Vientiane**
62 B1 **Viangphoukha** Laos
111 C3 **Viannos** Greece
154 C1 **Vianópolis** Brazil
108 B2 **Viareggio** Italy
93 E4 **Viborg** Denmark
109 C3 **Vibo Valentia** Italy
107 D1 **Vic** Spain

144 A1	Vicente Guerrero Mex.	
108 B1	Vicenza Italy	
105 C2	Vichy France	
142 B2	Vicksburg U.S.A.	
155 D3	Viçosa Brazil	
52 A3	Victor Harbor Austr.	
50 C1	Victoria r. Austr.	
52 B3	Victoria state Austr.	
128 B3	Victoria Can.	
153 A4	Victoria Chile	
113 K7	Victoria Seychelles	
139 D3	Victoria U.S.A.	
119 D3	Victoria, Lake Africa	
52 B2	Victoria, Lake Austr.	
62 A1	Victoria, Mount Myanmar	
59 D3	Victoria, Mount P.N.G.	
120 B2	Victoria Falls Zambia/Zimbabwe	
126 D2	Victoria Island Can.	
50 C1	Victoria River Downs Austr.	
122 B3	Victoria West S. Africa	
135 C4	Victorville U.S.A.	
110 C2	Videle Romania	
92 ⬜A2	Viðidalsá Iceland	
153 B5	Viedma Arg.	
153 A5	Viedma, Lago l. Arg.	
100 B2	Vielsalm Belgium	
101 E2	Vienenburg Ger.	
103 D2	Vienna Austria	
105 C2	Vienne France	
104 C2	Vienne r. France	
62 B2	Vientiane Laos	
100 C2	Viersen Ger.	
104 C2	Vierzon France	
144 B2	Viesca Mex.	
109 C2	Vieste Italy	
62 B2	Vietnam country Asia	
62 B1	Việt Trì Vietnam	
64 B1	Vigan Phil.	
108 A1	Vigevano Italy	
106 B1	Vigo Spain	
73 C3	Vijayawada India	
92 ⬜B3	Vík Iceland	
129 C2	Viking Can.	
150 C3	Vila Bela da Santíssima Trindade Brazil	
106 B2	Vila Franca de Xira Port.	
106 B1	Vilagarcía de Arousa Spain	
106 B1	Vilalba Spain	
106 B1	Vila Nova de Gaia Port.	
107 D1	Vilanova i la Geltrú Spain	
106 B1	Vila Real Port.	
106 B1	Vilar Formoso Port.	
155 D2	Vila Velha Brazil	
150 A3	Vilcabamba, Cordillera mts Peru	
92 G3	Vilhelmina Sweden	
150 B3	Vilhena Brazil	
88 C2	Viljandi Estonia	
88 B3	Vilkaviškis Lith.	
83 H1	Vil'kitskogo, Proliv str. Russia	
144 B1	Villa Ahumada Mex.	
106 B1	Villablino Spain	
102 C2	Villach Austria	
144 B2	Villa de Cos Mex.	
152 B4	Villa Dolores Arg.	
145 C3	Villa Flores Mex.	
145 C2	Villagrán Mex.	
145 C3	Villahermosa Mex.	
144 A2	Villa Insurgentes Mex.	
152 B4	Villa María Arg.	
152 B3	Villa Montes Bol.	
144 B2	Villanueva Mex.	
106 B2	Villanueva de la Serena Spain	
106 C2	Villanueva de los Infantes Spain	
152 C3	Villa Ocampo Arg.	
108 A3	Villaputzu Sardinia Italy	
152 C3	Villarrica Para.	
106 C2	Villarrobledo Spain	
152 B3	Villa Unión Arg.	
144 B2	Villa Unión Durango Mex.	
144 B2	Villa Unión Sinaloa Mex.	
150 A1	Villavicencio Col.	
152 B3	Villazon Bol.	
107 C2	Villena Spain	
104 C3	Villeneuve-sur-Lot France	
142 B2	Ville Platte U.S.A.	
105 C2	Villeurbanne France	
102 B2	Villingen Ger.	
88 C3	Vilnius Lith.	
91 C2	Vil'nohirs'k Ukr.	
91 D2	Vil'nyans'k Ukr.	
100 B2	Vilvoorde Belgium	
88 C3	Vilyeyka Belarus	
83 J2	Vilyuy r. Russia	
93 G4	Vimmerby Sweden	
153 A4	Viña del Mar Chile	
107 D1	Vinaròs Spain	
140 B3	Vincennes U.S.A.	
55 S3	Vincennes Bay Antarctica	
141 D3	Vineland U.S.A.	
62 B2	Vinh Vietnam	
63 B2	Vinh Long Vietnam	
139 D1	Vinita U.S.A.	
90 B2	Vinnytsya Ukr.	
55 O2	Vinson, Mount mt. Antarctica	
93 E3	Vinstra Norway	
108 B1	Vipiteno Italy	
64 B1	Virac Phil.	
129 D3	Virden Can.	
104 B2	Vire France	
120 A2	Virei Angola	
155 D1	Virgem da Lapa Brazil	
138 A1	Virgin r. U.S.A.	
123 C2	Virginia S. Africa	
137 E1	Virginia U.S.A.	
141 D3	Virginia state U.S.A.	
141 D3	Virginia Beach U.S.A.	
135 C3	Virginia City U.S.A.	
147 D3	Virgin Islands (U.K.) terr. West Indies	
147 D3	Virgin Islands (U.S.A.) terr. West Indies	
63 B2	Virôchey Cambodia	
109 C1	Virovitica Croatia	
100 B3	Virton Belgium	
88 B2	Virtsu Estonia	
109 C2	Vis i. Croatia	
88 C2	Visaginas Lith.	
135 C3	Visalia U.S.A.	
74 B2	Visavadar India	
64 B1	Visayan Sea Phil.	
93 G4	Visby Sweden	
126 D2	Viscount Melville Sound sea chan. Can.	
151 D2	Viseu Brazil	
106 B1	Viseu Port.	
73 C3	Vishakhapatnam India	
88 C2	Viški Latvia	
109 C2	Visoko Bos. & Herz.	
103 D1	Vistula r. Pol.	
108 B2	Viterbo Italy	
49 F4	Viti Levu i. Fiji	
83 I3	Vitim r. Russia	
155 D2	Vitória Brazil	
151 D3	Vitória da Conquista Brazil	
106 C1	Vitoria-Gasteiz Spain	
104 B2	Vitré France	
105 C2	Vitry-le-François France	
89 D2	Vitsyebsk Belarus	
105 D2	Vittel France	
108 B3	Vittoria Sicily Italy	
108 B1	Vittorio Veneto Italy	
106 B1	Viveiro Spain	
144 A2	Vizcaíno, Sierra mts Mex.	
110 C2	Vize Turkey	
73 C3	Vizianagaram India	
100 B2	Vlaardingen Neth.	
87 D4	Vladikavkaz Russia	
89 F2	Vladimir Russia	
66 B2	Vladivostok Russia	
109 D2	Vlasotince Serbia	
100 B1	Vlieland i. Neth.	
100 A2	Vlissingen Neth.	
109 C2	Vlorë Albania	
102 C2	Vöcklabruck Austria	
101 D2	Vogelsberg hills Ger.	
119 D3	Voi Kenya	
105 D2	Voiron France	
131 D1	Voisey's Bay Can.	
109 C1	Vojvodina prov. Serbia	
92 J3	Voknavolok Russia	
	Volcano Bay Japan see Uchiura-wan	
89 E2	Volga Russia	
89 F2	Volga r. Russia	
87 D4	Volgodonsk Russia	
87 D4	Volgograd Russia	
89 D2	Volkhov Russia	
89 D1	Volkhov r. Russia	
101 E2	Volkstedt Ger.	
91 D2	Volnovakha Ukr.	
90 B2	Volochys'k Ukr.	
91 D2	Volodars'ke Ukr.	
90 B1	Volodars'k-Volyns'kyy Ukr.	
90 B1	Volodymyrets' Ukr.	
90 A1	Volodymyr-Volyns'kyy Ukr.	
89 E2	Vologda Russia	
89 E2	Volokolamsk Russia	
91 D1	Volokonovka Russia	
111 B3	Volos Greece	
88 C2	Volosovo Russia	
89 D2	Volot Russia	
89 E3	Volovo Russia	
87 D3	Vol'sk Russia	
114 C4	Volta, Lake resr Ghana	
155 D2	Volta Redonda Brazil	
87 D4	Volzhskiy Russia	
92 ⬜C2	Vopnafjörður Iceland	
88 C3	Voranava Belarus	
86 F2	Vorkuta Russia	
88 B2	Vormsi i. Estonia	
89 E3	Voronezh Russia	
91 E1	Voronezh r. Russia	
89 E1	Vorontsovka Russia	
91 C2	Vorskla r. Russia	
88 C2	Võrtsjärv l. Estonia	
88 C2	Võru Estonia	
122 B3	Vosburg S. Africa	
105 D2	Vosges mts France	
89 E2	Voskresensk Russia	
93 E3	Voss Norway	
83 H3	Vostochnyy Sayan mts Russia	
86 F3	Vostok Russia	
154 C2	Votuporanga Brazil	
100 B3	Vouziers France	
92 J2	Voynitsa Russia	
91 C2	Voznesens'k Ukr.	
93 E4	Vrådal Norway	
66 B2	Vrangel' Russia	
	Vrangelya, Ostrov i. Russia see Wrangel Island	
109 D2	Vranje Serbia	
110 C2	Vratnik pass Bulg.	
110 B2	Vratsa Bulg.	
109 C1	Vrbas r. Bos. & Herz.	
109 C1	Vrbas Serbia	
122 A3	Vredenburg S. Africa	
122 A3	Vredendal S. Africa	
100 B3	Vresse-sur-Semois Belgium	
109 D1	Vršac Serbia	
122 B3	Vryburg S. Africa	
123 D2	Vryheid S. Africa	

89 D1 **Vsevolozhsk** Russia
Vučitrn Kosovo *see* **Vushtrri**
109 C1 **Vukovar** Croatia
86 E2 **Vuktyl** Russia
123 C2 **Vukuzakhe** S. Africa
109 B3 **Vulcano, Isola** *i.* Italy
63 B2 **Vung Tau** Vietnam
109 D2 **Vushtrri** Kosovo
74 B2 **Vyara** India
Vyatka Russia *see* **Kirov**
89 D2 **Vyaz'ma** Russia
93 I3 **Vyborg** Russia
88 C2 **Vyerkhnyadzvinsk** Belarus
87 D3 **Vyksa** Russia
90 B2 **Vylkove** Ukr.
90 A2 **Vynohradiv** Ukr.
89 D2 **Vypolzovo** Russia
89 D2 **Vyritsa** Russia
91 D2 **Vyselki** Russia
90 C1 **Vyshhorod** Ukr.
89 D2 **Vyshniy-Volochek** Russia
103 D2 **Vyškov** Czech Rep.
86 C2 **Vytegra** Russia

W

114 B3 **Wa** Ghana
100 B2 **Waal** *r.* Neth.
100 B2 **Waalwijk** Neth.
128 C2 **Wabasca** *r.* Can.
128 C2 **Wabasca-Desmarais** Can.
140 B3 **Wabash** *r.* U.S.A.
129 E2 **Wabowden** Can.
143 D3 **Waccasassa Bay** U.S.A.
101 D2 **Wächtersbach** Ger.
139 D2 **Waco** U.S.A.
115 D2 **Waddān** Libya
Waddeneilanden *is* Neth. *see* **West Frisian Islands**
100 B1 **Waddenzee** *sea chan.* Neth.
128 B2 **Waddington, Mount** Can.
100 B1 **Waddinxveen** Neth.
129 D2 **Wadena** Can.
137 D1 **Wadena** U.S.A.
74 A2 **Wadh** Pak.
116 B2 **Wadi Halfa** Sudan
116 B3 **Wad Madani** Sudan
70 C2 **Wafangdian** China
100 B2 **Wageningen** Neth.
127 F2 **Wager Bay** Can.
53 C3 **Wagga Wagga** Austr.
137 D2 **Wahoo** U.S.A.
137 D1 **Wahpeton** U.S.A.
54 B2 **Waiau** *r.* N.Z.
59 C3 **Waigeo** *i.* Indon.
58 B3 **Waikabubak** Indon.
52 A2 **Waikerie** Austr.
54 B2 **Waimate** N.Z.
75 B3 **Wainganga** *r.* India
58 C3 **Waingapu** Indon.
129 C2 **Wainwright** Can.
54 C1 **Waiouru** N.Z.
54 B2 **Waipara** N.Z.
54 C1 **Waipawa** N.Z.
54 B2 **Wairau** *r.* N.Z.
54 C1 **Wairoa** N.Z.
54 B2 **Waitaki** *r.* N.Z.
54 B1 **Waitara** N.Z.
54 B1 **Waiuku** N.Z.
119 E2 **Wajir** Kenya
67 C3 **Wakasa-wan** *b.* Japan
54 A3 **Wakatipu, Lake** N.Z.
129 D2 **Wakaw** Can.
67 C4 **Wakayama** Japan
136 D3 **WaKeeney** U.S.A.
54 B2 **Wakefield** N.Z.
98 C2 **Wakefield** U.K.
48 F1 **Wake Island** N. Pacific Ocean
66 D1 **Wakkanai** Japan
123 D2 **Wakkerstroom** S. Africa
103 D1 **Wałbrzych** Pol.
53 D2 **Walcha** Austr.
100 C1 **Walchum** Ger.
103 D1 **Wałcz** Pol.
99 B2 **Wales** *admin. div.* U.K.
53 C2 **Walgett** Austr.
119 C3 **Walikale** Dem. Rep. Congo
135 C3 **Walker Lake** U.S.A.
52 A2 **Wallaroo** Austr.
98 B2 **Wallasey** U.K.
134 C1 **Walla Walla** U.S.A.
101 D3 **Walldürn** Ger.
53 C2 **Wallendbeen** Austr.
49 G3 **Wallis and Futuna Islands** *terr.* S. Pacific Ocean
96 □ **Walls** U.K.
98 B1 **Walney, Isle of** *i.* U.K.
99 C2 **Walsall** U.K.
136 C3 **Walsenburg** U.S.A.
101 D1 **Walsrode** Ger.
143 D2 **Walterboro** U.S.A.
120 A3 **Walvis Bay** Namibia
119 C2 **Wamba** Dem. Rep. Congo
52 B1 **Wanaaring** Austr.
54 A2 **Wanaka** N.Z.
54 A2 **Wanaka, Lake** N.Z.
130 B2 **Wanapitei Lake** Can.
154 B3 **Wanda** Arg.
66 B1 **Wanda Shan** *mts* China
62 A1 **Wanding** China
54 C1 **Wanganui** N.Z.
53 C3 **Wangaratta** Austr.
65 B1 **Wangqing** China
62 A1 **Wan Hsa-la** Myanmar
71 B4 **Wanning** China
100 B2 **Wanroij** Neth.
99 C3 **Wantage** U.K.
70 A2 **Wanyuan** China
70 A2 **Wanzhou** China
73 B3 **Warangal** India
101 D2 **Warburg** Ger.
50 B2 **Warburton** Austr.
52 A1 **Warburton** *watercourse* Austr.
75 B2 **Wardha** India
96 C1 **Ward Hill** U.K.
128 B2 **Ware** Can.
101 F1 **Waren (Müritz)** Ger.
100 C2 **Warendorf** Ger.
53 D1 **Warialda** Austr.
122 A2 **Warmbad** Namibia
135 C3 **Warm Springs** U.S.A.
134 C2 **Warner Lakes** U.S.A.
143 D2 **Warner Robins** U.S.A.
152 B2 **Warnes** Bol.
52 B3 **Warracknabeal** Austr.
53 C2 **Warrego** *r.* Austr.
53 C2 **Warren** Austr.
142 B2 **Warren** *AR* U.S.A.
140 C2 **Warren** *OH* U.S.A.
141 D2 **Warren** *PA* U.S.A.
97 C1 **Warrenpoint** U.K.
137 E3 **Warrensburg** U.S.A.
122 B2 **Warrenton** S. Africa
115 C4 **Warri** Nigeria
98 B2 **Warrington** U.K.
52 B3 **Warrnambool** Austr.
103 E1 **Warsaw** Pol.
Warszawa Pol. *see* **Warsaw**
103 C1 **Warta** *r.* Pol.
53 D1 **Warwick** Austr.
99 C2 **Warwick** U.K.
135 C3 **Wasco** U.S.A.
136 C1 **Washburn** U.S.A.
141 D3 **Washington** *DC* U.S.A.
137 E2 **Washington** *IA* U.S.A.
140 B3 **Washington** *IN* U.S.A.
137 E3 **Washington** *MO* U.S.A.
143 E1 **Washington** *NC* U.S.A.
140 C2 **Washington** *PA* U.S.A.
134 B1 **Washington** *state* U.S.A.
141 E2 **Washington, Mount** U.S.A.
140 C3 **Washington Court House** U.S.A.
74 A2 **Washuk** Pak.
130 C1 **Waskaganish** Can.
129 E2 **Waskaiowaka Lake** Can.
122 A2 **Wasser** Namibia
130 C2 **Waswanipi, Lac** *l.* Can.
58 C3 **Watampone** Indon.
141 E2 **Waterbury** U.S.A.
129 D2 **Waterbury Lake** Can.
97 C2 **Waterford** Ireland
97 C2 **Waterford Harbour** Ireland
137 E2 **Waterloo** U.S.A.
123 C1 **Waterpoort** S. Africa
141 D2 **Watertown** *NY* U.S.A.
137 D2 **Watertown** *SD* U.S.A.
140 B2 **Watertown** *WI* U.S.A.
141 F2 **Waterville** U.S.A.
99 C3 **Watford** U.K.
136 C1 **Watford City** U.S.A.
129 D2 **Wathaman** *r.* Can.
139 D1 **Watonga** U.S.A.
129 D2 **Watrous** Can.
119 C2 **Watsa** Dem. Rep. Congo
140 B2 **Watseka** U.S.A.
118 C3 **Watsi Kengo** Dem. Rep. Congo
128 B1 **Watson Lake** Can.
135 B3 **Watsonville** U.S.A.
59 C3 **Watubela, Kepulauan** *is* Indon.
59 D3 **Wau** P.N.G.
117 A4 **Wau** South Sudan
53 D2 **Wauchope** Austr.
140 B2 **Waukegan** U.S.A.
139 D2 **Waurika** U.S.A.
140 B2 **Wausau** U.S.A.
99 D2 **Waveney** *r.* U.K.
137 E2 **Waverly** U.S.A.
143 D2 **Waycross** U.S.A.
137 D2 **Wayne** U.S.A.
143 D2 **Waynesboro** *GA* U.S.A.
141 D3 **Waynesboro** *VA* U.S.A.
143 D1 **Waynesville** U.S.A.
74 B1 **Wazirabad** Pak.
98 C1 **Wear** *r.* U.K.
139 D2 **Weatherford** U.S.A.
134 B2 **Weaverville** U.S.A.
130 B1 **Webequie** Can.
117 C4 **Webi Shabeelle** *r.* Somalia
137 D1 **Webster** U.S.A.
137 E2 **Webster City** U.S.A.
55 Q2 **Weddell Sea** Antarctica
100 B2 **Weert** Neth.
53 C2 **Wee Waa** Austr.
100 C2 **Wegberg** Ger.
103 E1 **Węgorzewo** Pol.
101 F3 **Weiden in der Oberpfalz** Ger.
70 B2 **Weifang** China
70 C2 **Weihai** China
53 C1 **Weilmoringle** Austr.
101 E2 **Weimar** Ger.
70 A2 **Weinan** China
51 D1 **Weipa** Austr.
53 C1 **Weir** *r.* Austr.
140 C2 **Weirton** U.S.A.
62 B1 **Weishan** China
101 E2 **Weißenfels** Ger.
102 C2 **Weißkugel** *mt.* Austria/Italy
103 D1 **Wejherowo** Pol.
128 C1 **Wekweètì** Can.

140 C3 **Welch** U.S.A.
117 B3 **Weldiya** Eth.
123 C2 **Welkom** S. Africa
99 C2 **Welland** *r.* U.K.
51 C1 **Wellesley Islands** Austr.
53 C2 **Wellington** Austr.
54 B2 **Wellington** N.Z.
122 A3 **Wellington** S. Africa
137 D3 **Wellington** U.S.A.
153 A5 **Wellington, Isla** *i.* Chile
53 C3 **Wellington, Lake** Austr.
128 B2 **Wells** Can.
99 B3 **Wells** U.K.
134 D2 **Wells** U.S.A.
50 B2 **Wells, Lake** *imp. l.* Austr.
54 B1 **Wellsford** N.Z.
98 D2 **Wells-next-the-Sea** U.K.
102 C2 **Wels** Austria
99 B2 **Welshpool** U.K.
123 C2 **Wembesi** S. Africa
130 C1 **Wemindji** Can.
134 B1 **Wenatchee** U.S.A.
154 C2 **Wenceslau Braz** Brazil
71 B4 **Wencheng** China
114 B4 **Wenchi** Ghana
70 C2 **Wendeng** China
101 E1 **Wendisch Evern** Ger.
117 B4 **Wendo** Eth.
134 D2 **Wendover** U.S.A.
71 B3 **Wengyuan** China
71 C3 **Wenling** China
71 A3 **Wenshan** China
52 B2 **Wentworth** Austr.
71 C3 **Wenzhou** China
123 C2 **Wepener** S. Africa
122 B2 **Werda** Botswana
101 F2 **Werdau** Ger.
101 F1 **Werder (Havel)** Ger.
101 F3 **Wernberg-Köblitz** Ger.
101 E2 **Wernigerode** Ger.
101 D2 **Werra** *r.* Ger.
53 D2 **Werris Creek** Austr.
101 D3 **Wertheim** Ger.
100 C2 **Wesel** Ger.
101 E1 **Wesendorf** Ger.
101 D1 **Weser** *r.* Ger.
101 D1 **Weser** *sea chan.* Ger.
51 C1 **Wessel, Cape** Austr.
51 C1 **Wessel Islands** Austr.
123 C2 **Wesselton** S. Africa
80 B2 **West Bank** *disp. terr.* Asia
140 B2 **West Bend** U.S.A.
99 C2 **West Bromwich** U.K.
100 C2 **Westerburg** Ger.
100 C1 **Westerholt** Ger.
50 B2 **Western Australia** *state* Austr.
122 B3 **Western Cape** *prov.* S. Africa
116 A2 **Western Desert** Egypt
73 B3 **Western Ghats** *mts* India
114 A2 **Western Sahara** *disp. terr.* Africa
100 A2 **Westerschelde** *est.* Neth.
100 C1 **Westerstede** Ger.
100 C2 **Westerwald** *hills* Ger.
153 B6 **West Falkland** *i.* Falkland Is
140 B3 **West Frankfort** U.S.A.
100 B1 **West Frisian Islands** *is* Neth.
147 D2 **West Indies** *is* Caribbean Sea
96 A1 **West Loch Roag** *b.* U.K.
128 C2 **Westlock** Can.
100 B2 **Westmalle** Belgium
142 B1 **West Memphis** U.S.A.
140 C3 **Weston** U.S.A.
99 B3 **Weston-super-Mare** U.K.
143 D3 **West Palm Beach** U.S.A.
137 E3 **West Plains** U.S.A.
137 D2 **West Point** U.S.A.
97 B2 **Westport** Ireland
54 B2 **Westport** N.Z.
129 D2 **Westray** Can.
96 C1 **Westray** *i.* U.K.
100 B1 **West-Terschelling** Neth.
140 C3 **West Virginia** *state* U.S.A.
53 C2 **West Wyalong** Austr.
134 D2 **West Yellowstone** U.S.A.
59 C3 **Wetar** *i.* Indon.
128 C2 **Wetaskiwin** Can.
101 D2 **Wetzlar** Ger.
59 D3 **Wewak** P.N.G.
97 C2 **Wexford** Ireland
129 D2 **Weyakwin** Can.
129 D3 **Weyburn** Can.
101 D1 **Weyhe** Ger.
99 B3 **Weymouth** U.K.
54 C1 **Whakatane** N.Z.
129 E1 **Whale Cove** Can.
96 □ **Whalsay** *i.* U.K.
54 B1 **Whangarei** N.Z.
98 C2 **Wharfe** *r.* U.K.
139 D3 **Wharton** U.S.A.
128 C1 **Whatî** Can.
136 B2 **Wheatland** U.S.A.
138 B1 **Wheeler Peak** *NM* U.S.A.
135 D3 **Wheeler Peak** *NV* U.S.A.
140 C2 **Wheeling** U.S.A.
98 B1 **Whernside** *h.* U.K.
128 B2 **Whistler** Can.
98 C1 **Whitby** U.K.
142 B2 **White** *r.* U.S.A.
50 B2 **White, Lake** *imp. l.* Austr.
131 E2 **White Bay** Can.
136 C1 **White Butte** *mt.* U.S.A.
52 B2 **White Cliffs** Austr.
128 C2 **Whitecourt** Can.
134 D1 **Whitefish** U.S.A.
98 B1 **Whitehaven** U.K.
97 D1 **Whitehead** U.K.
128 A1 **Whitehorse** Can.
142 B3 **White Lake** U.S.A.
135 C3 **White Mountain Peak** U.S.A.
116 B3 **White Nile** *r.* South Sudan/Uganda
86 C2 **White Sea** Russia
134 D1 **White Sulphur Springs** U.S.A.
143 E2 **Whiteville** U.S.A.
114 B3 **White Volta** *r.* Burkina Faso/Ghana
138 B2 **Whitewater Baldy** *mt.* U.S.A.
130 B1 **Whitewater Lake** Can.
129 D2 **Whitewood** Can.
96 B3 **Whithorn** U.K.
54 C1 **Whitianga** N.Z.
135 C3 **Whitney, Mount** U.S.A.
51 D2 **Whitsunday Island** Austr.
52 A2 **Whyalla** Austr.
100 A2 **Wichelen** Belgium
137 D3 **Wichita** U.S.A.
139 D2 **Wichita Falls** U.S.A.
96 C1 **Wick** U.K.
138 A2 **Wickenburg** U.S.A.
97 C2 **Wicklow** Ireland
97 D2 **Wicklow Head** Ireland
97 C2 **Wicklow Mountains** Ireland
98 B2 **Widnes** U.K.
101 D1 **Wiehengebirge** *hills* Ger.
100 C2 **Wiehl** Ger.
103 D1 **Wieluń** Pol.
Wien Austria *see* **Vienna**
103 D2 **Wiener Neustadt** Austria
100 B1 **Wieringerwerf** Neth.
100 D2 **Wiesbaden** Ger.
100 C1 **Wiesmoor** Ger.
103 D1 **Wieżyca** *h.* Pol.
99 C3 **Wight, Isle of** *i.* U.K.
96 B3 **Wigtown** U.K.
100 B2 **Wijchen** Neth.
52 B2 **Wilcannia** Austr.
123 C3 **Wild Coast** S. Africa
123 C2 **Wilge** *r.* S. Africa
100 D1 **Wilhelmshaven** Ger.
141 D2 **Wilkes-Barre** U.S.A.
55 H3 **Wilkes Land** *reg.* Antarctica
129 D2 **Wilkie** Can.
138 B2 **Willcox** U.S.A.
100 B2 **Willebroek** Belgium
147 D3 **Willemstad** Curaçao
52 B3 **William, Mount** Austr.
52 A1 **William Creek** Austr.
138 A1 **Williams** U.S.A.
141 D3 **Williamsburg** U.S.A.
128 B2 **Williams Lake** Can.
140 C3 **Williamson** U.S.A.
141 D2 **Williamsport** U.S.A.
143 E1 **Williamston** U.S.A.
122 B3 **Williston** S. Africa
136 C1 **Williston** U.S.A.
128 B2 **Williston Lake** Can.
135 B3 **Willits** U.S.A.
137 D1 **Willmar** U.S.A.
122 B3 **Willowmore** S. Africa
123 C3 **Willowvale** S. Africa
50 B2 **Wills, Lake** *imp. l.* Austr.
52 A3 **Willunga** Austr.
52 A2 **Wilmington** Austr.
141 D3 **Wilmington** *DE* U.S.A.
143 E2 **Wilmington** *NC* U.S.A.
100 D2 **Wilnsdorf** Ger.
143 E1 **Wilson** U.S.A.
53 C3 **Wilson's Promontory** *pen.* Austr.
100 B3 **Wiltz** Lux.
50 B2 **Wiluna** Austr.
99 C3 **Winchester** U.K.
140 C3 **Winchester** *KY* U.S.A.
141 D3 **Winchester** *VA* U.S.A.
98 B1 **Windermere** *l.* U.K.
120 A3 **Windhoek** Namibia
137 D2 **Windom** U.S.A.
51 D2 **Windorah** Austr.
136 B2 **Wind River Range** *mts* U.S.A.
53 D2 **Windsor** Austr.
130 B2 **Windsor** Can.
147 D3 **Windward Islands** Caribbean Sea
147 C3 **Windward Passage** Cuba/Haiti
137 D3 **Winfield** U.S.A.
100 A2 **Wingene** Belgium
53 D2 **Wingham** Austr.
130 B1 **Winisk** *r.* Can.
130 B1 **Winisk (abandoned)** Can.
130 B1 **Winisk Lake** Can.
63 A2 **Winkana** Myanmar
129 E3 **Winkler** Can.
114 B4 **Winneba** Ghana
140 B2 **Winnebago, Lake** U.S.A.
134 C2 **Winnemucca** U.S.A.
136 C2 **Winner** U.S.A.
142 B2 **Winnfield** U.S.A.
137 E1 **Winnibigoshish, Lake** U.S.A.
129 E3 **Winnipeg** Can.
129 E2 **Winnipeg** *r.* Can.
129 E2 **Winnipeg, Lake** Can.
129 D2 **Winnipegosis, Lake** Can.
141 E2 **Winnipesaukee, Lake** U.S.A.
142 B2 **Winnsboro** U.S.A.
137 E2 **Winona** *MN* U.S.A.
142 C2 **Winona** *MS* U.S.A.
100 C1 **Winschoten** Neth.
101 D1 **Winsen (Aller)** Ger.
101 E1 **Winsen (Luhe)** Ger.
138 A1 **Winslow** U.S.A.
143 D1 **Winston-Salem** U.S.A.
101 D2 **Winterberg** Ger.
143 D3 **Winter Haven** U.S.A.

100 C2 **Winterswijk** Neth.
105 D2 **Winterthur** Switz.
51 D2 **Winton** Austr.
54 A3 **Winton** N.Z.
52 A2 **Wirrabara** Austr.
99 D2 **Wisbech** U.K.
140 A2 **Wisconsin** *r.* U.S.A.
140 B2 **Wisconsin** *state* U.S.A.
140 B2 **Wisconsin Rapids** U.S.A.
Wisła *r.* Pol. *see* **Vistula**
101 E1 **Wismar** Ger.
98 D2 **Withernsea** U.K.
100 B1 **Witmarsum** Neth.
99 C3 **Witney** U.K.
123 D2 **Witrivier** S. Africa
101 F2 **Wittenberg, Lutherstadt** Ger.
101 E1 **Wittenberge** Ger.
101 E1 **Wittenburg** Ger.
101 E1 **Wittingen** Ger.
100 C3 **Wittlich** Ger.
100 C1 **Wittmund** Ger.
101 F1 **Wittstock/Dosse** Ger.
120 A3 **Witvlei** Namibia
101 D2 **Witzenhausen** Ger.
103 D1 **Władysławowo** Pol.
103 D1 **Włocławek** Pol.
53 C3 **Wodonga** Austr.
59 C3 **Wokam** *i.* Indon.
99 C3 **Woking** U.K.
101 F2 **Wolfen** Ger.
101 E1 **Wolfenbüttel** Ger.
101 D2 **Wolfhagen** Ger.
136 B1 **Wolf Point** U.S.A.
101 E1 **Wolfsburg** Ger.
100 C3 **Wolfstein** Ger.
131 D2 **Wolfville** Can.
102 C1 **Wolgast** Ger.
102 C1 **Wolin** Pol.
129 D2 **Wollaston Lake** Can.
129 D2 **Wollaston Lake** Can.
126 D2 **Wollaston Peninsula** Can.
53 D2 **Wollongong** Austr.
101 E2 **Wolmirsleben** Ger.
101 E1 **Wolmirstedt** Ger.
100 C1 **Wolvega** Neth.
99 B2 **Wolverhampton** U.K.
65 B2 **Wonju** S. Korea
128 B2 **Wonowon** Can.
65 B2 **Wŏnsan** N. Korea
53 C3 **Wonthaggi** Austr.
52 A2 **Woocalla** Austr.
51 C1 **Woodah, Isle** *i.* Austr.
99 D2 **Woodbridge** U.K.
136 B3 **Woodland Park** U.S.A.
50 C2 **Woodroffe, Mount** Austr.
51 C1 **Woods, Lake** *imp. l.* Austr.
129 E3 **Woods, Lake of the** Can./U.S.A.
53 C3 **Woods Point** Austr.
131 D2 **Woodstock** *N.B.* Can.
140 C2 **Woodstock** *Ont.* Can.
54 C2 **Woodville** N.Z.
139 D1 **Woodward** U.S.A.
53 D2 **Woolgoolga** Austr.
52 A2 **Woomera** Austr.
140 C2 **Wooster** U.S.A.
122 A3 **Worcester** S. Africa
99 B2 **Worcester** U.K.
141 E2 **Worcester** U.S.A.
102 C2 **Wörgl** Austria
98 B1 **Workington** U.K.
98 C2 **Worksop** U.K.
136 B2 **Worland** U.S.A.
101 D3 **Worms** Ger.
99 C3 **Worthing** U.K.
137 D2 **Worthington** U.S.A.
58 C3 **Wotu** Indon.
59 C3 **Wowoni** *i.* Indon.
83 N2 **Wrangel Island** Russia
128 A2 **Wrangell** U.S.A.
96 B1 **Wrath, Cape** U.K.
136 C2 **Wray** U.S.A.
122 A2 **Wreck Point** S. Africa
98 B2 **Wrexham** U.K.
136 B2 **Wright** U.S.A.
63 A2 **Wrightmyo** India
128 B1 **Wrigley** Can.
103 D1 **Wrocław** Pol.
103 D1 **Września** Pol.
70 B2 **Wu'an** China
70 A2 **Wuhai** China
70 B2 **Wuhan** China
70 B2 **Wuhu** China
Wujin China *see* **Changzhou**
115 C4 **Wukari** Nigeria
62 B1 **Wuliang Shan** *mts* China
59 C3 **Wuliaru** *i.* Indon.
71 A3 **Wumeng Shan** *mts* China
130 B1 **Wunnummin Lake** Can.
101 F2 **Wunsiedel** Ger.
101 D1 **Wunstorf** Ger.
62 A1 **Wuntho** Myanmar
100 C2 **Wuppertal** Ger.
122 A3 **Wuppertal** S. Africa
101 E2 **Wurzbach** Ger.
101 D3 **Würzburg** Ger.
101 F2 **Wurzen** Ger.
68 C2 **Wuwei** China
70 A2 **Wuxi** *Chongqing* China
70 C2 **Wuxi** *Jiangsu* China
Wuxing China *see* **Huzhou**
71 A3 **Wuxuan** China
Wuyang China *see* **Zhenyuan**
69 E1 **Wuyiling** China
71 B3 **Wuyishan** China
71 B3 **Wuyi Shan** *mts* China
70 A1 **Wuyuan** China
71 A4 **Wuzhishan** China
70 A2 **Wuzhong** China
71 B3 **Wuzhou** China
53 C2 **Wyangala Reservoir** Austr.
52 B3 **Wycheproof** Austr.
99 B3 **Wye** *r.* U.K.
50 B1 **Wyndham** Austr.
142 B1 **Wynne** U.S.A.
129 D2 **Wynyard** Can.
136 B2 **Wyoming** *state* U.S.A.
103 E1 **Wyszków** Pol.
140 C3 **Wytheville** U.S.A.

X

117 D3 **Xaafuun** Somalia
62 B2 **Xaignabouli** Laos
121 C3 **Xai-Xai** Moz.
145 C3 **Xalapa** Mex.
70 A1 **Xamba** China
62 A1 **Xamgyi'nyilha** China
62 B1 **Xam Nua** Laos
120 A2 **Xangongo** Angola
81 C2 **Xankändi** Azer.
111 B2 **Xanthi** Greece
150 B3 **Xapuri** Brazil
107 C2 **Xàtiva** Spain
Xiaguan China *see* **Dali**
71 B3 **Xiamen** China
70 A2 **Xi'an** China
70 A3 **Xianfeng** China
Xiangfan China *see* **Xiangyang**
Xianggang *aut. reg.* China *see* **Hong Kong**
Xianggang China *see* **Hong Kong**
Xiangjiang China *see* **Huichang**
71 B3 **Xiangtan** China
70 B2 **Xiangyang** China
70 B2 **Xiantao** China
70 A2 **Xianyang** China
70 B2 **Xiaogan** China
69 E1 **Xiao Hinggan Ling** *mts* China
70 C2 **Xiaoshan** China
62 B1 **Xichang** China
145 C2 **Xicohténcatl** Mex.
71 A3 **Xifeng** China
75 C2 **Xigazê** China
83 I4 **Xilinhot** China
70 B2 **Xincai** China
68 C2 **Xinghai** China
70 B2 **Xinghua** China
71 B3 **Xingning** China
70 A2 **Xingping** China
70 B2 **Xingtai** China
151 C2 **Xingu** *r.* Brazil
71 A3 **Xingyi** China
71 B3 **Xinhua** China
68 C2 **Xining** China
Xinjiang Uygur Autonomous Region *aut. reg.* China *see* **Xinjiang Uygur Zizhiqu**
75 C1 **Xinjiang Uygur Zizhiqu** *aut. reg.* China
Xinjing China *see* **Jingxi**
69 D2 **Xinkou** China
65 A1 **Xinmin** China
71 B3 **Xinning** China
70 B2 **Xintai** China
70 B2 **Xinxiang** China
70 B2 **Xinyang** China
70 B2 **Xinyi** China
71 C3 **Xinying** Taiwan
71 B3 **Xinyu** China
70 B2 **Xinzhou** China
71 C3 **Xinzhu** Taiwan
106 B1 **Xinzo de Limia** Spain
Xiongshan China *see* **Zhenghe**
68 C2 **Xiqing Shan** *mts* China
151 D3 **Xique-Xique** Brazil
70 A1 **Xishanzui** China
71 A3 **Xiushan** China
71 B3 **Xiuying** China
70 B2 **Xixia** China
Xixón Spain *see* **Gijón/Xixón**
68 B2 **Xizang Zizhiqu** *aut. reg.* China
76 B2 **Xo'jayli** Uzbek.
70 B2 **Xuancheng** China
71 A3 **Xuanwei** China
70 B2 **Xuchang** China
Xucheng China *see* **Xuwen**
117 C4 **Xuddur** Somalia
Xujiang China *see* **Guangchang**
71 B3 **Xun Jiang** *r.* China
71 B3 **Xunwu** China
107 C2 **Xúquer, Riu** *r.* Spain
71 B3 **Xuwen** China
71 A3 **Xuyong** China
70 B2 **Xuzhou** China

Y

117 B4 **Yabēlo** Eth.
69 D1 **Yablonovyy Khrebet** *mts* Russia
143 D1 **Yadkin** *r.* U.S.A.
75 C2 **Yadong** China

Yagnitsa

89 E2 Yagnitsa Russia
118 B1 Yagoua Cameroon
128 C3 Yahk Can.
144 B2 Yahualica Mex.
80 B2 Yahyalı Turkey
67 C4 Yaizu Japan
134 B1 Yakima U.S.A.
114 B3 Yako Burkina Faso
66 D2 Yakumo Japan
126 B3 Yakutat U.S.A.
83 J2 Yakutsk Russia
91 D2 Yakymivka Ukr.
63 B3 Yala Thai.
53 C3 Yallourn Austr.
111 C2 Yalova Turkey
91 C3 Yalta Ukr.
65 A1 Yalu Jiang r. China/N. Korea
67 D3 Yamagata Japan
67 B4 Yamaguchi Japan
 Yamal, Poluostrov pen. Russia
 see Yamal Peninsula
86 F1 Yamal Peninsula Russia
53 D1 Yamba Austr.
117 A4 Yambio South Sudan
110 C2 Yambol Bulg.
86 G2 Yamburg Russia
62 A1 Yamethin Myanmar
114 B4 Yamoussoukro Côte d'Ivoire
90 B2 Yampil' Ukr.
75 C2 Yamuna r. India
75 D2 Yamzho Yumco l. China
83 K2 Yana r. Russia
70 A2 Yan'an China
150 A3 Yanaoca Peru
78 B4 Yanbu' al Baḥr Saudi Arabia
70 C2 Yancheng China
50 A3 Yanchep Austr.
70 B2 Yangcheng China
71 B3 Yangchun China
65 B2 Yangdok N. Korea
71 B3 Yangjiang China
 Yangôn Myanmar see
 Rangoon
70 B2 Yangquan China
71 B3 Yangshuo China
70 C2 Yangtze r. China
69 E2 Yangtze, Mouth of the China
70 B2 Yangzhou China
65 B1 Yanji China
137 D2 Yankton U.S.A.
83 K2 Yano-Indigirskaya
 Nizmennost' lowland Russia
83 K2 Yanskiy Zaliv g. Russia
53 C1 Yantabulla Austr.
70 C2 Yantai China
118 B2 Yaoundé Cameroon
59 D2 Yap i. Micronesia
59 D3 Yapen i. Indon.
59 D3 Yapen, Selat sea chan. Indon.
144 A2 Yaqui r. Mex.
51 D2 Yaraka Austr.
86 D3 Yaransk Russia
48 F2 Yaren Nauru
78 B3 Yarim Yemen
77 D3 Yarkant He r. China
75 D2 Yarlung Zangbo r. China
131 D2 Yarmouth Can.
86 F2 Yarono Russia
89 E2 Yaroslavl' Russia
66 B2 Yaroslavskiy Russia
53 C3 Yarram Austr.
89 D2 Yartsevo Russia
89 E3 Yasnogorsk Russia
63 B2 Yasothon Thai.
53 C2 Yass Russia
111 C3 Yatağan Turkey
129 E1 Yathkyed Lake Can.
67 B4 Yatsushiro Japan

150 B2 Yavarí r. Brazil/Peru
75 B2 Yavatmal India
90 A2 Yavoriv Ukr.
67 B4 Yawatahama Japan
81 D2 Yazd Iran
142 B2 Yazoo City U.S.A.
111 B3 Ydra i. Greece
63 A2 Ye Myanmar
77 D3 Yecheng China
144 B2 Yécora Mex.
89 E3 Yefremov Russia
89 E2 Yegor'yevsk Russia
117 B4 Yei South Sudan
86 F3 Yekaterinburg Russia
77 D1 Yekibastuz Kazakh.
89 E3 Yelets Russia
114 A3 Yélimané Mali
96 □ Yell i. U.K.
128 C1 Yellowknife Can.
70 B2 Yellow River r. China
69 E2 Yellow Sea N. Pacific Ocean
136 C1 Yellowstone r. U.S.A.
136 A2 Yellowstone Lake U.S.A.
88 C3 Yel'sk Belarus
78 B3 Yemen country Asia
86 E2 Yemva Russia
91 D2 Yenakiyeve Ukr.
62 A1 Yenangyaung Myanmar
62 B1 Yên Bai Vietnam
114 B4 Yendi Ghana
111 C3 Yenice Turkey
111 C3 Yenifoça Turkey
68 C1 Yenisey r. Russia
65 B2 Yeongdeok S. Korea
53 C2 Yeoval Austr.
99 B3 Yeovil U.K.
51 E2 Yeppoon Austr.
83 J2 Yerbogachen Russia
81 C1 Yerevan Armenia
77 D1 Yereymentau Kazakh.
87 D3 Yershov Russia
 Yertis r. Kazakh./Russia see
 Irtysh
65 B2 Yesan S. Korea
77 C1 Yesil' Kazakh.
77 D1 Yesil' r. Kazakh./Russia
111 C3 Yeşilova Turkey
99 A3 Yes Tor h. U.K.
53 D1 Yetman Austr.
62 A1 Ye-U Myanmar
104 B2 Yeu, Île d' i. France
91 C2 Yevpatoriya Ukr.
91 D2 Yeya r. Russia
91 D2 Yeysk Russia
88 C2 Yezyaryshcha Belarus
71 A3 Yibin China
70 B2 Yichang China
69 E1 Yichun Heilong. China
71 B3 Yichun Jiangxi China
110 C2 Yıldız Dağları mts Turkey
80 B2 Yıldızeli Turkey
70 A2 Yinchuan China
65 B1 Yingchengzi China
71 B3 Yingde China
70 C1 Yingkou China
70 B2 Yingshan Hubei China
70 A2 Yingshan Sichuan China
71 B3 Yingtan China
77 E2 Yining China
117 B4 Yirga Alem Eth.
70 B2 Yishui China
68 C2 Yiwu China
71 B3 Yiyang China
71 A3 Yizhou China
92 I2 Yli-Kitka l. Fin.
92 H2 Ylitornio Fin.
92 H3 Ylivieska Fin.
61 C2 Yogyakarta Indon.

118 B2 Yoko Cameroon
67 C3 Yokohama Japan
115 D4 Yola Nigeria
72 A3 Yonezawa Japan
71 B3 Yong'an China
71 B3 Yongchun China
70 A2 Yongdeng China
71 C3 Yongkang China
62 B1 Yongsheng China
71 B3 Yongzhou China
150 A1 Yopal Col.
50 A3 York Austr.
98 C2 York U.K.
137 D2 York NE U.S.A.
141 D3 York PA U.S.A.
51 D1 York, Cape Austr.
52 A3 Yorke Peninsula Austr.
52 A3 Yorketown Austr.
98 C2 Yorkshire Wolds hills U.K.
129 D2 Yorkton Can.
87 D3 Yoshkar-Ola Russia
97 C3 Youghal Ireland
53 C2 Young Austr.
52 A3 Younghusband Peninsula
 Austr.
140 C2 Youngstown U.S.A.
114 B3 Youvarou Mali
77 E2 Youyi Feng mt. China/Russia
80 B2 Yozgat Turkey
134 B2 Yreka U.S.A.
76 C2 Yrgyz Kazakh.
 Yr Wyddfa mt. U.K. see
 Snowdon
105 C2 Yssingeaux France
93 F4 Ystad Sweden
 Ysyk-Köl Kyrg. see Balykchy
77 D2 Ysyk-Köl salt l. Kyrg.
92 □A3 Ytri-Rangá r. Iceland
71 B3 Yuanjiang China
62 B1 Yuan Jiang r. China
62 B1 Yuanmou China
135 B3 Yuba City U.S.A.
67 D3 Yūbari Japan
145 C3 Yucatán pen. Mex.
126 A2 Yucatán r. Can./U.S.A.
146 B2 Yucatan Channel Cuba/Mex.
50 C2 Yuendumu Austr.
71 B3 Yueyang China
86 F2 Yugorsk Russia
83 L2 Yukagirskoye Ploskogor'ye
 plat. Russia
128 A1 Yukon admin. div. Can.
126 A2 Yukon r. Can./U.S.A.
71 B3 Yulin Guangxi China
70 A2 Yulin Shaanxi China
62 B1 Yulong Xueshan mt. China
138 A2 Yuma AZ U.S.A.
136 C2 Yuma CO U.S.A.
 Yumen China see
 Laojunmiao
80 B2 Yunak Turkey
70 B2 Yuncheng China
71 B3 Yunfu China
62 B1 Yunnan prov. China
52 A2 Yunta Austr.
70 B2 Yunyang China
62 D3 Yurihonjō Japan
150 A2 Yurimaguas Peru
77 E1 Yurungkax He r. China
71 C3 Yu Shan mt. Taiwan
68 C2 Yushu China
 Yushuwan China see
 Huaihua
81 C1 Yusufeli Turkey
77 E3 Yutian China
62 B1 Yuxi China
66 D2 Yuzhno-Kuril'sk Russia
83 K3 Yuzhno-Sakhalinsk Russia
91 C2 Yuzhnoukrayins'k Ukr.